APPLICATIONS OF DIGITAL SIGNAL PROCESSING TO AUDIO AND ACOUSTICS

THE KLUWER INTERNATIONAL SERIES
IN ENGINEERING AND COMPUTER SCIENCE

APPLICATIONS OF DIGITAL SIGNAL PROCESSING TO AUDIO AND ACOUSTICS

edited by

Mark Kahrs
Rutgers University
Piscataway, New Jersey, USA

Karlheinz Brandenburg
Fraunhofer Institut Integrierte Schaltungen
Erlangen, Germany

KLUWER ACADEMIC PUBLISHERS
Boston / Dordrecht / London

Distributors for North, Central and South America:
Kluwer Academic Publishers
101 Philip Drive
Assinippi Park
Norwell, Massachusetts 02061 USA

Distributors for all other countries:
Kluwer Academic Publishers Group
Distribution Centre
Post Office Box 322
3300 AH Dordrecht, THE NETHERLANDS

Library of Congress Cataloging-in-Publication Data

A C.I.P. Catalogue record for this book is available
from the Library of Congress.

Copyright © 1998 by Kluwer Academic Publishers

All rights reserved. No part of this publication may be reproduced, stored in a retrieval system or transmitted in any form or by any means, mechanical, photo-copying, recording, or otherwise, without the prior written permission of the publisher, Kluwer Academic Publishers, 101 Philip Drive, Assinippi Park, Norwell, Massachusetts 02061

Printed on acid-free paper.

Printed in the United States of America

The cover depicts a wave of great power seen from the coast near Kanagawa. This famous print from Hokusai ("Kanagawa oki nami-ura") was first published around 1830; it forms part of the "Thirty-six views of Mt. Fuji". Reprinted with the kind permission of Heibonsha Ltd., Tokyo, Japan from *The Thirty-six Views of Mt. Fuji*, (c) 1966.

Contents

List of Figures — xiii

List of Tables — xxi

Contributing Authors — xxiii

Introduction — xxix
Karlheinz Brandenburg and Mark Kahrs

1
Audio quality determination based on perceptual measurement techniques — 1
John G. Beerends

    1.1   Introduction — 1
    1.2   Basic measuring philosophy — 2
    1.3   Subjective versus objective perceptual testing — 6
    1.4   Psychoacoustic fundamentals of calculating the internal sound representation — 8
    1.5   Computation of the internal sound representation — 13
    1.6   The perceptual audio quality measure (PAQM) — 17
    1.7   Validation of the PAQM on speech and music codec databases — 20
    1.8   Cognitive effects in judging audio quality — 22
    1.9   ITU Standardization — 29
           1.9.1   ITU-T, speech quality — 30
           1.9.2   ITU-R, audio quality — 35
    1.10 Conclusions — 37

2
Perceptual Coding of High Quality Digital Audio — 39
Karlheinz Brandenburg

    2.1   Introduction — 39

2.2	Some Facts about Psychoacoustics		42
	2.2.1	Masking in the Frequency Domain	42
	2.2.2	Masking in the Time Domain	44
	2.2.3	Variability between listeners	45
2.3	Basic ideas of perceptual coding		47
	2.3.1	Basic block diagram	48
	2.3.2	Additional coding tools	49
	2.3.3	Perceptual Entropy	50
2.4	Description of coding tools		50
	2.4.1	Filter banks	50
	2.4.2	Perceptual models	59
	2.4.3	Quantization and coding	63
	2.4.4	Joint stereo coding	68
	2.4.5	Prediction	72
	2.4.6	Multi-channel: to matrix or not to matrix	73
2.5	Applying the basic techniques: real coding systems		74
	2.5.1	Pointers to early systems (no detailed description)	74
	2.5.2	MPEG Audio	75
	2.5.3	MPEG-2 Advanced Audio Coding (MPEG-2 AAC)	79
	2.5.4	MPEG-4 Audio	81
2.6	Current Research Topics		82
2.7	Conclusions		83

3
Reverberation Algorithms 85
William G. Gardner

3.1	Introduction		85
	3.1.1	Reverberation as a linear filter	86
	3.1.2	Approaches to reverberation algorithms	87
3.2	Physical and Perceptual Background		88
	3.2.1	Measurement of reverberation	89
	3.2.2	Early reverberation	90
	3.2.3	Perceptual effects of early echoes	93
	3.2.4	Reverberation time	94
	3.2.5	Modal description of reverberation	95
	3.2.6	Statistical model for reverberation	97
	3.2.7	Subjective and objective measures of late reverberation	98
	3.2.8	Summary of framework	100
3.3	Modeling Early Reverberation		100
3.4	Comb and Allpass Reverberators		105
	3.4.1	Schroeder's reverberator	105
	3.4.2	The parallel comb filter	108
	3.4.3	Modal density and echo density	109
	3.4.4	Producing uncorrelated outputs	111
	3.4.5	Moorer's reverberator	112
	3.4.6	Allpass reverberators	113
3.5	Feedback Delay Networks		116

	3.5.1	Jot's reverberator	119
	3.5.2	Unitary feedback loops	121
	3.5.3	Absorptive delays	122
	3.5.4	Waveguide reverberators	123
	3.5.5	Lossless prototype structures	125
	3.5.6	Implementation of absorptive and correction filters	128
	3.5.7	Multirate algorithms	128
	3.5.8	Time-varying algorithms	129
3.6	Conclusions		130

4
Digital Audio Restoration 133
Simon Godsill, Peter Rayner and Olivier Cappé

4.1	Introduction	134
4.2	Modelling of audio signals	135
4.3	Click Removal	137
	4.3.1 Modelling of clicks	137
	4.3.2 Detection	141
	4.3.3 Replacement of corrupted samples	144
	4.3.4 Statistical methods for the treatment of clicks	152
4.4	Correlated Noise Pulse Removal	155
4.5	Background noise reduction	163
	4.5.1 Background noise reduction by short-time spectral attenuation	164
	4.5.2 Discussion	177
4.6	Pitch variation defects	177
	4.6.1 Frequency domain estimation	179
4.7	Reduction of Non-linear Amplitude Distortion	182
	4.7.1 Distortion Modelling	183
	4.7.2 Non-linear Signal Models	184
	4.7.3 Application of Non-linear models to Distortion Reduction	186
	4.7.4 Parameter Estimation	188
	4.7.5 Examples	190
	4.7.6 Discussion	190
4.8	Other areas	192
4.9	Conclusion and Future Trends	193

5
Digital Audio System Architecture 195
Mark Kahrs

5.1	Introduction	195
5.2	Input/Output	196
	5.2.1 Analog/Digital Conversion	196
	5.2.2 Sampling clocks	202
5.3	Processing	203
	5.3.1 Requirements	204
	5.3.2 Processing	207
	5.3.3 Synthesis	208

viii APPLICATIONS OF DSP TO AUDIO AND ACOUSTICS

	5.3.4 Processors	209
5.4	Conclusion	234

6 Signal Processing for Hearing Aids
James M. Kates

6.1	Introduction	236
6.2	Hearing and Hearing Loss	237
	6.2.1 Outer and Middle Ear	238
6.3	Inner Ear	239
	6.3.1 Retrocochlear and Central Losses	247
	6.3.2 Summary	248
6.4	Linear Amplification	248
	6.4.1 System Description	249
	6.4.2 Dynamic Range	251
	6.4.3 Distortion	252
	6.4.4 Bandwidth	253
6.5	Feedback Cancellation	253
6.6	Compression Amplification	255
	6.6.1 Single-Channel Compression	256
	6.6.2 Two-Channel Compression	260
	6.6.3 Multi-Channel Compression	261
6.7	Single-Microphone Noise Suppression	263
	6.7.1 Adaptive Analog Filters	263
	6.7.2 Spectral Subtraction	264
	6.7.3 Spectral Enhancement	266
6.8	Multi-Microphone Noise Suppression	267
	6.8.1 Directional Microphone Elements	267
	6.8.2 Two-Microphone Adaptive Noise Cancellation	268
	6.8.3 Arrays with Time-Invariant Weights	269
	6.8.4 Two-Microphone Adaptive Arrays	269
	6.8.5 Multi-Microphone Adaptive Arrays	271
	6.8.6 Performance Comparison in a Real Room	273
6.9	Cochlear Implants	275
6.10	Conclusions	276

7 Time and Pitch scale modification of audio signals
Jean Laroche

7.1	Introduction	279
7.2	Notations and definitions	282
	7.2.1 An underlying sinusoidal model for signals	282
	7.2.2 A definition of time-scale and pitch-scale modification	282
7.3	Frequency-domain techniques	285
	7.3.1 Methods based on the short-time Fourier transform	285
	7.3.2 Methods based on a signal model	293
7.4	Time-domain techniques	293

		7.4.1 Principle	293
		7.4.2 Pitch independent methods	294
		7.4.3 Periodicity-driven methods	298
	7.5	Formant modification	302
		7.5.1 Time-domain techniques	302
		7.5.2 Frequency-domain techniques	302
	7.6	Discussion	303
		7.6.1 Generic problems associated with time or pitch scaling	303
		7.6.2 Time-domain vs frequency-domain techniques	308

8
Wavetable Sampling Synthesis   311
Dana C. Massie

	8.1	Background and Introduction	311
		8.1.1 Transition to Digital	312
		8.1.2 Flourishing of Digital Synthesis Methods	313
		8.1.3 Metrics: The Sampling - Synthesis Continuum	314
		8.1.4 Sampling vs. Synthesis	315
	8.2	Wavetable Sampling Synthesis	318
		8.2.1 Playback of digitized musical instrument events.	318
		8.2.2 Entire note - not single period	318
		8.2.3 Pitch Shifting Technologies	319
		8.2.4 Looping of sustain	331
		8.2.5 Multi-sampling	337
		8.2.6 Enveloping	338
		8.2.7 Filtering	338
		8.2.8 Amplitude variations as a function of velocity	339
		8.2.9 Mixing or summation of channels	339
		8.2.10 Multiplexed wavetables	340
	8.3	Conclusion	341

9
Audio Signal Processing Based on Sinusoidal Analysis/Synthesis   343
T.F. Quatieri and R.J. McAulay

	9.1	Introduction	344
	9.2	Filter Bank Analysis/Synthesis	346
		9.2.1 Additive Synthesis	346
		9.2.2 Phase Vocoder	347
		9.2.3 Motivation for a Sine-Wave Analysis/Synthesis	350
	9.3	Sinusoidal-Based Analysis/Synthesis	351
		9.3.1 Model	352
		9.3.2 Estimation of Model Parameters	352
		9.3.3 Frame-to-Frame Peak Matching	355
		9.3.4 Synthesis	355
		9.3.5 Experimental Results	358
		9.3.6 Applications of the Baseline System	362
		9.3.7 Time-Frequency Resolution	364
	9.4	Source/Filter Phase Model	366

	9.4.1	Model	367
	9.4.2	Phase Coherence in Signal Modification	368
	9.4.3	Revisiting the Filter Bank-Based Approach	381
9.5	Additive Deterministic/Stochastic Model		384
	9.5.1	Model	385
	9.5.2	Analysis/Synthesis	387
	9.5.3	Applications	390
9.6	Signal Separation Using a Two-Voice Model		392
	9.6.1	Formulation of the Separation Problem	392
	9.6.2	Analysis and Separation	396
	9.6.3	The Ambiguity Problem	399
	9.6.4	Pitch and Voicing Estimation	402
9.7	FM Synthesis		403
	9.7.1	Principles	404
	9.7.2	Representation of Musical Sound	407
	9.7.3	Parameter Estimation	409
	9.7.4	Extensions	411
9.8	Conclusions		411

10
Principles of Digital Waveguide Models of Musical Instruments 417
Julius O. Smith III

10.1	Introduction	418
	10.1.1 Antecedents in Speech Modeling	418
	10.1.2 Physical Models in Music Synthesis	420
	10.1.3 Summary	422
10.2	The Ideal Vibrating String	423
	10.2.1 The Finite Difference Approximation	424
	10.2.2 Traveling-Wave Solution	426
10.3	Sampling the Traveling Waves	426
	10.3.1 Relation to Finite Difference Recursion	430
10.4	Alternative Wave Variables	431
	10.4.1 Spatial Derivatives	431
	10.4.2 Force Waves	432
	10.4.3 Power Waves	434
	10.4.4 Energy Density Waves	435
	10.4.5 Root-Power Waves	436
10.5	Scattering at an Impedance Discontinuity	436
	10.5.1 The Kelly-Lochbaum and One-Multiply Scattering Junctions	439
	10.5.2 Normalized Scattering Junctions	441
	10.5.3 Junction Passivity	443
10.6	Scattering at a Loaded Junction of N Waveguides	446
10.7	The Lossy One-Dimensional Wave Equation	448
	10.7.1 Loss Consolidation	450
	10.7.2 Frequency-Dependent Losses	451
10.8	The Dispersive One-Dimensional Wave Equation	451
10.9	Single-Reed Instruments	455

		457
	10.9.1 Clarinet Overview	457
	10.9.2 Single-Reed Theory	458
10.10	Bowed Strings	462
	10.10.1 Violin Overview	463
	10.10.2 The Bow-String Scattering Junction	464
10.11	Conclusions	466

References 467

Index 535

List of Figures

1.1	Basic philosophy used in perceptual audio quality determination	4
1.2	Excitation pattern for a single sinusoidal tone	9
1.3	Excitation pattern for a single click	10
1.4	Excitation pattern for a short tone burst	11
1.5	Masking model overview	12
1.6	Time-domain smearing as a function of frequency	15
1.7	Basic auditory transformations used in the PAQM	18
1.8	Relation between MOS and PAQM, ISO/MPEG 1990 database	19
1.9	Relation between MOS and PAQM, ISO/MPEG 1991 database	21
1.10	Relation between MOS and PAQM, ITU-R 1993 database	22
1.11	Relation between MOS and PAQM, ETSI GSM full rate database	23
1.12	Relation between MOS and PAQM, ETSI GSM half rate database	24
1.13	Basic approach used in the development of $PAQM_C$	25
1.14	Relation between MOS and $PAQM_C$, ISO/MPEG 1991 database	28
1.15	Relation between MOS and $PAQM_C$, ITU-R 1993 database	29
1.16	Relation between MOS and $PAQM_C$, ETSI GSM full rate database	30
1.17	Relation between MOS and $PAQM_C$, ETSI GSM half rate database	31
1.18	Relation between MOS and PSQM, ETSI GSM full rate database	32
1.19	Relation between MOS and PSQM, ETSI GSM half rate database	33
1.20	Relation between MOS and PSQM, ITU-T German speech database	34
1.21	Relation between MOS and PSQM, ITU-T Japanese speech database	35
1.22	Relation between Japanese and German MOS values	36
2.1	Masked thresholds: Masker: narrow band noise at 250 Hz, 1 kHz, 4 kHz	44
2.2	Example of pre-masking and post-masking	45

2.3	Masking experiment as reported in [Spille, 1992]	46
2.4	Example of a pre-echo	47
2.5	Block diagram of a perceptual encoding/decoding system	48
2.6	Basic block diagram of an n-channel analysis/synthesis filter bank with downsampling by k	51
2.7	Window function of the MPEG-1 polyphase filter bank	54
2.8	Frequency response of the MPEG-1 polyphase filter bank	55
2.9	Block diagram of the MPEG Layer 3 hybrid filter bank	57
2.10	Window forms used in Layer 3	58
2.11	Example sequence of window forms	59
2.12	Example for the bit reservoir technology (Layer 3)	67
2.13	Main axis transform of the stereo plane	69
2.14	Basic block diagram of M/S stereo coding	70
2.15	Signal flow graph of the M/S matrix	70
2.16	Basic principle of intensity stereo coding	71
2.17	ITU Multichannel configuration	73
2.18	Block diagram of an MPEG-1 Layer 3 encode	77
2.19	Transmission of MPEG-2 multichannel information within an MPEG-1 bitstream	78
2.20	Block diagram of the MPEG-2 AAC encoder	80
2.21	MPEG-4 audio scaleable configuration	82
3.1	Impulse response of reverberant stairwell measured using ML sequences.	90
3.2	Single wall reflection and corresponding image source A'.	91
3.3	A regular pattern of image sources occurs in an ideal rectangular room.	91
3.4	Energy decay relief for occupied Boston Symphony Hall	96
3.5	Canonical direct form FIR filter with single sample delays.	101
3.6	Combining early echoes and late reverberation	102
3.7	FIR filter cascaded with reverberator	102
3.8	Associating absorptive and directional filters with early echoes.	103
3.9	Average head-related filter applied to a set of early echoes	104
3.10	Binaural early echo simulator	104
3.11	One-pole, DC-normalized lowpass filter.	104
3.12	Comb filter response	106
3.13	Allpass filter formed by modification of a comb filter	106
3.14	Schroeder's reverberator consisting of a parallel comb filter and a series allpass filter [Schroeder, 1962].	108
3.15	Mixing matrix used to form uncorrelated outputs	112

3.16	Controlling IACC in binaural reverberation	112
3.17	Comb filter with lowpass filter in feedback loop	113
3.18	Lattice allpass structure.	115
3.19	Generalization of figure 3.18.	115
3.20	Reverberator formed by adding absorptive losses to an allpass feedback loop	115
3.21	Dattorro's plate reverberator based on an allpass feedback loop	117
3.22	Stautner and Puckette's four channel feedback delay network	118
3.23	Feedback delay network as a general specification of a reverberator containing N delays	120
3.24	Unitary feedback loop	121
3.25	Associating an attenuation with a delay.	122
3.26	Associating an absorptive filter with a delay.	123
3.27	Reverberator constructed with frequency dependent absorptive filters	124
3.28	Waveguide network consisting of a single scattering junction to which N waveguides are attached	124
3.29	Modification of Schroeder's parallel comb filter to maximize echo density	126
4.1	Click-degraded music waveform taken from 78 rpm recording	138
4.2	AR-based detection, P=50. (a) Prediction error filter (b) Matched filter.	138
4.3	Electron micrograph showing dust and damage to the grooves of a 78rpm gramophone disc.	139
4.4	AR-based interpolation, P=60, classical chamber music, (a) short gaps, (b) long gaps	147
4.5	Original signal and excitation (P=100)	150
4.6	LSAR interpolation and excitation ($P = 100$)	150
4.7	Sampled AR interpolation and excitation (P=100)	151
4.8	Restoration using Bayesian iterative methods	155
4.9	Noise pulse from optical film sound track ('silent' section)	157
4.10	Signal waveform degraded by low frequency noise transient	157
4.11	Degraded audio signal with many closely spaced noise transients	161
4.12	Estimated noise transients for figure 4.11	161
4.13	Restored audio signal for figure 4.11 (different scale)	162
4.14	Modeled restoration process	164
4.15	Background noise suppression by short-time spectral attenuation	165
4.16	Suppression rules characteristics	168
4.17	Restoration of a sinusoidal signal embedded in white noise	169
4.18	Probability density of the relative signal level for different mean values	172

4.19	Short-time power variations	175
4.20	Frequency tracks generated for example 'Viola'	179
4.21	Estimated (full line) and true (dotted line) pitch variation curves generated for example 'Viola'	180
4.22	Frequency tracks generated for example 'Midsum'	180
4.23	Pitch variation curve generated for example 'Midsum'	181
4.24	Model of the distortion process	184
4.25	Model of the signal and distortion process	186
4.26	Typical section of AR-MNL Restoration	191
4.27	Typical section of AR-NAR Restoration	191
5.1	DSP system block diagram	196
5.2	Successive Approximation Converter	198
5.3	16 Bit Floating Point DAC (from [Kriz, 1975])	202
5.4	Block diagram of Moore's FRMbox	210
5.5	Samson Box block diagram	211
5.6	diGiugno 4A processor	213
5.7	IRCAM 4B data path	214
5.8	IRCAM 4C data path	215
5.9	IRCAM 4X system block diagram	216
5.10	Sony DAE-1000 signal processor	217
5.11	Lucasfilm ASP ALU block diagram	218
5.12	Lucasfilm ASP interconnect and memory diagram	219
5.13	Moorer's update queue data path	219
5.14	MPACT block diagram	222
5.15	Rossum's cached interpolator	226
5.16	Sony OXF DSP block diagram	227
5.17	DSP.* block diagram	228
5.18	Gnusic block diagram	229
5.19	Gnusic core block diagram	230
5.20	Sony SDP-1000 DSP block diagram	232
5.21	Sony's OXF interconnect block diagram	233
6.1	Major features of the human auditory system	238
6.2	Features of the cochlea: transverse cross-section of the cochlea	239
6.3	Features of the cochlea: the organ of Corti	240
6.4	Sample tuning curves for single units in the auditory nerve of the cat	241
6.5	Neural tuning curves resulting from damaged hair cells	242
6.6	Loudness level functions	244
6.7	Mean results for unilateral cochlear impairments	246

6.8	Simulated neural response for the normal ear	247
6.9	Simulated neural response for impaired outer cell function	248
6.10	Simulated neural response for 30 dB of gain	249
6.11	Cross-section of an in-the-ear hearing aid	250
6.12	Block diagram of an ITE hearing aid inserted into the ear canal	251
6.13	Block diagram of a hearing aid incorporating signal processing for feedback cancellation	255
6.14	Input/output relationship for a typical hearing-aid compression amplifier	256
6.15	Block diagram of a hearing aid having feedback compression	257
6.16	Compression amplifier input/output curves derived from a simplified model of hearing loss.	260
6.17	Block diagram of a spectral-subtraction noise-reduction system.	265
6.18	Block diagram of an adaptive noise-cancellation system.	268
6.19	Block diagram of an adaptive two-microphone array.	270
6.20	Block diagram of a time-domain five-microphone adaptive array.	271
6.21	Block diagram of a frequency-domain five-microphone adaptive array.	274
7.1	Duality between Time-scaling and Pitch-scaling operations	285
7.2	Time stretching in the time-domain	293
7.3	A modified tape recorder for analog time-scale or pitch-scale modification	294
7.4	Pitch modification with the sampling technique	295
7.5	Output elapsed time versus input elapsed time in the sampling method for Time-stretching	296
7.6	Time-scale modification of a sinusoid	297
7.7	Output elapsed time versus input elapsed time in the optimized sampling method for Time-stretching	300
7.8	Pitch-scale modification with the PSOLA method	301
7.9	Time-domain representation of a speech signal showing shape invariance	305
7.10	Time-domain representation of a speech signal showing loss of shape-invariance	306
8.1	Expressivity vs. Accuracy	316
8.2	Sampling tradeoffs	316
8.3	Labor costs for synthesis techniques	317
8.4	Rudimentary sampling	320
8.5	"Drop Sample Tuning" table lookup sampling playback oscillator	323
8.6	Classical sample rate conversion chain	325
8.7	Digital Sinc function	326

8.8	Frequency response of at linear interpolation sample rate converter	327
8.9	A sampling playback oscillator using high order interpolation	329
8.10	Traditional ADSR amplitude envelope	331
8.11	Backwards forwards loop at a loop point with even symmetry	333
8.12	Backwards forwards loop at a loop point with odd symmetry	333
8.13	Multisampling	337
9.1	Signal and spectrogram from a trumpet	345
9.2	Phase vocoder based on filter bank analysis/synthesis.	349
9.3	Passage of single sine wave through one bandpass filter.	350
9.4	Sine-wave tracking based on frequency-matching algorithm	356
9.5	Block diagram of baseline sinusoidal analysis/synthesis	358
9.6	Reconstruction of speech waveform	359
9.7	Reconstruction of trumpet waveform	360
9.8	Reconstruction of waveform from a closing stapler	360
9.9	Magnitude-only reconstruction of speech	361
9.10	Onset-time model for time-scale modification	370
9.11	Transitional properties of frequency tracks with adaptive cutoff	372
9.12	Estimation of onset times for time-scale modification	374
9.13	Analysis/synthesis for time-scale modification	375
9.14	Example of time-scale modification of trumpet waveform	376
9.15	Example of time-varying time-scale modification of speech waveform	376
9.16	KFH phase dispersion using the sine-wave preprocessor	380
9.17	Comparison of original waveform and processed speech	381
9.18	Time-scale expansion ($x2$) using subband phase correction	383
9.19	Time-scale expansion ($x2$) of a closing stapler using filter bank/overlap-add	385
9.20	Block diagram of the deterministic plus stochastic system.	389
9.21	Decomposition example of a piano tone	391
9.22	Two-voice separation using sine-wave analysis/synthesis and peak-picking	393
9.23	Properties of the STFT of $x(n) = x_a(n) + x_b(n)$	396
9.24	Least-squared error solution for two sine waves	397
9.25	Demonstration of two-lobe overlap	400
9.26	$\mathbf{H}$ matrix for the example in Figure 9.25	401
9.27	Demonstration of ill conditioning of the $\mathbf{H}$ matrix	402
9.28	FM Synthesis with different carrier and modulation frequencies	405
9.29	Spectral dynamics of FM synthesis with linearly changing modulation index	406

9.30	Comparison of Equation (9.82) and (9.86) for parameter settings $\omega_c = 2000$, $\omega_m = 200$, and $I = 5.0$	407
9.31	Spectral dynamics of trumpet-like sound using FM synthesis	408
10.1	The ideal vibrating string.	423
10.2	An infinitely long string, "plucked" simultaneously at three points.	427
10.3	Digital simulation of the ideal, lossless waveguide with observation points at $x = 0$ and $x = 3X = 3cT$.	429
10.4	Conceptual diagram of interpolated digital waveguide simulation.	429
10.5	Transverse force propagation in the ideal string.	433
10.6	A waveguide section between two partial sections. a) Physical picture indicating traveling waves in a continuous medium whose wave impedance changes from R_0 to R_1 to R_2. b) Digital simulation diagram for the same situation.	437
10.7	The Kelly-Lochbaum scattering junction.	439
10.8	The one-multiply scattering junction.	440
10.9	The normalized scattering junction.	441
10.10	A three-multiply normalized scattering junction.	443
10.11	Four ideal strings intersecting at a point to which a lumped impedance is attached.	446
10.12	Discrete simulation of the ideal, lossy waveguide.	449
10.13	Discrete-time simulation of the ideal, lossy waveguide.	450
10.14	Section of a stiff string where allpass filters play the role of unit delay elements.	453
10.15	Section of a stiff string where the allpass delay elements are consolidated at two points, and a sample of pure delay is extracted from each allpass chain.	454
10.16	A schematic model for woodwind instruments.	455
10.17	Waveguide model of a single-reed, cylindrical-bore woodwind, such as a clarinet.	457
10.18	Schematic diagram of mouth cavity, reed aperture, and bore.	458
10.19	Normalized reed impedance $G(p_\Delta) \triangleq R_b u_m(p_\Delta)$ overlaid with the "bore load line" $p_\Delta^+ - p_\Delta = R_b u_b$.	459
10.20	Simple, qualitatively chosen reed table for the digital waveguide clarinet.	461
10.21	A schematic model for bowed-string instruments.	463
10.22	Waveguide model for a bowed string instrument, such as a violin.	464
10.23	Simple, qualitatively chosen bow table for the digital waveguide violin.	465

List of Tables

2.1	Critical bands according to [Zwicker, 1982]	43
2.2	Huffman code tables used in Layer 3	66
5.1	Pipeline timing for Samson box generators	212
6.1	Hearing thresholds, descriptive terms, and probable handicaps (after Goodman, 1965)	236

Acknowledgments

Mark Kahrs would like to acknowledge the support of J.L. Flanagan. He would also like to acknowledge the the assistance of Howard Trickey and S.J. Orfanidis. Jean Laroche has helped out with the production and served as a valuable forcing function. The patience of Diane Litman has been tested numerous times and she has offered valuable advice.

Karlheinz Brandenburg would like to thank Mark for his patience while he was always late in delivering his parts.

Both editors would like to acknowledge the patience of Bob Holland, our editor at Kluwer.

Contributing Authors

John G. Beerends was born in Millicent, Australia, in 1954. He received a degree in electrical engineering from the HTS (Polytechnic Institute) of The Hague, The Netherlands, in 1975. After working in industry for three years he studied physics and mathematics at the University of Leiden where he received the degree of M.Sc. in 1984. In 1983 he was awarded a prize of DFl 45000,- by Job Creation, for an innovative idea in the field of electro-acoustics. During the period 1984 to 1989 he worked at the Institute for Perception Research where he received a Ph.D. from the Technical University of Eindhoven in 1989. The main part of his Ph.D. work, which deals with pitch perception, was patented by the N.V. Philips Gloeilampenfabriek. In 1989 he joined the audio group of the KPN research lab in Leidschendam where he works on audio quality assessment. Currently he is also involved in the development of an objective video quality measure.

Karlheinz Brandenburg received M.S. (Diplom) degrees in Electrical Engineering in 1980 and in Mathematics in 1982 from Erlangen University. In 1989 he earned his Ph.D. in Electrical Engineering, also from Erlangen University, for work on digital audio coding and perceptual measurement techniques. From 1989 to 1990 he was with AT&T Bell Laboratories in Murray Hill, NJ, USA. In 1990 he returned to Erlangen University to continue the research on audio coding and to teach a course on digital audio technology. Since 1993 he is the head of the Audio/Multimedia department at the Fraunhofer Institute for Integrated Circuits (FhG-IIS). Dr. Brandenburg is a member of the technical committee on Audio and Electroacoustics of the IEEE Signal Processing Society. In 1994 he received the AES Fellowship Award for his work on perceptual audio coding and psychoacoustics.

Olivier Cappé was born in Villeurbanne, France, in 1968. He received the M.Sc. degree in electrical engineering from the Ecole Supérieure d'Electricité (ESE), Paris, in 1990, and the Ph.D. degree in signal processing from the Ecole Nationale Supérieure des Télécommunications (ENST), Paris, in 1993. His Ph.D. thesis dealt with noise-reduction for degraded audio recordings. He is currently with the Centre National de la Recherche Scientifique (CNRS) at ENST, Signal department. His research interests are in statistical signal processing for telecomunications and speech/audio processing. Dr. Cappé received the IEEE Signal Processing Society's Young Author Best Paper Award in 1995.

Bill Gardner was born in 1960 in Meriden, CT, and grew up in the Boston area. He received a bachelor's degree in computer science from MIT in 1982 and shortly thereafter joined Kurzweil Music Systems as a software engineer. For the next seven years, he helped develop software and signal processing algorithms for Kurzweil synthesizers. He left Kurzweil in 1990 to enter graduate school at the MIT Media Lab, where he recently completed his Ph.D. on the topic of 3-D audio using loudspeakers. He was awarded a Motorola Fellowship at the Media Lab, and was recipient of the 1997 Audio Engineering Society Publications Award. He is currently an independent consultant working in the Boston area. His research interests are spatial audio, reverberation, sound synthesis, realtime signal processing, and psychoacoustics.

Simon Godsill studied for the B.A. in Electrical and Information Sciences at the University of Cambridge from 1985-88. Following graduation he led the technical development team at the newly-formed CEDAR Audio Ltd., researching and developing DSP algorithms for restoration of degraded sound recordings. In 1990 he took up a post as Research Associate in the Signal Processing Group of the Engineering Department at Cambridge and in 1993 he completed his doctoral thesis: The Restoration of Degraded Audio Signals. In 1994 he was appointed as a Research Fellow at Corpus Christi College, Cambridge and in 1996 as University Lecturer in Signal Processing at the Engineering Department in Cambridge. Current research topics include: Bayesian and statistical methods in signal processing, modelling and enhancement of speech and audio signals, source signal separation, non-linear and non-Gaussian techniques, blind estimation of communications channels and image sequence analysis.

Mark Kahrs was born in Rome, Italy in 1952. He received an A.B. from Revelle College, University of California, San Diego in 1974. He worked intermittently for Tymshare, Inc. as a Systems Programmer from 1968 to 1974. During the summer of 1975 he was a Research Intern at Xerox PARC and then from 1975 to 1977 was a Research Programmer at the Center for Computer Research in Music and

Acoustics (CCRMA) at Stanford University. He was a chercheur at the Institut de Recherche et Coordination Acoustique Musique (IRCAM) in Paris during the summer of 1977. He received a PhD. in Computer Science from the University of Rochester in 1984. He worked and consulted for Bell Laboratories from 1984 to 1996. He has been an Assistant Professor at Rutgers University from 1988 to the present where he taught courses in Computer Architecture, Digital Signal Processing and Audio Engineering. In 1993 he was General Chair of the *IEEE Workshop on Applications of Signal Processing to Audio and Acoustics* ("Mohonk Workshop"). Since 1993 he has chaired the Technical Committee on Audio And Electroacoustics in the Signal Processing Society of the IEEE.

James M. Kates was born in Brookline, Massachusetts, in 1948. He received the degrees of BSEE and MSEE from the Massachusetts Institute of Technology in 1971 and the professional degree of Electrical Engineer from MIT in 1972. He is currently Senior Scientist at AudioLogic in Boulder, Colorado, where he is developing signal processing for a new digital hearing aid. Prior to joining AudioLogic, he was with the Center for Research in Speech and Hearing Sciences of the City University of New York. His research interests at CUNY included directional microphone arrays for hearing aids, feedback cancellation strategies, signal processing for hearing aid test and evaluation, procedures for measuring sound quality in hearing aids, speech enhancement algorithms for the hearing-impaired, new procedures for fitting hearing aids, and modeling normal and impaired cochlear function. He also held an appointment as an Adjunt Assistant Professor in the Doctoral Program in Speech and Hearing Sciences at CUNY, where he taught a course in modeling auditory physiology and perception. Previously, he has worked on applied research for hearing aids (Siemens Hearing Instruments), signal processing for radar, speech, and hearing applications (SIGNATRON, Inc.), and loudspeaker design and signal processing for audio applications (Acoustic Research and CBS Laboratories). He has over three dozen published papers and holds eight patents.

Jean Laroche was born in Bordeaux, France, in 1963 He earned a degree in Mathematics and Sciences from the Ecole Polytechnique in 1986, and a Ph.D. degree in Digital Signal Processing from the Ecole Nationale des Télécommunications in 1989. He was a post-doc student at the Center for Music Experiment at UCSD in 1990, and came back to the Ecole Nationale des Télécommunications in 1991 where he taught audio DSP, and acoustics. Since 1996 he has been a researcher in audio/music DSP at the Joint Emu/Creative Technology Center in Scotts Valley, CA.

Robert J. McAulay was born in Toronto, Ontario, Canada on October 23, 1939. He received the B.A.Sc. degree in Engineering Physics with honors from the University of Toronto, in 1962; the M.Sc. degree in Electrical Engineering from the University of Illinois, Urbana in 1963; and the Ph.D. degree in Electrical Engineering from the University of California, Berkeley, in 1967. He joined the Radar Signal Processing Group of the Massachusetts Institute of Technology, Lincoln Laboratory, Lexington, MA, where he worked on problems in estimation theory and signal/filter design using optimal control techniques. From 1970 until 1975, he was a member of the Air Traffic Control Division at Lincoln Laboratory, and worked on the development of aircraft tracking algorithms, optimal MTI digital signal processing and on problems of aircraft direction finding for the Discrete Address Beacon System. On a leave of absence from Lincoln Laboratory during the winter and spring of 1974, he was a Visiting Associate Professor at McGill University, Montreal, P.Q., Canada. From 1975 until 1996, he was a member of the Speech Systems Technology Group at Lincoln Laboratory, where he was involved in the development of robust narrowband speech vocoders. In 1986 he served on the National Research Council panel that reviewed the problem of the removal of noise from speech. In 1987 he was appointed to the position of Lincoln Laboratory Senior Staff. On retiring from Lincoln Laboratory in 1996, he accepted the position of Senior Scientist at Voxware to develop high-quality speech products for the Internet. In 1978 he received the M. Barry Carlton Award for the best paper published in the IEEE Transactions on Aerospace and Electronic Systems for the paper "Interferometer Design for Elevation Angle Estimation". In 1990 he received the IEEE Signal Processing Society's Senior Award for the paper "Speech Analysis/Synthesis Based on a Sinusoidal Representation", published in the IEEE Transactions on Acoustics, Speech and Signal Processing.

Dana C. Massie studied electronic music synthesis and composition at Virginia Commonwealth University in Richmond Virginia, and electrical engineering at Virginia Polytechnic Institute and State University in Blacksburg, VA. He worked in professional analog recording console and digital telecom systems design at Datatronix, Inc., in Reston, VA from 1981 through 1983. He then moved to E-mu Systems, Inc., in California, to design DSP algorithms and architectures for electronic music. After brief stints at NeXT Computer, Inc. and WaveFrame, Inc., developing MultiMedia DSP applications, he returned to E-mu Systems to work in digital filter design, digital reverberation design, and advanced music synthesis algorithms. He is now the Director of the Joint E-mu/Creative Technology Center, in Scotts Valley, California. The "Tech Center" develops advanced audio technologies for both E-mu Systems and Creative Technology, Limited in Singapore, including VLSI designs, advanced music synthesis algorithms, 3D audio algorithms, and software tools.

Thomas F. Quatieri was born in Somerville, Massachusetts on January 31, 1952. He received the B.S. degree from Tufts University, Medford, Massachusetts in 1973, and the S.M., E.E., and Sc.D. degrees from the Massachusetts Institute of Technology (M.I.T.), Cambridge, Massachusetts in 1975, 1977, and 1979, respectively. He is currently a senior research staff member at M.I.T. Lincoln Laboratory, Lexington, Massachusetts. In 1980, he joined the Sensor Processing Technology Group of M.I.T., Lincoln Laboratory, Lexington, Massachusetts where he worked on problems in multi-dimensional digital signal processing and image processing. Since 1983 he has been a member of the Speech Systems Technology Group at Lincoln Laboratory where he has been involved in digital signal processing for speech and audio applications, underwater sound enhancement, and data communications. He has contributed many publications to journals and conference proceedings, written several patents, and co-authored chapters in numerous edited books including: *Advanced Topics in Signal Processing* (Prentice Hall, 1987), *Advances in Speech Signal Processing* (Marcel Dekker, 1991), and *Speech Coding and Synthesis* (Elsevier, 1995). He holds the position of Lecturer at MIT where he has developed the graduate course *Digital Speech Processing*, and is active in advising graduate students on the MIT campus. Dr. Quatieri is the recipient of the 1982 Paper Award of the IEEE Acoustics, Speech and Signal Processing Society for the paper, "Implementation of 2-D Digital Filters by Iterative Methods". In 1990, he received the IEEE Signal Processing Society's Senior Award for the paper, "Speech Analysis/Synthesis Based on a Sinusoidal Representation", published in the *IEEE Transactions on Acoustics, Speech and Signal Processing*, and in 1994 won this same award for the paper "Energy Separation in Signal Modulations with Application to Speech Analysis" which was also selected for the 1995 IEEE W.R.G. Baker Prize Award. He was a member of the IEEE Digital Signal Processing Technical Committee, from 1983 to 1992 served on the steering committee for the bi-annual Digital Signal Processing Workshop, and was Associate Editor for the *IEEE Transactions on Signal Processing* in the area of nonlinear systems.

Peter J.W. Rayner received the M.A. degree from Cambridge University, U.K., in 1968 and the Ph. D. degree from Aston University in 1969. Since 1968 he has been with the Department of Engineering at Cambridge University and is Head of the Signal Processing and Communications Research Group. In 1990 he was appointed to an ad-hominem Readership in Information Engineering. He teaches course in random signal theory, digital signal processing, image processing and communication systems. His current research interests include image sequence restoration, audio restoration, non-linear estimation and detection and time series modelling and classification.

Julius O. Smith received the B.S.E.E. degree from Rice University, Houston, TX, in 1975. He received the M.S. and Ph.D. degrees from Stanford University, Stanford, CA, in 1978 and 1983, respectively. His Ph.D. research involved the application of digital signal processing and system identification techniques to the modeling and synthesis of the violin, clarinet, reverberant spaces, and other musical systems. From 1975 to 1977 he worked in the Signal Processing Department at ESL in Sunnyvale, CA, on systems for digital communications. From 1982 to 1986 he was with the Adaptive Systems Department at Systems Control Technology in Palo Alto, CA, where he worked in the areas of adaptive filtering and spectral estimation. From 1986 to 1991 he was employed at NeXT Computer, Inc., responsible for sound, music, and signal processing software for the NeXT computer workstation. Since then he has been an Associate Professor at the Center for Computer Research in Music and Acoustics (CCRMA), Stanford University, teaching courses in signal processing and music technology, and pursuing research in signal processing techniques applied to musical instrument modeling, audio spectral modeling, and related topics.

INTRODUCTION
Karlheinz Brandenburg and Mark Kahrs

With the advent of multimedia, digital signal processing (DSP) of sound has emerged from the shadow of bandwidth-limited speech processing. Today, the main applications of audio DSP are high quality audio coding and the digital generation and manipulation of music signals. They share common research topics including perceptual measurement techniques and analysis/synthesis methods. Smaller but nonetheless very important topics are hearing aids using signal processing technology and hardware architectures for digital signal processing of audio. In all these areas the last decade has seen a significant amount of application oriented research.

The topics covered here coincide with the topics covered in the biannual workshop on "Applications of Signal Processing to Audio and Acoustics". This event is sponsored by the IEEE Signal Processing Society (Technical Committee on Audio and Electroacoustics) and takes place at Mohonk Mountain House in New Paltz, New York.

A short overview of each chapter will illustrate the wide variety of technical material presented in the chapters of this book.

John Beerends: Perceptual Measurement Techniques. The advent of perceptual measurement techniques is a byproduct of the advent of digital coding for both speech and high quality audio signals. Traditional measurement schemes are bad estimates for the subjective quality after digital coding/decoding. Listening tests are subject to statistical uncertainties and the basic question of repeatability in a different environment. John Beerends explains the reasons for the development of perceptual measurement techniques, the psychoacoustic fundamentals which apply to both perceptual measurement and perceptual coding and explains some of the more advanced techniques which have been developed in the last few years. Completed and ongoing standardization efforts concludes his chapter. This is recommended reading not only to people interested in perceptual coding and measurement but to anyone who wants to know more about the psychoacoustic fundamentals of digital processing of sound signals.

Karlheinz Brandenburg: Perceptual Coding of High Quality Digital Audio.
High quality audio coding is rapidly progressing from a research topic to widespread applications. Research in this field has been driven by a standardization process within the Motion Picture Experts Group (MPEG). The chapter gives a detailed introduction of the basic techniques including a study of filter banks and perceptual models. As the main example, MPEG Audio is described in full detail. This includes a description of the new MPEG-2 Advanced Audio Coding (AAC) standard and the current work on MPEG-4 Audio.

William G. Gardner: Reverberation Algorithms. This chapter is the first in a number of chapters devoted to the digital manipulation of music signals. Digitally generated reverb was one of the first application areas of digital signal processing to high quality audio signals. Bill Gardner gives an in depth introduction to the physical and perceptual aspects of reverberation. The remainder of the chapter treats the different types of artificial reverberators known today. The main quest in this topic is to generate natural sounding reverb with low cost. Important milestones in the research, various historic and current types of reverberators are explained in detail.

Simon Godsill, Peter Rayner and Olivier Cappé: Digital Audio Restoration.
Digital signal processing of high quality audio does not stop with the synthesis or manipulation of new material: One of the early applications of DSP was the manipulation of sounds from the past in order to restore them for recording on new or different media. The chapter presents the different methods for removing clicks, noise and other artifacts from old recordings or film material.

Mark Kahrs: Digital Audio System Architecture. An often overlooked part of the processing of high quality audio is the system architecture. Mark Kahrs introduces current technologies both for the conversion between analog and digital world and the processing technologies. Over the years there is a clear path from specialized hardware architectures to general purpose computing engines. The chapter covers specialized hardware architectures as well as the use of generally available DSP chips. The emphasis is on high throughput digital signal processing architectures for music synthesis applications.

James M. Kates: Signal Processing for Hearing Aids. A not so obvious application area for audio signal processing is the field of hearing aids. Nonetheless this field has seen continuous research activities for a number of years and is another field where widespread application of digital technologies is under preparation today. The chapter contains an in-depth treatise of the basics of signal processing for hearing aids including the description of different types of hearing loss, simpler amplification and compression techniques and current research on multi-microphone techniques and cochlear implants.

Jean Laroche: Time and Pitch Scale Modification of Audio Signals. One of the conceptionally simplest problems of the manipulation of audio signals is difficult enough to warrant ongoing research for a number of years: Jean Laroche explains the basics of time and pitch scale modification of audio signals for both speech and musical signals. He discusses both time domain and frequency domain methods including methods specially suited for speech signals.

Dana C. Massie: Wavetable Sampling Synthesis. The most prominent example today of the application of high quality digital audio processing is wavetable sampling synthesis. Tens of millions of computer owners have sound cards incorporating wavetable sampling synthesis. Dana Massie explains the basics and modern technologies employed in sampling synthesis.

T.F. Quatieri and R.J. McAulay: Audio Signal Processing Based on Sinusoidal Analysis/Synthesis. One of the basic paradigms of digital audio analysis, coding (i.e. analysis/synthesis) and synthesis systems is the sinusoidal model. It has been used for many systems from speech coding to music synthesis. The chapter contains the unified view of both the basics of sinusoidal analysis/synthesis and some of the applications.

Julius O. Smith III: Principles of Digital Waveguide Models of Musical Instruments. This chapter describes a recent research topic in the synthesis of music instruments: Digital waveguide models are one method of physical modeling. As in the case of the Vocoder for speech, a model of an existing or hypothetical instrument is used for the sound generation. In the tutorial the vibrating string is taken as the principle illustrative example. Another example using the same underlying principles is the acoustic tube. Complicated instruments are derived by adding signal scattering and reed-bore or bow-string interactions.

Summary This book was written to serve both as a text book for an advanced graduate course on digital signal processing for audio or as a reference book for the practicing engineer. We hope that this book will stimulate further research and interest in this fascinating and exciting field.

1 AUDIO QUALITY DETERMINATION BASED ON PERCEPTUAL MEASUREMENT TECHNIQUES

John G. Beerends

Royal PTT Netherlands N.V.
KRN Research, P. Box 421, AK Leidenham
The Netherlands
J.G.Beerends@research.kpn.com

Abstract: A new, perceptual, approach to determine audio quality is discussed. The method does not characterize the audio system under test but characterizes the perception of the output signal of the audio system. By comparing the degraded output with the ideal (reference), using a model of the human auditory system, predictions can be made about the subjectively perceived audio quality of the system output using any input signal. A perceptual model is used to calculate the internal representations of both the degraded output and reference. A simple cognitive model interprets differences between the internal representations. The method can be used for quality assessment of wideband music codecs as well as for telephone-band (300-3400 Hz) speech codecs. The correlation between subjective and objective results is above 0.9 for a wide variety of databases derived from subjective quality evaluations of music and speech codecs. For the measurement of quality of telephone-band speech codecs a simplified method is given. This method was standardized by the International Telecommunication Union (Telecom sector) as recommendation P.861.

1.1 INTRODUCTION

With the introduction and standardization of new, perception based, audio (speech and music) codecs, [ISO92st, 1993], [ISO94st, 1994], [ETSIstdR06, 1992], [CCIT-

TrecG728, 1992], [CCITTrecG729, 1995], classical methods for measuring audio quality, like signal to noise ratio and total harmonic distortion, became useless.
During the standardization process of these codecs the quality of the different proposals was therefore assessed only subjectively (see e.g. [Natvig, 1988], [ISO90, 1990] and [ISO91, 1991]). Subjective assessments are however time consuming, expensive and difficult to reproduce.

A fundamental question is whether objective methods can be formulated that can be used for prediction of the subjective quality of such perceptual coding techniques in a reliable way. A difference with classical approaches to audio quality assessment is that system characterizations are no longer useful because of the time varying, signal adaptive, techniques that are used in these codecs. In general the quality of modern audio codecs is dependent on the input signal. The newly developed method must therefore be able to measure the quality of the codec using any audio signal, that is speech, music and test signals. Methods that rely on test signals only, either with or without making use of a perceptual model, can not be used.

This chapter will present a general method for measuring the quality of audio devices including perception based audio codecs. The method uses the concept of the internal sound representation, the representation that matches as close as possible the one that is used by subjects in their quality judgement. The input and output of the audio device are mapped onto the internal signal representation and the difference in this representation is used to define a perceptual audio quality measure (PAQM). It will be shown that this PAQM has a high correlation with the subjectively perceived audio quality especially when differences in the internal representation are interpreted, in a context dependent way, by a cognitive module. Furthermore a simplified method, derived from PAQM, for measuring the quality of telephone-band (300-3400 Hz) speech codecs is presented. This method was standardized by the ITU-T (International Telecommunication Union - Telecom sector) as recommendation P.861 [ITUTrecP861, 1996].

1.2 BASIC MEASURING PHILOSOPHY

In the literature on measuring the quality of audio devices one mostly finds measurement techniques that characterize the audio device under test. The characterization either has build in knowledge of human auditory perception or the characterization has to be interpreted with knowledge of human auditory perception.

For linear, time-invariant systems a complete characterization is given by the impulse or complex frequency response [Papoulis, 1977]. With perceptual interpretation of this characterization one can determine the audio quality of the system under test. If the design goal of the system under test is to be transparent (no audible differences between input and output) then quality evaluation is simple and brakes down to the requirement of a flat amplitude and phase response (within a specified template) over the audible frequency range (20-20000 Hz).

For systems that are nearly linear or time-variant, the concept of the impulse (complex frequency) response is still applicable. For weakly non-linear systems the characterization can be extended by including measurements of the non-linearity (noise, distortion, clipping point). For time-variant systems the characterization can be extended by including measurements of the time dependency of the impulse response. Some of the additional measurements incorporate knowledge of the human auditory system which lead to system characterizations that have a direct link to the perceived audio quality (e.g. the perceptually weighted signal to noise ratio).

The advantage of the system characterization approach is that it is (or better that it should be) largely independent of the test signals that are used. The characterizations can thus be measured with standardized signals and measurement procedures. Although the system characterization is mostly independent of the signal the subjectively perceived quality in most cases depends on the audio signal that is used. If we take e.g. a system that adds white noise to the input signal then the perceived audio quality will be very high if the input signal is wideband. The same system will show a low audio quality if the input signal is narrowband. For a wideband input signal the noise introduced by the audio system will be masked by the input signal. For a narrowband input signal the noise will be clearly audible in frequency regions where there is no input signal energy. System characterizations therefore do not characterize the perceived quality of the output signal.

A disadvantage of the system characterization approach is that although the characterization is valid for a wide variety of input signals it can only be measured on the basis of knowledge of the system. This leads to system characterizations that are dependent on the type of system that is tested. A serious drawback in the system characterization approach is that it is extremely difficult to characterize systems that show a non-linear and time-variant behavior.

An alternative approach to the system characterization, valid for any system, is the perceptual approach. In the context of this chapter a perceptual approach is defined as an approach in which aspects of human perception are modelled in order to make measurements on audio signals that have a high correlation with the subjectively perceived quality of these signals and that can be applied to any signal, that is, speech, music and test signals.

In the perceptual approach one does not characterize the system under test but one characterizes the audio quality of the output signal of the system under test. It uses the ideal signal as a reference and an auditory perception model to determine the audible differences between the output and the ideal. For audio systems that should be transparent the ideal signal is the input signal. An overview of the basic philosophy used in perceptual audio quality measurement techniques is given in Fig. 1.1.

4   APPLICATIONS OF DSP TO AUDIO AND ACOUSTICS

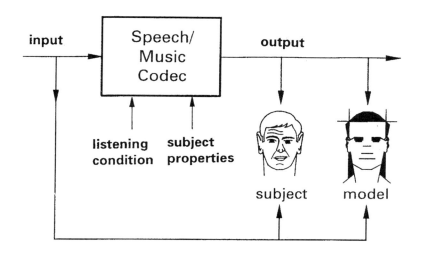

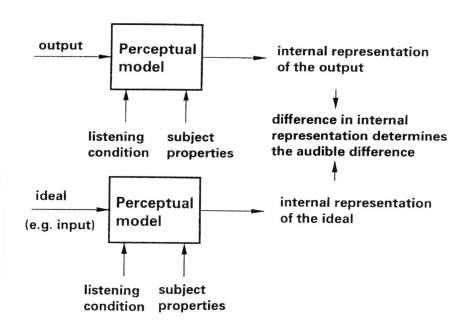

Figure 1.1   Overview of the basic philosophy used in the development of perceptual audio quality measurement techniques. A computer model of the subject is used to compare the output of the device under test (e.g. a speech codec or a music codec) with the ideal, using any audio signal. If the device under test must be transparent then the ideal is equal to the input.

The big advantage of the perceptual approach is that it is system independent and can be applied to any system, including systems that show a non-linear and time-variant behavior. A disadvantage is that for the characterization of the audio quality of a system one needs a large set of relevant test signals (speech and music signals).

If the perceptual approach is used for the prediction of subjectively perceived audio quality of the output of a linear, time-invariant system then the system characterization approach and the perceptual approach must lead to the same answer. In the system characterization approach one will first characterize the system and then interpret the results using knowledge of both the auditory system and the input signal for which one wants to determine the quality. In the perceptual approach one will characterize the perceptual quality of the output signals with the input signals as a reference.

Until recently several perceptual measurement techniques have been proposed but most of them are either focussed on speech codec quality [Gray and Markel, 1976], [Schroeder et al., 1979], [Gray et al., 1980], [Nocerino et al., 1985], [Quackenbush et al., 1988], [Hayashi and Kitawaki, 1992], [Halka and Heute, 1992], [Wang et al., 1992], [Ghitza, 1994] [Beerends and Stemerdink, 1994b] or on music codec quality [Paillard et al., 1992], [Brandenburg and Sporer, 1992], [Beerends and Stemerdink, 1992] [Colomes et al., 1994]. Although one would expect that a model for the measurement of the quality of wide band music codecs can be applied to telephone-band speech codecs, recent investigations show that this is rather difficult [Beerends, 1995].

In this chapter an overview is presented of the perceptual audio quality measure (PAQM) [Beerends and Stemerdink, 1992] and it will be shown that the PAQM approach can be used for the measurement of the quality of music and speech codecs. The PAQM method is currently under study within the ITU-R (International Telecommunication Union - Radio sector) [ITURsg10con9714, 1997], [ITURsg10con9719, 1997] for future standardization of a perception based audio quality measurement method. A simplified method, derived from PAQM, for measuring the quality of telephone-band (300-3400 Hz) speech codecs was standardized by the ITU-T (International Telecommunication Union - Telecom sector) as recommendation P.861 [ITUTrecP861, 1996] [ITUTsg12rep31.96, 1996]. Independent validation of this simplified method, called perceptual speech quality measure (PSQM), showed superior correlation between objective and subjective results, when compared to several other methods [ITUTsg12con9674, 1996].

A general problem in the development of perceptual measurement techniques is that one needs audio signals for which the subjective quality, when compared to a reference, is known. Creating databases of audio signals and their subjective quality is by no means trivial and many of the problems that are encountered in subjective testing have a direct relation to problems in perceptual measurement techniques. High correlations between objective and subjective results can only be obtained when the objective and subjective evaluation are closely related. In the next section some important points of discussion are given concerning the relation between subjective and objective perceptual testing.

1.3 SUBJECTIVE VERSUS OBJECTIVE PERCEPTUAL TESTING

In the development of perceptual measurement techniques one needs databases with reliable quality judgements, preferably using the same experimental setup and the same common subjective quality scale.

All the subjective results that will be used in this chapter come from large ITU databases for which subjects were asked to give their opinion on the quality of an audio fragment using a five point rating scale. The average of the quality judgements of the subjects gives a so called mean opinion score (MOS) on a five point scale. Subjective experiments in which the quality of telephone-band speech codecs (300-3400 Hz) or wideband music codecs (20-20000 Hz compact disc quality) were evaluated are used. For both, speech and music codec evaluation, the five point ITU MOS scale is used but the procedures in speech codec evaluation [CCITTrecP80, 1994] are different from the experimental procedures in music codec evaluation [CCIRrec562, 1990], [ITURrecBS1116, 1994].

In the speech codec evaluations, absolute category rating (ACR) was carried out with quality labels ranging from bad (MOS=1.0) to excellent (MOS=5.0) [CCITTrecP80, 1994]. In ACR experiments subjects do not have access to the original uncoded audio signal. In music codec evaluations a degradation category rating (DCR) scale was employed with quality labels ranging from "difference is audible and very annoying" (MOS=1.0) to "no perceptible difference" (MOS=5.0). The music codec databases used in this paper were all derived from DCR experiments where subjects had a known and a hidden reference [ITURrecBS1116, 1994].

In general it is not allowed to compare MOS values obtained in different experimental contexts. A telephone-band speech fragment may have a MOS that is above 4.0 in a certain experimental context while the same fragment may have a MOS that is lower than 2.0 in another context. Even if MOS values are obtained within the same experimental context but within a different cultural environment large differences in MOS values can occur [Goodman and Nash, 1982]. It is therefore impossible to develop a perceptual measurement technique that will predict correct MOS values under all conditions.

Before one can start predicting MOS scores several problems have to be solved. The first one is that different subjects have different auditory systems leading to a large range of possible models. If one wants to determine the quality of telephone-band speech codecs (300-3400 Hz) differences between subjects are only of minor importance. In the determination of the quality of wideband music codecs (compact disc quality, 20-20000 Hz) differences between subjects are a major problem, especially if the codec shows dynamic band limiting in the range of 10-20 kHz. Should an objective perceptual measurement technique use an auditory model that represents the best available (golden) ear, just model the average subject, or use an individual model for each subject [Treurniet, 1996]. The answer depends on the application. For prediction of mean opinion scores one has to adapt the auditory model to the average subject. In this chapter all perceptual measurements were done with a threshold of an average subject with an age between 20 and 30 years and an upper frequency audibility limit of 18 kHz. No accurate data on the subjects were available.

Another problem in subjective testing is that the way the auditory stimulus is presented has a big influence on the perceived audio quality. Is the presentation is in a quiet room or is there some background noise that masks small differences? Are the stimuli presented with loudspeakers that introduce distortions, either by the speaker itself or by interaction with the listening room? Are subjects allowed to adjust the volume for each audio fragment? Some of these differences, like loudness level and background noise, can be modelled in the perceptual measurement fairly easy, whereas for others it is next to impossible. An impractical solution to this problem is to make recordings of the output signal of the device under test and the reference signal (input signal) at the entrance of the ear of the subjects and use these signals in the perceptual evaluation.

In this chapter all objective perceptual measurements are done directly on the electrical output signal of the codec using a level setting that represents the average listening level in the experiment. Furthermore the background noise present during the listening experiments was modelled using a steady state Hoth noise [CCITTsup13, 1989]. In some experiments subjects were allowed to adjust the level individually for each audio fragment which leads to correlations that are possibly lower than one would get if the level in the subjective experiment would be fixed for all fragments. Correct setting of the level turned out be very important in the perceptual measurements.

It is clear that one can only achieve high correlations between objective measurements and subjective listening results when the experimental context is known and can be taken into account correctly by the perceptual or cognitive model.

The perceptual model as developed in this chapter is used to map the input and output of the audio device onto internal representations that are as close as possible to the internal representations used by the subject to judge the quality of the audio device. It is shown that the difference in internal representation can form the basis of a perceptual audio quality measure (PAQM) that has a high correlation with the subjectively perceived audio quality. Furthermore it is shown that with a simple cognitive module that interprets the difference in internal representation the correlation between objective and subjective results is always above 0.9 for both wideband music and telephone-band speech signals. For the measurement of the quality of telephone-band speech codecs a simplified version of the PAQM, the perceptual speech quality measure (PSQM), is presented.

Before introducing the method for calculating the internal representation the psychoacoustic fundamentals of the perceptual model is explained in the next chapter.

1.4 PSYCHOACOUSTIC FUNDAMENTALS OF CALCULATING THE INTERNAL SOUND REPRESENTATION

In thinking about how to calculate the internal representation of a signal one could dream of a method where all the transformation characteristics of the individual elements of the human auditory system would be measured and modelled. In this exact approach one would have the, next to impossible, task of modelling the ear, the transduction mechanism and the neural processing at a number of different abstraction levels.

Literature provides examples of the exact approach [Kates, 1991b], [Yang et al., 1992], [Giguère and Woodland, 1994a], [Giguère and Woodland, 1994b] but no results on large subjective quality evaluation experiments have been published yet. Preliminary results on using the exact approach to measure the quality of speech codecs have been published (e.g. [Ghitza, 1994]) but show rather disappointing results in terms of correlation between objective and subjective measurements. Apparently it is very difficult to calculate the correct internal sound representation on the basis of which subjects judge sound quality. Furthermore it may not be enough to just calculate differences in internal representations, cognitive effects may dominate quality perception.

One can doubt whether it is necessary to have an exact model of the lower abstraction levels of the auditory system (outer-, middle-, inner ear, transduction). Because audio quality judgements are, in the end, a cognitive process a crude approximation of the internal representation followed by a crude cognitive interpretation may be more appropriate then having an exact internal representation without cognitive interpretation of the differences.

In finding a suitable internal representation one can use the results of psychoacoustic experiments in which subjects judge certain aspects of the audio signal in terms of psychological quantities like loudness and pitch. These quantities already include a certain level of subjective interpretation of physical quantities like intensity and frequency. This psychoacoustic approach has led to a wide variety of models that can predict certain aspects of a sound e.g. [Zwicker and Feldtkeller, 1967], [Zwicker, 1977], [Florentine and Buus, 1981], [Martens, 1982], [Srulovicz and Goldstein, 1983], [Durlach et al., 1986], [Beerends, 1989], [Meddis and Hewitt, 1991]. However, if one wants to predict the subjectively perceived quality of an audio device a large range of the different aspects of sound perception has to be modelled. The most important aspects that have to be modelled in the internal representation are masking, loudness of partially masked time-frequency components and loudness of time-frequency components that are not masked.

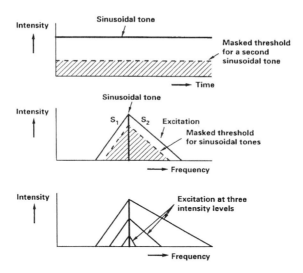

Figure 1.2 From the masking pattern it can be seen that the excitation produced by a sinusoidal tone is smeared out in the frequency domain. The right hand slope of the excitation pattern is seen to vary as a function of masker intensity (steep slope at low and flat slope at high intensities).
(Reprinted with permission from [Beerends and Stemerdink, 1992], ©Audio Engineering Society, 1992)

For stationary sounds the internal representation is best described by means of a spectral representation. The internal representation can be measured using a test signal having a small bandwidth. A schematic example for a single sinusoidal tone (masker) is given in Fig. 1.2 where the masked threshold of such a tone is measured with a second sinusoidal probe tone (target). The masked threshold can be interpreted as resulting from an internal representation that is given in Fig. 1.2 as an excitation pattern. Fig. 1.2 also gives an indication of the level dependence of the excitation pattern of a single sinusoidal tone. This level dependence makes interpretations in terms of filterbanks doubtful.

For non-stationary sounds the internal representation is best described by means of a temporal representation. The internal representation can be measured by means of a test signal of short duration. A schematic example for a single click (masker) is given in Fig. 1.3 where the masked threshold of such a click is measured with a second click (target). The masked threshold can be interpreted as the result of an internal, smeared out, representation of the puls (Fig. 1.3, excitation pattern).

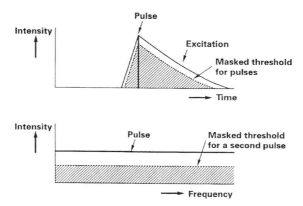

Figure 1.3   From the masking pattern it can be seen that the excitation produced by a click is smeared out in the time domain.
(Reprinted with permission from [Beerends and Stemerdink, 1992], ©Audio Engineering Society, 1992)

An example of a combination of time and frequency-domain masking, using a tone burst, is given in Fig. 1.4.

For the examples given in Figs. 1.2-1.4 one should realize that the masked threshold is determined with a target signal that is a replica of the masker signal. For target signals that are different from the masker signal (e.g. a sine that masks a band of noise) the masked threshold looks different, making it impossible to talk about *the masked threshold of a signal*. The masked threshold of a signal depends on the target, while the internal representation and the excitation pattern do not depend on the target.

In Figs. 1.2-1.4 one can see that any time-frequency component in the signal is smeared out along both the time and frequency axis. This smearing of the signal results in a limited time-frequency resolution of the auditory system. Furthermore it is known that two smeared out time-frequency components in the excitation domain do not add up to a combined excitation on the basis of energy addition. Therefore the smearing consists of two parts, one part describing how the energy at one point in the time-frequency domain results in excitation at another point, and a part that describes how the different excitations at a certain point, resulting from the smearing of the individual time-frequency components, add up.

Until now only time-frequency smearing of the audio signal by the ear, which leads to an excitation representation, has been described. This excitation representation is generally measured in dB SPL (Sound Pressure Level) as a function of time and frequency. For the frequency scale one does, in most cases, not use the linear Hz scale but the non-linear Bark scale. This Bark scale is a pitch scale representing the

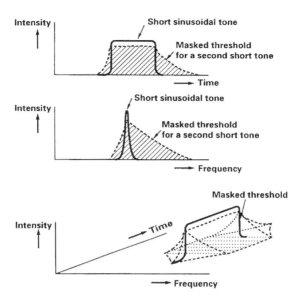

Figure 1.4 Excitation pattern for a short tone burst. The excitation produced by a short tone burst is smeared out in the time and frequency domain.
(Reprinted with permission from [Beerends and Stemerdink, 1992], ©Audio Engineering Society, 1992)

psychophysical equivalent of frequency. Although smearing is related to an important property of the human auditory system, viz. time-frequency domain masking, the resulting representation in the form of an excitation pattern is not very useful yet. In order to obtain an internal representation that is as close as possible to the internal representation used by subjects in quality evaluation one needs to compresses the excitation representation in a way that reflects the compression as found in the inner ear and in the neural processing.

The compression that is used to calculate the internal representation consists of a transformation rule from the excitation density to the compressed Sone density as formulated by Zwicker [Zwicker and Feldtkeller, 1967]. The smearing of energy is mostly the result of peripheral processes [Viergever, 1986] while compression is a more central process [Pickles, 1988]. With the two simple mathematical operations, smearing and compression, it is possible to model the masking properties of the auditory system not only at the masked threshold, but also the partial masking [Scharf, 1964] *above* masked threshold (see Fig. 1.5).

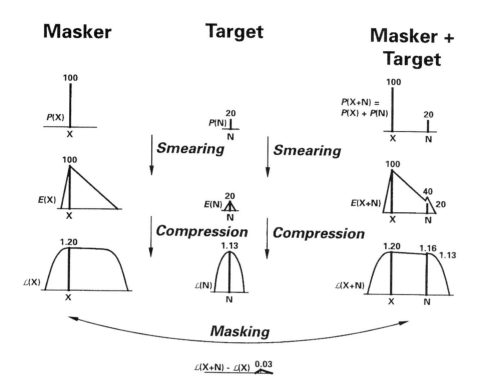

Figure 1.5 Overview on how masking is modelled in the internal representation model. Smearing and compression with $\mathcal{L} = E^{0.04}$ results in masking. The first representation (top) is in terms of power P and may represent clicks in the time domain or sines in the frequency domain. X represents the signal, or masker, and N the noise, or target. The left side shows transformations of the masker, in the middle the transformation of the target in isolation. The right side deals with the transformation of the composite signal (masker + target). The second representation is in terms of excitation E and shows the excitation as a function of time or frequency. The third representation is the internal representation using a simple compression $\mathcal{L} = E^{0.04}$. The bottom line shows the effect of masking, the internal representation of the target in isolation, $\mathcal{L}(N)$, is significantly larger than the internal representation of the target in the presence of a strong masker $\mathcal{L}(X+N) - \mathcal{L}(X)$.
(Reprinted with permission from [Beerends, 1995], ©Audio Engineering Society, 1995)

1.5 COMPUTATION OF THE INTERNAL SOUND REPRESENTATION

As a start in the quantification of the two mathematical operations, smearing and compression, used in the internal representation model one can use the results of psychoacoustic experiments on time-frequency masking and loudness perception. The frequency smearing can be derived from frequency domain masking experiments where a single steady-state narrow-band masker and a single steady-state narrow-band target are used to measure the slopes of the masking function [Scharf and Buus, 1986], [Moore, 1997]. These functions depend on the level and frequency of the masker signal. If one of the signals is a small band of noise and the other a pure tone then the slopes can be approximated by Eq. (1.1) (see Terhardt 1979, [Terhardt, 1979]):

$$S_1 = 31 \text{ dB/Bark, target frequency} < \text{masker frequency;} \qquad (1.1)$$

$$S_2 = (22 + \min(230/f, 10) - 0.2L) \text{ dB/Bark,}$$

target frequency > masker frequency;

with f the masker frequency in Hz and L the level in dB SPL. A schematic example of this frequency-domain masking is shown in Fig. 1.2. The masked threshold can be interpreted as resulting from a smearing of the narrow band signals in the frequency domain (see Fig. 1.2). The slopes as given in Eq. (1.1) can be used as an approximation of the smearing of the excitation in the frequency domain in which case the masked threshold can be interpreted as a fraction of the excitation.

If more than one masker is present at the same time the masked energy threshold of the composite signal $M_{\text{composite}}$ is not simply the sum of the n individual masked energy thresholds M_i but is given approximately by:

$$M_{\text{composite}} = \left(\sum_{i=1}^{n} M_i^\alpha\right)^{1/\alpha} \quad \alpha < 1. \qquad (1.2)$$

This addition rule holds for simultaneous (frequency-domain) [Lufti, 1983], [Lufti, 1985] and non-simultaneous (time-domain) [Penner, 1980], [Penner and Shiffrin, 1980] masking [Humes and Jesteadt, 1989] although the value of the compression power α may be different along the frequency (α_{freq}) and time (α_{time}) axis.

In the psychoacoustic model that is used in this chapter no masked threshold is calculated explicitly in any form. Masking is modelled by a combination of smearing and compression as explained in Fig. 5. Therefore the amount of masking is dependent on the parameters α_{freq} and α_{time} which determine, together with the slopes S_1 and S_2, the amount of smearing. However the values for α_{freq} and α_{time} found in literature were optimized with respect to the masked threshold and can thus not be used in our

14   APPLICATIONS OF DSP TO AUDIO AND ACOUSTICS

model. Therefore these two $\alpha's$ will be optimized in the context of audio quality measurements.

In the psychoacoustic model the physical time-frequency representation is calculated using a FFT with a 50% overlapping Hanning (sin^2) window of approximately 40 ms, leading to a time resolution of about 20 ms. Within this window the frequency components are smeared out according to Eq. (1.1) and the excitations are added according to Eq. (1.2). Due to the limited time resolution only a rough approximation of the time-domain smearing can be implemented.

From masking data found in the literature [Jesteadt et al., 1982] an estimate was made how much energy is left in a frame from a preceding frame using a shift of half a window (50% overlap). This fraction can be expressed as a time constant τ in the expression:

$$\Delta E(\Delta t) = e^{-\Delta t/\tau}, \tag{1.3}$$

with Δt = time distance between two frames = T_f. The fraction of the energy present in the next window depends on the frequency and therefore a different τ was used for each frequency band. This energy fraction also depends on the level of the masker [Jesteadt et al., 1982] but this level-dependency of τ yielded no improvement in the correlation and was therefore omitted from the model. At frequencies above 2000 Hz the smearing is dominated by neural processes and remains about the same [Pickles, 1988]. The values of τ are given in Fig. 1.6 and give an exponential approximation of time-domain masking using window shifts in the neighborhood of 20 ms.

An example of the decomposition of a sinusoidal tone burst in the time-frequency domain is given in Fig. 1.4. It should be realised that these time constants τ only give an exponential approximation, at the distance of half a window length, of the time-domain masking functions.

After having applied the time-frequency smearing operation one gets an excitation pattern representation of the audio signal in (dB$_{exc}$, seconds, Bark). This representation is then transformed to an internal representation using a non-linear compression function. The form of this compression function can be derived from loudness experiments.

Scaling experiments using steady-state signals have shown that the loudness of a sound is a non-linear function of the intensity. Extensive measurements on the relationship between intensity and loudness have led to the definition of the Sone. A steady-state sinusoid of 1 kHz at a level of 40 dB SPL is defined to have a loudness of one Sone. The loudness of other sounds can be estimated in psychoacoustic experiments. In a first approximation towards calculating the internal representation one would map the physical representation in dB/Bark onto a representation in Sone/Bark:

$$\mathcal{L} = k(P - P_0)^\gamma, \tag{1.4}$$

in which k is a scaling constant (about 0.01), P the level of the tone in μPa, P_0 the absolute hearing threshold for the tone in μPa, and γ the compression parameter, in

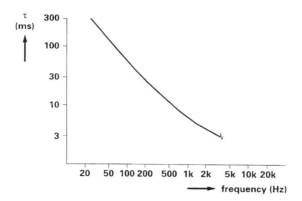

Figure 1.6 Time constant τ, that is used in the time-domain smearing, as a function of frequency. This function is only valid for window shifts of about 20 ms and only allows a crude estimation of the time-domain smearing, using a α_{time} of 0.6.
(Reprinted with permission from [Beerends and Stemerdink, 1992], ©Audio Engineering Society, 1992)

the literature estimated to be about 0.6 [Scharf and Houtsma, 1986]. This compression relates a physical quantity (acoustic pressure P) to a psychophysical quantity (loudness $\mathcal{L}$).

The Eqs (1.1), (1.2) and (1.4) involve quantities that can be measured directly. After application of Eq. (1.1) to each time frequency component and addition of all the individual excitation contributions using (1.2), the resulting excitation pattern forms the basis of the internal representation. (The exact method to calculate the excitation pattern is given in Appendix A, B and C of [Beerends and Stemerdink, 1992] while a compact algorithm is given in Appendix D of [Beerends and Stemerdink, 1992]).

Because Eq. (1.4) maps the physical domain directly to the internal domain it has to be replaced by a mapping from the excitation to the internal representation. Zwicker gave such a mapping (eq. 52,17 in [Zwicker and Feldtkeller, 1967]):

$$\mathcal{L} = k(\frac{E_0}{s})^\gamma \left[(1 - s + s\frac{E}{E_0})^\gamma - 1 \right], \quad (1.5)$$

in which k is an arbitrary scaling constant, E the excitation level of the tone, E_0 the excitation at the absolute hearing threshold for the tone, s the "schwell" factor as defined by Zwicker [Zwicker and Feldtkeller, 1967] and γ a compression parameter that was fitted to loudness data. Zwicker found an optimal value γ of about 0.23.

Although the γ of 0.23 may be optimal for the loudness scale it will not be appropriate for the subjective quality model which needs an internal representation that is

16   APPLICATIONS OF DSP TO AUDIO AND ACOUSTICS

as close as possible to the representation that is used by subjects to base their quality judgements on. Therefore γ is taken as a parameter which can be fitted to the masking behavior of the subjects in the context of audio quality measurements. The scaling k has no influence on the performance of the model. The parameter γ was fitted to the ISO/MPEG 1990 (International Standards Organization/Motion Picture Expert Group) database [ISO90, 1990] in terms of maximum correlation (minimum deviation) between objective and subjective results.

The composite operation, smearing followed by compression, results in partial masking (see Fig. 1.5). The advantage of this method is that the model automatically gives a prediction of the behavior of the auditory system when distortions are above masked threshold.

Summarizing, the model uses the following transformations (see Fig. 1.7):

- The input signal $x(t)$ and output signal $y(t)$ are transformed to the frequency domain, using an FFT with a Hanning ($\sin^2$) window $w(t)$ of about 40 ms. This leads to the physical signal representations $P_x(t, f)$ and $P_y(t, f)$ in (dB, seconds, Hz) with a time-frequency resolution that is good enough as a starting point for the time-frequency smearing.

- The frequency scale f (in Hz) is transformed to a pitch scale z (in Bark) and the signal is filtered with the transfer function $a_0(z)$ from outer to inner ear (free or diffuse field). This results in the power-time-pitch representations $p_x(t, z)$ and $p_y(t, z)$ measured in (dB, seconds, Bark). A more detailed description of this transformation is given in Appendix A of [Beerends and Stemerdink, 1992].

- The power-time-pitch representations $p_x(t, z)$ and $p_y(t, z)$ are multiplied with a frequency-dependent fraction $e^{-T_f/\tau(z)}$ using Eq. (1.3) and Fig. 1.6, for addition with α_{time} within the next frame (T_f = time shift between two frames $\approx$ 20 ms). This models the time-domain smearing of $x(t)$ and $y(t)$.

- The power-time-pitch representations $p_x(t, z)$ and $p_y(t, z)$ are convolved with the frequency-smearing function Λ, as can be derived from Eq. (1.1), leading to excitation-time-pitch (dB$_{exc}$, seconds, Bark) representations $E_x(t, z)$ and $E_y(t, z)$ (see Appendices B, C, D of [Beerends and Stemerdink, 1992]). The form of the frequency-smearing function depends on intensity and frequency, and the convolution is carried out in a non-linear way using Eq. (1.2) (see Appendix C of [Beerends and Stemerdink, 1992]) with parameter α_{freq}.

- The excitation-time-pitch representations $E_x(t, z)$ and $E_y(t, z)$ (dB$_{exc}$, seconds, Bark) are transformed to compressed loudness-time-pitch representations $\mathcal{L}_x(t, z)$ and $\mathcal{L}_y(t, z)$ (compressed Sone, seconds, Bark) using Eq. (1.5) with parameter γ (see Appendix E of [Beerends and Stemerdink, 1992]).

- The compressed loudness-time-pitch representation $\mathcal{L}_y(t,z)$ of the output of the audio device is scaled independently in three different pitch ranges with bounds at 2 and 22 Bark. This operation performs a global pattern matching between input and output representations and already models some of the higher, cognitive, levels of sound processing.

In psychoacoustic literature many experiments on masking behavior can be found for which the internal representation model should, in theory, be able to predict the behavior of subjects. One of these effects is the sharpening of the excitation pattern after switching off an auditory stimulus [Houtgast, 1977], which is partly modelled implicitly here in the form of the dependence of the slope S_2 in Eq. (1.1) on intensity. After "switching off" the masker the representation in the next frame in the model is a "sharpened version of the previous frame".

Another important effect is the asymmetry of masking between a tone masking a band of noise versus a noiseband masking a tone [Hellman, 1972]. In models using the masked threshold this effect has to be modelled explicitly by making the threshold dependent on the type of masker e.g. by calculating a tonality index as performed within the psychoacoustic models used in the ISO/MPEG audio coding standard [ISO92st, 1993]. Within the internal representation approach this effect is accounted for by the nonlinear addition of the individual time frequency components in the excitation domain.

1.6 THE PERCEPTUAL AUDIO QUALITY MEASURE (PAQM)

After calculation of the internal loudness-time-pitch representations of the input and output of the audio device the perceived quality of the output signal can be derived from the difference between the internal representations. The density functions $\mathcal{L}_x(t,z)$ (loudness density $\mathcal{L}$ as a function of time and pitch for the input x) and scaled $\mathcal{L}_y(t,z)$ are subtracted to obtain a noise disturbance density function $\mathcal{L}_n(t,z)$. This $\mathcal{L}_n(t,z)$ is integrated over frequency resulting in a momentary noise disturbance $\mathcal{L}_n(t)$ (see Fig. 1.7).

The momentary noise disturbance is averaged over time to obtain the noise disturbance $\mathcal{L}_n$. We will not use the term noise loudness because the value of γ is taken such that the subjective quality model is optimized; in that case $\mathcal{L}_n$ does not necessarily represent noise loudness. The logarithm ($\log_{10}$) of the noise disturbance is defined as the perceptual audio quality measure (PAQM).

The optimization of α_{freq}, α_{time} and γ is performed using the subjective audio quality database that resulted from the ISO/MPEG 1990 audio codec test [ISO90, 1990]. The optimization used the standard error of the estimated MOS from a third order regression line fitted through the PAQM, MOS datapoints. The optimization was carried out by minimization of the standard error of the estimated MOS as a function of α_{freq}, α_{time}, γ.

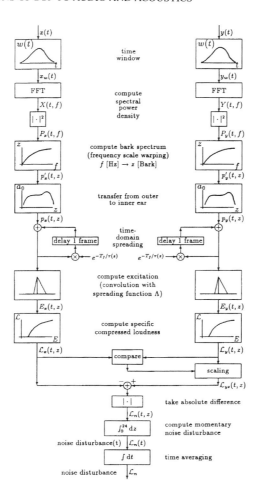

Figure 1.7 Overview of the basic transformations which are used in the development of the PAQM (Perceptual Audio Quality Measure). The signals $x(t)$ and $y(t)$ are windowed with a window $w(t)$ and then transformed to the frequency domain. The power spectra as function of time and frequency, $P_x(t, f)$ and $P_y(t, f)$ are transformed to power spectra as function of time and pitch, $p_x(t, z)$ and $p_y(t, z)$ which are convolved with the smearing function resulting in the excitations as a function of pitch $E_x(t, z)$ and $E_y(t, z)$. After transformation with the compression function we get the internal representations $\mathcal{L}_x(t, z)$ and $\mathcal{L}_y(t, z)$ from which the average noise disturbance $\mathcal{L}_n$ over the audio fragment can be calculated.
(Reprinted with permission from [Beerends and Stemerdink, 1992], ©Audio Engineering Society, 1992)

The optimal values of the parameters α_{freq} and α_{time} depend on the sampling of the time-frequency domain. For the values used in our implementation, $\Delta z = 0.2$ Bark and $\Delta t = 20$ ms (total window length is about 40 ms), the optimal values of the parameters in the model were found to be $\alpha_{freq} = 0.8$, $\alpha_{time} = 0.6$ and $\gamma = 0.04$. The dependence of the correlation on the time-domain masking parameter α_{time} turned out to be small.

Because of the small γ that was found in the optimization the resulting density as function of pitch (in Bark) and time does not represent the loudness density but a compressed loudness density. The integrated difference between the density functions of the input and the output therefore does not represent the loudness of the noise but the compressed loudness of the noise.

The relationship between the objective (PAQM) and subjective quality measure (MOS) in the optimal settings of α_{freq}, α_{time} and γ, for the ISO/MPEG 1990 database [ISO90, 1990], is given in Fig. 1.8. [1]

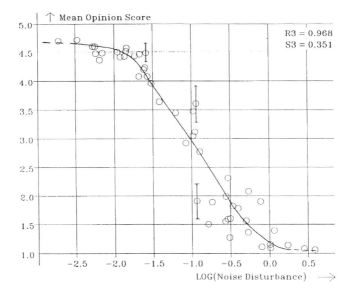

Figure 1.8 Relation between the mean opinion score and the perceptual audio quality measure (PAQM) for the 50 items of the ISO/MPEG 1990 codec test [ISO90, 1990] in loudspeaker presentation.
(Reprinted with permission from [Beerends and Stemerdink, 1992], ©Audio Engineering Society, 1992)

The internal representation of any audio signal can now be calculated by using the transformations given in the previous section. The quality of an audio device can thus be measured with test signals (sinusoids, sweeps, noise etc) as well as "real life" signals (speech, music). Thus the method is universally applicable. In general audio devices are tested for transparency (i.e. the output must resemble the input as closely as possible) in which case the input and output are both mapped onto their internal representations and the quality of the audio device is determined by the difference between these input (the reference) and output internal representations.

1.7 VALIDATION OF THE PAQM ON SPEECH AND MUSIC CODEC DATABASES

The optimization of the PAQM that is described in the previous section results in a PAQM that shows a good correlation between objective and subjective results. In this section the PAQM is validated using the results of the second ISO/MPEG audio codec test (ISO/MPEG 1991 [ISO91, 1991]) and the results of the ITU-R TG10/2 1993 [ITURsg10cond9343, 1993] audio codec test. In this last test several tandeming conditions of ISO/MPEG Layer II and III were evaluated subjectively while three different objective evaluation models presented objective results.

This section also gives a validation of the PAQM on databases that resulted from telephone-band (300-3400 Hz) speech codec evaluations.

The result of the validation using the ISO/MPEG 1991 database is given in Fig. 1.9. A good correlation (R3=0.91) and a reasonable low standard error of the estimate (S3=0.48) between the objective PAQM and the subjective MOS values was found.

A point of concern is that for the same PAQM values sometimes big deviations in subjective scores are found (see Fig. 1.9). [2]

The result of the validation using the ITU-R TG10/2 1993 database (for the Contribution Distribution Emission test) is given in Fig. 1.10 [3] and shows a good correlation and low standard error of the estimate (R3=0.83 and S3=0.29) between the objective PAQM and the subjective MOS. These results were verified by the Swedish Broadcasting Corporation [ITURsg10cond9351, 1993] using a software copy that was delivered before the ITU-R TG10/2 test was carried out.

The two validations that were carried out both use databases in which the subjective quality of the output signals of music codecs was evaluated. If the PAQM is really a universal audio quality measure it should also be applicable to speech codec evaluation. Although speech codecs generally use a different approach towards data reduction of the audio bitstream than music codecs the quality judgement of both is always carried with the same auditory system. A universal objective perceptual approach towards quality measurement of speech and music codecs must thus be feasible. When looking into the literature one finds a large amount of information on how to measure the quality of speech codecs (e.g. [Gray and Markel, 1976], [Schroeder et al., 1979], [Gray et al.,

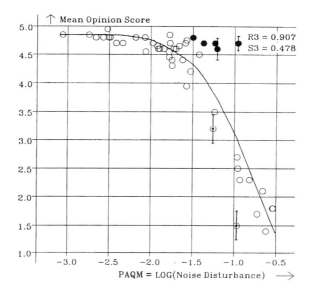

Figure 1.9 Relation between the mean opinion score (MOS) and the perceptual audio quality measure (PAQM) for the 50 items of the ISO/MPEG 1991 codec test [ISO91, 1991] in loudspeaker presentation. The filled circles are items whose quality was judged significantly lower by the model than by the subjects.
(Reprinted with permission from [Beerends and Stemerdink, 1992], ©Audio Engineering Society, 1992)

1980], [Nocerino et al., 1985], [Quackenbush et al., 1988], [Hayashi and Kitawaki, 1992], [Halka and Heute, 1992], [Wang et al., 1992], [Ghitza, 1994] [Beerends and Stemerdink, 1994b]), but non of the methods can be used for both narrowband speech and wideband music codecs.

To test whether the PAQM can be applied to evaluation of speech codec quality a validation was setup using subjective test results on the ETSI GSM (European Telecommunications Standards Institute, Global System for Mobile communications) candidate speech codecs.   Both the GSM full rate (13 kbit/s, [Natvig, 1988]) and half rate (6 kbit/s, [ETSI91tm74, 1991]) speech codec evaluations were used in the validation. In these experiments the speech signals were judged in quality over a standard telephone handset [CCITTrecP48, 1989].   Consequently in validating the PAQM both the reference input speech signal and the degraded output speech signal were filtered using the standard telephone filter characteristic [CCITTrecP48, 1989]. Furthermore the speech quality evaluations were carried out in a controlled noisy

22 APPLICATIONS OF DSP TO AUDIO AND ACOUSTICS

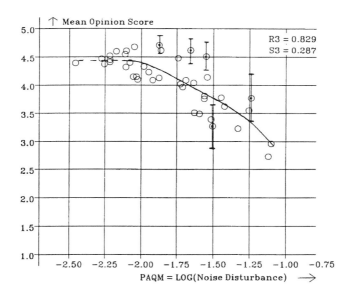

Figure 1.10 Relation between MOS and PAQM for the 43 ISO layer II tandeming conditions of the ITU-R TG10/2 1993 [ITURsg10cond9343, 1993] audio codec test (Reprinted with permission from [Beerends and Stemerdink, 1994a], ©Audio Engineering Society, 1994)

environment using Hoth noise as a masking background noise. Within the PAQM validation this masking noise was modelled by adding the correct spectral level of Hoth noise [CCITTsup13, 1989] to the power-time-pitch representations of input and output speech signal.

The results of the validation on speech codecs are given in Figs. 1.11 and 1.12. One obvious difference between this validation and the one carried out using music codecs is the distribution of the PAQM values. For music the PAQM values are all below -0.5 (see Figs. 1.9, 1.10) while for speech they are mostly above -0.5 (see Figs. 1.11, [4] 1.12 [5]). Apparently the distortions in these databases are significantly larger than those in the music databases. Furthermore the correlation between objective and subjective results of this validation are worse then those of the validation using music codecs.

1.8 COGNITIVE EFFECTS IN JUDGING AUDIO QUALITY

Although the results of the validation of the PAQM on the music and speech codec databases showed a rather good correlation between objective and subjective results, improvements are still necessary. The reliability of the MOS predictions is not good

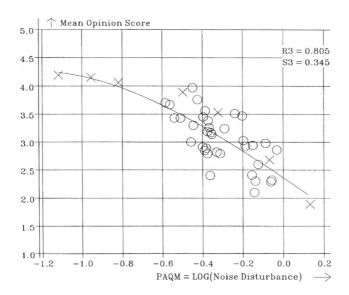

Figure 1.11 Relation between the MOS and the PAQM for the ETSI GSM full rate speech database. Crosses represent data from the experiment based on the modulated noise reference unit, circles represent data from the speech codecs.

enough for the selection of the speech or music codec with the highest audio quality. As stated in the section on the psychoacoustic fundamentals of the method, it may be more appropriate to have crude perceptual model combined with a crude cognitive interpretation then having an exact perceptual model. Therefore the biggest improvement is expected to come from a better modelling of cognitive effects. In the PAQM approach as presented until now, the only cognitive effect that is modelled is the overall timbre matching in three different frequency regions. This section will focus on improvements in the cognitive domain and the basic approach as given in Fig. 1.1 is modified slightly (see Fig. 1.13) by incorporating a central module which interprets differences in the internal representation.

Possible central, cognitive, effects that are important in subjective audio quality assessment are:

1. **Informational masking**, where the masked threshold of a complex target masked by a complex masker may decrease after training by more than 40 dB [Leek and Watson, 1984].

24   APPLICATIONS OF DSP TO AUDIO AND ACOUSTICS

2. **Separation of linear from non-linear distortions**. Linear distortions of the input signal are less objectionable than non-linear distortions.

3. **Auditory scene analysis**, in which decisions are made as to which parts of an auditory event integrate into one percept [Bregman, 1990].

4. **Spectro-temporal weighting**. Some spectro-temporal regions in the audio signal carry more information, and may therefore be more important, than others. For instance one expects that silent intervals in speech carry no information are therefore less important.

1) Informational masking can be modelled by defining a spectro-temporal complexity, entropy like, measure. The effect is most probably dependent on the amount of training that subjects are exposed to before the subjective evaluation is carried out. In general, quality evaluations of speech codecs are performed by naive listeners [CCITTrecP80, 1994], while music codecs are mostly evaluated by expert listeners [CCIRrec562, 1990], [ITURrecBS1116, 1994].

For some databases the informational masking effect plays a significant role and modelling this effect turned out to be mandatory for getting high correlations between objective and subjective results [Beerends et al., 1996]. The modelling can best be done by calculating a local complexity number over a time window of about 100 ms. If

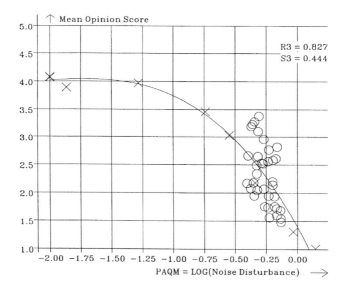

Figure 1.12   The same as Fig. 1.11 but for the ETSI GSM half rate speech database.

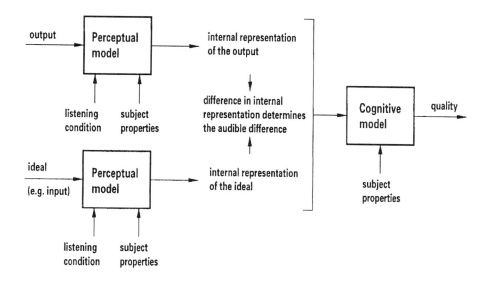

Figure 1.13 Basic approach used in the development of PAQM$_C$, the cognitive corrected PAQM. Differences in internal representation are judged by a central cognitive module.
(Reprinted with permission from [Beerends, 1995], ©Audio Engineering Society, 1995)

this local complexity is high then distortions within this time window are more difficult to hear then when the local complexity is low [Beerends et al., 1996].

Although the modelling of informational masking gives higher correlations for some databases, other databases may show a decrease in correlation. No general formulation was found yet that could be used to model informational masking in a satisfactory, general applicable, way. Modelling of this effect is therefore still under study and not taken into account here.

2) Separation of linear from non-linear distortions can be implemented fairly easy by using adaptive inverse filtering of the output signal. However it gave no significant improvement in correlation between objective and subjective results using the available databases (ISO/MPEG 1990, ISO/MPEG 1991, ITU-R 1993, ETSI GSM full rate 1988, ETSI GSM half rate 1991).

Informal experiments however showed that this separation is important when the output signal contains severe linear distortions.

26   APPLICATIONS OF DSP TO AUDIO AND ACOUSTICS

3) Auditory scene analysis is a cognitive effect that describes how subjects separate different auditory events and group them into different objects. Although a complete model of auditory scene analysis is beyond the scope of this chapter the effect was investigated in more detail. A pragmatic approach as given in [Beerends and Stemerdink, 1994a] turned out to be very successful in quantifying an auditory scene analysis effect. The idea in this approach is that if a time-frequency component is not coded by a codec, the remaining signal still forms one coherent auditory scene while introduction of a new unrelated time-frequency component leads to two different percepts. Because of the split in two different percepts the distortion will be more objectionable then one would expect on the basis of the loudness of the newly introduced distortion component. This leads to a perceived asymmetry between the disturbance of a distortion that is caused by not coding a time-frequency component versus the disturbance caused by the introduction of a new time-frequency component.

In order to be able to model this cognitive effect it was necessary to quantify to what extent a distortion, as found by the model, resulted from leaving out a time-frequency component or from the introduction of a new time-frequency component in the signal. One problem was that when a distortion is introduced in the signal at a certain time-frequency point there will in general already be a certain power level at that point. Therefore a time-frequency component will never be completely new. A first approach to quantify the asymmetry was to use the power ratio between output and input at a certain time-frequency point to quantify the "newness" of this component. When the power ratio between the output y and input x, p_y/p_x at a certain time-frequency point is larger than 1.0 an audible distortion is assumed more annoying than when this ratio is less than 1.0.

In the internal representation model the time-frequency plane is divided in cells with a resolution of 20 ms along in the time axis (time index m) and of 0.2 Bark along the frequency axis (frequency index l). A first approach was to use the power ratio between the output y and input x, p_y/p_x in every $(\Delta t, \Delta f)$ cell (m, l) as a correction factor for the noise disturbance $\mathcal{L}_n(m, l)$ in that cell (nomenclature is chosen to be consistent with [Beerends and Stemerdink, 1992]).

A better approach turned out to be to average the power ratio p_y/p_x between the output y and input x over a number of consecutive time frames. This implies that if a codec introduces a new time-frequency component this component will be more annoying if it is present over a number of consecutive frames. The general form of the cognitive correction is defined as:

$$\mathcal{L}_{Cn}(m,l) = \begin{cases} C(m,l)^\lambda \, \mathcal{L}_n(m,l) & \text{if } C(m,l) < 5 \\ 5^\lambda \, \mathcal{L}_n(m,l) & \text{if } C(m,l) \geq 5 \end{cases}$$

with

$$C(m,l) = \sum_{i=0}^{4} \frac{p_y(m-i,l)}{p_x(m-i,l)} \quad (1.6)$$

and an additional clipping of the noise disturbance in each time window $\mathcal{L}_{Cn}(m) = \sum_{l=1}^{l=maximum} \mathcal{L}_{Cn}(m,l)$ at a level of 20.

The simple modelling of auditory scene analysis with the asymmetry factor $C(m,l)$ gave significant improvements in correlation between objective and subjective results. However it was found that for maximal correlation the amount of correction, as quantified by the parameter λ, was different for speech and music. When applied to music databases the optimal corrected noise disturbance was found for $\lambda = 1.4$ (PAQM$_{C1.4}$) whereas for speech databases the optimal λ was around 4.0 (PAQM$_{C4.0}$).

The results for music codec evaluations are given in Fig. 1.14[6] (ISO/MPEG 1991) and Fig. 1.15[7] (ITU-R TG10/2 1993) and show a decrease in the standard error of the MOS estimate of more than 25%. For the ISO/MPEG 1990 database no improvement was found. For speech the improvement in correlation was slightly less but as it turned out the last of the listed cognitive effects, spectro-temporal weighting, dominates subjective speech quality judgements. The standard error of the MOS estimate in the speech databases could be decreased significantly more when both the asymmetry and spectro-temporal weighting are modelled simultaneously.

4) Spectro-temporal weighting was found to be important only in quality judgements on speech codecs. Probably in music all spectro-temporal components in the signal, even silences, carry information, whereas for speech some spectro-temporal components, like formants, clearly carry more information then others, like silences. Because speech databases used in this paper are all telephone-band limited spectral weighting turned out to be only of minor importance and only the weighting over time had to be modelled.

This weighting effect over time was modelled in a very simple way, the speech frames were categorized in two sets, one set of speech active frames and one set of silent frames. By weighting the noise disturbance occurring in silent frames with a factor W_{sil} between 0.0 (silences are not taken into account) and 0.5 (silences are equally important as speech) the effect was quantified.

A problem in quantifying the silent interval behavior is that the influence of the silent intervals depends directly on the length of these intervals. If the input speech does not contain any silent intervals the influence is zero. If the input speech signal contains a certain percentage of silent frames the influence is proportional to this percentage. Using a set of trivial boundary conditions with $\mathcal{L}_{spn}$ the average noise disturbance over speech active frames and $\mathcal{L}_{siln}$ the average noise disturbance over silent frames one can show that the correct weighting is:

$$\mathcal{L}_{Wn} = \frac{W_{sp} \cdot p_{sp}}{W_{sp} \cdot p_{sp} + p_{sil}} \mathcal{L}_{spn} + \frac{p_{sil}}{W_{sp} \cdot p_{sp} + p_{sil}} \mathcal{L}_{siln} \qquad (1.7)$$

with:

$\mathcal{L}_{Wn}$ the noise disturbance corrected with a weight factor W_{sil},
$W_{sp} = (1 - W_{sil})/W_{sil}$,
p_{sil} the fraction of silent frames,
p_{sp} the fraction of speech active frames ($p_{sil} + p_{sp} = 1.0$).

When both the silent intervals and speech active intervals are equally important, such as found in music codec testing, the weight factor W_{sil} is equal to 0.5 and Eq. (1.7) brakes down to $\mathcal{L}_{Wn} = p_{sp}.\mathcal{L}_{spn} + p_{sil}.\mathcal{L}_{siln}$. For both of the speech databases the weight factor for silent interval noise for maximal correlation between objective and subjective results was found to be 0.1 showing that noise in silent intervals is less disturbing than equally loud noise during speech activity.

When *both* the asymmetry effect, resulting from the *auditory scene analysis*, and the *temporal weighting* are quantified correctly, the correlation between subjective and objective results for both of the speech databases improves significantly. Using $\lambda = 4.0$ (asymmetry modelling) and a silent interval weighting of 0.1 (denoted as

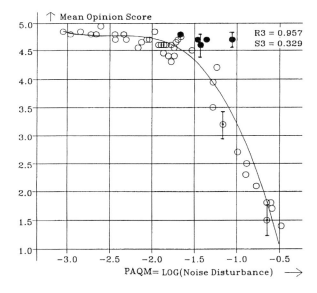

Figure 1.14 Relation between the mean opinion score (MOS) and the cognitive corrected PAQM (PAQM$_{C1.4}$) for the 50 items of the ISO/MPEG 1991 codec test [ISO91, 1991] in loudspeaker presentation.
(Reprinted with permission from [Beerends and Stemerdink, 1992], ©Audio Engineering Society, 1992)

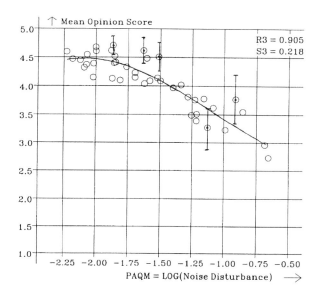

Figure 1.15 Relation between MOS and cognitive corrected PAQM (PAQM$_{C1.4}$) for the 43 ISO layer II tandeming conditions of the ITU-R TG10/2 1993 [ITURsg10cond9343, 1993] audio codec test.
(Reprinted with permission from [Beerends and Stemerdink, 1994a], ©Audio Engineering Society, 1994)

PAQM$_{C4.0,W0.1}$) the decrease in the standard error of the MOS estimate is around 40% for both the ETSI GSM full rate (see Fig. 1.16)[8] and half rate database (see Fig. 1.17[9]).

One problem of the resulting two cognitive modules is that predicting the subjectively perceived quality is dependent on the experimental context. One has to set values for the asymmetry effect and the weighting of the silent intervals in advance.

1.9 ITU STANDARDIZATION

Within the ITU several study groups deal with audio quality measurements. However, only two groups specifically deal with objective perceptual audio quality measurements. ITU-T Study Group 12 deals with the quality of telephone-band (300-3400 Hz) and wide-band speech signals, while ITU-R Task Group 10/4 deals with the quality of speech and music signals in general.

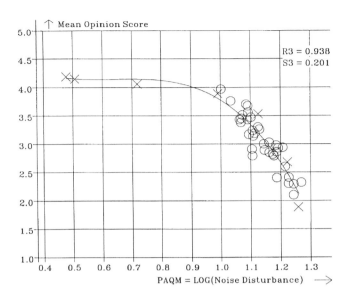

Figure 1.16 Relation between the MOS and the cognitive corrected PAQM (PAQM$_{C4.0,W0.1}$) for the ETSI GSM full rate speech database. Crosses represent data from the experiment based on the modulated noise reference unit, circles represent data from the speech codecs.

1.9.1 ITU-T, speech quality

Within ITU-T Study Group 12 five different methods for measuring the quality of telephone-band (300-3400 Hz) speech signals were proposed.

The first method, the cepstral distance, was developed by the NTT (Japan). It uses the cepstral coefficients [Gray and Markel, 1976] of the input and output signal of the speech codec.

The second method, the coherence function, was developed by Bell Northern Research (Canada). It uses the coherent (signal) and non-coherent (noise) powers to derive a quality measure [CCITT86sg12con46, 1986].

The third method was developed by the Centre National D'Etudes des Télécommunication (France) and is based on the concept of mutual information. It is called the information index and is described in the ITU-T series P recommendations [CCITTsup3, 1989] (supplement 3, pages 272-281).

The fourth method is a statistical method that uses multiple voice parameters and a non linear mapping to derive a quality measure via a training procedure on a training set

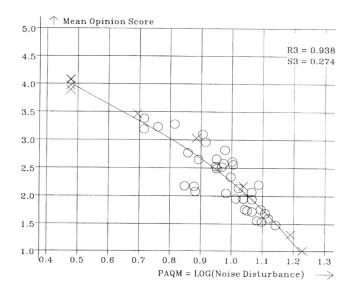

Figure 1.17 The same as Fig. 1.16 but for the ETSI GSM half rate speech database using PAQM$_{C4.0, W0.1}$.

of data. It is an expert pattern recognition technique and was developed by the National Telecommunication Information Administration (USA) [Kubichek et al., 1989].

The last method that was proposed is the perceptual speech quality measure (PSQM), a method derived from the PAQM as described in this chapter. It uses a simplified internal representation without taking into account masking effects that are caused by time-frequency smearing. Because of the band limitation used in telephone-band speech coding and because distortions are always rather large, masking effects as modelled in the PAQM are less important. In fact it has been shown that when cognitive effects as described in the previous chapter are not taken into account the modelling of masking behavior caused by time-frequency smearing may even lead to lower correlations [Beerends and Stemerdink, 1994b]. Within the PSQM masking is only taken into account when two time-frequency components coincide in both the time and frequency domain. The time frequency mapping that is used in the PSQM is exactly the same as the one used in the PAQM. Further simplifications used in the PSQM are:

32   APPLICATIONS OF DSP TO AUDIO AND ACOUSTICS

- No outer ear transfer function $a_0(z)$.

- A simplified mapping from intensity to loudness.

- A simplified cognitive correction for modelling the asymmetry effect.

An exact description of the PSQM method is given in [ITUTrecP861, 1996].

Although the PSQM uses a rather simple internal representation model the correlation with the subjectively perceived speech quality is very high. For the two speech quality databases that were used in the PAQM validation the method even gives a minor improvement in correlation. Because of a difference in the mapping from intensity to loudness a different weighting for the silent intervals has to be used (compare Figs. 1.16, 1.17 with 1.18,[10] 1.19[11]).

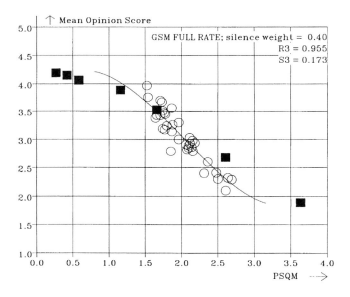

Figure 1.18   Relation between the MOS and the PSQM for the ETSI GSM full rate speech database. Squares represent data from the experiment based on the modulated noise reference unit, circles represent data from the speech codecs.

Within ITU-T Study Group 12 a benchmark was carried out by the NTT (Japan) on the five different proposals for measuring the quality of telephone-band speech codecs. The results showed that the PSQM was superior in predicting the subjective MOS values. The correlation on the unknown benchmark database was 0.98 [ITUTsg12con9674, 1996]. In this benchmark the asymmetry value λ for the PSQM

AUDIO QUALITY DETERMINATION USING PERCEPTUAL MEASUREMENT

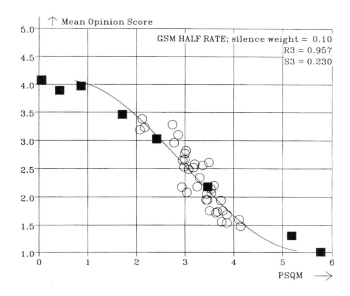

Figure 1.19 The same as Fig. 1.18 but for the ETSI GSM half rate speech database using the PSQM.

was fixed and three different weighting factors for the silent intervals were evaluated. The PSQM method was standardized by the ITU-T as recommendation P.861 [ITUTrecP861, 1996], objective quality measurement of telephone-band (300-3400 Hz) speech codecs.

A problem in the prediction of MOS values in speech quality evaluations is the weight factor of the silent intervals which depends on the experimental context. Within the ITU-T Study Group 12 benchmark the overall best performance was found for a weight factor of 0.4. However as can be seen in Fig. 1.19 the optimum weight factor can be significantly lower. In recommendation P.861 this weight factor of the silent intervals is provisionally set to 0.2. An argument for a low setting of the silent interval weight factor is that in real life speech codecs are mostly used in conversational contexts. When one is talking over a telephone connection the noise on the line present during talking is masked by ones own voice. Only when both parties are not talking this noise becomes apparent. In the subjective listening test however this effect does not occur because subjects are only required to listen. In all ITU-T and ETSI speech codec tests the speech material contained about 50% speech activity, leading to an overestimation of the degradation caused by noise in silent intervals.

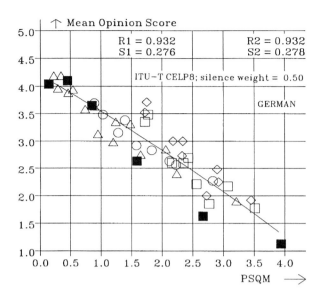

Figure 1.20 Relation between the PSQM and the MOS in experiment 2 of the ITU-T 8 kbit/s 1993 speech codec test for the German language. The silent intervals are weighted with the optimal weighting factor (0.5). Squares represent data from the experiment based on the modulated noise reference unit, the other symbols represent data from the speech codecs.

When the silent interval weighting in an experiment is known the PSQM has a very high correlation with the subjective MOS. In order to compare the reliability of subjective and objective measurements one should correlate two sets of subjective scores that are derived from the same set of speech quality degradations and compare this result with the correlation between the PSQM and subjective results. During the standardization of the G.729 speech codec [CCITTrecG729, 1995] a subjective test was performed at four laboratories with four different languages using the same set of speech degradations [ITUTsg12sq2.93, 1993], [ITUTsg12sq3.94, 1994]. The correlation between the subjective results and objective results, using the optimal weight factor, was between 0.91 and 0.97 for all four languages that were used [Beerends94dec, 1994]. The correlation between the subjective scores of the different languages varied between 0.85 and 0.96. For two languages, German and Japanese, the results are reproduced in Figs. 1.20[12], 1.21[13] and 1.22[14]. These results show that the PSQM is capable of predicting the correct mean opinion scores with an accuracy that is about the same as

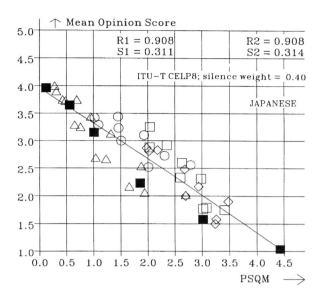

Figure 1.21 The same as Fig. 1.20 but for the Japanese language.

the accuracy obtained from a subjective experiment, once the experimental context is known.

1.9.2 ITU-R, audio quality

Within ITU-R Task Group 10/4 the following six methods for measuring the quality of audio signals were proposed:

- Noise to Mask Ratio (NMR, Fraunhofer Gesellschaft, Institut für Integrierte Schaltungen, Germany, [Brandenburg and Sporer, 1992])

- PERCeptual EVALuation method (PERCEVAL, Communications Research Centre,
  Canada [Paillard et al., 1992])

- Perceptual Objective Model (POM, Centre Commun d'Etudes de Télédiffusion et Télécommunication, France, [Colomes et al., 1994])

- Disturbance Index (DI, Technical University Berlin, [Thiede and Kabot, 1996])

- The toolbox (Institut für Rundfunk Technik, Germany)

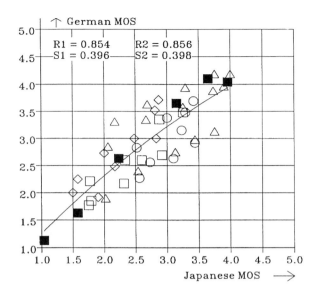

Figure 1.22 Relation between the Japanese and German MOS values using the subjective data of experiment 2 of the ITU-T 8 kbit/s 1993 speech codec test. Squares represent data from the experiment based on the modulated noise reference unit, the other symbols represent data from the speech codecs.

- Perceptual Audio Quality Measure (PAQM, Royal PTT Netherlands, [Beerends and Stemerdink, 1992], [Beerends and Stemerdink, 1994a])

The context in which these proposals were validated was much wider than the context used in the ITU-T Study Group 12 validation. Besides a number of audio codec conditions several types of distortions were used in the subjective evaluation. Because of this wide context each proponent was allowed to put in three different versions of his objective measurement method.

The wide validation context made it necessary to extend the PAQM method to include some binaural processing. Furthermore different implementations of the asymmetry effect were used and also a first attempt to model informational masking was included [Beerends et al., 1996].

Although the PAQM method showed highest correlation between objective and subjective results none of the eighteen (3*6) methods could be accepted as ITU-R recommendation [ITURsg10con9714, 1997]. Currently in a joint effort between the six proponents a new method is being developed, based on all eighteen proposals. [ITURsg10con9719, 1997].

1.10 CONCLUSIONS

A method for measuring audio quality, based on the internal representation of the audio signal, has been presented. The method does not characterize the audio system, but the perception of the output signal of the audio system. It can be applied to measurement problems where a reference and a degraded output signal are available. For measurement of audio codec quality the input signal to the codec is used as a reference and the assumption is made that all differences that are introduced by the codec lead to a degradation in quality.

In the internal representation approach the quality of an audio device is measured by mapping the reference and output of the device from the physical signal representation (measured in dB, seconds, Hertz) onto a psychoacoustic (internal) representation (measured in compressed Sone, seconds, Bark). From the difference in internal representation the perceptual audio quality measure (PAQM) can be calculated which shows good correlation with the subjectively perceived audio quality.

The PAQM is optimized using the ISO/MPEG music codec test of 1990 and validated with several speech and music databases. The PAQM can be improved significantly by incorporation of two cognitive effects. The first effect deals with the asymmetry between the disturbance of a distortion that is caused by not coding a time-frequency component versus the disturbance caused by the introduction of a new time-frequency component. The second effect deals with the difference in perceived disturbance between noise occurring in silent intervals and noise occurring during the presence of audio signals. This last correction is only relevant in quality measurements on speech codecs. When both cognitive effects are modelled correctly the correlations between objective and subjective results are above 0.9 using three different music codec databases and two different speech codec databases.

For measurement of the quality of telephone-band speech codecs a simplified method, the perceptual speech quality measure (PSQM), is presented. The PSQM was benchmarked together with four other speech quality measurement methods within ITU-T Study Group 12 by the NTT (Japan). It showed superior performance in predicting subjective mean opinion scores. The correlation on the unknown benchmark database was 0.98 [ITUTsg12con9674, 1996]. The PSQM method was standardized by the ITU-T as recommendation P.861 [ITUTrecP861, 1996], objective quality measurement of telephone-band (300-3400 Hz) speech codecs.

Notes

1. The 95% confidence intervals of the MOS lie in the range of 0.1-0.4. For some items, which differ significantly from the fitted curve, the confidence intervals are given. The correlation and standard error of the estimate (R3=0.97 and S3=0.35) are derived from the third order regression line that is drawn using a NAG curve fitting routine.

2. The 95% confidence intervals of the MOS lie in the range of 0.1-0.4. For some items, which differ significantly from the fitted curve, the confidence intervals are given. The correlation and standard error of the estimate (R3=0.91 and S3=0.48) are derived from the third order regression line that is drawn using a NAG curve fitting routine.

3. The 95% confidence intervals of the MOS lie in the range of 0.1-0.5. For some items, which differ significantly from the fitted curve, the confidence intervals are given. The correlation and standard error of the estimate (R3=0.83 and S3=0.29) are derived from the third order regression line that is drawn using a NAG curve fitting routine. The result as given in this figure was validated by the Swedish Broadcasting Corporation [ITURsg10cond9351, 1993].

4. The correlation and standard error of the estimate (R3=0.81 and S3=0.35) are derived from the third order regression line that is drawn using a NAG curve fitting routine.

5. The correlation and standard error of the estimate (R3=0.83 and S3=0.44) are derived from the third order regression line.

6. The 95% confidence intervals of the MOS lie in the range of 0.1-0.4. For some items, which differ significantly from the fitted curve, the confidence intervals are given. The filled circles are the same items as indicated in Fig. 1.9. The correlation and standard error of the estimate (R3=0.96 and S3=0.33) are derived from the third order regression line that is drawn using a NAG curve fitting routine.

7. The 95% confidence intervals of the MOS lie in the range of 0.1-0.5. For some items, which differ significantly from the fitted curve, the confidence intervals are given. The correlation and standard error of the estimate (R3=0.91 and S3=0.22) are derived from the third order regression line that is drawn using a NAG curve fitting routine. The result as given in this figure was validated by the Swedish Broadcasting Corporation [ITURsg10cond9351, 1993].

8. The correlation and standard error of the estimate (R3=0.94 and S3=0.20) are derived from the third order regression line that is drawn using a NAG curve fitting routine.

9. The correlation and standard error of the estimate (R3=0.94 and S3=0.27) are derived from the third order regression line.

10. The correlation and standard error of the estimate (R3=0.96 and S3=0.17) are derived from the third order regression line that is drawn using a NAG curve fitting routine.

11. The correlation and standard error of the estimate (R3=0.96 and S3=0.23) are derived from the third order regression line.

12. The correlations and standard errors that are given are derived from a first (R1, S1) and second (R2, S2) order regression line calculated with a NAG curve fitting routine. The second order regression line is drawn

13. The silent intervals are weighted with the optimal weighting factor (0.4). The correlations and standard errors that are given are derived from a first (R1, S1) and second (R2, S2) order regression line. The second order regression line is drawn. line.

14. The correlations and standard errors that are given are derived from a first (R1, S1) and second (R2, S2) order regression line calculated with a NAG curve fitting routine. The second order regression line is drawn.

2 PERCEPTUAL CODING OF HIGH QUALITY DIGITAL AUDIO

Karlheinz Brandenburg

Fraunhofer Institut Integrierte Schaltungen,
Audio/Multimedia Department, Erlangen, Germany
bdg@iis.fhg.de

Abstract: Perceptual coding of high quality digital audio signals or in short "audio compression" is one of the basic technologies of the multimedia age. This chapter introduces the basic ideas of perceptual audio coding and discusses the different options for the main building blocks of a perceptual coder. Several well known algorithms are described in detail.

2.1 INTRODUCTION

Perceptual coding of high quality digital audio is without doubt one of the most exciting chapters in applying signal processing to audio technology. The goal of this chapter is to describe the basic technologies and to introduce some of the refinements which are used to make decompressed sound perceptually equivalent to the original signal.

While the aggregate bandwidth for the transmission of audio (and video) signals is increasing every year, the demand increases even more. This leads to a large demand for compression technology. In the few years since the first systems and the first standardization efforts, perceptual coding of audio signals has found its way to a growing number of consumer applications. In addition, the technology has been used for a large number of low volume professional applications.

Application areas of audio coding. Current application areas include

- Digital Broadcasting: e.g. DAB (terrestrial broadcasting as defined by the European Digital Audio Broadcasting group), WorldSpace (satellite broadcasting).
- Accompanying audio for digital video: This includes all of digital TV.
- Storage of music including hard disc recording for the broadcasting environment.
- Audio transmission via ISDN, e.g. feeder links for FM broadcast stations
- Audio transmission via the Internet.

Requirements for audio coding systems. The target for the development of perceptual audio coding schemes can be defined along several criteria. Depending on the application, they are more or less important for the selection of a particular scheme.

- Compression efficiency: In many applications, to get a higher compression ratio at the same quality of service directly translates to cost savings. Therefore signal quality at a given bit-rate (or the bit-rate needed to achieve a certain signal quality) is the foremost criterion for audio compression technology.

- Absolute achievable quality: For a number of applications, high fidelity audio (defined as no audible difference to the original signal on CD or DAT) is required. Since no prior selection of input material is possible (everything can be called music), perceptual coding must be lossy in the sense that in most cases the original bits of a music signal cannot be recovered. Nonetheless it is important that, given enough bit-rate, the coding system is able to pass very stringent quality requirements.

- Complexity: For consumer applications, the cost of the decoding (and sometimes of the encoding, too) is relevant. Depending on the application, a different tradeoff between different kinds of complexity can be used. The most important criteria are:

    - Computational complexity: The most used parameter here is the signal processing complexity, i.e. the number of multiply-accumulate instructions necessary to process a block of input samples. If the algorithm is implemented on a general purpose computing architecture like a workstation or PC, this is the most important complexity figure.
    - Storage requirements: This is the main cost factor for implementations on dedicated silicon (single chip encoders/decoders). RAM costs are much higher than ROM cost, so RAM requirements are most important.

- Encoder versus decoder complexity: For most of the algorithms described below, the encoder is much more complex than the decoder. This asymmetry is useful for applications like broadcasting, where a one-to-many relation exists between encoders and decoders. For storage applications, the encoding can even be done off-line with just the decoder running in realtime.

As time moves along, complexity issues become less important. Better systems which use more resources are acceptable for more and more applications.

- Algorithmic delay: Depending on the application, the delay is or is not an important criterion. It is very important for two way communications applications and not relevant for pure storage applications. For broadcasting applications some 100 ms delay seem to be tolerable.

- Editability: For some applications, it is important to access the audio within a coded bitstream with high accuracy (down to one sample). Other applications demand just a time resolution in the order of one coder frame size (e.g. 24 ms) or no editability at all. A related requirement is break-in, i.e. the possibility to start decoding at any point in the bitstream without long synchronization times.

- Error resilience: Depending on the architecture of the bitstream, perceptual coders are more or less susceptible to single or burst errors on the transmission channel. This can be overcome by application of error-correction codes, but with more or less cost in terms of decoder complexity and/or decoding delay.

Source coding versus perceptual coding. In speech, video and audio coding the original data are analog values which have been converted into the digital domain using sampling and quantization. The signals have to be transmitted with a given fidelity, not necessarily without any difference on the signal part. The scientific notation for the "distortion which optimally can be achieved using a given data rate" is the rate distortion function ([Berger, 1971]). Near optimum results are normally achieved using a combination of removal of data which can be reconstructed (redundancy removal) and the removal of data which are not important (irrelevancy removal). It should be noted that in most cases it is not possible to distinguish between parts of an algorithm doing redundancy removal and parts doing irrelevancy removal.

In source coding the emphasis is on the removal of redundancy. The signal is coded using its statistical properties. In the case of speech coding a model of the vocal tract is used to define the possible signals that can be generated in the vocal tract. This leads to the transmission of parameters describing the actual speech signal together with some residual information. In this way very high compression ratios can be achieved.

For generic audio coding, this approach leads only to very limited success [Johnston and Brandenburg, 1992]. The reason for this is that music signals have no predefined method of generation. In fact, every conceivable digital signal may (and probably will by somebody) be called music and sent to a D/A converter. Therefore, classical source coding is not a viable approach to generic coding of high quality audio signals.

Different from source coding, in perceptual coding the emphasis is on the removal of only the data which are irrelevant to the auditory system, i.e. to the ear. The signal is coded in a way which minimizes noise *audibility*. This *can* lead to increased noise as measured by Signal-to-Noise-Ratio (SNR) or similar measures. The rest of the chapter describes how knowledge about perception can be applied to code generic audio in a very efficient way.

2.2 SOME FACTS ABOUT PSYCHOACOUSTICS

The main question in perceptual coding is: What amount of noise can be introduced to the signal without being audible? Answers to this question are derived from psychoacoustics. Psychoacoustics describes the relationship between acoustic events and the resulting auditory perceptions [Zwicker and Feldtkeller, 1967], [Zwicker and Fastl, 1990], [Fletcher, 1940].

The few basic facts about psychoacoustics given here are needed to understand the description of psychoacoustic models below. More about psychoacoustics can be found in John Beerend's chapter on perceptual measurement in this book and in [Zwicker and Fastl, 1990] and other books on psychoacoustics (e.g. [Moore, 1997]).

The most important keyword is 'masking'. It describes the effect by which a fainter, but distinctly audible signal (the maskee) becomes inaudible when a correspondingly louder signal (the masker) occurs simultaneously. Masking depends both on the spectral composition of both the masker and the maskee as well as on their variations with time.

2.2.1 Masking in the Frequency Domain

Research on the hearing process carried out by many people (see [Scharf, 1970]) led to a frequency analysis model of the human auditory system. The scale that the ear appears to use is called the critical band scale. The critical bands can be defined in various ways that lead to subdivisions of the frequency domain similar to the one shown in table 2.1. A critical band corresponds to both a constant distance on the cochlea and the bandwidth within which signal intensities are added to decide whether the combined signal exceeds a masked threshold or not. The frequency scale that is derived by mapping frequencies to critical band numbers is called the Bark scale. The critical band model is most useful for steady-state tones and noise.

Figure 2.1 (according to [Zwicker, 1982]) shows a masked threshold derived from the threshold in quiet and the masking effect of a narrow band noise (1 kHz, 60 dB sound pressure level; masker not indicated in the figure). All signals with a level below

Table 2.1 Critical bands according to [Zwicker, 1982]

z/Bark	f_u/Hz	f_o/Hz	Δf_G/Hz	f_m/Hz
0	0	100	100	50
1	100	200	100	150
2	200	300	100	250
3	300	400	100	350
4	400	510	110	450
5	510	630	120	570
6	630	770	140	700
7	770	920	150	840
8	920	1080	160	1000
9	1080	1270	190	1170
10	1270	1480	210	1370
11	1480	1720	240	1600
12	1720	2000	280	1850
13	2000	2320	320	2150
14	2320	2700	380	2500
15	2700	3150	450	2900
16	3150	3700	550	3400
17	3700	4400	700	4000
18	4400	5300	900	4800
19	5300	6400	1100	5800
20	6400	7700	1300	7000
21	7700	9500	1800	8500
22	9500	12000	2500	10500
23	12000	15500	3500	13500

the threshold are not audible. The masking caused by a narrow band noise signal is given by the spreading function. The slope of the spreading function is steeper towards lower frequencies. A good estimate is a logarithmic decrease in masking over a linear Bark scale (e.g., 27 dB / Bark). Its slope towards higher frequencies depends on the loudness of the masker, too. Louder maskers cause more masking towards higher frequencies, i.e., a less steep slope of the spreading function. Values of -6 dB / Bark for louder signals and -10 dB / Bark for signals with lower loudness have been reported [Zwicker and Fastl, 1990]. The masking effects are different depending on the tonality of the masker. A narrow band noise signal exhibits much greater 'masking ability' when masking a tone compared to a tone masking noise [Hellman, 1972].

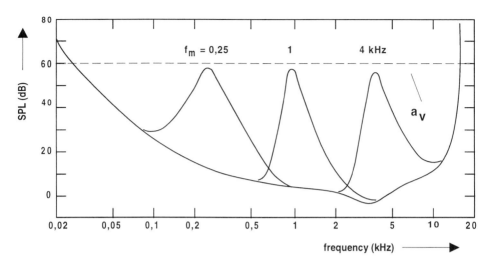

Figure 2.1 Masked thresholds: Masker: narrow band noise at 250 Hz, 1 kHz, 4 kHz (Reprinted from [Herre, 1995] ©1995, courtesy of the author)

Additivity of masking. One key parameter where there are no final answers from psychoacoustics yet is the additivity of masking. If there are several maskers and the single masking effects overlap, the combined masking is usually more than we expect from a calculation based on signal energies. More about this can be found in John Beerends chapter on perceptual measurement techniques in this book.

2.2.2 Masking in the Time Domain

The second main masking effect is masking in the time domain. As shown in Figure 2.2, the masking effect of a signal extends both to times after the masker is switched of (post-masking, also called forward masking) and to times before the masker itself is audible (pre-masking, also called backwards masking). This effect makes it possible to use analysis/synthesis systems with limited time resolution (e.g. high frequency resolution filter banks) to code high quality digital audio. The maximum negative time difference between masker and masked noise depends on the energy envelope of both signals. Experimental data suggest that backward masking exhibits quite a large variation between subjects as well as between different signals used as masker and maskee. Figure 2.3 (from [Spille, 1992]) shows the results of a masking experiment using a Gaussian-shaped impulse as the masker and noise with the same spectral density function as the test signal. The test subjects had to find the threshold of audibility for

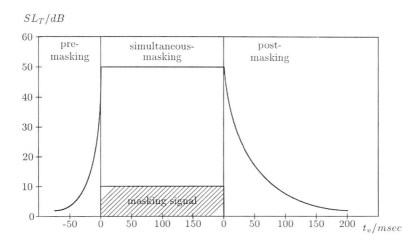

Figure 2.2 Example of pre-masking and post-masking (according to [Zwicker, 1982]) (Reprinted from [Sporer, 1998] ©1998, courtesy of the author)

the noise signal. As can be seen from the plot, the masked threshold approaches the threshold in quiet if the time differences between the two signals exceed 16 ms. Even for a time difference of 2 ms the masked threshold is already 25 dB below the threshold at the time of the impulse. The masker used in this case has to be considered a worst case (minimum) masker.

If coder-generated artifacts are spread in time in a way that they precede a time domain transition of the signal (e.g. a triangle attack), the resulting audible artifact is called "pre-echo". Since coders based on filter banks always cause a spread in time (in most cases longer than 4 ms) of the quantization error, pre-echoes are a common problem to audio coding systems.

2.2.3 Variability between listeners

One assumption behind the use of hearing models for coding is that "all listeners are created equal", i.e. between different listeners there are no or only small deviations in the basic model parameters. Depending on the model parameter, this is more or less true:

- Absolute threshold of hearing:
  It is a well known effect that the absolute threshold of hearing varies between listeners and even for the same listener over time with a general trend that the

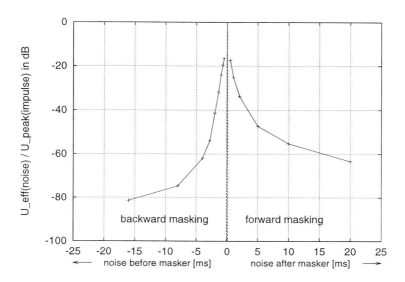

Figure 2.3 Masking experiment as reported in [Spille, 1992] (Reprinted from [Sporer, 1998] ©1998, courtesy of the author)

listening capabilities at high frequencies decrease with age. Hearing deficiencies due to overload of the auditory system further increase the threshold of hearing for part of the frequency range (see the chapter by Jim Kates) and can be found quite often. Perceptual models have to take a worst case approach, i.e. have to assume very good listening capabilities.

- Masked threshold:
  Fortunately for the designers of perceptual coding systems, variations for the actual masked thresholds in frequency domain are quite small. They are small enough to warrant one model of masking with a fixed set of parameters.

- Masking in time domain:
  The experiments described in [Spille, 1992] and other observations (including the author) show that there are large variations in the ability of test subjects to recognize small noise signals just before a loud masker (pre-echoes). It is known that the capability to recognize pre-echoes depends on proper training of the subjects, i.e. you might not hear it the first time, but will not forget the effect after you heard it for the 100th time. At present it is still an open question whether in addition to this training effect there is a large variation between different groups of listeners.

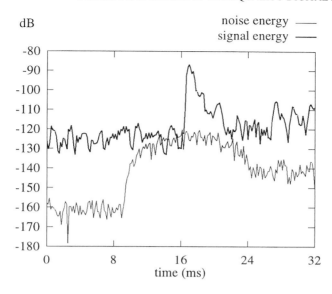

Figure 2.4 Example of a pre-echo. The lower curve (noise signal) shows the form of the analysis window

- Perception of imaging and imaging artifacts:
  This item seems to be related to the perception of pre-echo effects (test subjects who are very sensitive for pre-echoes in some cases are known to be very insensitive to imaging artifacts). Not much is known here, so this is a topic for future research.

As can be seen from the comments above, research on hearing is by no means a closed topic. Very simple models can be built very easily and can already be the base for reasonably good perceptual coding systems. If somebody tries to built advanced models, the limits of accuracy of the current knowledge about psychoacoustics are reached very soon.

2.3 BASIC IDEAS OF PERCEPTUAL CODING

The basic idea about perceptual coding of high quality digital audio signals is to hide the quantization noise below the signal dependent thresholds of hearing. Since the most important masking effects are described using a description in the frequency domain, but with stationarity ensured only for short time periods of around 15 ms, perceptual audio coding is best done in time/frequency domain. This leads to a basic structure of perceptual coders which is common to all current systems.

2.3.1 Basic block diagram

Figure 2.5 shows the basic block diagram of a perceptual encoding system.

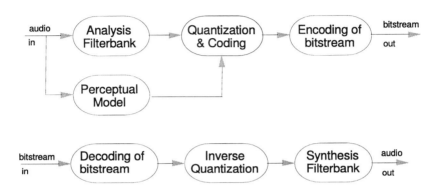

Figure 2.5   Block diagram of a perceptual encoding/decoding system (Reprinted from [Herre, 1995] ©1995, courtesy of the author)

- Filter bank:
  A filter bank is used to decompose the input signal into subsampled spectral components (time/frequency domain). Together with the corresponding filter bank in the decoder it forms an analysis/synthesis system.

- Perceptual model:
  Using either the time domain input signal or the output of the analysis filter bank, an estimate of the actual (time dependent) masked threshold is computed using rules known from psychoacoustics. This is called the perceptual model of the perceptual encoding system.

- Quantization and coding:
  The spectral components are quantized and coded with the aim of keeping the noise, which is introduced by quantizing, below the masked threshold. Depending on the algorithm, this step is done in very different ways, from simple block companding to analysis-by-synthesis systems using additional noiseless compression.

- Frame packing:
  A bitstream formatter is used to assemble the bitstream, which typically consists of the quantized and coded spectral coefficients and some side information, e.g. bit allocation information.

These processing blocks (in various ways of refinement) are used in every perceptual audio coding system.

2.3.2 Additional coding tools

Along the four mandatory main tools, a number of other techniques are used to enhance the compression efficiency of perceptual coding systems. Among these tools are:

- Prediction:
  Forward- or backward adaptive predictors can be used to increase the redundancy removal capability of an audio coding scheme. In the case of high resolution filter banks backward adaptive predictors of low order have been used with success [Fuchs, 1995].

- Temporal noise shaping:
  Dual to prediction in time domain (with the result of flattening the spectrum of the residual), applying a filtering process to parts of the spectrum has been used to control the temporal shape of the quantization noise within the length of the window function of the transform [Herre and Johnston, 1996].

- M/S stereo coding:
  The masking behavior of stereo signals is improved if a two-channel signal can be switched between left/right and sum/difference representation. Both broadband and critical band-wise switching has been proposed [Johnston, 1989a].

- Intensity stereo coding:
  For high frequencies, phase information can be discarded if the energy envelope is reproduced faithfully at each frequency. This is used in intensity stereo coding [van der Waal and Veldhuis, 1991, Herre et al., 1992].

- Coupling channel:
  In multichannel systems, a coupling channel is used as the equivalent to an n-channel intensity system. This system is also known under the names dynamic crosstalk or generalized intensity coding. Instead of n different channels, for part of the spectrum only one channel with added intensity information is transmitted [Fielder et al., 1996, Johnston et al., 1996].

- Stereo prediction:
  In addition to the intra-channel version, prediction from past samples of one channel to other channels has been proposed [Fuchs, 1995].

- Spectrum flattening:
  As a special version to enhance the efficiency of the quantization and coding module, an LPC analysis has been proposed to normalize the spectral values [Iwakami et al., 1995].

2.3.3 Perceptual Entropy

The term "Perceptual Entropy" (PE, see [Johnston, 1988]) is used to define the lowest data rate which is needed to encode some audio signal without any perceptual difference to the original. An estimate of the PE (there is not enough theory yet to calculate a "real" PE) can be used to determine how easy or how difficult it is to encode a given music item using a perceptual coder.

In practice, the calculation of the PE requires an analysis filter bank and a perceptual model. The PE is defined as

$$PE = \frac{1}{N} \sum_{f=f_l}^{f=f_u} max\left(0, \log_2 \frac{|\text{signal}(f)|}{\text{threshold}(f)}\right) \quad (2.1)$$

where N is the number of frequency components between f_l and f_u, f_l is the lower frequency limit (e.g. $f_l = 0$ Hz), f_u is the upper frequency limit (e.g. $f_u = 20000$ Hz), signal(f) is the amplitude of the frequency component f and threshold(f) is the estimated threshold level at the frequency f. This definition of the PE of course needs the existence of a concept of audibility resp. an auditory threshold. Examples for this are given later in this chapter.

2.4 DESCRIPTION OF CODING TOOLS

2.4.1 Filter banks

The filter bank is the deciding factor for the basic structure of a perceptual coding system. Figure 2.6 shows the basic block diagram of an static n-channel analysis/synthesis filter bank with downsampling by k. If $k = n$, it is called a filter bank with critical sampling. A number of basic parameters can be used to describe filter banks used for audio coding:

- Frequency resolution:
  Over the past years, two main types of filter banks have been used for high quality audio coding:

  - Low resolution filter banks (e.g. 32 subbands), normally combined with a quantization and coding module which works on blocks in time direction. These are frequently called *subband coders*.

- High frequency resolution filter banks (e.g. 512 subbands), normally combined with a quantization and coding module which works by combining adjacent frequency lines. These have traditionally been called *transform coders*.

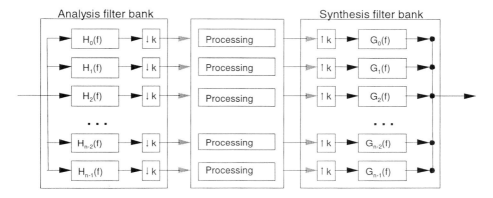

Figure 2.6 Basic block diagram of an n-channel analysis/synthesis filter bank with downsampling by k (Reprinted from [Herre, 1995] ©1995, courtesy of the author)

Mathematically, all transforms used in today's audio coding systems can be seen as filter banks. All uniform subband filter banks can be seen as transforms of L input samples into N spectral components as derived for example in [Edler, 1995, Temerinac and Edler, 1993]. There is no basic difference between both approaches, so any attempt to distinguish between subband coders and transform coders is against current scientific knowledge. Therefore, in the following we will use the term "filter bank" synonymously to "subband filter bank" and/or "transform".

A higher resolution filter bank for most signals exhibits a larger transform gain. For this reason, high frequency resolution filter banks are the tool of choice for audio coding systems built for maximum coding efficiency at low bit-rates.

- Perfect reconstruction:
Perfect reconstruction filter banks allow the lossless reconstruction of the input signal in an analysis–synthesis system without quantization. While not a necessary feature, the use of a perfect reconstruction filter bank simplifies the design of a coding system. While at some point other filter banks have been proposed for use in perceptual coders (e.g. wave digital filters, see [Sauvagerd, 1988]), all currently used filter banks are either perfect reconstruction or near perfect

reconstruction (very small reconstruction error in the absence of quantization of the spectral components).

- Prototype window:
  Especially in the case of low bit-rates (implying that a lot of quantization noise is introduced), the filter characteristics of the analysis and synthesis filters as determined by the prototype window / windowing function are a key factor for the performance of a coding system.

- Uniform or non-uniform frequency resolution:
  Both uniform or non-uniform frequency resolution filter banks have been proposed since the first work on high quality audio coding. While a non-uniform frequency resolution is closer to the characteristics of the human auditory system, in practical terms uniform frequency resolution filter banks have been more successful. This may be due to the fact that even at high frequencies for some signals the larger coding gain of a high frequency resolution is needed [Johnston, 1996].

- Static or adaptive filter bank:
  In an analysis/synthesis filter bank, all quantization errors on the spectral components show up on the time domain output signal as the modulated signal multiplied by the synthesis window. Consequently, the error is smeared in time over the length of the synthesis window / prototype filter. As described above, this may lead to audible errors if premasking is not ensured. This pre-echo effect (a somewhat misleading name, a better word would be pre-noise) can be avoided if the filter bank is not static, but switched between different frequency/time resolutions for different blocks of the overlap/add. An example of this technique called adaptive window switching is described below.

The following section gives a short overview of filter banks which are currently used for audio coding purposes.

QMF filter banks. Quadrature mirror filters (QMF, see [Esteban and Galand, 1977]) have often been proposed for audio coding. The most common configuration is the tree of filters with a two-way split. In one of the early examples [Theile et al., 1987] the '64d' filter design from [Johnston, 1980] has been used. The decomposition tree is set up so that the filter bands resemble critical bands. The QMF halfband filters are non-perfect reconstruction, but with perfect alias cancellation by design. The reconstruction error of the analysis/synthesis pair can be held at small amplitudes by increasing the filter length.

Instead of standard QMF filters, generalized QMF-techniques (GQMF) have been used as well [Edler, 1988].

The disadvantages of the QMF-tree technique are

- Non-perfect reconstruction: The passband ripple which is typical for QMF filter designs can lead to time domain artifacts which can be audible even if they are at very low amplitudes.

- Long system delay: The overall system delay can reach 250 ms, if the technique is used to design a filter bank with a frequency partitioning similar to critical bands.

- High computational complexity: The number of multiplications per sample which is needed to compute a QMF-tree filter bank using e.g. 64-tap filters is much higher compared to polyphase filter banks or FFTs.

Wavelet based filter banks. In the last few years a number of audio coding systems have been proposed using wavelet based filters [Sinha and Tewfik, 1993]. A thorough description of the theory of wavelet based filter banks can be found in [Vetterli and Kovačević, 1995].

Polyphase filter banks. Polyphase filter banks as used in audio coding have been introduced in [Rothweiler, 1983]. These are equally spaced filter banks which combine the filter design flexibility of generalized QMF banks with low computational complexity. Most current designs are based on the work in [Rothweiler, 1983].

The filter bank used in the MPEG/Audio coding system will be used as an example. A 511 tap prototype filter is used. Figure 2.7 shows the prototype filter (window function). It has been optimized for a very steep filter response and a stop band attenuation of better than 96 dB. Figure 2.8 shows the frequency response of the filter bank. In addition to the attenuation requirements it was designed as a reasonable tradeoff between time behavior and frequency localization [Dehery, 1991].

The advantage of polyphase filter banks as used for audio coding is the combination of the degrees of freedom for the prototype filter design and the relatively low complexity of the filter bank.

Only equally-spaced filter banks can be designed using this technique. This is the main disadvantage of polyphase filter banks.

Fourier Transform based filter banks (DFT, DCT): Some of the first work done in coding of high quality audio signals used DFT and DCT based transforms as known from image coding. The original idea of Adaptive Transform Coding (ATC) was to decorrelate the signal via the transform. This technique had been introduced for speech coding by [Zelinski and Noll, 1977] (see the description in [Jayant and Noll, 1982], too). The extension of this technique to high quality audio coding was first presented in [Brandenburg et al., 1982]. To reduce blocking artifacts, windowing and overlap/add techniques have been used. [Portnoff, 1980] gives a framework of FFT-based short time analysis/synthesis systems using windowing. The first perceptual

54   APPLICATIONS OF DSP TO AUDIO AND ACOUSTICS

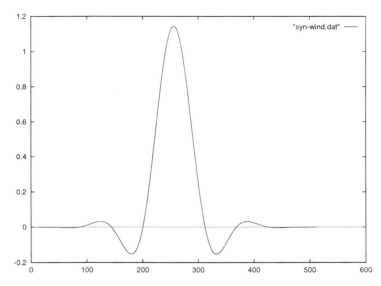

Figure 2.7   Window function of the MPEG-1 polyphase filter bank (Reprinted from [Sporer, 1998] ©1998, courtesy of the author)

transform coding systems all implemented an overlap of $1/16$ of the block length [Krahé, 1988, Brandenburg, 1987, Johnston, 1989b].

Another, now commonly used viewpoint is to look at a transform based and windowed analysis/synthesis system as a polyphase structure. The window takes the part of a prototype filter. The transform does the modulation of the filtered signal into the baseband.

In the last years all new high frequency resolution (transform) based coding systems use MDCT techniques (see below) instead of DFT or DCT.

The advantage of the transform-based approach is low computational complexity. An analysis/synthesis system with as much as 512 frequency components can be realized using, for example, 10 multiplications per time domain sample.

Time domain aliasing cancellation based filter banks.   The Modified Discrete Cosine Transform (MDCT) was first proposed in [Princen et al., 1987] as a sub-band/transform coding scheme using Time Domain Aliasing Cancellation (TDAC). It can be viewed as a dual to the QMF-approach doing frequency domain aliasing cancellation. The window is constructed in a way that satisfies the perfect reconstruction condition:

$$h(i)^2 + h(i + N/2)^2 = 1, \quad i = 0, ..., N/2 - 1 \tag{2.2}$$

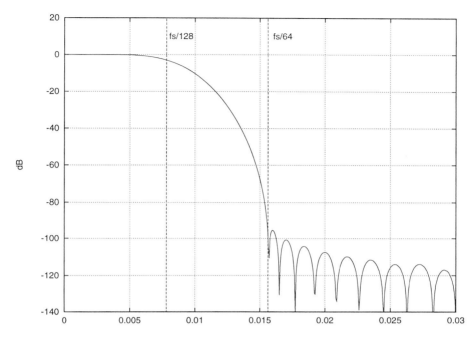

Figure 2.8 Frequency response of the MPEG-1 polyphase filter bank (Reprinted from [Sporer, 1998] ©1998, courtesy of the author)

where N is the window length. The equation above already assumes that analysis window and synthesis window are equal. While this is not a necessary condition, it is widely the case.

Normally, an overlap factor of two is used together with a sine window:

$$h(i) = sin(\pi \frac{i + 0.5}{N}), i = 0, ...N - 1 \qquad (2.3)$$

The transform kernel is a DCT with a time-shift component added:

$$X_t(m) = \sum_{k=0}^{N-1} h(k)x_t(k)cos[\frac{\pi}{2N}(2k + 1 + M)(2m + 1)] \qquad (2.4)$$

where N is the block length in time, $M = N/2$ is the block length in the frequency domain, $h(k), k = 0, ..., N - 1$ is the window, $x_t(k)$ are the samples of the tth block, and $X_t(m), m = 0, ..., M - 1$ are the frequency domain values.

56   APPLICATIONS OF DSP TO AUDIO AND ACOUSTICS

As can be seen from the equation above, there is frequency domain subsampling. As a result, the analysis/synthesis system does critical sampling of the input signal, that is the number of time/frequency components for transform block is equal to the update length of the input time domain sequence.

MDCT or similar schemes are used in several audio coding systems [Brandenburg, 1988, Mahieux et al., 1990, Brandenburg et al., 1991] [Davidson et al., 1990, Iwadare et al., 1992] because they combine critical sampling with the good frequency resolution provided by a sine window and the computational efficiency of a fast FFT-like algorithm. Typically, 128 to 2048 equally spaced bands are used.

As a further advantage of MDCT-like filter banks it should be noted that the time domain aliasing property needs to be valid for each half of the window independently from the other. Thus hybrid window forms (with different types of window functions for the first or second half) can be used. This leads to the realization of adaptive window switching systems ([Edler, 1989], see below).

The MDCT is known under the name Modulated Lapped Transform ([Malvar, 1990]) as well. Extensions using an overlap of more than a factor of two have been proposed [Vaupelt, 1991, Malvar, 1991] and used for coding of high quality audio [Vaupelt, 1991]. This type of filter banks can be described within the framework of cosine-modulated filter banks ([Koilpillai and Vaidyanathan, 1991][Ramstadt and Tanem, 1991, Malvar, 1992]).

Other window functions than the sine window have been proposed as well (see [Bosi et al., 1996b, Fielder et al., 1996]). Using Kaiser-Bessel-Derived window functions, a filter characteristic exhibiting better side-lobe suppression is possible. This is explained in [Fielder et al., 1996].

Hybrid filter banks.   Filter banks which consist of a cascade of different types of filter banks are called *hybrid filter banks*. They have been introduced in [Brandenburg and Johnston, 1990] to build an analysis/synthesis system which combines the different frequency resolution at different frequencies possible with QMF-tree structures with the computational efficiency of FFT-like algorithms. In the example of [Brandenburg and Johnston, 1990], the input signal is first subdivided into 4 bands using a QMF-tree. To avoid artifacts due to the QMF-filter bank, an 80-tap filter has been used. Each of the 4 bands is further subdivided into 64 or 128 frequency lines using an MDCT. A total of 320 frequency lines is generated. The time resolution for each line is between 21 ms for the lowest frequencies to 2.7 ms for the highest frequencies.

In ISO/MPEG Layer 3, a different approach to hybrid coding has been used (see Figure 2.9. To ensure compatibility to Layers 1 and 2, the same polyphase filter bank is used as the first filter in the hybrid filter bank. Each of the 32 polyphase subbands is normally further subdivided into 18 frequency lines using an MDCT. By using the window switching technique described below the subdivision can be switched to 6

PERCEPTUAL CODING OF HIGH QUALITY DIGITAL AUDIO 57

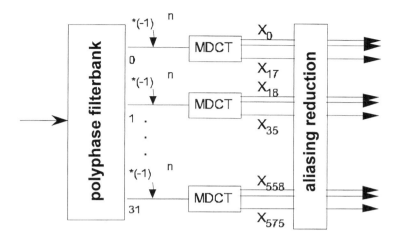

Figure 2.9 Block diagram of the MPEG Layer 3 hybrid filter bank (Reprinted from [Herre, 1995] ©1995, courtesy of the author)

lines for either all of the 32 polyphase subbands or for only the 30 higher polyphase subbands.

In summary it can be stated that hybrid filter banks allow increased flexibility in the design by including the possibility to have different frequency resolutions at different frequencies. Another degree of freedom which is gained by using hybrid filter banks is the adaptive switching of the filter bank to different time/frequency behavior. On the downside, a somewhat increased complexity compared to solutions based on cosine-modulated filter banks is necessary to implement adaptive hybrid systems.

Alias reduction for hybrid filter banks. One possible problem of all cascaded filter banks specific to hybrid filter banks needs to be mentioned. Since the frequency selectivity of the complete filter bank can be derived as the product of a single filter with the alias components folded in for each filter, there are spurious responses (alias components) possible at unexpected frequencies. Crosstalk between subbands over a distance of several times the bandwidth of the final channel separation can occur. The overall frequency response shows peaks within the stopbands.

In [Edler, 1992] a solution to this problem has been proposed. It is based on the fact that every frequency component of the input signal influences two subbands of the cascaded filter bank, one as a signal component and the other as an aliasing component. Since this influence is symmetric, a compensation can be achieved using a butterfly structure with the appropriate weighting factors. No complete cancellation of the additional alias terms can be achieved, but an optimization for the overall frequency response can be done. The resulting frequency response of the hybrid filter banks shows an improvement of the aliasing side lobes by about 5 – 10 dB.

Adaptive filter banks. In the basic configuration, all filter banks described above feature a time/frequency decomposition which is constant over time. As mentioned above, there are possibilities to switch the characteristics of a filter bank, going from one time/frequency decomposition to another one. We explain the basic principle using the example of MPEG Audio Layer 3:

The technique is based on the fact that alias terms which are caused by subsampling in the frequency domain of the MDCT are constrained to either half of the window. Adaptive window switching as used in Layer 3 is based on [Edler, 1989]. Figure 2.10 shows the different windows used in Layer 3, Figure 2.11 shows a typical sequence of window types if adaptive window switching is used. The function of the different

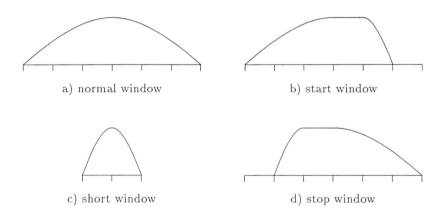

Figure 2.10   Window forms used in Layer 3 (Reprinted from [Sporer, 1998] ©1998, courtesy of the author)

window types is explained as follows:

- Long window:
  This is the normal window type used for stationary signals.

- Short window:
  The short window has basically the same form as the long window, but with 1/3 of the window length. It is followed by an MDCT of 1/3 length. The time

resolution is enhanced to 4 ms at 48 kHz sampling frequency. The combined frequency resolution of the hybrid filter bank in the case of short windows is 192 lines compared to 576 lines for the normal windows used in Layer 3.

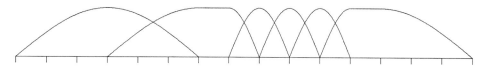

Figure 2.11 Example sequence of window forms (Reprinted from [Sporer, 1998] ©1998, courtesy of the author)

- Start window:
  In order to switch between the long and the short window type, this hybrid window is used. The left half has the same form as the left half of the long window type. The right half has the value one for 1/3 of the length and the shape of the right half of a short window for 1/3 of the length. The remaining 1/3 of the window is zero. Thus, alias cancellation can be obtained for the part which overlaps the short window.

- Stop window:
  This window type enables the switching from short windows back to normal windows. It is the time reverse of the start window.

A criterion when to switch the window form is necessary to control the adaptive block switching. One possible criterion to switch the filter bank is derived from the threshold calculation. If pre-echo control is implemented in the perceptual model as described below, pre-echo conditions result in a much increased estimated Perceptual Entropy (PE) [Johnston, 1988], i.e. in the amount of bits needed to encode the signal. If the demand for bits exceeds the average value by some extend, a pre-echo condition is assumed and the window switching logic is activated. Experimental data suggest that a big surge in PE is always due to pre-echo conditions. Therefore pre-echo detection via the threshold calculation works more reliable than purely time domain energy calculation based methods.

2.4.2 Perceptual models

As discussed above, the model of hearing built into a perceptual coding system forms the heart of the algorithm. A lot of systems (like MPEG Audio, see below) just define the transmission format, thus allowing changes and improvements to the perceptual model even after a standard is fixed and a lot of decoders are deployed at the customers.

60 APPLICATIONS OF DSP TO AUDIO AND ACOUSTICS

The main task of the perceptual model in a perceptual encoding system is to deliver accurate estimates of the allowed noise (just masked threshold) according to the time/frequency resolution of the coding system. Additional tasks include

- the control of adaptive block switching for adaptive filter banks,

- the control of a bit reservoir if applicable,

- the control of joint stereo coding tools like M/S and/or intensity coding

To solve these tasks, perceptual models often work directly on the time domain input data thus allowing a time and/or frequency resolution of the model which is better than the time and/or frequency resolution of the main filter bank of the perceptual coding system.

A trivial example. In the simplest case, a static model can be used. In the case of a frequency domain coding system, a worst-case SNR necessary for each band can be derived from the masking curves. Here a bit allocation strategy assigns the number of bits according to

$$\text{nbits}(i) = SNR_{\text{worst}(i)}/6.02 dB \qquad (2.5)$$

that is, the number of bits per band i is derived from the worst-case SNR $SNR_{\text{worst}(i)}$ for this band.

This model has been used in the earliest known digital perceptual audio coding system [Krasner, 1979]. Similar models have been used for the Low-Complexity-Adaptive Transform Coding (LC-ATC, [Seitzer et al., 1988]) and AC-2 ([Davidson et al., 1990]) systems.

More advanced models try to estimate a time-dependent Signal-to-Mask-Ratio (SMR) for each band used in the coder. Because the knowledge about masking effects is limited at this time and because different theories about additivity of masking or the effects of tonality exist, there is no such thing as 'the correct psychoacoustic model'.

Estimation of tonality. Following early research work by Scharf (see [Scharf, 1970]) and Hellman (see [Hellman, 1972]) one way to derive an estimate of the masked threshold is to distinguish between the masking estimates for noise maskers masking tonal signals and tone maskers masking noise signals. To do this, an estimate of the tonality of a signal is necessary. For complex signals we find that a tonality index $v(t,\omega)$ depending on time t and frequency ω leads to the best estimation of a masked threshold. To get such an estimate, a tonality measure using a simple polynomial predictor has been proposed in [Brandenburg and Johnston, 1990].

Two successive instances of magnitude and phase are used to predict magnitude and phase at each frequency line:

$$r(t, \omega) \quad \text{magnitude at time } t \text{ and frequency } \omega \qquad (2.6)$$

$$\Phi(t, \omega) \quad \text{phase at time } t \text{ and frequency } \omega \qquad (2.7)$$

The predicted values $\hat{r}$ and $\hat{\Phi}$ of r and Φ are calculated as:

$$\hat{r}(t, \omega) = r(t-1, \omega) + (r(t-1, \omega) - r(t-2, \omega)) \qquad (2.8)$$
$$\hat{\Phi}(t, \omega) = \Phi(t-1, \omega) + (\Phi(t-1, \omega) - \Phi(t-2, \omega)) \qquad (2.9)$$

The Euclidean distance between the predicted and the actual values is the unpredictability (sometimes called the 'chaos-measure') $c(t, \omega)$:

$$c(t, \omega) = \frac{dist\{[\hat{r}(t, \omega), \hat{\Phi}(t, \omega)], [r(t, \omega), \Phi(t, \omega)]\}}{r(t, \omega) + abs[\hat{r}(t, \omega)]} \qquad (2.10)$$

If the signal at frequency ω is very tonal, the prediction will be accurate and $c(t, \omega)$ will be very small. If the signal is noise-like, $c(t, \omega)$ assumes values up to 1 with a mean of 0.5. Therefore the chaos measure can be limited to the range 0.05 to .5 with 0.05 considered fully tonal and 0.5 considered fully noise-like:

$$c_l(t, \omega) = max\{0.05, min[0.5, c(t, \omega)]\} \qquad (2.11)$$

The chaos measure $c(t, \omega)$ can be mapped to a tonality measure $v(t, \omega)$ via a nonlinear mapping:

$$v(t, \omega) = -0.43 * \log_{10}(c_l(t, \omega)) - 0.299 \qquad (2.12)$$

The tonality index $v(t, \omega)$ denotes the final result of the tonality estimation and can be applied to a perceptual model as for example the one described below.

MPEG-1 perceptual model 2. As an example for actual perceptual models we give a description of the "perceptual model 2" as it is described in the informative annex of MPEG-1 audio [MPEG, 1992].

The frequency domain representation of the data is calculated via an FFT after applying a Hann window with a window length of 1024 samples. The calculation is done with a shift length equal to the block structure of the coding system. As described below, the shift length is 576 samples for Layer 3 of the ISO/MPEG-Audio system.

The separate calculation of the frequency domain representation is necessary because the filter bank output values (polyphase filter bank used in Layer 1/2 or hybrid filter bank used in Layer 3) can not easily be used to get a magnitude/phase representation of the input sequence as needed for the estimation of tonality.

The tonality estimation is based on the simple polynomial predictor as described above.

The magnitude values of the frequency domain representation are converted to a 1/3-critical band energy representation. This is done by adding the magnitude values within a threshold calculation partition.

$$e_b = \sum_{\omega_{low_b}}^{\omega_{high_b}} r(\omega)^2 \qquad (2.13)$$

where b is the threshold calculation partition index, low_b and $high_b$ are the lowest and the highest frequency line in b and $r(\omega)$ is the magnitude at frequency ω.

A weighted unpredictability c_b is derived from the unpredictability measure, $c(\omega)$, which has been computed according to the procedure described above.

$$c_b = \sum_{\omega=\omega_{low_b}}^{\omega_{high_b}} r(\omega)^2 c(\omega) \qquad (2.14)$$

A convolution of these values with the cochlear spreading function follows. Due to the non-normalized nature of the spreading function, the convolved versions of e_b and c_b should be renormalized. The convolved unpredictability, c_b, is mapped to the tonality index, t_b, using a log transform just as the unpredictability was mapped to the tonality index, $c(t, \omega)$, from equation (2.12).

The next step in the threshold estimation is the calculation of the just-masked noise level (also called masking level) in the cochlear domain using the tonality index and the convolved spectrum. This is done by first calculating the required signal to noise ratio SNR_b for each threshold calculation band b.

$$SNR_b = \max(\text{minval}_b, t_b * TMN_b + (1 - t_b) * NMT_b) \qquad (2.15)$$

where minval_b is a tabulated minimum value per threshold calculation band. TMN_b and NMT_b are estimates for the masking capabilities of tone masking noise and noise masking tone [Scharf, 1970, Hellman, 1972].

The final step to get the preliminary estimated threshold is the adjustment for the threshold in quiet. Since the sound pressure level of the final audio output is not known in advance, the threshold in quiet is assumed to be some amount below the LSB for the frequencies around 4 kHz.

If necessary, pre-echo control occurs at this point. This is done by using an actual threshold estimation which would be valid for the current block even if the sound which could cause pre-echo artifacts would be deleted from the signal. A good approximation for this hypothetical deletion is to use the data of the last block as an estimate for the current block. To have data of earlier blocks available, the preliminary estimated threshold is stored. It will be used for pre-echo control in the next input data block.

The preliminary threshold of the current block is then modified using the preliminary threshold of the last block.

$$thr_b = max(thrp_b, rpelev * throld_b) \qquad (2.16)$$

where thr_b is the final estimated masked threshold for the threshold calculation band b, $thrp_b$ is the preliminary threshold, $throld_b$ is the preliminary threshold of the last block of data and $rpelev$ is a constant. It introduces a weighting to the threshold data of the last block. We use $rpelev = 2$.

All calculations up to now have been done using the threshold calculation partitions, that is without any knowledge about the frequency partitioning used by the coding system. To map these values to coder partitions, all the estimated threshold values are first mapped to spectral densities. From there, using again the magnitude values $r(\omega)$, the signal to mask ratios SMR_n are derived. n denotes the coder partition or coder subband number.

$$SMR_n = 10 * \log_{10}(\frac{e_n}{thr_n}) \qquad (2.17)$$

with e_n denoting the signal energy in the coder partition or coder subband n and thr_n describing the estimated masked threshold for the coder partition n.

The values SMR_n can be used either directly in the case of coding systems using noise allocation or to control a bit allocation algorithm.

2.4.3 Quantization and coding

The quantization and coding tools in an encoder do the main data-reduction work. As in the case of filter banks, a number of design options are possible and have been explored.

- Quantization alternatives:
  Most systems apply uniform quantization. One exception to this rule is the application of non-uniform quantization with a power law in MPEG-1 and MPEG-2 audio.

- Coding alternatives:
  The quantized spectral components are stored and/or transmitted either directly as quantized values according to a bit allocation strategy (including bit packing) or as entropy coded words.

- Quantization and coding control structures:
  The two approaches currently in wide use are

    - Bit allocation (direct structure):
      A bit allocation algorithm driven either by data statistics or by a perceptual

model decides how many bits are assigned to each spectral component. This is performed before the quantization is done.
   – Noise allocation (indirect structure):
   The data are quantized with possible modifications to the quantization step sizes according to a perceptual model. A count of how many bits are used for each component can only be done after the process is completed.
- Tools to improve quantization and coding:
  A lot of small variations to the basic ideas have been applied to further remove redundancy of the quantized values. Examples can be found e.g. in the documents describing standardized systems. [MPEG, 1992, MPEG, 1994a, MPEG, 1997a, ATSC, 1995].

The following sections describe some of the widely used tools for quantization and coding in more detail.

Block companding. This method is also known under the name "block floating point". A number of values, ordered either in time domain (successive samples) or in frequency domain (adjacent frequency lines) are normalized to a maximum absolute value. The normalization factor is called the scalefactor (or, in some cases, exponent). All values within one block are then quantized with a quantization step size selected according to the number of bits allocated for this block. A bit allocation algorithm is necessary to derive the number of bits allocated for each block from the perceptual model. In some cases, a simple bit allocation scheme without an explicit perceptual model (but still obeying masking rules) is used.

Non-uniform scalar quantization. While usually non-uniform scalar quantization is applied to reduce the mean squared quantization errors like in the well known MAX quantizer, another possibility is to implement some default noise shaping via the quantizer step size. This is explained using the example of the quantization formula for MPEG Layer 3 or MPEG-2 Advanced Audio Coding:

The basic formula is

$$is(i) = nint\left(\left(\frac{|xr(i)|}{quant}\right)^{0.75} - 0.0946\right) \quad (2.18)$$

where $xr(i)$ is the value of the frequency line at index i, $quant$ is the actual quantizer step size, $nint$ is the 'nearest integer' function and $is(i)$ is the quantized absolute value at index i.

The quantization is of the mid-tread type, i.e. values around zero get quantized to zero and the quantizer is symmetric.

In this case, bigger values are quantized less accurately than smaller values thus implementing noise shaping by default.

Vector quantization. In vector quantization, not the individual filter bank output samples are quantized, but n-tuples of values. This technique is used in most current speech and video coding techniques. Recently, vector quantization has been applied in a scheme called TWIN-VQ ([Iwakami et al., 1995]). This system has been proposed for MPEG-4 audio coding (see [MPEG, 1997b]).

Noise allocation followed by scalar quantization and Huffman coding. In this method, no explicit bit allocation is performed. Instead, an amount of allowed noise equal to the estimated masked threshold is calculated for each scalefactor band. The scalefactors are used to perform a coloration of the quantization noise (i.e. they modify the quantization step size for all values within a scalefactor band) and are not the result of a normalization procedure. The quantized values are coded using Huffman coding. The whole process is normally controlled by one or more nested iteration loops. The technique is known as analysis-by-synthesis quantization control. It was first introduced for OCF [Brandenburg, 1987], PXFM [Johnston, 1989b] and ASPEC [Brandenburg et al., 1991]. In a practical application, the following computation steps are performed in an iterative fashion:

- Inner loop
  The quantization of the actual data is performed including the buffer control.

- Calculation of the actual quantization noise
  The quantization noise is calculated by subtracting the reconstructed from the unquantized signal values and summing the energies per scalefactor band.

- Scaling
  For each scalefactor band which violates the masked threshold as known from the calculation of the psychoacoustic model, the signal values are amplified. This corresponds to a decrease of the quantizer step size only for these bands.

- Check for termination of iteration loop
  If no scaling was necessary or another reason to terminate the loop applies, end the iterations. If not, continue with quantization using the modified signal values.

Huffman coding. One very successful tool for high quality audio coding is static Huffman coding applying different Huffman code tree tables according to the local statistics of the signal. As an example for refinements to the basic concept of Huffman coding, the following paragraph describes the noiseless coding techniques used within MPEG Layer 3.

Codes are transmitted only up to the highest numbered frequency line with a quantized value different from zero. The actually coded values are divided into one region called *big values*, where the frequency lines are coded with a 2-dimensional

66   APPLICATIONS OF DSP TO AUDIO AND ACOUSTICS

Huffman code and another region at higher frequencies (below the values which default to zero) containing only quantized values not exceeding magnitude 1. The values in the latter region are quantized using a 4-dimensional Huffman code. The *big values* region is split into 3 subregions. Each of them uses a separately selectable Huffman code table. A set of 16 possible Huffman code tables is used. For each section the

Table 2.2   Huffman code tables used in Layer 3

Table number	Quantization levels	Table size	Number of bits per pair of zeroes	ESC
0	0	0 x 0	0	
1	3	2 x 2	1	
2	5	3 x 3	1	
3	5	3 x 3	2	
4	not used			
5	7	4 x 4	1	
6	7	4 x 4	3	
7	11	6 x 6	1	
8	11	6 x 6	2	
9	11	6 x 6	3	
10	15	8 x 8	1	
11	15	8 x 8	2	
12	15	8 x 8	4	
13	31	16 x 16	1	
14	not used			
15	31	16 x 16	3	
16	33	16 x 16	1	*
24	33	16 x 16	4	*

table which is best adapted to the current signal statistics is searched. By individually adapting code tables to subregions coding efficiency is enhanced and simultaneously the sensitivity against transmission errors is decreased. The largest tables used in Layer 3 contain 16 by 16 entries. Larger values are coded using an escape mechanism. The table entry belonging to the largest value signals that the value is coded via a PCM-code.

The table numbers 17 to 23 and 25 to 31 are used to point to tables 16 resp. 24 but with different lengths of the codeword part which is coded using the escape mechanism.

PERCEPTUAL CODING OF HIGH QUALITY DIGITAL AUDIO 67

Short time buffering. While all systems described here are designed to work in a fixed bit-rate environment, it is desirable to support some locally varying bit-rates. Beyond the aim of smoothing out some local variations in the bit-rate demand, this is used to reduce the probability of audible pre-echoes even in systems where window switching is applied.

As described above, the pre-echo control in the perceptual model can lead to a PE signalling a bit-rate demand which is increased by a possibly large factor.

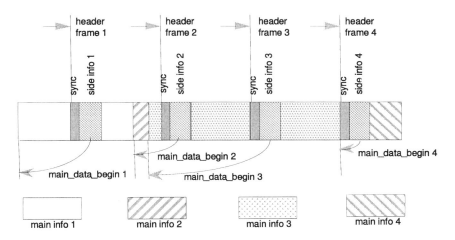

Figure 2.12   Example for the bit reservoir technology (Layer 3)

A buffer technique called *bit reservoir* was introduced to satisfy this additional need for bits. It can be described as follows:

The amount of bits corresponding to a frame is no longer constant, but varying with a constant long term average. To accommodate fixed rate channels, a maximum accumulated deviation of the actual bit-rate to the target (mean) bit-rate is allowed. The deviation is always negative, i.e. the actual mean bit-rate is never allowed to exceed the channel capacity. An additional delay in the decoder takes care of the maximum accumulated deviation from the target bit-rate.

If the actual accumulated deviation from the target bit-rate is zero, then (by definition) it holds that the actual bit-rate equals the target bit-rate. In this case the bit reservoir is called empty. If there is an accumulated deviation of n bits then the next frame may use up to n bits more than the average number without exceeding the mean bit-rate. In this case the bit reservoir is said to 'hold n bits'.

This is used in the following way in Layer 3: Normally the bit reservoir is kept at somewhat below the maximum number (accumulated deviation). If there is a surge in PE due to the pre-echo control then additional bits 'taken from the reservoir' are used to code this particular frame according to the PE demand. In the next few frames every frame is coded using some bits less than the average amount. The bit reservoir gets 'filled up' again.

Figure 2.12 shows an example (for Layer 3) of the succession of frames with different amounts of bits actually used. A pointer called main-data-begin is used to transmit the information about the actual accumulated deviation from the mean bit-rate to the decoder. The side information is still transmitted with the frame rate as derived from the channel capacity (mean rate). The main-data-begin pointer is used to find the main information in the input buffer of the decoder.

2.4.4 Joint stereo coding

As for the underlying audio coding methods itself the goal of joint stereo coding is to reduce the amount of information which is transmitted to the receiver without introducing audible artifacts. This is done by using the stereo redundancy and the irrelevancy of certain stereo coding artifacts.

Contrary to popular believe, for most signals there is not much correlation between the time signals corresponding to the left and right channel of a stereo source [Bauer and Seitzer, 1989b]. Only the power spectra of both channels are often highly correlated [Bauer and Seitzer, 1989a]. For binaural recordings this fact can easily be derived from a look at room acoustics and the way the signal is recorded. If the delay of some sound due to room acoustics is less than the time resolution of the filter bank, we find the resulting signals on both channels in the same sample period of the filter bank output. Generally it is true that stereo redundancy can be used more easily in high frequency resolution systems.

Looking for stereo irrelevancy we find that the ability of the human auditory system to discriminate the exact location of audio sources decreases at high frequencies [Blauert, 1983]. The cues to get spatial impression are mainly taken from the energy maxima in space at each frequency.

Pitfalls of stereo coding. Unfortunately for the coding system designer, in addition to the lack of redundancy between the stereo channels there are a number of issues which complicate stereo coding. In some cases, the necessary bit-rate for stereo coding exceeds the one for coding of two mono channels. Other effects forbit joint stereo coding for some classes of signals.

An especially interesting topic is the discussion of the "stereo unmasking effect". It describes the situation that certain coding artifacts which are masked in single channel coding can become audible when presented as a stereo signal coded by a dual mono coding system. The underlying psychoacoustic effects have been studied intensively by Blauert [Blauert, 1983]. The key parameter in the determination of stereo unmasking is the Binaural Masking Level Difference (BMLD). This effect is most pronounced at low frequencies. In any case the maximum masking is occurring when the direction of the virtual quantization noise source coincides with the direction of the main signal source.

The *precedence effect* describes the effect that sound sources are sometimes localized not according to the loudness of left versus right channel but on the origin of the first (not the loudest) wavefront. This time relationship between signals can be distorted by certain joint stereo coding techniques resulting in an altered stereo image.

General ideas. To apply the general findings to bit-rate reduction, the first idea is to rotate the stereo plane into the main axis direction (as shown in Figure 2.13. This has to be done independently for different frequencies, i.e. for each subband or each critical

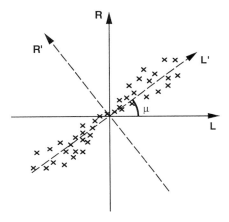

Figure 2.13 Main axis transform of the stereo plane (Reprinted from [Herre, 1995] ©1995, courtesy of the author)

band. The idea has not been implemented in any real world audio coding system because more bits are spent to transmit the direction information than are gained by this method. Two methods which have been used very successfully can be derived as simplifications of the main axis transform idea:

M/S stereo coding simplifies on the original idea by reducing the number of possible directions (to two).

Intensity stereo coding does not reduce the number of directions but keeps only the main channel information for each subband.

M/S stereo coding. M/S stereo coding was introduced to low bit-rate coding in [Johnston, 1989a]. A matrixing operation similar to the technique used in FM stereo transmission is used in the coder with the appropriate dematrixing in the decoder:

Instead of transmitting the left and right signal, the normalized sum and difference signals are handled (see Figure 2.14). They are referred to as the middle (M) and the

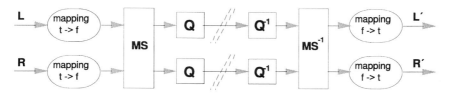

Figure 2.14 Basic block diagram of M/S stereo coding (Reprinted from [Herre, 1995] ©1995, courtesy of the author)

side (S) channel. The matrixing operation can be done in the time domain (i.e. before the analysis filter bank) as well as in the frequency domain (i.e. after the analysis filter bank). Figure 2.15 shows the matrix operation. M/S stereo coding can be seen as a special case of a main axis transform of the input signal (see [van der Waal and Veldhuis, 1991]).

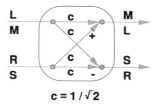

$c = 1/\sqrt{2}$

Figure 2.15 Signal flow graph of the M/S matrix (Reprinted from [Herre, 1995] ©1995, courtesy of the author)

The main features of M/S stereo processing can be described as follows [Herre et al., 1992]:

- Emphasis on redundancy removal
  The main focus of M/S joint stereo coding is on the redundancy removal for mono-like signals which often are critical for dual mono coding systems due to the stereo unmasking effects described below. The maximum gain is the theoretical gain of a main axis transform of a two-dimensional signal. However, stereo irrelevancy effects can be used in an M/S coding framework, too.

- Perfect reconstruction
  The matrixing done in M/S joint stereo coding is invertible. Without the quantization and coding of the matrix output the processing is completely transparent.

Therefore M/S coding is applicable to higher bit-rate very high quality coding, too.

- Signal dependend bit-rate gain
  The added coding efficiency of M/S stereo coding depends heavily on the actual signal. It varies from a maximum of nearly 50% if the left and right channel signals are equal (or exactly out of phase) to situations where M/S must not be used because of the possibility of new reverse unmasking effects.

- Useful for the whole spectral range
  Because M/S matrixing basically preserves the full spatial information, it may be applied to the full audio spectral range without the danger of the introduction of severe artifacts.

Intensity stereo coding. Intensity stereo coding is another simplified approximation to the general idea of directional transform coding. For each subband which is transmitted using intensity stereo modes, just the intensity information is retained. The directional information is transmitted via the coding of independent scalefactor values for the left and right channels. Thus, only the energy envelope is transmitted for both channels. Due to the irrelevancy of exact location information at high frequencies this

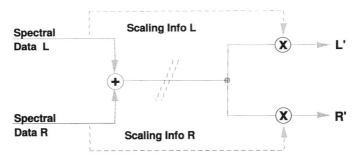

Figure 2.16 Basic principle of intensity stereo coding (Reprinted from [Herre, 1995] ©1995, courtesy of the author)

method is relatively successful. The main spatial cues are transmitted, however some details may be missing. It seems that this is especially obvious if the decoded signal is audited using headphones (see [MPEG, 1991]).

The main features of intensity stereo coding can be described as follows:

- Emphasis on irrelevancy reduction
  While signals with a large correlation of left versus right time domain signal still benefit from intensity stereo coding, the main emphasis is on the reduced spatial resolution at high frequencies.

- Not perfect reconstruction
  The signal components which are orthogonal in respect to the transmitted energy maximum are not transmitted, resulting in a loss of spatial information. The energy of the stereo signal is preserved, however. The potential loss of spatial information is considered to be less annoying than other coding artifacts. Therefore intensity stereo coding is mainly used at low bit-rates to prevent annoying coding artifacts.

- Saving of 50% of the sample data
  For the frequency range where intensity stereo coding is applied, only one channel of subband data has to be transmitted. If we assume that intensity stereo coding is applied for half of the spectrum, we can assume a saving of about 20% of the net bit-rate. The maximum saving is at about 40%.

- Useful only for the high frequency range
  As explained above, intensity stereo encoding is used only for part of the spectrum. Extending intensity stereo processing towards low frequencies can cause severe artifacts such as a major loss of directional information.

Coupling channels. In multichannel systems, a coupling channel is used as the equivalent to an n-channel intensity stereo system. This system is also known under the names dynamic crosstalk or generalized intensity coding. Instead of n different channels, for part of the spectrum only one channel with added intensity information is transmitted. Coupling channels are used in AC-3 ([Fielder et al., 1996]) and MPEG-2 AAC ([Johnston et al., 1996]).

In the coupling channel as used in MPEG-2 AAC [Johnston et al., 1996], the spectral data transmitted in the coupling element can be applied to any number of channels. Instead of replacing the data as in classical intensity coding, the coupling channel is added to the other channels. This enables coding of a residual signal in each of the channels.

2.4.5 Prediction

Prediction as a tool for high quality audio coding has been proposed a number of times (see for example [Edler, 1988, Singhal, 1990, Dimino and Parladori, 1995, Fuchs, 1995]). Prediction improves the redundancy reduction especially for near stationary signals. Dependent on the overall type of the coding system (low or high frequency resolution), different prediction strategies have found to be most efficient. The following example shows how prediction is used in a recent high frequency resolution coding system (MPEG-2 Advanced Audio Coding, the description below follows [Bosi et al., 1996b]).

PERCEPTUAL CODING OF HIGH QUALITY DIGITAL AUDIO   73

For high frequency resolution filter bank based coders the transmission of prediction coefficients would take a huge amount of additional side information. Therefore, a short (two tap) backward adaptive predictor is used. An attenuation factor is applied to the predictor to lower the long term impact of a connection loss. Prediction is switched on and off to ensure it is only used in blocks with an actual prediction gain. Additionally, the predictors are reset in certain intervals. In this way, small differences in the arithmetic accuracy between encoder and decoder do not lead to audible errors and can be tolerated.

2.4.6 Multi-channel: to matrix or not to matrix

A newer addition to perceptual encoding of high quality digital audio are systems which faithfully reproduce multichannel sound. The most common presentation structure is the 5 channel system as seen in Figure 2.17. A center channel is added to the usual left and right channels to increase the stability of the sound stage. With a center channel present, a sound source in the center (like a news speaker) stays in the center even if the listener is located slightly of-center. Two surround channels are added to give a much improved sound stage.

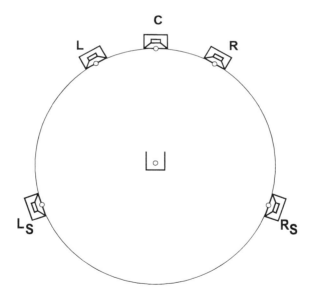

Figure 2.17   ITU Multichannel configuration

To enable a smooth transition between two-channel stereo and discrete surround sound transmission, different matrix systems have been proposed. One system is employed in the MPEG-2 backward compatible coding (see [MPEG, 1994a]) and uses an automatic downmix of the five original signals to yield a two-channel stereo signal which contains all parts of the original signal. Other systems ([ATSC, 1995],[MPEG, 1997a] propose to do a downmix of a 5-channel signal at the receiver end if a two channel presentation is needed. The following equation describes the mixdown of five to two channels. L, R, C, L_S and R_S are the left, right, center, left surround and right surround channels of the multichannel signal. L_C and R_C are the compatible left and right channels generated from the five channel signal. The matrix-mixdown coefficient a is usually selected to be one of $1/\sqrt{2}$, $1/2$, $1/(2\sqrt{2})$, 0.

$$L_C = \frac{1}{1 + 1/\sqrt{2} + a} \cdot [L + C/\sqrt{2} + a \cdot L_S] \qquad (2.19)$$

$$R_C = \frac{1}{1 + 1/\sqrt{2} + a} \cdot [R + C/\sqrt{2} + a \cdot R_S] \qquad (2.20)$$

Both compatibility matrixing in the encoder as well as downmix in the decoder have specific disadvantages. Encoder matrixing can lead to noise leakage in the decoder. This can be overcome, but at the expense of an increased bit-rate demand. Decoder matrixing can lead to some artifacts, too. However, this has not been observed to the same amount as the encoder matrixing artifacts. In both cases, the optimum two-channel mix is probably different from the automatic downmix from five channels.

2.5 APPLYING THE BASIC TECHNIQUES: REAL CODING SYSTEMS

As examples how to apply the basic techniques several well known perceptual coders are described below. The selection was based on the familiarity of the author with the schemes, not on the scientific or commercial importance of the systems. An overview containing more details about commercially available coding systems can be found in [Brandenburg and Bosi, 1997].

2.5.1 *Pointers to early systems (no detailed description)*

Instead of detailed descriptions it shall suffice to point to examples of early work on high quality audio coding. The first reference known to the author mentioning the idea of perceptual coding is [Blauert and Tritthart, 1975]. The original paper stimulating research on perceptual coding is [Schroeder et al., 1979]. Other references to early work on high quality audio coding are [Krasner, 1979, Schroeder and Voessing, 1986, Brandenburg et al., 1982].

2.5.2 MPEG Audio

Since 1988 ISO/IEC JTC1/SC29 WG11, called MPEG (Moving Pictures Experts Group) undertakes the standardization of compression techniques for video and audio. Three low bit-rate audio coding standards have been completed:

- MPEG-1 Audio [MPEG, 1992] became IS (International Standard) in 1992. It was designed to fit the demands of many applications including storage on magnetic tape, digital radio and the live transmission of audio via ISDN. A target system consisting of three modes called layers was devised. Layer 1 was originally optimized for a target bit-rate of 192 kbit/s per channel (as used in the Digital Compact Cassette, DCC), Layer 2 for a target bit-rate of 128 kbit/s per channel and Layer 3 for a target bit-rate of 64 kbit/s per channel. Sampling rates of 32 kHz, 44.1 kHz and 48 kHz are specified.

- MPEG-2 Audio [MPEG, 1994a] consists of two extensions to MPEG-1:
    - Backwards compatible multichannel coding adds the option of forward and backwards compatible coding of multichannel signals including the 5.1 channel configuration known from cinema sound.
    - Coding at lower sampling frequencies adds sampling frequencies of 16 kHz, 22.05 kHz and 24 kHz to the sampling frequencies supported by MPEG-1. This adds coding efficiency at very low bit-rates.

    Both extensions do not introduce new coding algorithms over MPEG-1 Audio.

- MPEG-2 Advanced Audio Coding [MPEG, 1997a] contains the definition of a second generation audio coding scheme for generic coding of stereo and multichannel signals including 5.1 and 7.1 configurations. This was formerly known under the name MPEG-2 NBC (non backwards-compatible coding).

MPEG-1 Layer 1 and Layer 2. The coding scheme contains the basic polyphase filter bank to map the digital audio input into 32 subbands, fixed segmentation to format the data into blocks, a psychoacoustic model to determine the adaptive bit allocation, and quantization using block companding and frame coding. The following description follows the lines of the basic block diagram of a perceptual coding system as shown in Figure 2.5.

The polyphase filter bank used in MPEG-1 uses a 511-tap prototype filter as described on page 53.

For each polyphase subband there are three main types of information to transmit:

- Bit allocation
  This determines the number of bits used to code each subband samples. The

quantizer is controlled by the bit allocation as well. In Layer 1 there are 4 bits used to transmit the bit allocation for each subband. In Layer 2 there are different possible bit allocation patterns depending on total bit-rate and sampling rate. This reduces the number of bits spent on bit allocation information at low bit-rates.

- Scalefactors
  A block floating point technique (block companding) is used to quantize the subband samples. The calculation of scalefactors is performed every 12 subband samples. The maximum absolute value of the 12 subband samples is quantized with a quantizer step size of 2 dB. With 6 bits allocated for the quantized scalefactors, the dynamic range can be up to 120 dB. Only scalefactors for subbands with a non-zero bit allocation are transmitted.

- Subband samples
  The subband samples are transmitted using the wordlength defined by the bit allocation for each subband. Uniform quantization and mid-tread quantizers are used.

Compared to Layer 1 (as described above), Layer 2 provides additional coding of bit allocation, scalefactors and samples. Different framing is used (24 ms versus 8 ms in Layer 1). The bit allocation is valid for the whole frame while scalefactors are used as exponents to blocks of 12 subband samples as in Layer 1. A scalefactor select information is used to flag whether a scalefactor is transmitted for each of the 3 blocks in a frame, for two of them or if one is valid for all 3 blocks. The scalefactor select information (scsfi) is coded using 2 bits per subband and frame. Whereas in Layer 1 the possible bit allocations are 0 and 2 to 15 bits, in Layer 2 additional fractional bit allocations are possible. They include quantizers using 3, 5, 7 and 9 quantization levels. Since many subbands are typically quantized with no more quantization levels, this results in a considerable bit-rate saving.

The bit allocation is derived from the SMR-values which have been calculated in the psychoacoustic model. This is done in an iterative fashion. The objective is to minimize the noise-to-mask ratio over every subband and the whole frame. In each iteration step the number of quantization levels is increased for the subband with the worst (maximum) noise-to-mask ratio. This is repeated until all available bits have been spent.

MPEG-1 Layer 3. Layer 3 combines some of the features of Layer 2 with the additional coding efficiency gained by higher frequency resolution and Huffman coding as found in ASPEC ([Brandenburg, 1991]). Figure 2.18 shows a block diagram.

Most of the features of Layer 3 have been described above.

The hybrid filter bank used in Layer 3 has been described on page 56. The filter bank is switchable with three possible selections corresponding to a 576 line, 216 line and 192 line frequency resolution.

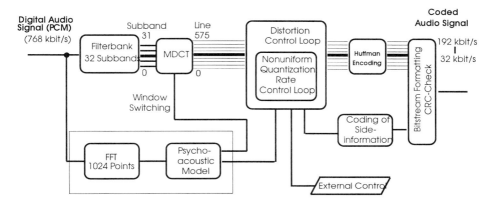

Figure 2.18 Block diagram of an MPEG-1 Layer 3 encode

Other coding tools in Layer 3 include a different (nonuniform) quantizer, analysis-by-synthesis control of the quantization noise and Huffman coding of the quantized values to increase the coding efficiency. All these have already been described earlier in this chapter.

In terms of joint stereo coding techniques, Layer 3 supports a combination of M/S coding (broad band) and intensity stereo coding (see [Herre et al., 1992]).

MPEG-2 Audio.   MPEG-2 audio coding contains two large additions to the MPEG-1 audio standard:

ISO/IEC IS 13818-3 is called "backward compatible MPEG-2 audio coding" (MPEG-2 BC) and contains extensions to MPEG-1 audio covering backwards compatible (matrixed) multichannel coding, bitstream definition extensions to cover multilingual services and the extension of all coding modes of MPEG-1 to lower sampling frequencies.

ISO/IEC IS 13818-7 is called "MPEG-2 Advanced Audio Coding" and covers a new, non backwards compatible audio coding system for flexible channel configurations including stereo and multichannel services.

Backwards compatible multichannel coding.   IS 13818-3 contains the definition of a backward-compatible multichannel coding system. The MPEG-1 L and R channels are replaced by the matrixed signals L_C and R_C according to equations (2.19) and (2.20), where L_C and R_C are encoded with an MPEG-1 encoder. Therefore an MPEG-

1 decoder can reproduce a comprehensive downmix of the full 5 channel information. The basic frame format is identical to the MPEG-1 bitstream format. The additional

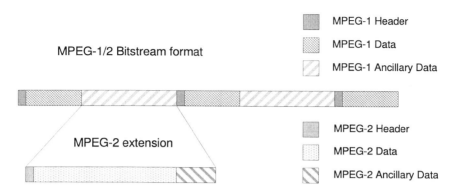

Figure 2.19 Transmission of MPEG-2 multichannel information within an MPEG-1 bitstream

channels e.g. C, L_S, and R_S are transmitted in the MPEG-1 ancillary data field.

During dematrixing in the decoder it can happen that the signal in a particular channel is derived from two channels with the signals being out of phase (cancelling each other). In this case, the corresponding quantization noise might not be out of phase and therefore survive the dematrixing. It then becomes audible as a dematrixing artifact. This way, quantization noise generated by coding of one channel can become audible in other channels.

As in the case of MPEG-1, there are three versions of the multichannel extension called Layer 1, Layer 2 and Layer 3. Layer 1 and Layer 2 MC extensions basically both use a bitstream syntax similar to Layer 2. As in the case of MPEG-1, Layer 3 is the most flexible system. As one special feature, MPEG-2 MC Layer 3 permits use of a flexible number of extension channels. While the original idea behind this was to alleviate the dematrixing artifact problem for some worst case items, this idea can be used to do simulcast of two-channel stereo and 5-channel extension without the artistic restrictions of a fixed compatibility matrix.

Coding at lower sampling frequencies. Another extension of MPEG-1 is the addition of modes using lower sampling frequencies, i.e. below 32 kHz. These modes are useful for the transmission of both wideband speech and medium quality audio at bit-rates between 64 and 16 kbit/s per channel, with applications for commentary as well as for Internet audio systems and whenever the bit-rate budget is very limited. The basic idea behind the addition of lower sampling frequencies (LSF) is the increase of coding gain for higher frequency resolution filter banks. Another advantage of LSF is an improved ratio of main information to side (esp. header) information. In a 1994 listening test [MPEG, 1994b] it was shown that 64 kb/s total bit-rate joint stereo Layer 3 at 24 kHz sampling frequencies approaches the quality (in reference to a 11 kHz signal) which was found in 1990 for the 64 kbit/s per channel ASPEC system.

2.5.3 MPEG-2 Advanced Audio Coding (MPEG-2 AAC)

MPEG-2 AAC has been designed to reduce the bit-rate where broadcast quality can be achieved as much as possible according to the state of the art. A number of new or improved coding tools have been introduced in order to improve the coding efficiency. This paragraph gives only a very short description of the main features. More information can be found in [Bosi et al., 1996b].

The block diagram of the MPEG-2 AAC encoder is shown in Figure 2.20. A brief description of the basic tools of the MPEG-2 AAC system follows:

Gain Control. A four-band polyphase quadrature filter bank (PQF) splits the input signal into four equally-spaced frequency bands. This tool is used for the scaleable sampling rate (SSR) profile only. Its time domain gain control component can be applied to reduce pre-echo effects.

Filterbank. A modified discrete cosine transform (MDCT/IMDCT) is used for the filter bank tool. The MDCT output consists of 1024 or 128 frequency lines. The window shape is selected between two alternative window shapes.

Temporal Noise Shaping (TNS). The TNS tool is used to control the temporal shape of the quantization noise within each window of the transform. This is done by applying a filtering process to parts of the spectral data.

Intensity Coding/Coupling. The intensity coding/coupling tool combines channel pairs or multiple channels and transmits only a single channel plus directional information for parts of the spectrum.

Prediction. Prediction is used to improve the redundancy reduction for stationary signals. This tool is implemented as a second order backward adaptive predictor.

M/S Stereo Coding. The M/S stereo coding tool allows to encode either Left and Right or Mid and Side of a channel pair for selected spectral regions in order to improve coding efficiency.

Scalefactors. The spectrum is divided in several groups of spectral coefficients called scalefactor bands which share one scalefactor. A scalefactor represents a gain value

80   APPLICATIONS OF DSP TO AUDIO AND ACOUSTICS

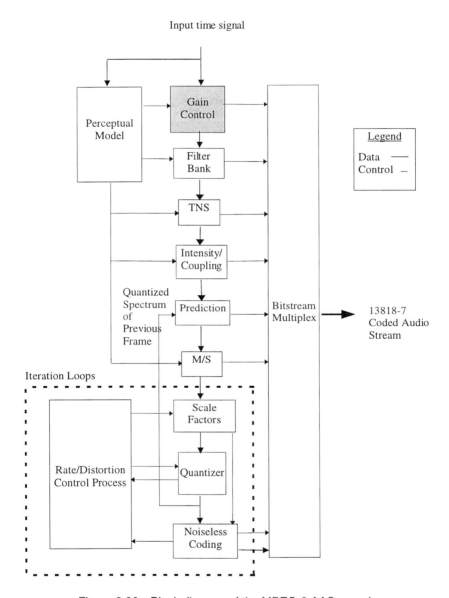

Figure 2.20   Block diagram of the MPEG-2 AAC encoder

which is used to change the amplitude of all spectral coefficients in that scalefactor band. This process provides shaping of the quantization noise according to the masked thresholds as estimated in the perceptual model.

Quantization. In the quantization tool a non-uniform quantizer (as in Layer 3) is used with a step size of 1.5 dB.

Noiseless Coding. Huffman coding is applied for the quantized spectrum, the differential scalefactors, and directional information. A total of 12 static Huffman codebooks are employed to code pairs or quadruples of spectral values.

Perceptual Model. A psychoacoustic model similar to IS 11172-3 psychoacoustic model II is employed.

2.5.4 MPEG-4 Audio

The newest coding system which is reported here is currently still under development. MPEG-4 audio, planned for completion in late 1998, will actually consist of a family of coding algorithms targeted for different bit-rates and different applications.

Bridging the gap between signal synthesis, speech coding and perceptual audio coding. The target bit-rates of MPEG-4 audio are from around 2 kbit/s up to 64 kbit/s per channel. Depending on the application, generic audio coding or speech coding is required. To fulfill this wide range of needs, MPEG-4 audio will contain a number of different algorithms. MPEG-4 audio will use MPEG-2 Advanced Audio Coding for the higher bit-rates and utilize coding tools based on MPEG-2 AAC as well as other proposals for lower bit-rates.

Scaleable audio coding. The main innovation of MPEG-4 audio besides the added flexibility is scaleability. In the context of MPEG-4 audio this is defined as the property that some part of a bitstream is still sufficient for decoding and generating a meaningful audio signal with lower fidelity, bandwidth or a selected content. Depending whether this embedded coding is realized as a number of large (e.g. 8 kbit/s) steps or with a fine granularity, it is called large step or small step scaleability. While scaleability can always be implemented via simulcast of different encoded versions of a signal, MPEG-4 audio calls for solutions with a small or no hit in coding efficiency due to the scaleability feature.

Figure 2.21 shows a block diagram of the configuration for scaleability of the planned MPEG-4 audio standard.

In the extreme case, scaleability can actually improve the coding efficiency at a certain bit-rate: If a good quality speech coder is used for the core layer the resulting

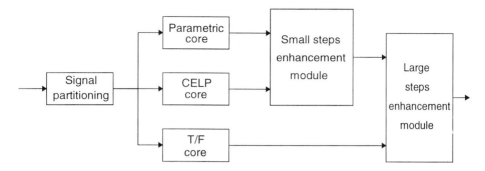

Figure 2.21 MPEG-4 audio scaleable configuration
Parametric core: Very low bit-rate coder based on parametric methods
CELP core: Speech coder
T/F-core: Time/frequency transform based perceptual coder

quality for speech signals may improve at combined bit-rates where this type of signal normally results in audible artifacts.

2.6 CURRENT RESEARCH TOPICS

Up to now, the most advanced perceptual coding systems have been built within the framework of the MPEG-Audio standardization effort. In parallel, research on alternative algorithms has been going on at universities and research institutes not involved in MPEG. The following paragraphs list just a few areas of active research on high quality audio coding.

Filterbanks. There is still continued research on filter banks for high quality audio coding. Topics include wavelet based filter banks, low delay filter banks [Schuller, 1995] or variable filter banks allowing a higher degree of variability than classic window switching [Princen and Johnston, 1995].

Perceptual Models. It seems that the search for more accurate psychoacoustic models will not be over for some time to come. Progress at very low bit-rates and for variable rate coding depends on the availability of better perceptual models. One area with promising results is the application of nonlinear models as proposed in [Baumgarte et al., 1995].

Quantization and Coding. While no major new ideas have been introduced for some time, refinements and variations on the currently used methods for quantization and coding are still an active research topic. Examples include experiments with arithmetic coding as conducted during the MPEG-2 NBC core experiment process [MPEG, 1996] or tools to improve the efficiency of currently used systems for some signal classes [Takamizawa et al., 1997].

Lossless and near lossless coding. For contribution and archiving purposes, the use of low bit-rate audio coding is dangerous to the final audio quality. If signals are coded in tandem at very low bit-rates, coding artifacts are accumulating and become audible. To overcome this problems, lossless and near lossless (high coding margin) systems have been proposed (see [Cellier, 1994, Brandenburg and Henke, 1993]). While no standardization is planned for such systems, there is ongoing work towards improved systems for lossless and near lossless coding.

2.7 CONCLUSIONS

The art of perceptual audio coding is still in between research (with a solid scientific foundation) and engineering (where it is important that things work even if nobody knows why). While the rate of innovation has somewhat slowed down, high quality audio coding is still a young research field with more results to be expected in the future.

Acknowledgments

The author would like to express his sincerest gratitude to Jürgen Herre, Thomas Sporer and Harald Popp for helping to prepare this paper. Part of the work at FhG-IIS has been supported by the Bavarian Ministry for Economy, Transportation and Technology and the German Federal Ministry for Education and Science.

3 REVERBERATION ALGORITHMS

William G. Gardner

MIT Media Laboratory, 20 Ames Street, Cambridge, MA 02139

billg@media.mit.edu

Abstract: This chapter discusses reverberation algorithms, with emphasis on algorithms that can be implemented for realtime performance. The chapter begins with a concise framework describing the physics and perception of reverberation. This includes a discussion of geometrical, modal, and statistical models for reverberation, the perceptual effects of reverberation, and subjective and objective measures of reverberation. Algorithms for simulating early reverberation are discussed first, followed by a discussion of algorithms that simulate late, diffuse reverberation. This latter material is presented in chronological order, starting with reverberators based on comb and allpass filters, then discussing allpass feedback loops, and proceeding to recent designs based on inserting absorptive losses into a lossless prototype implemented using feedback delay networks or digital waveguide networks.

3.1 INTRODUCTION

Our lives are for the most part spent in reverberant environments. Whether we are enjoying a musical performance in a concert hall, speaking to colleagues in the office, walking outdoors on a city street, or even in the woods, the sounds we hear are invariably accompanied by delayed reflections from many different directions. Rather than causing confusion, these reflections often go unnoticed, because our auditory system is well equipped to deal with them. If the reflections occur soon after the initial sound, the result is not perceived as separate sound events. Instead, the reflections modify the perception of the sound, changing the loudness, timbre, and most importantly, the spatial characteristics of the sound. Late reflections, common in very reverberant environments such as concert halls and cathedrals, often form a background ambience which is quite distinct from the foreground sound.

Interestingly, the presence of reverberation is clearly preferred for most sounds, particularly music. Music without reverberation sounds dry and lifeless. On the other hand, too much reverberation, or the wrong kind of reverberation, can cause a fine musical performance to sound muddy and unintelligible. Between these extremes is a beautiful reverberation appropriate for the music at hand, which adds fullness and a sense of space. Consequently, a number of concert halls have built reputations for having fine acoustics, based on the quality of the perceived reverberation.

The importance of reverberation in recorded music has resulted in the the creation of artificial reverberators, electro-acoustic devices that simulate the reverberation of rooms. Early devices used springs or steel plates equipped with transducers. The advent of digital electronics has replaced these devices with the modern digital reverberator, which simulates reverberation using a linear discrete-time filter. These devices are ubiquitous in the audio production industry. Almost every bit of audio that we hear from recordings, radio, television, and movies has had artificial reverberation added. Artificial reverberation has recently found another application in the field of virtual environments, where simulating room acoustics is critical for producing a convincing immersive experience.

The subject of this paper is the study of signal processing algorithms that simulate natural room reverberation. The emphasis will be on efficient algorithms that can be implemented for real-time performance.

3.1.1 Reverberation as a linear filter

From a signal processing standpoint, it is convenient to think of a room containing sound sources and listeners as a system with inputs and outputs, where the input and output signal amplitudes correspond to acoustic variables at points in the room. For example, consider a system with one input associated with a spherical sound source, and two outputs associated with the acoustical pressures at the eardrums of a listener. To the extent that the room can be considered a linear, time-invariant (LTI) system[1], a stereo transfer function completely describes the transformation of sound pressure from the source to the ears of a listener. We can therefore simulate the effect of the room by convolving an input signal with the *binaural impulse response* (BIR):

$$y_L(t) = \int_0^\infty h_L(\tau) x(t-\tau) d\tau \quad (3.1)$$

$$y_R(t) = \int_0^\infty h_R(\tau) x(t-\tau) d\tau$$

where $h_L(t)$ and $h_R(t)$ are the system impulse responses for the left and right ear, respectively; $x(t)$ is the source sound; and $y_L(t)$ and $y_R(t)$ are the resulting signals for the left and right ear, respectively. This concept is easily generalized to the case of multiple sources and multiple listeners.

3.1.2 Approaches to reverberation algorithms

We will speak of a *reverberation algorithm*, or more simply, a *reverberator*, as a linear discrete-time system that simulates the input-output behavior of a real or imagined room. The problem of designing a reverberator can be approached from a physical or perceptual point of view.

The physical approach. The physical approach seeks to simulate exactly the propagation of sound from the source to the listener for a given room. The preceding discussion of binaural impulse responses suggests an obvious way to do this, by simply measuring the binaural impulse response of an existing room, and then rendering the reverberation by convolution.

When the room to be simulated doesn't exist, we can attempt to predict its impulse response based on purely physical considerations. This requires detailed knowledge of the geometry of the room, properties of all surfaces in the room, and the positions and directivities of the sources and receivers. Given this prior information, it is possible to apply the laws of acoustics regarding wave propagation and interaction with surfaces to predict how the sound will propagate in the space. This technique has been termed *auralization* in the literature and is an active area of research [Kleiner et al., 1993]. Typically, an auralization system first computes the impulse response of the specified room, for each source-receiver pair. These finite impulse response (FIR) filters are then used to render the room reverberation.

The advantage of this approach is that it offers a direct relation between the physical specification of the room and the resulting reverberation. However, this approach is computationally expensive and rather inflexible. Compared to other algorithms we will study, real-time convolution with a large filter response is somewhat expensive, even using an efficient algorithm. Furthermore, there is no easy way to achieve real-time parametric control of the perceptual characteristics of the resulting reverberation without recalculating a large number of FIR filter coefficients.

The perceptual approach. The perceptual approach seeks to reproduce only the perceptually salient characteristics of reverberation. Let us assume that the space of all percepts caused by reverberation can be spanned by N independent dimensions, which correspond to independently perceivable attributes of reverberation. If each perceptual attribute can be associated with a physical feature of the impulse response, then we can attempt to construct a digital filter with N parameters that reproduces exactly these N attributes. In order to simulate the reverberation from a particular room, we can measure the room response, estimate the N parameters by analyzing the impulse response, and then plug the parameter estimates into our "universal" reverberator. The reverberator should then produce reverberation that is indistinguishable from the original, even though the fine details of the impulse responses may differ considerably.

This approach has many potential advantages:

- The reverberation algorithm can be based on efficient infinite impulse response (IIR) filters.

- The reverberation algorithm will provide real-time control of all the perceptually relevant parameters. The parameters do not need to be correlated as they often are in real rooms.

- Ideally, only one algorithm is required to simulate all reverberation.

- Existing rooms can be simulated using the analysis/synthesis approach outlined above.

One disadvantage of this method is that it doesn't necessarily provide an easy way to change a physical property of the simulated room.

The perceptually motivated method is essentially the approach that has been taken in the design of reverberation algorithms, with several caveats. First, there is a great deal of disagreement as to what the perceivable attributes of reverberation are, and how to measure these from an impulse response. Second, it is difficult to design digital filters to reproduce these attributes. Consequently, the emphasis has been to design reverberators that are *perceptually indistinguishable* from real rooms, without necessarily providing the reverberator with a complete set of independent perceptual controls.

In this paper, we will concentrate on the perceptually motivated method, because the resulting recursive algorithms are more practical and useful. We first present a concise physical and perceptual background for our study of reverberation, then discuss algorithms to simulate early reverberation, and conclude with a discussion of late reverberation algorithms.

3.2 PHYSICAL AND PERCEPTUAL BACKGROUND

The process of reverberation starts with the production of sound at a location within a room. The acoustic pressure wave expands radially outward, reaching walls and other surfaces where energy is both absorbed and reflected. Technically speaking, all reflected energy is *reverberation*. Reflection off large, uniform, rigid surfaces produces a reflection the way a mirror reflects light, but reflection off non-uniform surfaces is a complicated process, generally leading to a diffusion of the sound in various directions. The wave propagation continues indefinitely, but for practical purposes we can consider the propagation to end when the intensity of the wavefront falls below the intensity of the ambient noise level.

Assuming a direct path exists between the source and the listener, the listener will first hear the *direct sound*, followed by reflections of the sound off nearby surfaces, which are called *early echoes*. After a few hundred milliseconds, the number of reflected waves becomes very large, and the remainder of the reverberant decay is characterized by a dense collection of echoes traveling in all directions, whose intensity is relatively independent of location within the room. This is called *late reverberation* or *diffuse reverberation*, because there is equal energy propagating in all directions. In a perfectly diffuse soundfield, the energy lost due to surface absorption is proportional to the energy density of the soundfield, and thus diffuse reverberation decays exponentially with time. The time required for the reverberation level to decay to 60 dB below the initial level is defined as the *reverberation time*.

3.2.1 Measurement of reverberation

Measuring reverberation in a room usually consists of measuring an impulse response for a specific source and receiver. Pistol shots, balloon pops, and spark generators can be used as impulsive sources. Another possibility is to use an omnidirectional speaker driven by an electronic signal generator. Typical measurement signals include clicks, chirps (also known as time delay spectrometry [Heyser, 1967]), and various pseudo-random noise signals, such as maximum length (ML) sequences [Rife and Vanderkooy, 1987] and Golay codes [Foster, 1986]. The click (unit impulse) signal allows a direct measurement of the impulse response, but results in poor signal to noise ratio (SNR) because the signal energy is small for a given peak amplitude. The chirp and noise signals have significantly greater energy for a given peak amplitude, and allow the impulse response to be measured with improved SNR by deconvolving the impulse response from the recorded signal. The measurement signals are deliberately chosen to make the deconvolution easy to perform.

Figure 3.1 shows the impulse response of a concrete stairwell, plotting pressure as a function of time. The direct response is visible at the far left, followed by some early echoes, followed by the exponentially decaying late reverberation. The early echoes have greater amplitude than the direct response due to the directivities of the measurement speaker and microphone.

Rooms may contain a large number of sources with different positions and directivity patterns, each producing an independent signal. The reverberation created in a concert hall by a symphony orchestra cannot be characterized by a single impulse response. Fortunately, the statistical properties of late reverberation do not change significantly as a function of position. Thus, a point to point impulse response does characterize the late reverberation of the room, although the early echo pattern is dependent on the positions and directivities of the source and receiver.

90   APPLICATIONS OF DSP TO AUDIO AND ACOUSTICS

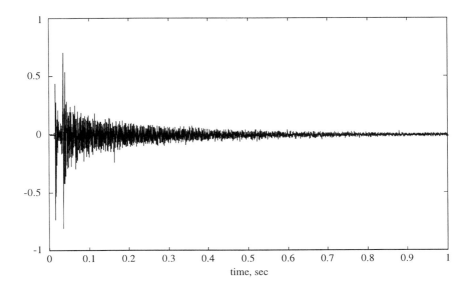

Figure 3.1   Impulse response of reverberant stairwell measured using ML sequences.

The fact that the early and late reverberation have different physical and perceptual properties permits us to logically split the study of reverberation into early and late reverberation.

3.2.2   Early reverberation

Early reverberation is most easily studied by considering a simple geometrical model of the room. These models depend on the assumption that the dimensions of reflective surfaces in the room are large compared to the wavelength of the sound. Consequently, the sound wave may be modeled as a ray that is normal to the surface of the wavefront and reflects specularly, like light bouncing off a mirror, when the ray encounters a wall surface. Figure 3.2 shows a wall reflection using the ray model. The source is at point A, and we are interested in how sound will propagate to a listener at point B.

The reflected ray may also be constructed by considering the mirror image of the source as reflected across the plane of the wall. In figure 3.2, the *image source* thus constructed is denoted A'. This technique of reflecting sources across wall surfaces is called the *source image method*. The method allows a source with reflective boundaries to be modeled as multiple sources with no boundaries.

The image source A' is a first order source, corresponding to a sound path with a single reflection. Higher order sources corresponding to sound paths with multiple

REVERBERATION ALGORITHMS

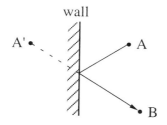

Figure 3.2 Single wall reflection and corresponding image source A'.

reflections are created by reflecting lower order sources across wall boundaries. Frequently the resulting sources are "invisible" to the listener position, and this condition must be tested explicitly for each source. When the room is rectangular, as shown in figure 3.3, the pattern of image sources is regular and trivial to calculate. Calculation of the image source positions in irregularly-shaped rooms is more difficult, but the problem has been solved in detail [Borish, 1984]. The number of image sources of order k is roughly N^k, where N is the number of wall surfaces. The source image method is impractical for studying late reverberation because the number of sources increases exponentially, and the simplified reflection model becomes inaccurate.

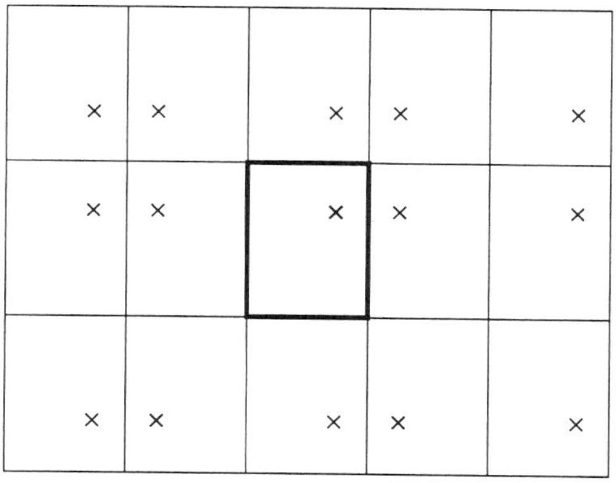

Figure 3.3 A regular pattern of image sources occurs in an ideal rectangular room.

In order to calculate the impulse response at the listener's position, the contributions from all sources are summed. Each source contributes a delayed impulse (echo), whose time delay is equal to the distance between the source and the listener divided by the speed of sound. The echo amplitude is inversely proportional to the distance travelled, to account for spherical expansion of the sound, and proportional to the product of the reflection coefficients of the surfaces encountered. This model ignores any frequency dependent absorption, which normally occurs during surface reflections and air propagation. A more accurate model uses linear filters to approximate these frequency dependent losses [Lehnert and Blauert, 1992], such that the spectrum of each echo reaching the listener is determined by the product of the transfer functions involved in the history of that echo:

$$A(\omega) = G(\omega) \prod_{j \in S} \Gamma_j(\omega) \tag{3.2}$$

where $A(\omega)$ is the spectrum of the echo, S is the set of walls encountered, $\Gamma_j(\omega)$ is the frequency dependent transfer function that models reflection with the jth wall, and $G(\omega)$ models the absorptive losses and time delay due to air propagation.

The simplifying assumptions that permit us to consider only specular reflections are no longer met when the wall surfaces contain features that are comparable in size to the wavelength of the sound. In this case, the reflected sound will be scattered in various directions, a phenomenon referred to as *diffusion*. The source image model cannot be easily extended to handle diffusion. Most auralization systems use another geometrical model, called *ray tracing* [Krokstad et al., 1968], to model diffuse reflections. A discussion of these techniques is beyond the scope of this paper.

The early response consists largely of discrete reflections that come from specific directions, and we now consider how to reproduce the directional information. It is well known that the auditory cues for sound localization are embodied in the transformation of sound pressure by the torso, head, and external ear (pinna) [Blauert, 1983]. A *head-related transfer function* (HRTF) is a frequency response that describes this transformation from a specific free field source position to the eardrum. HRTFs are usually measured using human subjects or dummy-head microphones, and consist of response pairs, for the left and right ears, corresponding to a large number of source positions surrounding the head. When computing the binaural transfer function of a room using the geometrical models just discussed, we must convolve each directional echo with the HRTF corresponding to the direction of the echo [Wightman and Kistler, 1989, Begault, 1994].

The binaural directional cues captured by HRTFs are primarily the interaural time difference (ITD) and interaural intensity difference (IID) which vary as a function of frequency. Echoes that arrive from lateral directions (i.e. from either side of the listener) are important for modifying the spatial character of the perceived reverberation. The ITD of a lateral sound source is well modeled by a delay corresponding to the difference in path lengths between the two ears. Similarly, the IID may be modeled as a lowpass filtering of the signal arriving at the opposite (contralateral) ear.

3.2.3 Perceptual effects of early echoes

The perceptual effects of early reflections can be studied by considering a simple soundfield consisting of the direct sound and a single delayed reflection. This situation is easy to reproduce in an anechoic chamber or with headphones. Using musical signals, when both the direct sound and reflection are presented frontally and the reflection delay is greater than about 80 msec, the reflection will be perceived as a distinct echo of the direct sound if it is sufficiently loud. As the reflection delay becomes smaller, the reflection and direct sound fuse into one sound, but with a tonal coloration attributed to the cancellation between the two signals at a periodic set of frequencies. The reflection can also increase the loudness of the direct sound. The delay and gain thresholds corresponding to the different percepts depend strongly on the source sound used for the experiment.

When the reflection comes from a lateral direction, the reflection can profoundly affect the spatial character of the sound. For small reflection delays (< 5 msec), the echo can cause the apparent location of the source to shift. Larger delays can increase the apparent size of the source, depending on its frequency content, or can create the sensation of being surrounded by sound.

In the literature, various terms are used to describe the spatial sensations attributed to lateral reflections, including *spaciousness*, *spatial impression*, *envelopment*, and *apparent source width* (ASW). Despite the lack of consistent terminology, it is generally accepted that spaciousness is a desirable attribute of reverberation [Beranek, 1992]. It has been hypothesized that lateral reflections affect the spatial character of the sound by directly influencing the localization mechanisms of the auditory system [Griesinger, 1992]; the presence of the lateral energy causes large interaural differences which would not otherwise occur in the presence of frontal (or medial) energy alone.

In Barron and Marshall's research into this phenomena using musical signals, it was determined that the degree of spatial impression was directly related to the sine of the reflection incidence angle, reaching a maximum for 90 degree (fully lateral) incidence [Barron and Marshall, 1981]. They proposed a simple acoustical measurement that predicted the spatial impression, called the lateral fraction (LF). LF is the ratio of early energy received by a dipole microphone (null axis facing forward) to the total early energy. A binaural acoustical measurement that has superceded LF for predicting spatial impression is the *interaural cross-correlation coefficient* (IACC) [Hidaka et al., 1995]:

$$IACF(\tau) = \frac{\int_{t_1}^{t_2} p_L(t) p_R(t+\tau) d\tau}{(\int_{t_1}^{t_2} p_L^2(t) dt \int_{t_1}^{t_2} p_R^2(t) dt)^{1/2}} \quad (3.3)$$

$$IACC = |IACF(\tau)|_{\max}, \text{ for } -1 < \tau < +1 \text{ ms}$$

where p_L and p_R are the pressures at the entrance to the left and right ear canals, respectively, and the integration limits t_1 and t_2 are chosen to be 0 and 80 msec, respectively, when the "early" $IACC_E$ is calculated. $IACF(\tau)$ is the normalized cross-correlation function of the left and right ear pressures with a time lag of τ, and IACC is the maximum of this function over a range of ± 1 msec, to account for the maximum interaural time delay. The time lag corresponding to the maximum value of IACF estimates the lateral direction of the source sound [Blauert and Cobben, 1978]. A broadening of the IACF, and consequently a lower IACC value, corresponds to increased spatial impression.

3.2.4 Reverberation time

Sabine's pioneering research started the field of modern room acoustics and established many important concepts, most notably the concept of *reverberation time* (RT) [Sabine, 1972]. His initial experiments consisted of measuring the reverberant decay time of a room, and observing the change in decay time as absorptive material was added to the room. Sabine determined that the reverberant decay time was proportional to the volume of the room and inversely proportional to the amount of absorption:

$$T_r \propto \frac{V}{A} \tag{3.4}$$

where T_r is the reverberation time required for the sound pressure to decay 60 dB, V is the volume of the room, and A is a measure of the total absorption of materials in the room. Because the absorptive properties of materials vary as a function of frequency, the reverberation time does as well. Most porous materials, such as carpeting and upholstery, are more absorptive at higher frequencies, and consequently the RT of most rooms decreases with increasing frequency.

Reverberation time can be measured by exciting a room to steady state with a noise signal, turning off the sound source, and plotting the resulting squared pressure as a function of time. The time required for the resulting *energy decay curve* (EDC) to decay 60 dB is defined as the RT. Narrowband noise centered at some frequency can be used to measure the RT at that frequency. The particular energy decay curve so obtained will depend on details of the noise signal used. By averaging many successive measurements using different noise signals, one obtains a more accurate estimate of the true energy decay curve.

Schroeder has shown that this averaging is unnecessary [Schroeder, 1965]. The true energy decay curve can be obtained by integrating the impulse response of the room as follows:

$$EDC(t) = \int_t^\infty h^2(\tau) d\tau \tag{3.5}$$

where $h(t)$ is the impulse response of the room which may be narrowband filtered to yield the EDC for some particular frequency. The integral (often called a Schroeder integral) computes the energy remaining in the impulse response after time t.

A useful way to display reverberation as a function of time and frequency is to start with the impulse response, bandpass filter it into frequency bands, compute the Schroeder integrals, and display the result as a 3-D surface. This has been proposed by several authors [Jot, 1992b, Jot, 1992a, Griesinger, 1995] and the concept has been formalized by Jot as the *energy decay relief*, $EDR(t, \omega)$, which is a time-frequency representation of the energy decay. Thus, $EDR(0, \omega)$ gives the power gain as a function of frequency and $EDR(t, \omega_0)$ gives the energy decay curve for some frequency ω_0. Figure 3.4 shows the energy delay relief of occupied Boston Symphony Hall displayed in third octave bands. As expected, the reverberation decays faster at higher frequencies.

The late portion of the EDR can be described in terms of the *frequency response envelope* $G(\omega)$ and the reverberation time $T_r(\omega)$, both functions of frequency [Jot, 1992b]. $G(\omega)$ is calculated by extrapolating the exponential decay backwards to time 0 to obtain a conceptual $EDR(0, \omega)$ of the late reverberation. For diffuse reverberation, which decays exponentially, $G(\omega) = EDR(0, \omega)$. In this case, the frequency response envelope $G(\omega)$ specifies the power gain of the room, and the reverberation time $T_r(\omega)$ specifies the energy decay rate. The smoothing of these functions is determined by the frequency resolution of the time-frequency distribution used.

3.2.5 Modal description of reverberation

When the room is highly idealized, for instance if it is perfectly rectangular with rigid walls, the reverberant behavior of the room can be described mathematically in closed form. This is done by solving the acoustical wave equation for the boundary conditions imposed by the walls of the room. This approach yields a solution based on the natural resonant frequencies of the room, called *normal modes*. For the case of a rectangular room shown in figure 3.3, the resonant frequencies are given by [Beranek, 1986]:

$$f_n = \frac{c}{2}\sqrt{\left(\frac{n_x}{L_x}\right)^2 + \left(\frac{n_y}{L_y}\right)^2 + \left(\frac{n_z}{L_z}\right)^2} \quad (3.6)$$

where:
$f_n = n$th normal frequency in Hz.
n_x, n_y, n_z = integers from 0 to ∞ that can be chosen separately.
L_x, L_y, L_z = dimensions of the room in meters.
c = speed of sound in m/sec.

The number N_f of normal modes below frequency f is approximately [Kuttruff, 1991]:

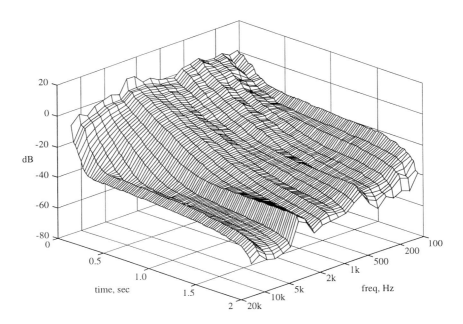

Figure 3.4 Energy decay relief for occupied Boston Symphony Hall. The impulse response was measured at 25 kHz sampling rate using a balloon burst source on stage and a dummy-head microphone in the 14th row. The Schroeder integrals are shown in third octave bands with 40 msec time resolution. At higher frequencies there is a substantial early sound component, and the reverberation decays faster. The frequency response envelope at time 0 contains the non-uniform frequency response of the balloon burst and the dummy-head microphone. The late spectral shape is a consequence of integrating measurement noise. The SNR of this measurement is rather poor, particularly at low frequencies, but the reverberation time can be calculated accurately by linear regression over a portion of the decay which is exponential (linear in dB).

$$N_f \approx \frac{4\pi V}{3c^3} f^3 \tag{3.7}$$

where V is the volume of the room ($V = L_x L_y L_z$). Differentiating with respect to f, we obtain the modal density as a function of frequency:

$$\frac{dN_f}{df} \approx \frac{4\pi V}{c^3} f^2 \tag{3.8}$$

The number of modes per unit bandwidth thus grows as the square of the frequency. For instance, consider a concert hall sized room with dimensions 44m x 25m x 17m whose volume is 18,700 m^3. Below 10,000 Hz, there are approximately 1.9×10^9 normal modes. At 1000 Hz, the modal density per Hz is approximately 5800, and thus the average spacing between modes is less than 0.0002 Hz.

When a sound source is turned on in an enclosure, it excites one or more of the normal modes of the room. When the source is turned off, the modes continue to resonate their stored energy, each decaying at a separate rate determined by the mode's damping constant, which depends on the absorption of the room. This is entirely analogous to an electrical circuit containing many parallel resonances [Beranek, 1986]. Each mode has a resonance curve associated with it, whose inxquality factor (Q) depends on the damping constant.

3.2.6 Statistical model for reverberation

The behavior of a large, irregularly shaped room can also be described in terms of its normal modes, even though a closed form solution may be impossible to achieve. It can be shown that equation 3.8 regarding modal density is generally true for irregular shapes [Kuttruff, 1991]. At high frequencies, the frequency response of the room is determined by a great many modes whose resonance curves overlap. At each frequency, the complex frequency response is a sum of the overlapping modal responses, which may be regarded as independent and randomly distributed. If the number of contributing terms is sufficiently large, the real and imaginary parts of the combined frequency response can be modeled as independent Gaussian random variables. Consequently, the resulting pressure magnitude response follows the well known Raleigh probability distribution. This yields a variety of statistical properties of reverberation in large rooms, including the average separation of maxima and the average height of a maximum [Schroeder, 1954, Schroeder, 1987, Schroeder and Kuttruff, 1962]. This statistical model for reverberation is justified for frequencies higher than:

$$f_g \approx 2000 \sqrt{\frac{T_r}{V}} \text{ Hz}. \tag{3.9}$$

where T_r is the reverberation time in seconds, and V is the volume of the room in m^3. The average separation of maxima in Hz is given by:

98   APPLICATIONS OF DSP TO AUDIO AND ACOUSTICS

$$\Delta f_{\max} \approx \frac{4}{T_r} \text{ Hz.} \qquad (3.10)$$

For example, a concert hall with volume of 18,700 m³ and an RT of 1.8 sec will have $\Delta f_{\max} = 2.2$ Hz, for frequencies greater than $f_g = 20$ Hz.

Another statistic of interest is the temporal density of echoes, which increases with time. This can be estimated by considering the source image model for reverberation, which for a rectangular room leads to a regular pattern of source images (figure 3.3). The number of echoes N_t that will occur before time t is equal to the number of image sources enclosed by a sphere with diameter ct centered at the listener [Kuttruff, 1991]. Since there is one image source per room volume, the number of image sources enclosed by the sphere can be estimated by dividing the volume of the sphere by the volume of the room:

$$N_t = \frac{4\pi(ct)^3}{3V} \qquad (3.11)$$

Differentiating with respect to t, we obtain the temporal density of echoes:

$$\frac{dN_t}{dt} = \frac{4\pi c^3}{V} t^2 \qquad (3.12)$$

Although this equation is not accurate for small times, it shows that the density of echoes grows as the square of time.

3.2.7   Subjective and objective measures of late reverberation

A practical consequence of architecture is to permit acoustical performances to large numbers of listeners by enclosing the sound source within walls. This dramatically increases the sound energy to listeners, particularly those far from the source, relative to free field conditions. A measure of the resulting frequency dependent gain of the room can be obtained from the EDR evaluated at time 0. This frequency response can be considered to be an equalization applied by the room, and is often easily perceived.

In the absence of any other information, the mid-frequency reverberation time is perhaps the best measure of the overall reverberant characteristics of a room. We expect a room with a long RT to sound more reverberant than a room with a short RT. However, this depends on the distance between the source and the listener, which affects the level of the direct sound relative to the level of the reverberation. The reverberant level varies little throughout the room, whereas the direct sound falls off inversely proportional to distance. Thus, the ratio of direct to reverberant level is an important perceptual cue for source distance [Blauert, 1983, Begault, 1992].

One acoustical measure of the direct to reverberant ratio is called the *clarity index*, and is defined as:

$$C = 10\log_{10}\left\{\frac{\int_0^{80\text{ ms}} p^2(t)dt}{\int_{80\text{ ms}}^{\infty} p^2(t)dt}\right\} \text{ dB.} \qquad (3.13)$$

where $p(t)$ is the impulse response of the room. This is essentially an early to late energy ratio, which correlates with the intelligibility of music or speech signals in reverberant environments. It is generally accepted that early energy perceptually fuses with the direct sound and thus increases intelligibility by providing more useful energy, whereas late reverberation tends to smear syllables and note phrases together.

When there is a sufficient amount of late reverberant energy, the reverberation forms a separately perceived background sound. The audibility of the reverberation depends greatly on the source sound as well as the EDR, due to masking of the reverberation by the direct sound [Gardner and Griesinger, 1994, Griesinger, 1995]. Consequently, the early portion of the reverberant decay, which is audible during the gaps between notes and syllables, contributes more to the perception of reverberance than does the late decay, which is only audible after complete stops in the sound. Existing measures for reverberance focus on the initial energy decay. The most used measure is the *early decay time* (EDT), typically calculated as the time required for the Schroeder integral to decay from 0 to -10 dB, multiplied by 6 to facilitate comparison with the RT.

Because late reverberation is spatially diffuse, the left and right ear signals will be largely uncorrelated. The resulting impression is that the listener is enveloped by the reverberation. A measurement of this, called $IACC_L$, is obtained by calculating IACC over the time limit $0.08 < t < 3$ sec [Hidaka et al., 1995] (see equation 3.3). We expect this to yield low values for nearly all rooms, and thus a high degree of spaciousness in the late reverberation.

Various experiments have been performed in an effort to determine an orthogonal set of perceptual attributes of reverberation, based on factor analysis or multidimensional scaling. A typical experiment presents subjects with pairs of reverberant stimuli, created by applying different reverberant responses to the same source sound, and the subjects estimate the subjective difference between each pair [Jullien et al., 1992]. The resulting distances are used to place the reverberant response data in an N-dimensional space such that error between the Cartesian distance and the subjective distance is minimized. The projection of the data points onto the axes can then be correlated with known objective or subjective properties of the data to assign meaning to the axes. A fundamental problem with this approach is that the number of dimensions is not known a priori, and it is difficult to assign relevance to higher dimensions which are added to improve the fit. In Jullien's experiments, 11 independent perceptual factors were found. The corresponding objective measures can be categorized as energy ratios or energy decay slopes calculated over different time-frequency regions of the EDR. Only one factor (a lateral energy measure) is not derivable from the EDR.

3.2.8 Summary of framework

The geometrical models allow the prediction of a room's early reverberant response, which will consist of a set of delayed and attenuated impulses. More accurate modeling of absorption and diffusion will tend to fill in the gaps with energy. Linear filters can be used to model absorption, and to a lesser extent diffusion, and allow reproduction of the directional properties of the early response.

The late reverberation is characterized by a dense collection of echoes traveling in all directions, in other words a diffuse sound field. The time decay of the diffuse reverberation can be broadly described in terms of the mid frequency reverberation time. A more accurate description considers the energy decay relief of the room. This yields the frequency response envelope and the reverberation decay time, both functions of frequency. The modal approach reveals that reverberation can be described statistically for sufficiently high frequencies. Thus, certain statistical properties of rooms, such as the mean spacing and height of frequency maxima, are independent of the shape of the room.

Early reverberation perceptually fuses with the direct sound, modifying its loudness, timbre, and spatial impression. Lateral reflections are necessary for the spatial modification of the direct sound. The level of the direct sound relative to the reverberation changes as a function of source distance, and serves as an important distance cue. Generally speaking, increased early energy relative to total energy contributes to the intelligibility of the signal, though this may not be subjectively preferred.

There are a large number of subjective attributes of reverberation which have been discussed in the literature. Most of these are monaural attributes directly correlated with acoustical measures that can be derived from the EDR. Consequently, it is convenient to think of the EDR as representative of all the monaural objective measures of a room impulse response. Presumably, the fine details of this shape are irrelevant, particularly in the late response, but no systematic study has been done to determine the resolution required to perceptually reproduce a reverberant response from its EDR.

Thus, in order to simulate a perceptually convincing room reverberation, it is necessary to simulate both the pattern of early echoes, with particular concern for lateral echoes, and the late energy decay relief. The latter can be parameterized as the frequency response envelope and the reverberation time, both of which are functions of frequency. The challenge is to design an artificial reverberator which has sufficient echo density in the time domain, sufficient density of maxima in the frequency domain, and a natural colorless timbre.

3.3 MODELING EARLY REVERBERATION

We are now prepared to discuss efficient algorithms that can render reverberation in real-time. For the case of early reverberation, the filter structures are fairly obvious. As we have already mentioned, convolution is a general technique that can be used to render a measured or predicted reverberant response. Implementing convolution using the direct form FIR filter (figure 3.5) is extremely inefficient when the filter size is large. Typical room responses are several seconds long, which at a 44.1 kHz sampling rate would translate to an 88,200 point filter for a 2 second response (for each channel). The early response consisting of the first 100 msec would require a 4410 point filter. These filter sizes are prohibitively large for direct form implementation. However, it is possible to implement convolution efficiently using a block processing algorithm based on the Fast Fourier transform (FFT) [Oppenheim and Schafer, 1989].

One problem with using block convolution methods for real-time processing is the input/output propagation delay inherent in block algorithms. Gardner has proposed a hybrid convolution algorithm that eliminates the propagation delay by segmenting the impulse response into blocks of exponentially increasing size [Gardner, 1995]. Convolution with the first block is computed using a direct form filter, and convolution with the remaining blocks is computed using frequency domain techniques. For large filter sizes, this hybrid algorithm is vastly more efficient than the direct form filter.

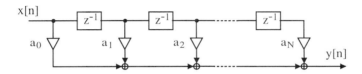

Figure 3.5   Canonical direct form FIR filter with single sample delays.

When the early response is derived from the source image model without any special provisions to model diffusion or absorption, the early response will be sparsely populated with delayed and attenuated impulses. Consequently, it is possible to efficiently implement this filter using a direct form structure with long delays between filter taps. An example of this is shown in figure 3.6, which is a structure proposed by Schroeder for generating a specific early echo pattern in addition to a late diffuse reverberation [Schroeder, 1970b]. The FIR structure is implemented with the set of delays m_i and tap gains a_i, and $R(z)$ is a filter that renders the late reverberation. Because this filter receives the delayed input signal, the FIR response will occur before the late response in the final output.

Moorer proposed a slightly different structure, shown in figure 3.7, where the late reverb is driven by the output of the early echo FIR filter [Moorer, 1979]. Moorer described this as a way of increasing the echo density of the late reverberation. The delays D_1 and D_2 can be adjusted so that the first pulse output from the late reverberator corresponds with the last pulse output from the FIR section. The gain g serves to balance the amount of late reverberation with respect to the early echoes. An important feature of this structure, apart from the early echo modeling, is the control

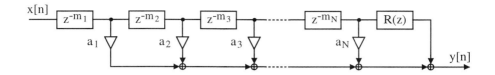

Figure 3.6 Combining early echoes and late reverberation [Schroeder, 1970b]. $R(z)$ is a reverberator.

it permits of the overall decay shape of the reverberation. For instance, if the FIR response has a wide rectangular envelope and the late reverberator has a relatively fast exponential decay, then the cascade response will have an flat plateau followed by a rapid decay. Such a multislope decay can be a useful and popular effect for musical signals [Griesinger, 1989].

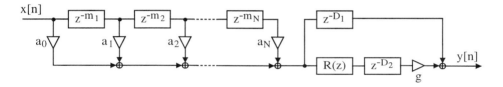

Figure 3.7 FIR filter cascaded with reverberator $R(z)$ [Moorer, 1979].

Modeling the early echoes using a sparse FIR filter results in an early response that can have an overly discrete sound quality, particularly with bright impulsive inputs. In practice it is necessary to associate some form of lowpass filtering with the early response to improve the sound quality. The simplest possible solution uses a single lowpass filter in series with the FIR filter [Gardner, 1992], where the filter response can be set empirically or by physical consideration of the absorptive losses.

In the structure shown in figure 3.6, Schroeder suggested replacing the gains a_i with frequency dependent filters $A_i(z)$. These filters can model the frequency dependent absorptive losses due to wall reflections and air propagation. Each filter is composed by considering the history of reflections for each echo, as given in equation 3.2. If the reverberator is intended for listening over headphones, we can also associate with each echo a directional filter intended to reproduce localization cues. This structure is shown in figure 3.8, where $A(z)$ is the transfer function which models absorptive losses, and $H_L(z)$ and $H_R(z)$ are the HRTFs corresponding to the direction of the echo.

Considering that the early echoes are not perceived as individual events, it seems unlikely that the spectral characteristics of each echo need to be modeled so carefully

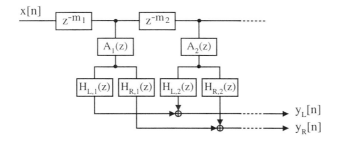

Figure 3.8 Associating absorptive and directional filters with early echoes.

[Begault, 1994, Bech, 1995, Jot et al., 1995]. It is far more efficient to sum sets of echoes together and process them with the same filter, such that all the echoes in a set have the same absorption and spatial location. Another possibility is to reproduce the interaural time and intensity difference separately for each echo, and lump the remaining spectral cues into an average directional filter for each set of echoes [Jot et al., 1995]. This is shown in figure 3.9. Each echo has an independent gain and interaural time and intensity difference, allowing for individual lateral locations. The final filter reproduces the remaining spectral features, obtained by a weighted average of the various HRTFs and absorptive filters.

If the reverberation is not presented binaurally, the early lateral echoes will not produce spatial impression, but will cause tonal coloration of the sound. In this case it may be preferable to omit the early echoes altogether. This is an important consideration in professional recording, and is the reason why orchestras are often moved to the concert hall floor when recording, to avoid the early stage reflections [Griesinger, 1989].

We conclude the section on early reverberation with an efficient algorithm that renders a convincing sounding early reverberation, particularly in regards to providing the sensation of spatial impression. Figure 3.10 shows Griesinger's binaural echo simulator that takes a monophonic input and produces stereo outputs intended for listening over headphones [Griesinger, 1997]. The algorithm simulates a frontally incident direct sound plus six lateral reflections, three per side. The echo times are chosen arbitrarily between 10 and 80 msec in order to provide a strong spatial impression, or may be derived from a geometrical model. The algorithm is a variation of the preceding structures: two sets of echoes are formed, and each set is processed through the same directional filter. Here the directional filter is modeled using a delay of 0.8 msec and a one-pole lowpass filter (figure 3.11) with a 2 kHz cutoff.

Various degrees of spatial impression can be obtained by increasing the gain of the echoes (via the g_e parameter). Whether the spatial impression is heard as a surrounding

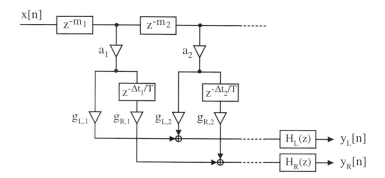

Figure 3.9 Average head-related filter applied to a set of early echoes lateralized using delays Δt_i and gains g_i [Jot et al., 1995]. T is the sampling period. If $\Delta t_i/T$ is non-integer, then an interpolated delay is required.

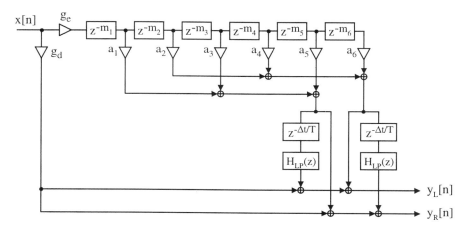

Figure 3.10 Binaural early echo simulator [Griesinger, 1997]. $\Delta t = 0.8$ msec. $H_{LP}(z)$ is a one-pole lowpass filter (figure 3.11) with $f_c = 2$ kHz.

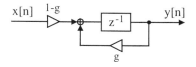

Figure 3.11 One-pole, DC-normalized lowpass filter.

spaciousness or as an increase in the source width depends on the input signal and the strength and timing of the reflections. This early echo simulator sounds very good with speech or vocal music as an input signal.

3.4 COMB AND ALLPASS REVERBERATORS

Now we discuss algorithms that reproduce late reverberation. The material is presented in roughly chronological order, starting with reverberators based on comb and allpass filters, and proceeding to more general methods based on feedback delay networks.

3.4.1 Schroeder's reverberator

The first artificial reverberators based on discrete-time signal processing were constructed by Schroeder in the early 1960's [Schroeder, 1962], and most of the important ideas about reverberation algorithms can be traced to his original papers. Schroeder's original proposal was based on *comb* and *allpass* filters. The comb filter is shown in figure 3.12 and consists of a delay whose output is recirculated to the input. The z transform of the comb filter is given by:

$$H(z) = \frac{z^{-m}}{1 - gz^{-m}} \qquad (3.14)$$

where m is the length of the delay in samples and g is the feedback gain. The time response of this filter is an exponentially decaying sequence of impulses spaced m samples apart. The system poles occur at the complex mth roots of g, and are thus harmonically spaced on a circle in the z plane. The frequency response is therefore shaped like a comb, with m periodic peaks that correspond to the pole frequencies.

Schroeder determined that the comb filter could be easily modified to provide a flat frequency response by mixing the input signal and the comb filter output as shown in figure 3.13. The resulting filter is called an allpass filter because its frequency response has unit magnitude for all frequencies. The z transform of the allpass filter is given by:

$$H(z) = \frac{z^{-m} - g}{1 - gz^{-m}} \qquad (3.15)$$

The poles of the allpass filter are thus the same as for the comb filter, but the allpass filter now has zeros at the conjugate reciprocal locations. The frequency response of the allpass filter can be written:

$$H(e^{j\omega}) = e^{j\omega m} \frac{1 - ge^{+j\omega m}}{1 - ge^{-j\omega m}} \qquad (3.16)$$

In this form it is easy to see that the magnitude response is unity, because the first term in the product, $e^{-j\omega m}$, has unit magnitude, and the second term is a quotient of complex conjugates, which also has unit magnitude. Thus,

106   APPLICATIONS OF DSP TO AUDIO AND ACOUSTICS

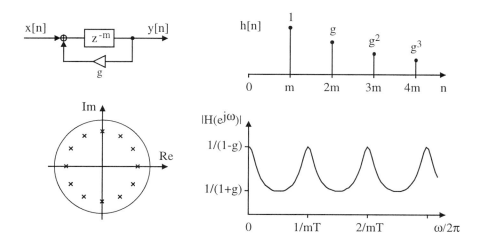

Figure 3.12   Comb filter: (clockwise from top-left) flow diagram, time response, frequency response, and pole diagram.

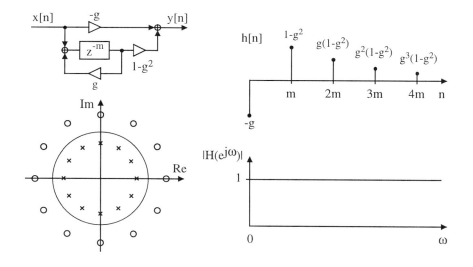

Figure 3.13   Allpass filter formed by modification of a comb filter: (clockwise from top-left) flow diagram, time response, frequency response, and pole-zero diagram.

$$|H(e^{j\omega})| = 1 \tag{3.17}$$

The phase response of the allpass filter is a non-linear function of frequency, leading to a smearing of the signal in the time domain.

Let us consider attempting to create a reverberator using a single comb or allpass filter. For the case of a comb filter, the reverberation time T_r is given by:

$$\frac{20\log_{10}(g_i)}{m_i T} = \frac{-60}{T_r} \tag{3.18}$$

where g_i is the gain of the comb filter, m_i is the length of the delay in samples, and T is the sampling period. For a desired reverberation time, we can choose the delay length and the feedback gain to tradeoff modal density for echo density. Of course, there are serious problems with using a single comb filter as a reverberator. For short delay times, which yield rapidly occurring echoes, the frequency response is characterized by widely spaced frequency peaks. These peaks correspond to the frequencies that will be reverberated, whereas frequencies falling between the peaks will decay quickly. When the peaks are widely spaced, the comb filter has a noticeable and unpleasant characteristic timbre. We can increase the density of peaks by increasing the delay length, but this causes the echo density to decrease in the time domain. Consequently, the reverberation is heard as a discrete set of echoes, rather than a smooth diffuse decay.

An allpass filter has a flat magnitude response, and we might expect it to solve the problem of timbral coloration attributed to the comb filter. However, the response of an allpass filter sounds quite similar to the comb filter, tending to create a timbral coloration. This is because our ears perform a short-time frequency analysis, whereas the mathematical property of the allpass filter is defined for an infinite time integration.

By combining two elementary filters in series, we can dramatically increase the echo density, because every echo generated by the first filter will create a set of echoes in the second. Comb filters are not good candidates for series connection, because the only frequencies that will pass are those that correspond to peaks in both comb filter responses. However, any number of allpass filters can be connected in series, and the combined response will still be allpass. Consequently, series allpass filters are useful for increasing echo density without affecting the magnitude response of the system.

A parallel combination of comb filters with incommensurate delays is also a useful structure, because the resulting frequency response contains peaks contributed by all of the individual comb filters. Moreover, the combined echo density is the sum of the individual densities. Thus, we can theoretically obtain arbitrary density of frequency peaks and time echoes by combining a sufficient number of comb filters in parallel.

Schroeder proposed a reverberator consisting of parallel comb filters and series allpass filters [Schroeder, 1962], shown in figure 3.14. The delays of the comb filters are chosen such that the ratio of largest to smallest is about 1.5 (Schroeder suggested a range of 30 to 45 msec). From equation 3.18, the gains g_i of the comb filters are set to give a desired reverberation time T_r according to

$$g_i = 10^{-3m_i T/T_r} \qquad (3.19)$$

The allpass delays t_5 and t_6 are much shorter than the comb delays, perhaps 5 and 1.7

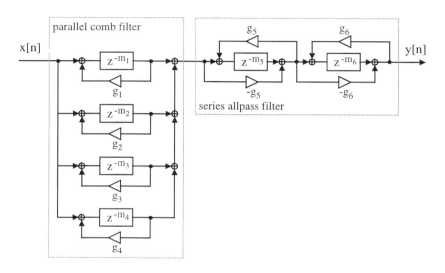

Figure 3.14 Schroeder's reverberator consisting of a parallel comb filter and a series allpass filter [Schroeder, 1962].

msec, with both allpass gains set to around 0.7. Consequently, the comb filters produce the long reverberant decay, and the allpass filters multiply the number of echoes output by the comb filters.

3.4.2 The parallel comb filter

The z transform of the parallel comb structure is given by [Jot and Chaigne, 1991]:

$$H(z) = \sum_{i=1}^{N} \frac{z^{-m_i}}{1 - g_i z^{-m_i}} \qquad (3.20)$$

where N is the number of comb filters. The poles are given by solutions to the following equation:

$$\prod_{i=1}^{N}(g_i - z^{m_i}) = 0 \qquad (3.21)$$

For each comb filter, the pole moduli are the same, and given by:

$$\gamma_i = \sqrt[m_i]{g_i} = 10^{-3T/T_r} \qquad (3.22)$$

Assuming all the gains g_i are set from the same reverberation time T_r, the pole moduli γ_i will be the same for all comb filters. Thus, all the resonant modes of the parallel comb structure will decay at the same rate. If the pole moduli were not all the same, the poles with the largest moduli would resonate the longest, and these poles would determine the tonal characteristic of the late decay [Moorer, 1979, Griesinger, 1989]. Consequently, to avoid tonal coloration in the late decay, it is important to respect condition 3.22 regarding the uniformity of pole modulus [Jot and Chaigne, 1991].

When the delay lengths of the comb filters are chosen to be incommensurate, in other words sharing no common factors, the pole frequencies will all be distinct (except at frequency 0). Furthermore, in the time response, the echoes from two comb filters i and k will not overlap until sample number $m_i m_k$.

3.4.3 Modal density and echo density

Two important criteria for the realism of a reverberation algorithm are the modal density and the echo density. The modal density of the parallel combs, expressed as the number of modes per Hz, is [Jot and Chaigne, 1991]:

$$D_m = \sum_{i=0}^{N-1} \tau_i = N \cdot \tau \qquad (3.23)$$

where τ_i is the length of delay i in seconds, and τ is the mean delay length. It is apparent that the modal density of the parallel combs is constant for all frequencies, unlike real rooms, whose modal density increases as the square of frequency (equation 3.8). However, in real rooms, once the modal density passes a certain threshold, the frequency response is characterized by frequency maxima whose mean spacing is constant (equations 3.9 and 3.10). It is therefore possible to approximate a room's frequency response by equating the modal density of the parallel comb filters with the density of frequency maxima in the room's response [Schroeder, 1962]. The total length of the comb delays, expressed in seconds, is equal to the modal density of the parallel comb filter, expressed as number of modes per Hz. Equating this to the density of frequency maxima of real rooms, we obtain the following relation between the total length of the delays and the maximum reverberation time we wish to simulate [Jot, 1992b]:

$$\sum_i \tau_i = D_m > D_f \approx \frac{T_{\max}}{4} \qquad (3.24)$$

where D_f is the density of frequency maxima according to the the statistical model for late reverberation (equal to the reciprocal of Δf_{max} in equation 3.10) and T_{max} is the maximum reverberation time desired.

Equation 3.24 specifies the *minimum* amount of total delay required. In practice, low modal density can lead to audible beating in response to narrowband signals. A narrowband signal may excite two neighboring modes which will beat at their difference frequency. To alleviate this, the mean spacing of modes can be chosen so that the average beat period is at least equal to the reverberation time [Stautner and Puckette, 1982]. This leads to the following relationship:

$$\sum_i \tau_i \geq T_{\max} \qquad (3.25)$$

For the case of the parallel comb filter, following constraint 3.22 guarantees that all the modes will decay at the same rate, but does not mean they will all have the same initial amplitude. As shown in figure 3.12, the height of a frequency peak is $1/(1-g)$. Following equation 3.18, longer delays will have smaller feedback gains, and hence smaller peaks. The modes of the parallel comb filter can be normalized by weighting the input of each comb filter with a gain proportional to its delay length [Jot, 1992b]. In practice, these normalizing gains are not necessary if the comb delay lengths are relatively similar.

The echo density of the parallel combs is the sum of the echo densities of the individual combs. Each comb filter i outputs one echo per time τ_i, thus the combined echo density, expressed as the number of echoes per second, is [Jot and Chaigne, 1991]:

$$D_e = \sum_{i=0}^{N-1} \frac{1}{\tau_i} \approx \frac{N}{\tau} \qquad (3.26)$$

This approximation is valid when the delays are similar. It is apparent that the echo density is constant as a function of time, unlike real rooms, whose echo density increases with the square of time (equation 3.12). Schroeder suggested that 1000 echoes per second was sufficient to sound indistinguishable from diffuse reverberation [Schroeder, 1962]. Griesinger has suggested that 10000 echoes per second may be required, and adds that this value is a function of the bandwidth of the system [Griesinger, 1989]. The mathematical definition of echo density includes all echoes regardless of amplitude, and does not consider system bandwidth. Jot has suggested the term *time density* to refer to the perceptual correlate of echo density, and he relates this to the crest factor of the impulse response [Jot, 1992b]. Griesinger obtains a measure of the time density of an impulse response by counting all echoes within 20 dB of the maximum echo in a 20 msec sliding window [Griesinger, 1989].

From equations 3.23 and 3.26, we can derive the number of comb filters required to achieve a given modal density D_m and echo density D_e [Jot and Chaigne, 1991]:

$$N \approx \sqrt{D_m D_e} \qquad (3.27)$$

Schroeder chose the parameters of his reverberator to have an echo density of 1000 echoes per second, and a frequency density of 0.15 peaks per Hz (one peak per 6.7 Hz). Strictly applying equation 3.27 using these densities would require 12 comb filters with a mean delay of 12 msec. However, this ignores the two series allpass filters, which will increase the echo density by approximately a factor of 10 [Schroeder, 1962]. Thus, only 4 comb filters are required with a mean delay of 40 msec.

3.4.4 Producing uncorrelated outputs

The reverberator in figure 3.14 is a monophonic reverberator with a single input and output. Schroeder suggested a way to produce multiple outputs by computing linear combinations of the comb filter outputs [Schroeder, 1962]. This requires that the allpass filters be placed in front of the comb filters, as shown in figure 3.15. A mixing matrix is then used to form multiple outputs, where the number of rows is equal to the number of comb filters, and the number of columns is equal to the number of outputs. Schroeder suggested that the coefficients of the mixing matrix have values of +1 or -1, and Jot suggests that the mixing matrix have orthogonal columns [Jot, 1992b]. The purpose of the linear combinations is to produce outputs which are mutually uncorrelated. For example, the mixing matrix

$$\begin{bmatrix} +1 & +1 \\ +1 & -1 \\ +1 & +1 \\ +1 & -1 \end{bmatrix} \qquad (3.28)$$

when used in the system of figure 3.15 produces a stereo reverberator which is quite spacious and enveloping when listened to over headphones.

Given two outputs $y_1(t)$ and $y_2(t)$ that are mutually uncorrelated, we can mix these signals to achieve any desired amount of interaural cross-correlation (see equation 3.3), as shown in equation 3.29 and figure 3.16 [Martin et al., 1993, Jot, 1992b]:

$$y_L(t) = \cos(\theta) y_1(t) + \sin(\theta) y_2(t) \qquad (3.29)$$

$$y_R(t) = \sin(\theta) y_1(t) + \cos(\theta) y_2(t)$$

112 APPLICATIONS OF DSP TO AUDIO AND ACOUSTICS

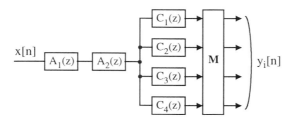

Figure 3.15 Mixing matrix **M** used to form uncorrelated outputs from parallel comb filters [Schroeder, 1962]. $A_i(z)$ are allpass filters, and $C_i(z)$ are comb filters.

$$\theta = \arcsin(IACC)/2$$

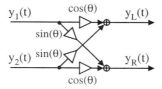

Figure 3.16 Controlling IACC in binaural reverberation [Martin et al., 1993, Jot, 1992b].

3.4.5 Moorer's reverberator

Schroeder's original reverberator sounds quite good, particularly for short reverberation times and moderate reverberation levels. For longer reverberation times or higher levels, some sonic deficiencies become noticeable and these have been described by various authors [Moorer, 1979, Griesinger, 1989, Jot and Chaigne, 1991]:

- The initial response sounds too discrete, leading to a grainy sound quality, particularly for impulsive input sounds, such as a snare drum.

- The amplitude of the late response, rather than decaying smoothly, can exhibit unnatural modulation, often described as a fluttering or beating sound.

- For longer reverberation times, the reverberation sounds tonally colored, usually referred to as a metallic timbre.

- The echo density is insufficient, and doesn't increase with time.

All reverberation algorithms are susceptible to one or more of these faults, which usually do not occur in real rooms, certainly not good sounding ones. In addition to these criticisms, there is the additional problem that Schroeder's original proposal does not provide a frequency dependent reverberation time.

Moorer later reconsidered Schroeder's reverberator and made several improvements [Moorer, 1979]. The first of these was to increase the number of comb filters from 4 to 6. This was necessary in order to effect longer reverberation times, while maintaining sufficient frequency and echo density according to equation 3.27. Moorer also inserted a one-pole lowpass filter into each comb filter feedback loop, as shown in figure 3.17. The cutoff frequencies of the lowpass filters were based on a physical consideration of the absorption of sound by air. Adding the lowpass filters caused the reverberation time to decrease at higher frequencies and Moorer noted that this made the reverberation sound more realistic. In addition, several other benefits were observed. The response to impulsive sounds was greatly improved, owing to the fact that the impulses are smoothed by the lowpass filtering. This improves the subjective quality of both the early response and the late response, which suffers less from a metallic sound quality or a fluttery decay.

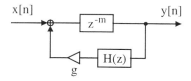

Figure 3.17 Comb filter with lowpass filter in feedback loop [Moorer, 1979].

Despite these improvements many problems remained. The frequency dependent reverberation time is the net result of the lowpass filtering, but it is not possible to specify a function $T_r(\omega)$ which defines the reverberation time as a function of frequency. Furthermore, the recurring problems of metallic sounding decay and fluttery late response are reduced but not entirely eliminated by this reverberator.

3.4.6 *Allpass reverberators*

We now study reverberators that are based on a series association of allpass filters[2]. Schroeder experimented with reverberators consisting of 5 allpass filters in series, with delays starting at 100 msec and decreasing roughly by factors of 1/3, and with gains of about 0.7 [Schroeder, 1962]. Schroeder noted that these reverberators were indistinguishable from real rooms in terms of coloration, which may be true with stationary input signals, but other authors have found that series allpass filters are extremely susceptible to tonal coloration, especially with impulsive inputs [Moorer, 1979, Gardner, 1992]. Moorer experimented with series allpass reverberators, and made the following comments [Moorer, 1979]:

- The higher the order of the system, the longer it takes for the echo density to build up to a pleasing level.
- The smoothness of the decay depends critically on the particular choice of the delay and gain parameters.
- The decay exhibits an annoying, metallic ringing sound.

The z transform of a series connection of N allpass filters is:

$$H(z) = \prod_{i=1}^{N} \frac{z^{-m_i} - g_i}{1 - g_i z^{-m_i}} \quad (3.30)$$

where m_i and g_i are the delay and gain, respectively, of allpass filter i. It is possible to ensure that the pole moduli are all the same, by basing the gains on the delay length as indicated by equation 3.19. However, this does not solve the problem of the metallic sounding decay.

Gardner has described reverberators based on a "nested" allpass filter, where the delay of an allpass filter is replaced by a series connection of a delay and another allpass filter [Gardner, 1992]. This type of allpass filter is identical to the lattice form shown in figure 3.18. Several authors have suggested using nested allpass filters for reverberators [Schroeder, 1962, Gerzon, 1972, Moorer, 1979]. The general form of such a filter is shown in figure 3.19, where the allpass delay is replaced with a system function $A(z)$, which is allpass. The transfer function of this form is written:

$$H(z) = \frac{A(z) - g}{1 - gA(z)} \quad (3.31)$$

The magnitude squared response of $H(z)$ is:

$$|H(z)|^2 = \frac{|A(z)|^2 - 2g\mathrm{Re}\{A(z)\} + g^2}{1 - 2g\mathrm{Re}\{A(z)\} + g^2|A(z)|^2} = 1 \text{ if } |A(z)| = 1 \quad (3.32)$$

which is verified to be allpass if $A(z)$ is allpass [Gardner, 1992, Jot, 1992b]. This filter is not realizable unless $A(z)$ can be factored into a delay in series with an allpass filter, otherwise a closed loop is formed without delay. The advantage of using a nested allpass structure can be seen in the time domain. Echoes created by the inner allpass filter are recirculated to itself via the outer feedback path. Thus, the echo density of a nested allpass filter increases with time, as in real rooms.

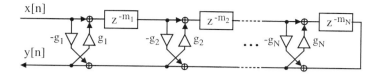

Figure 3.18 Lattice allpass structure.

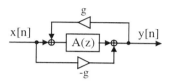

Figure 3.19 Generalization of figure 3.18.

A useful property of allpass filters is that no matter how many are nested or cascaded in series, the response is still allpass. This makes it very easy to verify the stability of the resulting system, regardless of complexity. Gardner suggested a general structure for a monophonic reverberator constructed with allpass filters, shown in figure 3.20 [Gardner, 1992]. The input signal flows through a cascade of allpass sections $A_i(z)$, and is then recirculated upon itself through a lowpass filter $H_{LP}(z)$ and an attenuating gain g. Gardner noted that when the output of the allpass filters was recirculated to the input through a sufficient delay, the characteristic metallic sound of the series allpass was greatly reduced.

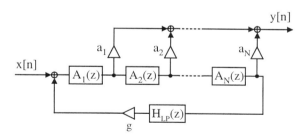

Figure 3.20 Reverberator formed by adding absorptive losses to an allpass feedback loop [Gardner, 1992].

The output is formed as a linear combination of the outputs of the allpass sections. The stability of the system is guaranteed, provided the magnitude of the loop gain is less than 1 for all frequencies (i.e. $|gH_{LP}(e^{j\omega})| < 1$ for all ω). The overall transfer function of this system is in general not allpass, due to phase cancellation between the output taps and also the presence of the outer feedback loop. As the input signal is diffused through the allpass filters, each tap outputs a different response shape. Consequently, it is possible to customize the amplitude envelope of the reverberant decay by adjusting the coefficients a_i. The reverberation time can be adjusted by changing the feedback gain g. The lowpass filter simulates frequency dependent absorptive losses, and lower cutoff frequencies generally result in a less metallic sounding, but duller, late response.

Figure 3.21 shows a complete schematic of an allpass feedback loop reverberator described by Dattorro [Dattorro, 1997], who attributes this style of reverberator to Griesinger. The circuit is intended to simulate an electro-acoustical plate reverberator, characterized by a rapid buildup of echo density followed by an exponential reverberant decay. The monophonic input signal passes through several short allpass filters, and then enters what Dattorro terms the reverberator "tank", consisting of two systems like that of figure 3.20 which have been cross-coupled. This is a useful structure for producing uncorrelated stereo outputs, which are obtained by forming weighted sums of taps within the tank. The reverberator incorporates a time varying delay element in each of the cross-coupled systems. The purpose of the time varying delays is to further decrease tonal coloration by dynamically altering the resonant frequencies.

There are many possible reverberation algorithms that can be constructed by adding absorptive losses to allpass feedback loops, and these reverberators can sound very good. However, the design of these reverberators has to date been entirely empirical. There is no way to specify in advance a particular reverberation time function $T_r(\omega)$, nor is there a deterministic method for choosing the filter parameters to eliminate tonal coloration.

3.5 FEEDBACK DELAY NETWORKS

Gerzon generalized the notion of unitary multichannel networks, which are N-dimensional analogues of the allpass filter [Gerzon, 1976]. An N-input, N-output LTI system is defined to be *unitary* if it preserves the total energy of all possible input signals. Similarly, a matrix $\mathbf{M}$ is unitary if $\|\mathbf{M}\mathbf{u}\| = \|\mathbf{u}\|$ for all vectors $\mathbf{u}$, which is equivalent to requiring that $\mathbf{M}^T\mathbf{M} = \mathbf{M}\mathbf{M}^T = \mathbf{I}$, where $\mathbf{I}$ is the identity matrix. It is trivial to show that the product of two unitary matrices is also unitary, and consequently the series cascade of two unitary systems is a unitary system. Simple unitary systems we have encountered include a set of N delay lines, and a set of N allpass filters. It is also easy to show that an N-channel unitary system and an M-channel unitary system can be combined to form an $(N + M)$ channel unitary system by diagonally juxtaposing their system matrices. Gerzon showed that a feedback modification can be

REVERBERATION ALGORITHMS 117

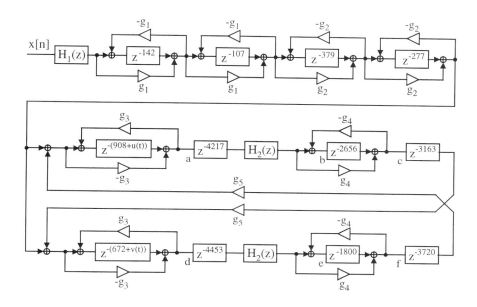

Figure 3.21 Dattorro's plate reverberator based on an allpass feedback loop, intended for 29.8 kHz sampling rate [Dattorro, 1997]. $H_1(z)$ and $H_2(z)$ are lowpass filters described in figure 3.11; $H_1(z)$ controls the bandwidth of signals entering the reverberator, and $H_2(z)$ controls the frequency dependent decay. Stereo outputs y_L and y_R are formed from taps taken from labelled delays as follows: $y_L = a[266] + a[2974] - b[1913] + c[1996] - d[1990] - e[187] - f[1066], y_R = d[353] + d[3627] - e[1228] + f[2673] - a[2111] - b[335] - c[121]$. In practice, the input is also mixed with each output to achieve a desired reverberation level. The time varying functions $u(t)$ and $v(t)$ are low frequency ($\approx$ 1 Hz) sinusoids that span 16 samples peak to peak. Typical coefficients values are $g_1 = 0.75, g_2 = 0.625, g_3 = 0.7, g_4 = 0.5, g_5 = 0.9$.

made to a unitary system without destroying the unitary property [Gerzon, 1976], in a form completely analogous to the feedback around the delay in an allpass filter. Gerzon applied these principles to the design of multichannel reverberators, and suggested the basic feedback topologies found in later work [Gerzon, 1971, Gerzon, 1972].

Stautner and Puckette proposed a four channel reverberator consisting of four delay lines with a feedback matrix [Stautner and Puckette, 1982], shown in figure 3.22. The feedback matrix allows the output of each delay to be recirculated to each delay input, with the matrix coefficients controlling the weights of these feedback paths. The structure can be seen as a generalization of Schroeder's parallel comb filter, which would arise using a diagonal feedback matrix. This structure is capable of much higher echo densities than the parallel comb filter, given a sufficient number of non-zero feedback coefficients and incommensurate delay lengths. The delays were chosen in accordance with Schroeder's suggestions.

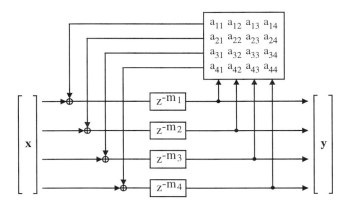

Figure 3.22 Stautner and Puckette's four channel feedback delay network [Stautner and Puckette, 1982].

Stautner and Puckette make a number of important points regarding this system:

- Stability is guaranteed if the feedback matrix $\mathbf{A}$ is chosen to be the product of a unitary matrix and a gain coefficient g, where $|g| < 1$. They suggest the matrix:

$$\mathbf{A} = g \frac{1}{\sqrt{2}} \begin{bmatrix} 0 & 1 & 1 & 0 \\ -1 & 0 & 0 & -1 \\ 1 & 0 & 0 & -1 \\ 0 & 1 & -1 & 0 \end{bmatrix} \qquad (3.33)$$

where g controls the reverberation time. If $|g| = 1$, $\mathbf{A}$ is unitary.

- The outputs will be mutually incoherent, and thus can be used in a four channel loudspeaker system to render a diffuse soundfield.

- Absorptive losses can be simulated by placing a lowpass filter in series with each delay line.

- The early reverberant response can be customized by injecting the input signal appropriately into the interior of the delay lines.

The authors note that fluttering and tonal coloration is present in the late decay of this reverberator. They attribute the fluttering to the beating of adjacent modes, and suggest that the beat period be made greater than the reverberation time by suitably reducing the mean spacing of modes according to equation 3.25. To reduce the tonal coloration, they suggest randomly varying the lengths of the delays.

3.5.1 Jot's reverberator

We now discuss the recent and important work by Jot, who has proposed a reverberator structure with two important properties [Jot, 1992b]:

- A reverberator can be designed with arbitrary time and frequency density while simultaneously guaranteeing absence of tonal coloration in the late decay.

- The resulting reverberator can be specified in terms of the desired reverberation time $T_r(\omega)$ and frequency response envelope $G(\omega)$.

This is accomplished by starting with an energy conserving system whose impulse response is perceptually equivalent to stationary white noise. Jot calls this a *reference filter*, but we will also use the term *lossless prototype*. Jot chooses lossless prototypes from the class of unitary feedback systems. In order to effect a frequency dependent reverberation time, absorptive filters are associated with each delay in the system. This is done in a way that eliminates coloration in the late response, by guaranteeing the local uniformity of pole modulus.

Jot generalizes the notion of a monophonic reverberator using the *feedback delay network* (FDN) structure shown in figure 3.23. The structure is a completely general specification of a linear system containing N delays.

Using vector notation and the z transform, the equations for the output of the system $y(z)$ and the delay lines $s_i(z)$ are [Jot and Chaigne, 1991]:

$$y(z) = \mathbf{c}^T \mathbf{s}(z) + dx(z) \tag{3.34}$$

$$\mathbf{s}(z) = \mathbf{D}(z)[\mathbf{A}\mathbf{s}(z) + \mathbf{b}x(z)] \tag{3.35}$$

where:

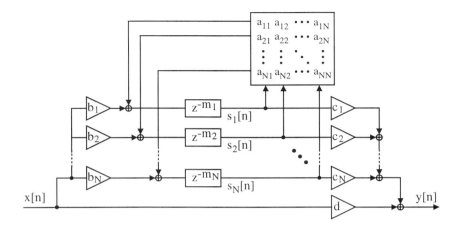

Figure 3.23 Feedback delay network as a general specification of a reverberator containing N delays [Jot and Chaigne, 1991]

$$\mathbf{s}(z) = \begin{bmatrix} s_1(z) \\ \vdots \\ s_N(z) \end{bmatrix} \quad \mathbf{b} = \begin{bmatrix} b_1 \\ \vdots \\ b_N \end{bmatrix} \quad \mathbf{c} = \begin{bmatrix} c_1 \\ \vdots \\ c_N \end{bmatrix} \quad (3.36)$$

$$\mathbf{D}(z) = \begin{bmatrix} z^{-m_1} & & 0 \\ & \ddots & \\ 0 & & z^{-m_N} \end{bmatrix} \quad \mathbf{A} = \begin{bmatrix} a_{11} & \cdots & a_{1N} \\ \vdots & \ddots & \vdots \\ a_{N1} & \cdots & a_{NN} \end{bmatrix} \quad (3.37)$$

The FDN can be extended to multiple inputs and outputs by replacing the vectors **b** and **c** with appropriate matrices. The system transfer function is obtained by eliminating $\mathbf{s}(z)$ from the preceding equations [Jot and Chaigne, 1991]:

$$H(z) = \frac{y(z)}{x(z)} = \mathbf{c}^T[\mathbf{D}(z^{-1}) - \mathbf{A}]^{-1}\mathbf{b} + d \quad (3.38)$$

The system zeros are given by [Rocchesso and Smith, 1994]:

$$\det[\mathbf{A} - \frac{\mathbf{bc}^T}{d} - \mathbf{D}(z^{-1})] = 0 \quad (3.39)$$

The system poles are given by those values of z that nullify the denominator of equation 3.38, in other words the solutions to the characteristic equation:

$$\det[\mathbf{A} - \mathbf{D}(z^{-1})] = 0 \tag{3.40}$$

Assuming **A** is a real matrix, the solutions to the characteristic equation 3.40 will either be real or complex-conjugate pole pairs. Equation 3.40 is not easy to solve in the general case, but for specific choices of **A** the solution is straightforward. For instance, when **A** is diagonal, the system represents Schroeder's parallel comb filter, and the poles are given by equation 3.21. More generally, when **A** is triangular, the matrix $\mathbf{A} - \mathbf{D}(z^{-1})$ is also triangular; and because the determinant of a triangular matrix is the product of the diagonal entries, equation 3.40 reduces to:

$$\prod_{i=1}^{N}(a_{ii} - z^{m_i}) = 0 \tag{3.41}$$

This is verified to be identical to equation 3.21. Any series combination of elementary filters – for instance, a series allpass filter – can be expressed as a feedback delay network with a triangular feedback matrix [Jot and Chaigne, 1991].

3.5.2 Unitary feedback loops

Another situation that interests us occurs when the feedback matrix A is chosen to be unitary, as suggested by Stautner and Puckette. Because the set of delay lines is also a unitary system, a unitary feedback loop is formed by the cascade of the two unitary systems. A general form of this situation is shown in figure 3.24, where $U_1(z)$ corresponds to the delay matrix, and $U_2(z)$ corresponds to the feedback matrix.

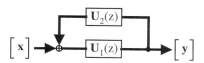

Figure 3.24   Unitary feedback loop [Jot, 1992b].

Because a unitary system preserves the energy of input signals, it is intuitively obvious that a unitary feedback loop will conserve energy. It can be shown that the system poles of a unitary feedback loop all have unit modulus, and thus the system response consists of non-decaying eigenmodes [Jot, 1992b].

Another way to demonstrate this is to consider the state variable description for the FDN shown in figure 3.23. It is straightforward to show that the resulting state transition matrix is unitary if and only if the feedback matrix **A** is unitary [Jot, 1992b, Rocchesso and Smith, 1997]. Thus, a unitary feedback matrix is sufficient to create a lossless

FDN prototype. However, we will later see that there are other choices for the feedback matrix that also yield a lossless system.

3.5.3 Absorptive delays

Jot has demonstrated that unitary feedback loops can be used to create lossless prototypes whose impulse responses are perceptually indistinguishable from stationary white noise [Jot and Chaigne, 1991]. Moorer previously noted that convolving source signals with exponentially decaying Gaussian white noise produces a very natural sounding reverberation [Moorer, 1979]. Consequently, by introducing absorptive losses into a suitable lossless prototype, we should obtain a natural sounding reverberator. Jot's method for introducing absorptive losses guarantees that the colorless quality of the lossless prototype is maintained. This is accomplished by associating a gain $k_i < 1$ with each delay i in the filter, as shown in figure 3.25.

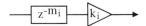

Figure 3.25 Associating an attenuation with a delay.

The logarithm of the the gain is proportional to the length of the delay:

$$k_i = \gamma^{m_i} \tag{3.42}$$

Provided all the delays are so modified, this has the effect of replacing z with z/γ in the expression for the system function $H(z)$, regardless of the filter structure. Starting from a lossless prototype whose poles are all on the unit circle, the above modification will cause all the poles to have a modulus equal to γ. Therefore, the lossless prototype response $h[n]$ will be multiplied by an exponential envelope γ^n where γ is the decay factor per sampling period [Jot and Chaigne, 1991, Jot, 1992b]. By maintaining the uniformity of pole modulus, we avoid the situation where the response in the neighborhood of a frequency is dominated by a few poles with relatively large moduli.

The decay envelope is made frequency dependent by specifying frequency dependent losses in terms of the reverberation time $T_r(\omega)$. This is accomplished by associating with each delay i an absorptive filter $h_i(z)$, as shown in figure 3.26. The filter is chosen such that the logarithm of its magnitude response is proportional to the delay length and inversely proportional to the reverberation time, as suggested by equation 3.19 [Jot and Chaigne, 1991]:

$$20 \log_{10} |h_i(e^{j\omega})| = \frac{-60T}{T_r(\omega)} m_i \tag{3.43}$$

This expression ignores the phase response of the absorptive filter, which has the effect

Figure 3.26 Associating an absorptive filter with a delay.

of slightly modifying the effective length of the delay. In practice, it is not necessary to take the phase delay into consideration [Jot and Chaigne, 1991]. By replacing each delay with an absorptive delay as described above, the poles of the prototype filter no longer appear on a circle centered at the origin, but now lie on a curve specified by the reverberation time $T_r(\omega)$.

A consequence of incorporating the absorptive filters into the lossless prototype is that the frequency response envelope of the reverberator will no longer be flat. For exponentially decaying reverberation, the frequency response envelope is proportional to the reverberation time at all frequencies. We can compensate for this effect by associating a correction filter $t(z)$ in series with the reference filter, whose squared magnitude is inversely proportional to the reverberation time [Jot, 1992b]:

$$|t(e^{j\omega})| \propto \frac{1}{\sqrt{T_r(\omega)}} \qquad (3.44)$$

After applying the correction filter, the frequency response envelope of the reverberator will be flat. This effectively decouples the reverberation time control from the overall gain of the reverberator. The final reverberator structure is shown in figure 3.27. Any additional equalization of the reverberant response, for instance, to match the frequency envelope of an existing room, can be effected by another filter in series with the correction filter.

3.5.4 Waveguide reverberators

Smith has proposed multichannel reverberators based on a *digital waveguide network* (DWN) [Smith, 1985]. Each waveguide is a bi-directional delay line, and junctions between multiple waveguides produce lossless signal scattering. Figure 3.28 shows an N-branch DWN which is isomorphic to the N-delay FDN shown in figure 3.23 [Smith and Rocchesso, 1994].

The waves travelling into the junction are associated with the FDN delay line outputs $s_i[n]$. The length of each waveguide is half the length of the corresponding FDN delay, because the waveguide signal must make a complete round trip to return to the scattering junction. An odd-length delay can be accommodated by replacing the non-inverting reflection with a unit sample delay.

124   APPLICATIONS OF DSP TO AUDIO AND ACOUSTICS

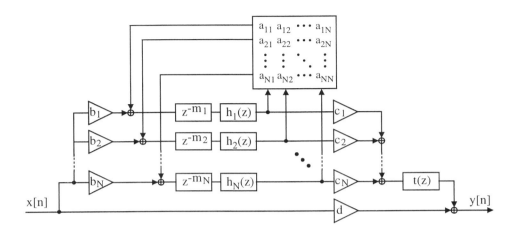

Figure 3.27   Reverberator constructed by associating a frequency dependent absorptive filter with each delay of a lossless FDN prototype filter [Jot and Chaigne, 1991].

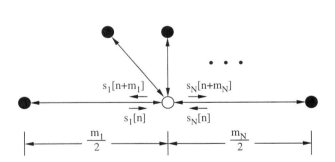

Figure 3.28   Waveguide network consisting of a single scattering junction to which N waveguides are attached. Each waveguide is terminated by an ideal non-inverting reflection, indicated by a black dot [Smith and Rocchesso, 1994].

The usual DWN notation defines the incoming and outgoing pressure variables as $p_i^+ = s_i[n]$ and $p_i^- = s_i[n+m_i]$, respectively, and therefore the operation of the scattering junction can be written in vector notation as

$$\mathbf{p}^- = \mathbf{A}\mathbf{p}^+ \tag{3.45}$$

where $\mathbf{A}$ is interpreted as a *scattering matrix* associated with the junction.

As we have already discussed, a lossless FDN results when the feedback matrix is chosen to be unitary. Smith and Rocchesso have shown that the waveguide interpretation leads to a more general class of lossless scattering matrices [Smith and Rocchesso, 1994]. This is due to the fact that each waveguide may have a different characteristic admittance. A scattering matrix is lossless if and only if the active complex power is scattering-invariant, i.e., if and only if

$$\begin{aligned} \mathbf{p}^{+*} \mathbf{\Gamma} \mathbf{p}^+ &= \mathbf{p}^{-*} \mathbf{\Gamma} \mathbf{p}^- \\ \Rightarrow \quad \mathbf{A}^* \mathbf{\Gamma} \mathbf{A} &= \mathbf{\Gamma} \end{aligned}$$

where $\mathbf{\Gamma}$ is a Hermitian, positive-definite matrix which can be interpreted as a generalized junction admittance. For the waveguide in figure 3.28, we have $\mathbf{\Gamma} = \mathrm{diag}(\Gamma_1, ... \Gamma_N)$, where Γ_i is the characteristic admittance of waveguide i. When $\mathbf{A}$ is unitary, we have $\mathbf{\Gamma} = \mathbf{I}$. Thus, unitary feedback matrices correspond to DWNs where the waveguides all have unit characteristic admittance, or where the signal values are in units of root power [Smith and Rocchesso, 1994].

Smith and Rocchesso have shown that a DWN scattering matrix (or a FDN feedback matrix) is lossless if and only if its eigenvalues have unit modulus and its eigenvectors are linearly independent. Therefore, lossless scattering matrices may be fully parameterized as

$$\mathbf{A} = \mathbf{T}^{-1} \mathbf{D} \mathbf{T} \tag{3.46}$$

where $\mathbf{D}$ is any unit modulus diagonal matrix, and $\mathbf{T}$ is any invertible matrix [Smith and Rocchesso, 1994]. This yields a larger class of lossless scattering matrices than given by unitary matrices. However, not all lossless scattering matrices can be interpreted as a physical junction of N waveguides (e.g., consider a permutation matrix).

3.5.5 Lossless prototype structures

Jot has described many lossless FDN prototypes based on unitary feedback matrices. A particularly useful unitary feedback matrix $\mathbf{A}_N$, which maximizes echo density while reducing implementation cost, is taken from the class of Householder matrices [Jot, 1992b]:

126   APPLICATIONS OF DSP TO AUDIO AND ACOUSTICS

$$\mathbf{A}_N = \mathbf{J}_N - \frac{2}{N}\mathbf{u}_N\mathbf{u}_N^T \qquad (3.47)$$

where $\mathbf{J}_N$ is an $N \times N$ permutation matrix, and $\mathbf{u}_N$ is an $N \times 1$ column vector of 1's. This unitary matrix contains only two different values, both nonzero, and thus it achieves maximum echo density when used in the structure of figure 3.27. Because $\mathbf{u}_N\mathbf{u}_N^T$ is a matrix containing all 1's, computation of $\mathbf{A}_N\mathbf{x}$ consists of permuting the elements of $\mathbf{x}$ according to $\mathbf{J}_N$, and adding to these the sum of the elements of $\mathbf{x}$ times the factor $-2/N$. This requires roughly $2N$ operations as opposed to the N^2 operations normally required. When $\mathbf{J}_N$ is the identity matrix $\mathbf{I}_N$, the resulting system is a modification of Schroeder's parallel comb filter which maximizes echo density as shown in figure 3.29.

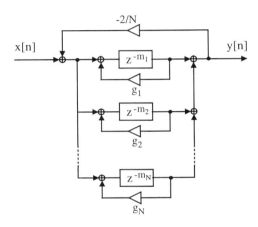

Figure 3.29   Modification of Schroeder's parallel comb filter to maximize echo density [Jot, 1992b].

Jot has discovered that this structure produces a periodic parasitic echo with period equal to the sum of the delay lengths. This is a result of constructive interference between the output signals of the delays, and can be eliminated by choosing the coefficients c_i in figure 3.27 such that every other channel undergoes a phase inversion (multiplication by -1) [Jot, 1992b]. Another interesting possibility proposed by Jot is choosing $\mathbf{J}_N$ to be a circular permutation matrix. This causes the delay lines to feed one another in series, which greatly simplifies the memory management in the final implementation.

Rocchesso and Smith have suggested using unitary circulant matrices for the feedback matrix of a FDN [Rocchesso and Smith, 1994]. Circulant matrices have the form:

$$\mathbf{A} = \begin{bmatrix} a[0] & a[1] & a[2] & \cdots & a[N-1] \\ a[N-1] & a[0] & a[1] & \cdots & a[N-2] \\ a[N-2] & a[N-1] & a[0] & \cdots & a[N-3] \\ \vdots & \vdots & \vdots & \ddots & \vdots \\ a[1] & a[2] & a[3] & \cdots & a[0] \end{bmatrix} \tag{3.48}$$

Multiplication by a circulant matrix implements circular convolution of a column vector with the first row of the matrix. A circulant matrix $\mathbf{A}$ can be factored as shown in equation 3.46 where $\mathbf{T}$ is the discrete Fourier transform (DFT) matrix and $\mathbf{D}$ is a diagonal matrix whose elements are the DFT of the first row of $\mathbf{A}$. The diagonal elements of $\mathbf{D}$ are the eigenvalues of $\mathbf{A}$. A circulant matrix is thus lossless (and unitary) when its eigenvalues (the spectrum of the first row) have unit modulus. The advantages of using a circulant matrix are that the eigenvalues can be explicitly specified, and computation of the product can be accomplished in $O(N \log(N))$ time using the Fast Fourier transform (FFT).

All of the late reverberator structures we have studied can be seen as an energy conserving system with absorptive losses inserted into the structure. When the absorptive losses are removed, the structure of the lossless prototype is revealed. This is true for Schroeder's parallel comb filter when the feedback coefficients are unity, which corresponds to a FDN feedback matrix equal to the identity matrix. The allpass feedback loop reverberator in figure 3.20 consists of a unitary feedback loop when absorptive losses are removed. Stautner and Puckette's FDN reverberator is also a unitary feedback loop when $|g| = 1$ (see equation 3.33). However, the method shown for adding the absorptive losses in these reverberators does not necessarily prevent coloration in the late decay. This can be accomplished by associating an absorptive filter with each delay in the reverberator according to equation 3.43.

The parameters of the reference structure are the number of delays N, the lengths of the delays m_i, and the feedback matrix coefficients. If a large number of inputs or outputs is desired, this can also affect the choice of the reference structure. The total length of the delays in seconds, equal to the modal density, should be greater than the density of frequency maxima for the room to be simulated. Thus, the minimum total length required is $T_r/4$, after equation 3.24. A total delay of 1 to 2 seconds is sufficient to produce a reference filter response that is perceptually indistinguishable from white noise [Jot, 1992b], which gives an upper bound on the total delay required for infinite reverberation times with broadband input signals. To improve the quality of the reverberation in response to narrowband input signals, one may wish to use a total delay at least equal to the maximum reverberation time desired, after equation 3.25. The number of delays and the lengths of the delays, along with the choice of feedback matrix, determines the buildup of echo density. These decisions must be made empirically by evaluating the quality of the reference filter response.

3.5.6 Implementation of absorptive and correction filters

Once a lossless prototype has been chosen, the absorptive filters and the correction filter need to be implemented based on a desired reverberation time curve. Jot has specified a simple solution using first order IIR filters for the absorptive filters, whose transfer functions are written [Jot, 1992b]:

$$h_i(z) = k_i \frac{1 - \beta_i}{1 - \beta_i z^{-1}} \tag{3.49}$$

Remarkably, this leads to a correction filter which is first order FIR:

$$t(z) = g \frac{1 - \beta z^{-1}}{1 - \beta} \tag{3.50}$$

The filter parameters are based on the reverberation time at zero frequency and the Nyquist frequency, notated $T_r(0)$ and $T_r(\pi)$, respectively:

$$k_i = 10^{-3\tau_i/T_r(0)}, \; \beta_i = 1 - \frac{2}{1 + k_i^{(1+1/\epsilon)}} \tag{3.51}$$

$$g = \sqrt{\frac{\sum \tau_i}{T_r(0)}}, \; \beta = \frac{1 - \sqrt{\epsilon}}{1 + \sqrt{\epsilon}}, \; \epsilon = \frac{T_r(\pi)}{T_r(0)}$$

The derivation of these parameters is detailed in the reference [Jot, 1992b]. The family of reverberation time curves obtained from first order filters is limited, but leads to natural sounding reverberation. Jot also describes methods for creating higher order absorption and correction filters by combining first order sections.

3.5.7 Multirate algorithms

Jot's method of incorporating absorptive filters into a lossless prototype yields a system whose poles lie on a curve specified by the reverberation time. An alternative method to obtain the same pole locus is to combine a bank of bandpass filters with a bank of comb filters, such that each comb filter processes a different frequency range. The feedback gain of each comb filter then determines the reverberation time for the corresponding frequency band.

This approach has been extended to a multirate implementation by embedding the bank of comb filters in the interior of a multirate analysis/synthesis filterbank [Zoelzer et al., 1990]. A multirate implementation reduces the memory requirements for the comb filters, and also allows the use of an efficient polyphase analysis/synthesis filterbank [Vaidyanathan, 1993].

3.5.8 Time-varying algorithms

There are several reasons why one might want to incorporate time variation into a reverberation algorithm. One motivation is to reduce coloration and fluttering in the reverberant response by varying the resonant frequencies. Another use of time variation is to reduce feedback when the reverberator is coupled to an electro-acoustical sound reinforcement system, as is the case in reverberation enhancement systems [Griesinger, 1991]. The time variation should always be implemented so as to yield a natural sounding reverberation free from audible amplitude or frequency modulations. There are several ways to add time variation to an existing algorithm:

- Modulate the lengths of the delays, e.g., as shown in figure 3.21.

- Vary the coefficients of the feedback matrix in the reference filter while maintaining the energy conserving property, or similarly vary the allpass gains of an allpass feedback loop reverberator.

- Modulate the output tap gains of an allpass feedback loop structure such as in figure 3.20, or similarly vary the mixing matrix shown in equation 3.28.

There are many ways to implement variable delay lines [Laakso et al., 1996]. A simple linear interpolator works well, but for better high frequency performance, it may be preferable to use a higher order Lagrangian interpolator. Dattorro has suggested using allpass interpolation, which is particularly suitable because the required modulation rate is low [Dattorro, 1997]. Obviously, modulating the delay length causes the signal passing through the delay to be frequency modulated. If the depth or rate of the modulation is too great, the modulation will be audible in the resulting reverberation. This is particularly easy to hear with solo piano music. The maximum detune should be restricted to a few cents, and the modulation rate should be on the order of 1 Hz.

The notion of changing the filter coefficients while maintaining an energy conserving system has been suggested by Smith [Smith, 1985], who describes the result as placing the signal in a changing lossless maze. Smith suggests that all coefficient modulation be done at sub-audio rates to avoid sideband generation, and warns of an "undulating" sound that can occur with slow modulation that is too deep.

Although many commercial reverberators use time variation to reduce tonal coloration, very little has been published on time-varying techniques. There is no theory which relates the amount and type of modulation to the reduction of tonal coloration in the late response, nor is there a way to predict whether the modulation will be noticeable. Consequently, all the time-varying methods are completely empirical in nature.

3.6 CONCLUSIONS

This paper has discussed algorithms for rendering reverberation in real-time. A straightforward method for simulating room acoustics is to sample a room impulse response and render the reverberation using convolution. Synthetic impulse responses can be created using auralization techniques. The availability of efficient, zero delay convolution algorithms make this a viable method for real-time room simulation. The drawback of this method is the lack of parameterized control over perceptually salient characteristics of the reverberation. This can be a problem when we attempt to use these systems in interactive virtual environments.

Reverberators implemented using recursive filters offer parameterized control due to the small number of filter coefficients. The problem of designing efficient, natural sounding reverberation algorithms has always been to avoid unpleasant coloration and fluttering in the decay. In many ways, Jot's work has revolutionized the state of the art, because it is now possible to design colorless reverberators without resorting to solely empirical design methods. It is possible to specify in advance the reverberation time curve of the reverberator, permitting an analysis/synthesis method for reverberator design which concentrates on reproducing the energy decay relief of the target room. Interestingly, many of the fundamental ideas can be traced back to Schroeder's original work, which is now more than thirty years old.

There are still problems to be solved. Reproducing a complicated reverberation time curve using Jot's method requires associating a high order filter with each delay in the lossless prototype, and this is expensive. It is an open question whether the constraint of uniform pole modulus necessarily requires one absorptive filter per delay line (Jean-Marc Jot, personal communication, 1994). Many of the commercially available reverberators probably use time-variation to reduce tonal coloration, yet the study of time-varying algorithms has received almost no attention in the literature. A general theory of tonal coloration in reverberation is needed to explain why certain algorithms sound good and others sound bad.

The study of reverberation has been fertile ground for many acousticians, psychologists, and electrical engineers. There is no doubt it will continue to be so in the future.

Acknowledgments

The author would like to express his sincerest gratitude to David Griesinger for making this paper possible, to Jean-Marc Jot for his inspiring work, to Jon Dattorro for his useful comments, and to the author's colleagues at the Machine Listening Group at the MIT Media Lab – in particular, Eric Scheirer, Keith Martin, and Dan Ellis – for helping to prepare this paper.

Notes

1. Rooms are very linear but they are not time-invariant due to the motion of people and air. For practical purposes we consider them to be LTI systems.

2. There are many ways to implement allpass filters [Moorer, 1979, Jot, 1992b]; two methods are shown in Figures 3.13 and 3.14.

4 DIGITAL AUDIO RESTORATION

Simon Godsill and Peter Rayner

Dept. of Engineering
University of Cambridge,
Cambridge, U.K.
{sjg,pjwr}@eng.cam.ac.uk

Olivier Cappé

CNRS / ENST
Paris, France
cappe@sig.enst.fr

Abstract: This chapter is concerned with the application of modern signal processing techniques to the restoration of degraded audio signals. Although attention is focussed on gramophone recordings, film sound tracks and tape recordings, many of the techniques discussed have applications in other areas where degraded audio signals occur, such as speech transmission, telephony and hearing aids.

We aim to provide a wide coverage of existing methodology while giving insight into current areas of research and future trends.

4.1 INTRODUCTION

The introduction of high quality digital audio media such as Compact Disk (CD) and Digital Audio Tape (DAT) has dramatically raised general awareness and expectations about sound quality in all types of recordings. This, combined with an upsurge in interest in historical and nostalgic material, has led to a growing requirement for restoration of degraded sources ranging from the earliest recordings made on wax cylinders in the nineteenth century, through disc recordings (78 rpm, LP, etc.) and finally magnetic tape recording technology, which has been available since the 1950's. Noise reduction may occasionally be required even in a contemporary digital recording if background noise is judged to be intrusive.

Degradation of an audio source will be considered as any undesirable modification to the audio signal which occurs as a result of (or subsequent to) the recording process. For example, in a recording made direct-to-disc from a microphone, degradations could include noise in the microphone and amplifier as well as noise in the disc cutting process. Further noise may be introduced by imperfections in the pressing material, transcription to other media or wear and tear of the medium itself. We do not strictly consider any noise present in the recording environment such as audience noise at a musical performance to be degradation, since this is part of the 'performance'. Removal of such performance interference is a related topic which is considered in other applications, such as speaker separation for hearing aid design. An ideal restoration would then reconstruct the original sound source exactly as received by the transducing equipment (microphone, acoustic horn, etc.). Of course, this ideal can never be achieved perfectly in practice, and methods can only be devised which come close according to some suitable error criterion. This should ideally be based on the perceptual characteristics of the human listener.

Analogue restoration techniques have been available for at least as long as magnetic tape, in the form of manual cut-and-splice editing for clicks and frequency domain equalization for background noise (early mechanical disk playback equipment will also have this effect by virtue of its poor response at high frequencies). More sophisticated electronic click reducers were based upon high pass filtering for detection of clicks, and low pass filtering to mask their effect [Carrey and Buckner, 1976, Kinzie, Jr. and Gravereaux, 1973].[1] None of these methods was sophisticated enough to perform a significant degree of noise reduction without interfering with the underlying signal quality. Digital methods allow for a much greater degree of flexibility in processing, and hence greater potential for noise removal, although indiscriminate application of inappropriate digital methods can be more disastrous than analogue processing!

Some of the earliest digital signal processing work for audio restoration involved deconvolution for enhancement of a solo voice (Caruso) from an acoustically recorded source (see Miller [Miller, 1973] and Stockham *et al.* [Stockham et al., 1975]). Since then, research groups at Cambridge, Le Mans, Paris and elsewhere have worked in the area, developing sophisticated techniques for treatment of degraded audio. The results of this research are summarized and referenced later in the chapter.

There are several distinct types of degradation common in audio sources. These can be broadly classified into two groups: *localized* degradations and *global* degradations. Localized degradations are discontinuities in the waveform which affect only certain samples, including clicks, crackles, scratches, breakages and clipping. Global degradations affect all samples of the waveform and include background noise, wow and flutter and certain types of non-linear distortion. Mechanisms by which all of these defects can occur are discussed later.

The chapter is organized as follows. We firstly describe models which are suitable for audio signal restoration, in particular those which are used in later work. Subsequent sections describe individual restoration problems separately, considering the alternative methods available to the restorer. A concluding section summarizes the work and discusses future trends.

4.2 MODELLING OF AUDIO SIGNALS

Many signal processing techniques will be model-based, either explicitly or implicitly, and this certainly applies to most of the audio restoration algorithms currently available. The quality of processing will depend largely on how well the modelling assumptions fit the data. For an audio signal, which might contain speech, music and general acoustical noises the model must be quite general and robust to deviations from the assumptions. It should also be noted that most audio signals are non-stationary in nature, although practical modelling will often assume short-term stationarity of the signal. We now discuss some models which are appropriate for audio signals.

A model which has found application in many areas of time series processing, including audio restoration (see sections 4.3 and 4.7), is the autoregressive (AR) or all-pole model (see Box and Jenkins [Box and Jenkins, 1970], Priestley [Priestley, 1981] and also Makhoul [Makhoul, 1975] for an introduction to linear predictive analysis) in which the current value of a signal is represented as a weighted sum of P previous signal values and a white noise term:

$$s[n] = \sum_{i=1}^{P} s[n-i]a_i + e[n]. \tag{4.1}$$

The AR model is a reasonable representation for many stationary linear processes, allowing for noise-like signals (poles close to origin) and near-harmonic signals (poles close to unit circle). A more appropriate model for many situations might be the autoregressive moving-average (ARMA) model which allows zeros as well as poles. However, the AR model offers far greater analytic flexibility than the ARMA model, so a high order AR model will often be used in practice to approximate an ARMA signal (it is well known that an infinite order AR model can represent any finite-order

ARMA model (see, e.g. [Therrien, 1992])). Model order for the autoregressive process will reflect the complexity of the signal under consideration. For example, a highly complex musical signal can require a model order of $P > 100$ to represent the waveform adequately, while simpler signals may be modelled by an order 30 system. Strictly, any processing procedure should thus include a model order selection strategy. For many applications, however, it is sufficient to fix the model order to a value high enough for representation of the most complex signal likely to be encountered. Clearly no audio signal is truly stationary, so it will be necessary to implement the model in a block-based or adaptive fashion. Suitable block lengths and adaptation rates will depend upon the signal type, but block lengths between 500 and 2000 samples at the 44.1kHz sampling rate are generally found to be appropriate.

There are many well-known methods for estimating AR models, including maximum likelihood/least-squares [Makhoul, 1975] and methods robust to noise [Huber, 1981, Spath, 1991]. Adaptive parameter estimation schemes are reviewed in [Haykin, 1991]. The class of methods robust to noise, both block-based and adaptive, will be of importance to many audio restoration applications, since standard parameter estimation schemes can be heavily biased in the presence of noise, in particular impulsive noise such as is commonly encountered in click-degraded audio. A standard approach to this problem is the M-estimator [Huber, 1981, Spath, 1991]. This method achieves robustness by iteratively re-weighting excitation values in the least-squares estimator using a non-linear function such as Huber's psi-function [Huber, 1964] or Tukey's bisquare function [Mosteller and Tukey, 1977]. Applications of these methods to parameter estimation, detection of impulses and robust filtering include [Martin, 1981, Arakawa et al., 1986, Efron and Jeen, 1992].

Another model which is a strong candidate for musical signals is the sinusoidal model, which has been used effectively for speech applications ([McAulay and Quatieri, 1986b] and chapter 9 of this book). A constrained form of the sinusoidal model is implicitly at the heart of short-time spectral attenuation (STSA) methods of noise reduction (see section 4.5.1). The model is also a fundamental assumption of the pitch variation algorithms presented in section 4.6. In its general form the signal can be expressed as:

$$s[n] = \sum_{i=1}^{P_n} a_i[n] \sin\left(\int_0^{nT} \omega_i(t)dt + \phi_i\right). \quad (4.2)$$

This is quite a general model, allowing for frequency and amplitude modulation (by allowing $a_i[n]$ and $\omega_i(t)$ to vary with time) as well as the 'birth' and 'death' of individual components (by allowing P_n to vary with time). However, parameter estimation for such a general model is difficult, and restrictive constraints must typically be placed upon the amplitude and frequency variations. The sinusoidal model is not suited to modelling noise-like signals, although an acceptable representation can be achieved by using a large number of sinusoids in the expansion.

Other models include adaptations to the basic AR/ARMA models to allow for speech-like periodic excitation pulses [Rabiner and Schafer, 1978a] and non-linearity (see section 4.7). Further 'non-parametric' modelling possibilities arise from other basis function expansions which might be more appropriate for audio signal analysis, including Wavelets [Akansu and Haddad, 1992] and signal dependent transforms which employ principal component-based analysis [Gerbrands, 1981]. Choice of model will in general involve a compromise between prior knowledge of signal characteristics, computational power and how critical the accuracy of the model is to the application.

4.3 CLICK REMOVAL

The term 'clicks' is used here to refer to a generic localized type of degradation which is common to many audio media. We will classify all finite duration defects which occur at random positions in the waveform as clicks. Clicks are perceived in a number of ways by the listener, ranging from tiny 'tick' noises which can occur in any recording medium, including modern digital sources, through the characteristic 'scratch' and 'crackle' noise associated with most analogue disc recording methods. For example, a poor quality 78 rpm record might typically have around 2,000 clicks per second of recorded material, with durations ranging from less than $20\mu s$ up to 4ms in extreme cases. See figure 4.1 for a typical example of a recorded music waveform degraded by localized clicks. In most examples at least 90% of samples remain undegraded, so it is reasonable to hope that a convincing restoration can be achieved.

There are many mechanisms by which clicks can occur. Typical examples are specks of dirt and dust adhering to the grooves of a gramophone disc (see figure 4.3 [2].) or granularity in the material used for pressing such a disc. Further click-type degradation may be caused through damage to the disc in the form of small scratches on the surface. Similar artifacts are encountered in other analogue media, including optical film sound tracks and early wax cylinder recordings, although magnetic tape recordings are generally free of clicks. Ticks can occur in digital recordings as a result of poorly concealed digital errors and timing problems.

Peak-related distortion, occurring as a result either of overload during recording or wear and tear during playback, can give rise to a similar perceived effect to clicks, but is really a different area which should receive separate attention (see section 4.7), even though click removal systems can often go some way towards alleviating the worst effects.

4.3.1 Modelling of clicks

Localized defects may be modelled in many different ways. For example, a defect may be additive to the underlying audio signal, or it may replace the signal altogether for some short period. An additive model has been found to be acceptable for most surface

138 APPLICATIONS OF DSP TO AUDIO AND ACOUSTICS

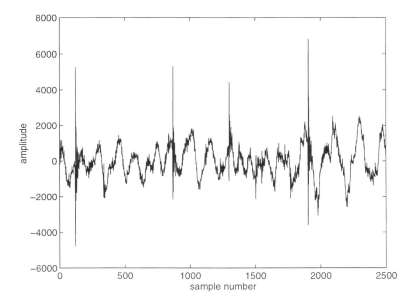

Figure 4.1   Click-degraded music waveform taken from 78 rpm recording

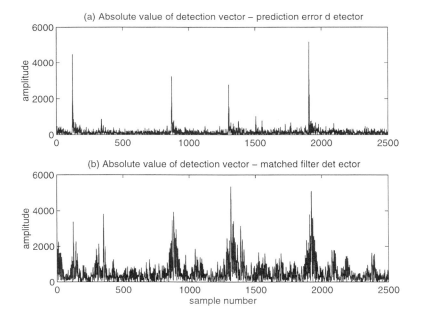

Figure 4.2   AR-based detection, P=50. (a) Prediction error filter (b) Matched filter.

DIGITAL AUDIO RESTORATION 139

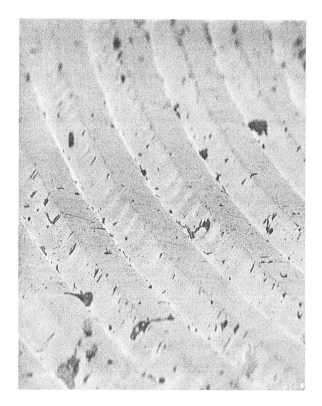

Figure 4.3 Electron micrograph showing dust and damage to the grooves of a 78rpm gramophone disc.

defects in recording media, including small scratches, dust and dirt. A replacement model may be appropriate for very large scratches and breakages which completely obliterate any underlying signal information, although such defects usually excite long-term resonances in mechanical playback systems and must be treated differently (see section 4.4). Here we will consider primarily the additive model, although many of the results are at least robust to replacement noise.

An additive model for localized degradation can be expressed as:

$$x[n] = s[n] + i[n]v[n] \qquad (4.3)$$

where $s[n]$ is the underlying audio signal, $v[n]$ is a corrupting noise process and $i[n]$ is a 0/1 'switching' process which takes the value 1 only when the localized degradation is present. Clearly the value of $v[n]$ is irrelevant to the output when the switch is in position 0. The statistics of the switching process $i[n]$ thus govern which samples are degraded, while the statistics of $v[n]$ determine the amplitude characteristics of the corrupting process.

This model is quite general and can account for a wide variety of noise characteristics encountered in audio recordings. It does, however, assume that the degradation process does not interfere with the timing content of the original signal, as observed in $x[n]$. This is reasonable for all but very severe degradations, which might temporarily upset the speed of playback, or actual breakages in the medium which have been mechanically repaired (such as a broken disc recording which has been glued back together).

Any procedure which is designed to remove localized defects in audio signals must take account of the typical characteristics of these artifacts. Some important features which are common to many click-degraded audio media include:

- Degradation tends to occur in contiguous 'bursts' of corrupted samples, starting at random positions in the waveform and of random duration (typically between 1 and 200 samples at 44.1 kHz sampling rates). Thus there is strong dependence between successive samples of the switching process $i[n]$, and the noise cannot be assumed to follow a classical impulsive noise pattern in which single impulses occur independently of each other (the Bernoulli model). It is considerably more difficult to treat clusters of impulsive disturbance than single impulses, since the effects of adjacent impulses can cancel each other in the detection space ('missed detections') or add constructively to give the impression of more impulses ('false alarms').

- The amplitude of the degradation can vary greatly within the same recorded extract, owing to a range of size of physical defects. For example, in many recordings the largest click amplitudes will be well above the largest signal amplitudes, while the smallest audible defects can be more than 40dB below the local signal level (depending on psychoacoustical masking by the signal and the amount of background noise). This leads to a number of difficulties. In particular, large amplitude defects will tend to bias any parameter estimation and threshold determination procedures, leaving smaller defects undetected. As we shall see in section 4.3.2, threshold selection for some detection schemes becomes a difficult problem in this case.

Many approaches are possible for the restoration of such defects. It is clear, however, that the ideal system will process only on those samples which are degraded, leaving the others untouched in the interests of fidelity to the original source. Two tasks can thus be identified for a successful click restoration system. The first is a detection procedure in which we estimate the process $i[n]$, that is decide which samples are degraded. The second is an estimation procedure in which we attempt to reconstruct the underlying audio data when corruption is present. A method which assumes that no useful information about the underlying signal is contained in the degraded samples will involve a pure interpolation of the audio data using the undegraded samples, while more sophisticated techniques will attempt in addition to extract extra information from samples degraded with noise using some degree of noise modelling.

4.3.2 Detection

Click detection for audio signals involves the identification of samples which are not drawn from the underlying audio signal; in other words they are drawn from a spurious 'outlier' distribution. We will see a close relationship between click detection and work in robust parameter estimation and treatment of outliers, from fields as diverse as medical signal processing, underwater signal processing and statistical data analysis. In the statistical field in particular there has been a vast amount of work in the treatment of outliers (see e.g. [Beckman and Cook, 1983, Barnett and Lewis, 1984] for extensive review material, and further references in section 4.3.4). Various criteria for detection are possible, including minimum probability of error, P_E, and related concepts, but strictly speaking the aim of any audio restoration is to remove only those artifacts which are audible to the listener. Any further processing is not only unnecessary but will increase the chance of distorting the perceived signal quality. Hence a truly optimal system should take into account the trade-off between the audibility of artifacts and perceived distortion as a result of processing, and will involve consideration of complex psychoacoustical effects in the human ear (see e.g. [Moore, 1997]). Such an approach, however, is difficult both to formulate and to realize, so we will limit discussion here only to criteria which are well understood in a mathematical sense.

The simplest click detection methods involve a high-pass filtering operation on the signal, the assumption being that most audio signals contain little information at high frequencies, while impulses have spectral content at all frequencies. Clicks are thus enhanced relative to the signal by the high-pass filtering operation and can easily be detected by thresholding the filtered output. The method has the advantage of being simple to implement and having no unknown system parameters (except for a detection threshold). This principle is the basis of most analogue de-clicking equipment [Carrey and Buckner, 1976, Kinzie, Jr. and Gravereaux, 1973] and some simple digital click detectors [Kasparis and Lane, 1993]. Of course, the method will fail if the audio signal has strong high frequency content or the clicks are band-limited. Along similar lines, wavelets and multiresolution methods in general [Akansu and Haddad, 1992, Chui, 1992a, Chui, 1992b] have useful localization properties for singularities in signals (see e.g. [Mallat and Hwang, 1992]), and a Wavelet filter at a fine resolution can be used for the detection of clicks. Such methods have been studied and demonstrated successfully by Montresor, Valière *et al.* [Valière, 1991, Montresor et al., 1990].

Other methods attempt to incorporate prior information about signal and noise into a model-based detection procedure. Techniques for detection and removal of impulses from autoregressive signals have been developed from robust filtering principles (see section 4.2 and [Arakawa et al., 1986, Efron and Jeen, 1992]). These methods apply non-linear functions to the autoregressive excitation sequence, and can be related to the click detection methods of Vaseghi and Rayner [Vaseghi and Rayner, 1988, Vaseghi, 1988, Vaseghi and Rayner, 1990], which are now discussed. See also section 4.3.4 for recent detection methods based on statistical decision theory.

Autoregressive (AR) model-based Click Detection. In this method ([Vaseghi and Rayner, 1988, Vaseghi, 1988, Vaseghi and Rayner, 1990]) the underlying audio data $s[n]$ is assumed to be drawn from a short-term stationary autoregressive (AR) process (see equation (4.1)). The AR model parameters **a** and the excitation variance σ_e^2 are estimated from the corrupted data $x[n]$ using some procedure robust to impulsive noise, such as the M-estimator (see section 4.2).

The corrupted data $x[n]$ is filtered using the prediction error filter $H(z) = (1 - \sum_{i=1}^{P} a_i z^{-i})$ to give a detection signal $e_d[n]$:

$$e_d[n] = x[n] - \sum_{i=1}^{P} x[n-i]a_i. \qquad (4.4)$$

Substituting for $x[n]$ from (4.3) and using (4.1) gives:

$$e_d[n] = e[n] + i[n]v[n] - \sum_{i=1}^{P} i[n-i]v[n-i]a_i \qquad (4.5)$$

which is composed of the signal excitation $e[n]$ and a weighted sum of present and past impulsive noise values. If $s[n]$ is zero mean and has variance σ_s^2 then $e[n]$ is white noise with variance $\sigma_e^2 = 2\pi \dfrac{\sigma_s^2}{\int_{-\pi}^{\pi} \frac{1}{|H(e^{j\theta})|^2} d\theta}$. The reduction in power here from signal to excitation can be 40dB or more for highly correlated audio signals. Consideration of (4.5), however, shows that a single impulse contributes the impulse response of the prediction error filter, weighted by the impulse amplitude, to the detection signal $e_d[n]$, with maximum amplitude corresponding to the maximum in the impulse response. This means that considerable amplification of the impulse relative to the signal can be achieved for all but uncorrelated, noise-like signals. It should be noted, however, that this amplification is achieved at the expense of localization in time of the impulse, whose effect is now spread over $P+1$ samples of the detection signal $e_d[n]$. This will have adverse consequences when a number of impulses is present in the same vicinity, since their impulse responses may cancel one another out or add constructively to give false detections. More generally, threshold selection will be troublesome when impulses of widely differing amplitudes are present, since a low threshold which is appropriate for very small clicks will lead to false detections in the P detection values which follow a large impulse.

Detection can then be performed by thresholding $e_d[n]^2$ to identify likely impulses. Choice of threshold will depend upon the AR model, the variance of $e[n]$ and the size of impulses present (see [Godsill, 1993] for optimal thresholds under Gaussian signal and noise assumptions), and will reflect trade-offs between false and missed detection rates. See figure 4.2(a) for a typical example of detection using this method, which shows how the impulsive interference is strongly amplified relative to the signal component.

An adaptation of this method, also devised by Vaseghi and Rayner, considers the impulse detection problem from a matched filtering perspective [VanTrees, 1968]. The 'signal' is the impulse itself, while the autoregressive audio data is regarded as coloured additive noise. The prediction error filter described above can then be viewed as a pre-whitening stage for the autoregressive noise, and the full matched filter is given by $H(z)H(z^{-1})$, a non-causal filter with $2P+1$ coefficients which can be realized with P samples of lookahead. The matched filtering approach provides additional amplification of impulses relative to the signal, but further reduces localization of impulses for a given model order. Choice between the two methods will thus depend on the range of click amplitudes present in a particular recording and the degree of separation of individual impulses in the waveform. See figure 4.2(b) for an example of detection using the matched filter. Notice that the matched filter has high-lighted a few additional impulse positions, but at the expense of a much more 'smeared' response which will make accurate localization very awkward. Hence the prediction-error detector is usually preferred in practice.

Both the prediction error detection algorithm and the matched filtering algorithm are efficient to implement and can be operated in real time using DSP microprocessors. Results of a very high standard can be achieved if a careful strategy is adopted for extracting the precise click locations from the detection signal. Iterative schemes are also possible which re-apply the detection algorithms to the restored data (see section 4.3.3) in order to achieve improved parameter estimates and to ensure that any previously undetected clicks are detected.

4.3.3 Replacement of corrupted samples

Once clicks have been detected, a replacement strategy must be devised to mask their effect. It is usually appropriate to assume that clicks have in no way interfered with the timing of the material, so the task is then to fill in the 'gap' with appropriate material of identical duration to the click. As discussed above, this amounts to an interpolation or generalized prediction problem, making use of the good data values surrounding the corruption and possibly taking account of signal information which is buried in the corrupted section. An effective technique will have the ability to interpolate gap lengths from one sample up to at least 100 samples at a sampling rate of 44.1kHz.

The replacement problem may be formulated as follows. Consider N samples of audio data, forming a vector s. The corresponding click-degraded data vector is x, and the (known) vector of detection values $i[n]$ is i. The audio data s may be partitioned into two sub-vectors, one containing elements whose value is known (i.e. $i[n] = 0$), denoted by $s_\mathcal{K}$, and the second containing unknown elements which are corrupted by noise ($i[n] = 1$), denoted by $s_\mathcal{U}$. Vectors x and i are partitioned in a similar fashion. The replacement problem requires the estimation of the unknown data $s_\mathcal{U}$, given the observed (corrupted) data x. This will be a statistical estimation procedure for audio signals, which are stochastic in nature, and estimation methods might be chosen to satisfy criteria such as minimum mean-square error (MMSE), maximum likelihood (ML), maximum *a posteriori* (MAP) or perceptual features.

Numerous methods have been developed for the interpolation of corrupted or missing samples in speech and audio signals. The 'classical' approach is perhaps the median filter [Tukey, 1971, Pitas and Venetsanopoulos, 1990] which can replace corrupted samples with a median value while retaining detail in the signal waveform. A suitable system is described in [Kasparis and Lane, 1993], while a hybrid autoregressive prediction/ median filtering method is presented in [Nieminen et al., 1987]. Median filters, however, are too crude to deal with gap lengths greater than a few samples. Other techniques 'splice' uncorrupted data from nearby into the gap [Lockhart and Goodman, 1986, Platte and Rowedda, 1985] in such a manner that there is no signal discontinuity at the start or end of the gap. These methods rely on the periodic nature of many speech and music signals and also require a reliable estimate of pitch period.

The most effective and flexible methods to date have been model-based, allowing for the incorporation of reasonable prior information about signal characteristics. A good coverage is given by Veldhuis [Veldhuis, 1990], and a number of interpolators suited to speech and audio signals is presented. These are based on minimum variance estimation under various modelling assumptions, including sinusoidal, autoregressive, and periodic. The autoregressive interpolator, originally derived in [Janssen et al.,

1986], was later developed by Vaseghi and Rayner [Vaseghi, 1988, Vaseghi and Rayner, 1988, Vaseghi and Rayner, 1990] for the restoration of gramophone recordings. This interpolator and other developments based on autoregressive modelling are discussed in the next section.

Autoregressive interpolation. An interpolation procedure which has proved highly successful is the Least Squares AR-based (LSAR) method [Janssen et al., 1986, Veldhuis, 1990], devised originally for the concealment of uncorrectable errors in CD systems. Corrupted data is considered truly 'missing' in that no account is taken of its value in making the interpolation. We present the algorithm in a matrix/vector notation in which the locations of degraded samples can be arbitrarily specified within the data block through the detection vector $\mathbf{i}$.

Consider a block of N data samples $\mathbf{s}$ which are drawn from a short-term stationary AR process with parameters $\mathbf{a}$. Equation 4.1 can be re-written in matrix/vector notation as:

$$\mathbf{e} = \mathbf{A}\mathbf{s} \qquad (4.6)$$

where $\mathbf{A}$ is an $((N-P) \times N)$ matrix, whose $(j-P)$th row is constructed so as to generate the prediction error, $e[j] = s[j] - \sum_{i=1}^{P} s[j-i]a_j$. Elements on the right hand side of this equation can be partitioned into known and unknown sections as described above, with $\mathbf{A}$ being partitioned by column. The least squares solution is then obtained by minimizing the sum of squares $E = \mathbf{e}^T\mathbf{e}$ w.r.t. the unknown data segment, to give the solution:

$$\mathbf{s}_\mathcal{U}^{\mathrm{LS}} = -(\mathbf{A}_\mathcal{U}{}^T\mathbf{A}_\mathcal{U})^{-1}\mathbf{A}_\mathcal{U}{}^T\mathbf{A}_\mathcal{K}\,\mathbf{s}_\mathcal{K}. \qquad (4.7)$$

This interpolator has useful properties, being the minimum-variance unbiased estimator for the missing data [Veldhuis, 1990]. Viewed from a probabilistic perspective, it corresponds to maximization of $p(\mathbf{s}_\mathcal{U} \mid \mathbf{s}_\mathcal{K}, a, \sigma_e^2)$ under Gaussian assumptions,[3] and is hence also the maximum *a posteriori* (MAP) estimator [Godsill, 1993, Veldhuis, 1990]. In cases where corruption occurs in contiguous bursts separated by at least P 'good' samples, the interpolator leads to a Toeplitz system of equations which can be efficiently solved using the Levinson-Durbin recursion [Durbin, 1959]. See figure 4.4 for examples of interpolation using the LSAR method. A succession of interpolations has been performed, with increasing numbers of missing samples from left to right in the data (gap lengths increase from 25 samples up to more than 100). The autoregressive model order is 60. The shorter length interpolations are almost indistinguishable from the true signal (left-hand side of figure 4.4(a)), while the interpolation is much poorer as the number of missing samples becomes large (right-hand side of figure 4.4(b)). This is to be expected of any interpolation scheme when the data is drawn from a random process, but the situation can often be improved by use of a higher order autoregressive model. Despite poor accuracy of the interpoland for longer gap lengths, good continuity is maintained at the start and end of the missing data blocks, and the signal appears to have the right 'character'. Thus effective removal of click artifacts in typical audio sources can usually be achieved.

The basic formulation given in (4.7) assumes that the AR parameters are known *a priori*. In practice we may have a robust estimate of the parameters obtained during the detection stage (see section 4.3.2). This, however, is strictly sub-optimal and we should perhaps consider interpolation methods which treat the parameters as unknown. Minimization of the term $E = \mathbf{e}^T\mathbf{e}$ w.r.t. both $\mathbf{s}_u$ and $\mathbf{a}$ corresponds to the joint least squares estimator for the parameters and the missing data, and also to the approximate joint ML estimator.[4] E, however, contains fourth-order terms in the unknowns and cannot be minimized analytically. Janssen, Veldhuis and Vries [Janssen et al., 1986] propose an alternating variables iteration which performs linear maximizations w.r.t. data and parameters in turn, and is guaranteed to converge at least to a local maximum of the likelihood. The true likelihood for the missing data, $p(\mathbf{s}_\kappa \mid \mathbf{s}_u)$, can be maximized using the expectation-maximize (EM) algorithm [Dempster et al., 1977], an approach which has been investigated by Ó Ruanaidh and Fitzgerald [ÓRuanaidh, 1994, ÓRuanaidh and Fitzgerald, 1993]. Convergence to local maxima of the likelihood is also a potential difficulty with this method.

The LSAR approach to interpolation performs well in most cases. However, certain classes of signal which do not fit the modelling assumptions (such as periodic pulse-driven voiced speech) and very long gap lengths can lead to an audible 'dulling' of the signal or unsatisfactory masking of the original corruption. Increasing the order of the AR model will usually improve the results; however, several developments to the method are now outlined which can lead to better performance.

Vaseghi and Rayner [Vaseghi and Rayner, 1990] propose an extended AR model to take account of signals with long-term correlation structure, such as voiced speech, singing or near-periodic music. The model, which is similar to the long term prediction schemes used in some speech coders, introduces extra predictor parameters around the pitch period T, so that equation 4.1 becomes:

$$s[n] = \sum_{i=1}^{P} s[n-i]a_i + \sum_{j=-Q}^{Q} s[n-T-j]b_j + e[n], \qquad (4.8)$$

where Q is typically smaller than P. Least squares/ML interpolation using this model is of a similar form to equation 4.7, and parameter estimation is straightforwardly derived as an extension of standard AR parameter estimation methods (see section 4.2). The method gives a useful extra degree of support from adjacent pitch periods which can only be obtained using very high model orders in the standard AR case. As a result, the 'under-prediction' sometimes observed when interpolating long gaps is improved. Of course, an estimate of T is required, but results are quite robust to errors in this. Veldhuis [Veldhuis, 1990][chapter 4] presents a special case of this

DIGITAL AUDIO RESTORATION   147

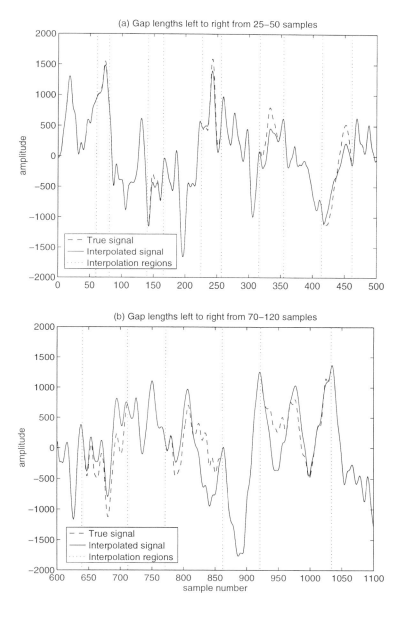

Figure 4.4 AR-based interpolation, $P=60$, classical chamber music, (a) short gaps, (b) long gaps

interpolation method in which the signal is modelled by one single 'prediction' element at the pitch period (i.e. $Q = 0$ and $P = 0$ in the above equation).

A second modification to the LSAR method is concerned with the characteristics of the excitation signal. We notice that the LSAR procedure (4.7) seeks to minimize the excitation energy of the signal, irrespective of its time domain autocorrelation. This is quite correct, and desirable mathematical properties result (see above). However, figure 4.6 shows that the resulting excitation signal corresponding to the corrupted region can be correlated and well below the level of surrounding excitation. As a result, the 'most probable' interpolands may under-predict the true signal levels and be over-smooth compared with the surrounding signal. In other words, ML/MAP procedures do not necessarily generate interpolands which are *typical* for the underlying model, which is an important factor in the *perceived* effect of the restoration. Rayner and Godsill [Rayner and Godsill, 1991] have devised a method which addresses this problem. Instead of minimizing the excitation energy, we consider interpolands with constant excitation energy. The excitation energy may be expressed as:

$$E = (\mathbf{s}_u - \mathbf{s}_u^{\text{LS}})^T \mathbf{A}_u^T \mathbf{A}_u (\mathbf{s}_u - \mathbf{s}_u^{\text{LS}}) + E_{\text{LS}}, \qquad E > E_{\text{LS}}, \qquad (4.9)$$

where E_{LS} is the excitation energy corresponding to the LSAR estimate $\mathbf{s}_u^{\text{LS}}$. The positive definite matrix $\mathbf{A}_u^T \mathbf{A}_u$ can be factorized into 'square roots' by Cholesky or any other suitable matrix decomposition [Golub and Van Loan, 1989] to give $\mathbf{A}_u^T \mathbf{A}_u = \mathbf{M}^T \mathbf{M}$, where $\mathbf{M}$ is a non-singular square matrix. A transformation of variables $\mathbf{u} = \mathbf{M}(\mathbf{s}_u - \mathbf{s}_u^{\text{LS}})$ then serves to de-correlate the missing data samples, simplifying equation (4.9) to:

$$E = \mathbf{u}^T \mathbf{u} + E_{\text{LS}}, \qquad (4.10)$$

from which it can be seen that the (non-unique) solutions with constant excitation energy correspond to vectors $\mathbf{u}$ with constant L_2-norm. The resulting interpoland can be obtained by the inverse transformation $\mathbf{s}_u = \mathbf{M}^{-1}\mathbf{u} + \mathbf{s}_u^{\text{LS}}$. One suitable criterion for selecting $\mathbf{u}$ might be to minimize the autocorrelation at non-zero lags of the resulting excitation signal, since the excitation is assumed to be white noise. This, however, requires a non-linear minimization, and a practical alternative is to generate $\mathbf{u}$ as Gaussian white noise with variance $(E - E_{\text{LS}})/l$, where l is the number of corrupted samples. The resulting excitation will have approximately the desired energy and uncorrelated character. A suitable value for E is the expected excitation energy for the AR model, provided this is greater than E_{LS}, i.e. $E = max(E_{\text{LS}}, N\sigma_e^2)$. Viewed within a probabilistic framework, the case when $E = E_{\text{LS}} + l\sigma_e^2$, where l is the number of unknown sample values, is equivalent to drawing a sample from the posterior density for the missing data, $p(\mathbf{s}_u \mid \mathbf{s}_\kappa, \mathbf{a}, \sigma_e^2)$. Figures 4.5-4.7 illustrate the principles involved in this sampled interpolation method. A short section taken from a modern solo vocal recording is shown in figure 4.5, alongside its estimated autoregressive excitation. The waveform has a fairly 'noise-like' character, and the corresponding excitation is noise-like as expected. The standard LSAR interpolation and corresponding excitation is shown in figure 4.6. The interpolated section (between the dotted vertical lines) is reasonable, but has lost the random noise-like quality of the original. Examination of the excitation signal shows that the LSAR interpolator has done 'too good' a job of minimizing the excitation energy, producing an interpolant which, while optimal in a mean-square error sense, cannot be regarded as typical of the autoregressive process. This might be heard as a momentary change in sound quality at the point of interpolation. The sampling-based interpolator is shown in figure 4.7.

Its waveform retains the random quality of the original signal, and likewise the excitation signal in the gap matches the surrounding excitation. Hence the sub-optimal interpolant is likely to sound more convincing to the listener than the LSAR reconstruction.

Ó Ruanaidh and Fitzgerald [ÓRuanaidh and Fitzgerald, 1994, ÓRuanaidh, 1994] have successfully extended the idea of sampled interpolates to a full Gibbs' Sampling framework [Geman and Geman, 1984, Gelfand and Smith, 1990] in order to generate typical interpolates from the marginal posterior density $p(\mathbf{s}_\mathcal{U} \mid \mathbf{s}_\mathcal{K})$. The method is iterative and involves sampling from the conditional posterior densities of $\mathbf{s}_\mathcal{U}$, $\mathbf{a}$ and σ_e^2 in turn, with the other unknowns fixed at their most recent sampled values. Once convergence has been achieved, the interpolation used is the last sampled estimate from $p(\mathbf{s}_\mathcal{U} \mid \mathbf{s}_\mathcal{K}, \mathbf{a}, \sigma_e^2)$.

Other methods. Several transform-domain methods have been developed for click replacement. Montresor, Valiére and Baudry [Montresor et al., 1990] describe a simple method for interpolating wavelet coefficients of corrupted audio signals, which involves substituting uncorrupted wavelet coefficients from nearby signal according to autocorrelation properties. This, however, does not ensure continuity of the restored waveform and is not a localized operation in the signal domain. An alternative method, based in the discrete Fourier domain, which is aimed at restoring long sections of missing data is presented by Maher [Maher, 1994]. In a similar manner to the sinusoidal coding algorithms of McAulay and Quatieri [McAulay and Quatieri, 1986b], this technique assumes that the signal is composed as a sum of sinusoids with slowly varying frequencies and amplitudes (see equation 4.2). Spectral peak 'tracks' from either side of the gap are identified from the Discrete Fourier Transform (DFT) of successive data blocks and interpolated in frequency and amplitude to generate estimated spectra for the intervening material. The inverse DFTs of the missing data blocks are then inserted back into the signal. The method is reported to be successful for gap lengths of up to 30ms, or well over 1000 samples at audio sampling rates. A method for interpolation of signals represented by the multiple sinusoid model is given in [Veldhuis, 1990][Chapter 6].

150   APPLICATIONS OF DSP TO AUDIO AND ACOUSTICS

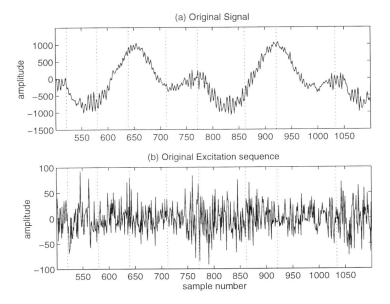

Figure 4.5   Original signal and excitation (P=100)

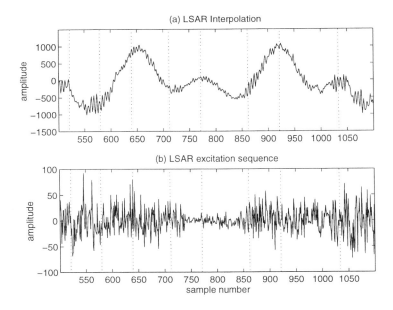

Figure 4.6   LSAR interpolation and excitation ($P = 100$)

DIGITAL AUDIO RESTORATION 151

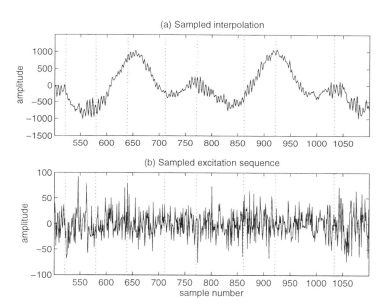

Figure 4.7   Sampled AR interpolation and excitation (P=100)

Godsill and Rayner [Godsill and Rayner, 1993a, Godsill, 1993] have derived an interpolation method which operates in the DFT domain. This can be viewed as an alternative to the LSAR interpolator (see section 4.3.3) in which power spectral density (PSD) information is directly incorporated in the frequency domain. Real and imaginary DFT components are modelled as independent Gaussians with variance proportional to the PSD at each frequency. These assumptions of independence are shown to hold exactly for random periodic processes [Therrien, 1989], so the method is best suited to musical signals with strongly tonal content. The method can, however, also be used for other stationary signals provided that a sufficiently long block length is used (e.g. 500-2000 samples) since the assumptions also improve as block length increases [Papoulis, 1991]. The Maximum *a posteriori* solution is of a similar form and complexity to the LSAR interpolator, and is particularly useful as an alternative to the other method when the signal has a quasi-periodic or tonal character. A robust estimate is required for the PSD, and this can usually be obtained through averaged DFTs of the surrounding data, although iterative methods are also possible, as in the case of the LSAR estimator.

Recent statistical model-based detection and interpolation methods are discussed in the next section.

4.3.4 Statistical methods for the treatment of clicks

The detection and replacement techniques described in the preceding sections can be combined to give very successful click concealment, as demonstrated by a number of research and commercial systems which are now used for the re-mastering of old recordings. However, some of the difficulties outlined above concerning the 'masking' of smaller defects by large defects in the detection process, the poor time localization of some detectors in the presence of impulse 'bursts' and the inadequate performance of existing interpolation methods for certain signal categories, has led to further research which considers the problem from a more fundamental statistical perspective.

In [Godsill and Rayner, 1992, Godsill and Rayner, 1995a, Godsill, 1993] click detection is studied within a model-based Bayesian framework (see e.g. [Box and Tiao, 1973, Bernardo and Smith, 1994]). The Bayesian approach is a simple and elegant framework for performing decision and estimation within complex signal and noise modelling problems such as this, and relevant Bayesian approaches to the related problem of outlier detection in statistical data can be found in [Box and Tiao, 1968, Abraham and Box, 1979, McCulloch and Tsay, 1994]. Detection is formulated explicitly as estimation of the noise 'switching' process $i[n]$ (see section 4.3.1) conditional upon the corrupted data $x[n]$. The switching process can be regarded as a random discrete (1/0) process for which a posterior probability is calculated. Detection is then achieved by determining the switching values which minimize risk according to some appropriate cost function. In the most straightforward case, this will involve selecting switch values which maximize the posterior probability, leading to the maximum *a posteriori* (MAP) detection. The posterior detection probability for a block of N data points may be expressed using Bayes' rule as:

$$P(\mathbf{i} \mid \mathbf{x}) = \frac{p(\mathbf{x} \mid \mathbf{i}) P(\mathbf{i})}{p(\mathbf{x})} \qquad (4.11)$$

where all terms are implicitly conditional upon the prior modelling assumptions, $\mathcal{M}$. The prior detection probability $P(\mathbf{i})$ reflects any prior knowledge about the switching process. In the case of audio clicks this might, for example, incorporate the knowledge that clicks tend to occur as short 'bursts' of consecutive impulses, while the majority of samples are uncorrupted. A suitable prior which expresses this time dependence is the discrete Markov chain prior (see [Godsill and Rayner, 1995a, Godsill, 1993] for discussion this point). The term $p(\mathbf{x})$ is constant for any given set of observations, and so can be ignored as a constant scale factor. Attention will thus focus on $p(\mathbf{x} \mid \mathbf{i})$, the detection-conditioned likelihood for a particular detection vector $\mathbf{i}$. It is shown in [Godsill and Rayner, 1995a, Godsill, 1993, Godsill and Rayner, 1992] that within the additive noise modelling framework of (4.3), the likelihood term is given by

$$p(\mathbf{x} \mid \mathbf{i}) = \int_{\mathbf{s}_\mathcal{U}} p_{\mathbf{v}_\mathcal{U} \mid \mathbf{i}}(\mathbf{x}_\mathcal{U} - \mathbf{s}_\mathcal{U} \mid \mathbf{i}) \, p_{\mathbf{s}}(\mathbf{s}) \mid_{\mathbf{s}_\mathcal{K} = \mathbf{x}_\mathcal{K}} d\mathbf{s}_\mathcal{U} \qquad (4.12)$$

where $p_{\mathbf{v}_\mathcal{U} \mid \mathbf{i}}$ is the probability density function for the corrupting noise values and $p_\mathbf{s}$ is the density for the underlying audio data. This formulation holds for any random additive noise process which is independent of the signal. In particular, the calculation of (4.12) is analytic in the case of linear Gaussian models. In [Godsill and Rayner, 1992, Godsill and Rayner, 1995a, Godsill, 1993] the autoregressive signal model with Gaussian corruption is studied in detail.

In order to obtain the MAP detection estimate from the posterior probability expression of equation (4.11) an exhaustive search over all 2^N possible configurations of the (1/0) vector $\mathbf{i}$ is necessary. This is clearly infeasible for any useful value of N, so alternative strategies must be devised. A sequential approach is developed in [Godsill and Rayner, 1995a, Godsill, 1993] for the Gaussian AR case. This is based around a recursive calculation of the likelihood (4.12), and hence posterior probability, as each new data sample is presented. The sequential algorithm performs a reduced binary tree search through possible configurations of the detection vector, rejecting branches which have low posterior probability and thus making considerable computational savings compared with the exhaustive search. The method has been evaluated experimentally in terms of detection error probabilities and perceived quality of restoration and found to be a significant improvement over the autoregressive detection methods described in section 4.3.2, although more computationally intensive.

Click detection within a Bayesian framework has introduced the concept of an explicit model for the corrupting noise process through the noise density $p_{\mathbf{v}_\mathcal{U} \mid \mathbf{i}}$. Effective noise modelling can lead to improvements not only in click detection, but also in replacement, since it allows useful signal information to be extracted from the corrupted data values. This information is otherwise discarded as irrevocably lost, as in the interpolators described in earlier sections. In fact, it transpires that an intrinsic part of the likelihood calculation in the Bayesian detection algorithm (equation 4.12) is calculation of the MAP estimate for the unknown data conditional upon the detection vector **i**. This MAP interpolation can be used as the final restored output after detection, without resort to other interpolation methods. The form of this 'interpolator' is closely related to the LSAR interpolator (section 4.3.3) and may be expressed as:

$$\mathbf{s}_u^{\text{MAP}} = -\left(\mathbf{A}_u^T \mathbf{A}_u + \frac{\sigma_e^2}{\sigma_v^2}\mathbf{I}\right)^{-1}\left(\mathbf{A}_u^T \mathbf{A}_\kappa \mathbf{s}_\kappa - \frac{\sigma_e^2}{\sigma_v^2}\mathbf{x}_u\right), \qquad (4.13)$$

(see [Godsill and Rayner, 1995a][equations (12-14)]), where σ_v^2 is the variance of the corrupting noise, which is assumed independent and Gaussian. Of course, the quality of the restored output is now dependent on the validity of the assumed noise statistics. The Bayesian detector itself shows considerable robustness to errors in these assumptions [Godsill and Rayner, 1995a, Godsill, 1993], but the interpolator is less tolerant. This will be particularly noticeable when the true noise distributions are more 'heavy-tailed' than the Gaussian, a scenario for which there is strong evidence in many degraded audio signals. The noise modelling can in fact be generalized to a more realistic class of distributions by allowing the individual noise components $v[n]$ to have separate, unknown variances and even unknown correlation structure. We are essentially then modelling noise sources as continuous scale mixtures of Gaussians:

$$p(v[n]) = \int N(0, \lambda) g(\lambda) d\lambda$$

where $N(\mu, \lambda)$ is the Gaussian distribution with mean μ and variance λ, and $g(\lambda)$ is a continuous 'mixing' density [West, 1984]. These extensions allow for non-Gaussian defects with of widely varying magnitude and also for the noise correlation which might be expected when the signal has been played through a mechanical pick-up system followed by equalization circuitry. This noise modelling framework can be used to develop highly robust interpolators, and a Bayesian approach which requires no prior knowledge of AR parameters or noise statistics is presented in [Godsill and Rayner, 1995b], using an iterative EM-based solution. Similar noise modelling principles can be used to extend the Bayesian detection algorithms, and Markov chain Monte Carlo (MCMC) methods [Hastings, 1970, Geman and Geman, 1984, Gelfand and Smith, 1990] are presented for the solution of this problem in [Godsill and Rayner, 1998, Godsill and Rayner, 1996b]. An example of these Bayesian iterative restoration methods for removal of clicks is shown in figure 4.8 for a typical 78 rpm recording. The same framework may be extended to perform joint removal of clicks and background

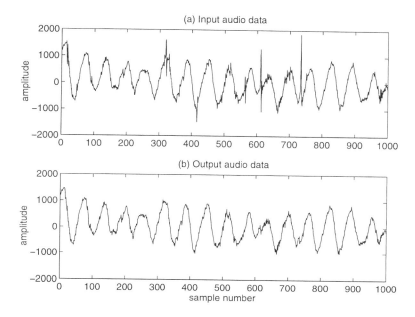

Figure 4.8 Restoration using Bayesian iterative methods

noise in one single procedure, and some recent work on this problem can be found in [Godsill and Rayner, 1996a] for autoregressive signals and in [Godsill, 1997a, Godsill, 1997b] for autoregressive moving-average (ARMA) signals.

The statistical methods described here provide a highly flexible framework for audio restoration and signal enhancement in general. Solution for these complex models is usually of significantly higher computational complexity than the techniques described in earlier sections, but this is unlikely to be problematic for applications where restoration quality is the highest priority. The methods are still in their infancy, but we believe that future research work in the field will require sophisticated statistical modelling of signals and noise, with associated increases in solution complexity, in order to achieve improved fidelity of restoration. The Bayesian methods discussed here are likely to find application in many other areas of audio processing (see later sections).

4.4 CORRELATED NOISE PULSE REMOVAL

A further problem which is common to several recording media including gramophone discs and optical film sound tracks is that of low frequency noise pulses. This form of degradation is typically associated with large scratches or even breakages in the

156   APPLICATIONS OF DSP TO AUDIO AND ACOUSTICS

surface of a gramophone disc. The precise form of the noise pulse depends upon the mechanical and electrical characteristics of the playback system, but a typical result is shown in figure 4.9. A large discontinuity is observed followed by a decaying low frequency transient. The noise pulses appear to be additively superimposed on the undistorted signal waveform (see figure 4.10).

Low frequency noise pulses appear to be the response of the playback system to extreme step-like or impulsive stimuli caused by breakages in the groove walls of gramophone discs or large scratches on an optical film sound track. The audible effect of this response is a percussive 'pop' indexNoise, Pop noise or 'thump' in the recording. This type of degradation is often the most disturbing artifact present in a given extract. It is thus highly desirable to eliminate noise pulses as a first stage in the restoration process.

The effects of the noise pulse are quite long-term, as can be seen from figure 4.9, and thus a straightforward interpolation using the methods of section 4.3.3 is not a practical proposition. Since the majority of the noise pulse is of very low frequency it might be thought that some kind of high pass filtering operation would remove the defect. Unfortunately this does not work well either, since the discontinuity at the start of the pulse has significant high frequency content. Some success has been achieved with a combination of localized high pass filtering , followed by interpolation to remove discontinuities. However it is generally found that significant artifacts remain after processing or that the low frequency content of the signal has been damaged.

It should be noted that the problem of transient noise pulses can in principle be circumvented by use of suitable playback technology. For example, in the case of gramophone disks the use of a laser-based reader should eliminate any mechanical resonance effects and thus reduce the artifact to a large click which can be restored using the methods of previous sections. Of course, this does not help in the many cases where the original source medium has been discarded after transcription using standard equipment to another medium such as magnetic tape!

Template-based methods. The first digital approach to this problem was devised by Vaseghi and Rayner [Vaseghi, 1988, Vaseghi and Rayner, 1988]. This technique, which employs a 'template' for the noise pulse waveform, has been found to give good results for many examples of broken gramophone discs. The observation was made that the resonant sections (i.e. after the initial discontinuity) of successive noise pulses in the same recording were nearly identical in shape (to within a scale factor). This would correspond with the idea that noise pulses are simply the step response of a linear time-invariant (LTI) mechanical system. Given the waveform of the repetitive section of the noise pulse (the 'template' $t[n]$) it is then possible to subtract appropriately scaled versions from the corrupted signal $x[n]$ wherever pulses are detected. The position M and scaling G of the noise pulse are estimated by cross-correlating the template with

DIGITAL AUDIO RESTORATION   157

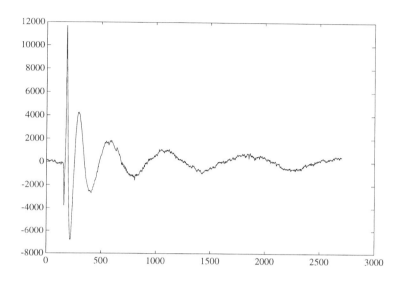

Figure 4.9   Noise pulse from optical film sound track ('silent' section)

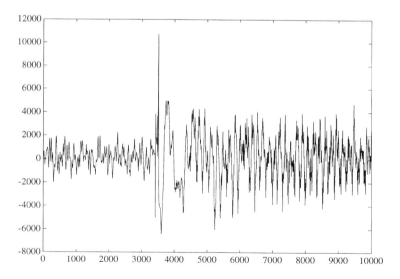

Figure 4.10   Signal waveform degraded by low frequency noise transient

the corrupted waveform, and the restored signal is then obtained as:

$$y[n] = x[n] - Gt[n - M], \quad M \leq n < M + N_t \quad (4.14)$$

where N_t is the length of the template. Any remaining samples close to the start of the pulse which are irrevocably distorted can then be interpolated using a method such as the LSAR interpolator discussed earlier (see section 4.3.3).

The template $t[n]$ is obtained by long term averaging of many such pulses from the corrupted signal. Alternatively, a noise-free example of the pulse shape may be available from a 'silent' section of the recording or the lead-in groove of a gramophone disc.

The template method has been very successful in the restoration of many recordings. However, it is limited in several important ways which hinder the complete automation of pulse removal. While the assumption of constant template shape is good for short extracts with periodically recurring noise pulses (e.g. in the case of a broken gramophone disc) it is not a good assumption for many other recordings. Even where noise pulses do correspond to a single radial scratch or fracture on the record the pulse shape is often found to change significantly as the recording proceeds, while much more variety is found where pulses correspond to randomly placed scratches and breakages on the recording. Further complications arise where several pulses become superimposed as is the case for several closely spaced scratches. These effects may be partly due to the time-varying nature of the mechanical system as the stylus moves towards the centre of the disk, but also non-linearity in the playback apparatus. There is some evidence for the latter effect in optical film sound track readers [Godsill, 1993], where the frequency of oscillation can be observed to decrease significantly as the response decays.

Correct detection can also be a challenge. This may seem surprising since the defect is often very large relative to the signal. However, audible noise pulses do occur in high amplitude sections of the signal. In such cases the cross-correlation method of detection can give false alarms from low frequency components in the signal; in other circumstances noise pulses can be missed altogether. This is partly as a result of the correlated nature of the signal which renders the cross-correlation procedure suboptimal. A true matched filter for the noise pulse would take into account the signal correlations (see e.g. [Van Trees, 1968]) and perhaps achieve some improvements in detection. This issue is not addressed here, however, since other restoration methods are now available.

Model-based separation methods. A full study of the noise pulse mechanism would involve physical modelling of the (possibly non-linear) playback system for both gramophone systems and optical sound track readers. A full description is beyond the scope of this article (see [Roys, 1978] for some more detail), but can be used to shed further light upon this and other audio restoration areas including click removal and background noise reduction.

A linear modelling approach to noise pulse removal is presented in [Godsill, 1993]. In this it is assumed that the corrupted waveform **x** consists of a linear sum of the underlying audio waveform **s** and resonant noise pulses **v**:

$$\mathbf{x} = \mathbf{s} + \mathbf{v}. \tag{4.15}$$

We note that **s** and **v** are the responses of the playback system, including mechanical components and amplification/ equalization circuitry, to the recorded audio and noise signals, respectively. The assumption of a linear system allows the overall response **x** to be written as the linear superposition of individual responses to signal and noise components.

Here the noise pulses are modelled by a low order autoregressive process which is driven by a low level white noise excitation with variance σ_{v0}^2 most of the time, and bursts of high level impulsive excitation with variance $\sigma_{v1}^2 \gg \sigma_{v0}^2$ at the initial discontinuity of the noise transient. We can define a binary noise switching process $i[n]$ to switch between low and high variance components in a similar way to the click generation model of section 4.3. This modelling approach is quite flexible in that it allows for variations in the shape of individual noise pulses as well as for the presence of many superimposed pulses within a short period. The restoration task is then one of separating the two superimposed responses, **s** and **v**. If the audio signal's response is also modelled as an autoregressive process then the MAP estimator for **s** under Gaussian assumptions is obtained from:

$$\left(\frac{\mathbf{A}^T \mathbf{A}}{\sigma_e^2} + \mathbf{A}_v{}^T \mathbf{\Lambda}_v{}^{-1} \mathbf{A}_v \right) \mathbf{s}^{\mathrm{MAP}} = \mathbf{A}_v{}^T \mathbf{\Lambda}_v{}^{-1} \mathbf{A}_v \, \mathbf{x}. \tag{4.16}$$

Terms of this equation are defined similarly to those for the LSAR interpolator of section 4.3.3, with subscript 'v' referring to the autoregressive process for the noise pulses. $\mathbf{\Lambda}_v$ is a diagonal matrix whose mth diagonal element $\lambda_v[m]$ is the variance of the mth noise excitation component, i.e.

$$\lambda_v[m] = \sigma_{v0}^2 + i[m](\sigma_{v1}^2 - \sigma_{v0}^2). \tag{4.17}$$

This signal separation algorithm requires knowledge of both AR systems, including noise variances and actual parameter values, as well as the switching vector **i** which indicates which noise samples have impulsive excitation and which have low level excitation. These can be treated as unknowns within a similar iterative statistical framework to that outlined for click removal in section 4.3.4, and this could form a useful piece of future research. In practice, however, these unknowns can usually be estimated by simpler means. The switching process can be estimated much as clicks are detected (see section 4.3.2), with a higher threshold selected to indicate large disturbances which are likely to be noise pulses. The autoregressive system for the noise can often be estimated from a noise pulse captured during a 'silent' section of the recording or from a similar type of pulse taken from another recording, and the very large (or even infinite) value chosen for the high level excitation variance σ_{v1}^2. The autoregressive system for the underlying audio data is then estimated from uncorrupted data in the vicinity of the noise pulses, usually in the section just prior to the start of the section to be restored.

Even with suitable estimation schemes for the unknown parameters, the separation formula of equation (4.16) is of relatively high computational complexity, since the noise process can affect thousands of samples following the initial impulsive discontinuity. This problem can be partially overcome by restoring samples which are fairly distant from the initial transient using a simple linear phase high-pass filter. The separation algorithm is then constrained to give continuity with this filtered signal at either end of the restored section in much the same way as the LSAR interpolator (section 4.3.3). Further computational savings can be achieved by working with a sub-sampled version of the noise pulse waveform, since it is typically over-sampled by a factor of at least one hundred for the much of its duration. This sub-sampling can be incorporated into the separation algorithm by use of an analytic interpolation operator such as the second order spline. An alternative scheme, which takes advantage of the Markovian nature of the AR models, is based on Kalman filtering [Anderson and Moore, 1979]. This is currently being investigated and results will be reported in future publications.

Results from the model-based separation approach have demonstrated much more generality of application and ease of automation than the templating technique, which can be a highly operator-intensive procedure, and the perceived quality of output is certainly at least as good as the templating method. Figures 4.11-4.13 show the restoration of a particularly badly degraded 78 rpm recording which exhibits many closely spaced noise transients. A second order autoregressive model was found to be adequate for modelling the noise transients, while the signal was modelled to order 80. The restored signal (shown on a different scale) shows no trace of the original corruption, and the perceptual results are very effective.

Summary. Two principal methods for removal of low frequency noise transients are currently available. The model-based separation approach has shown more flexibility and generality, but is computationally rather intensive. It is felt that future work in the area should consider the problem from a realistic physical modelling perspective, which takes into account linear and non-linear characteristics of gramophone and film sound playback systems, in order to detect and correct these artifacts more effectively. Such an approach could involve both experimental work with playback systems and sophisticated non-linear modelling techniques. Statistical approaches related to those outlined in the click removal work (section 4.3.4) may be applicable to this latter task.

DIGITAL AUDIO RESTORATION

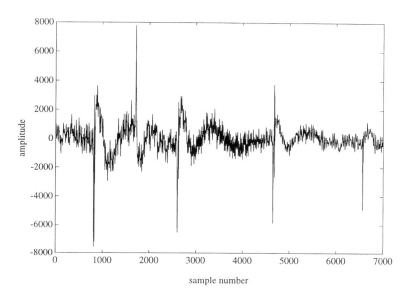

Figure 4.11 Degraded audio signal with many closely spaced noise transients

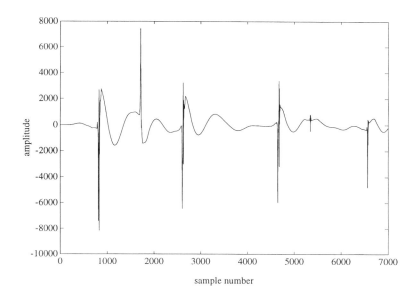

Figure 4.12 Estimated noise transients for figure 4.11

162 APPLICATIONS OF DSP TO AUDIO AND ACOUSTICS

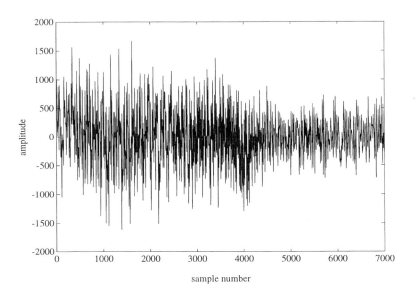

Figure 4.13 Restored audio signal for figure 4.11 (different scale)

4.5 BACKGROUND NOISE REDUCTION

Random, additive background noise is a form of degradation common to all analogue measurement, storage and recording systems. In the case of audio signals the noise, which is generally perceived as 'hiss' by the listener, will be composed of electrical circuit noise, irregularities in the storage medium and ambient noise from the recording environment. The combined effect of these sources will generally be treated as one single noise process, although we note that a pure restoration should strictly not treat the ambient noise, which might be considered as a part of the original 'performance'. Random noise generally has significant components at all audio frequencies, and thus simple filtering and equalization procedures are inadequate for restoration purposes.

Analogue tape recordings typically exhibit noise characteristics which are stationary and for most purposes white. At the other end of the scale, many early 78 rpm and cylinder recordings exhibit highly non-stationary noise characteristics, such that the noise can vary considerably within each revolution of the playback system. This results in the characteristic 'swishing' effect associated with some early recordings. In recording media which are also affected by local disturbances, such as clicks and low frequency noise resonances, standard practice is to restore these defects prior to any background noise treatment.

Noise reduction has been of great importance for many years in engineering disciplines. The classic least-squares work of Norbert Wiener[Wiener, 1949] placed noise reduction on a firm analytic footing, and still forms the basis of many noise reduction methods. In the field of speech processing a large number of techniques has been developed for noise reduction, and many of these are more generally applicable to noisy audio signals. We do not attempt here to describe every existing method in detail, since these are well covered in speech processing texts (see for example [Lim and Oppenheim, 1979, Lim, 1983, Boll, 1991]). We do, however, discuss some standard approaches which are appropriate for general audio signals and emerging techniques which are likely to be of use in future work. It is worth mentioning that where methods are derived from speech processing techniques, as in for example the spectral attenuation methods of section 4.5.1, sophisticated modifications to the basic schemes are required in order to match the stringent fidelity requirements and signal characteristics of an audio restoration system.

Certainly the most popular methods for noise reduction in audio signals to date are based upon short-time Fourier processing. These methods, which can be derived from non-stationary adaptations to the frequency-domain Wiener filter, are discussed fully in section 4.5.1.

Within a model-based framework, Lim and Oppenheim [Lim and Oppenheim, 1978] studied noise reduction using an autoregressive signal model, deriving iterative MAP and ML procedures. These methods are computationally intensive, although the signal estimation part of the iteration is shown to have a simple frequency-domain

Wiener filtering interpretation (see also [Paliwal and Basu, 1987, Koo et al., 1989] for Kalman filtering realizations of the signal estimation step). It is felt that new and more sophisticated model-based procedures may provide noise reducers which are competitive with the well-known short-time Fourier based methods. In particular, modern statistical methodology for solution of complex problems (for example, the Markov chain Monte-Carlo (MCMC) methods discussed in section 4.3.4 for click removal) allows for more realistic signal and noise modelling, including non-Gaussianity, non-linearity and non-stationarity. Such a framework can also be used to perform joint restoration of both clicks and random noise in one single process. A Bayesian approach to this joint problem using an autoregressive signal model is described in [Godsill, 1993][section 4.3.2] and [Godsill and Rayner, 1996a] and in [Godsill, 1997a, Godsill, 1997b] for the more general autoregressive moving average (ARMA) model,. In addition, [Niedźwiecki, 1994, Niedźwiecki and Cisowski, 1996] present an extended Kalman filter for joint removal of noise and clicks from AR- and ARMA-modelled audio signals.

Other methods which are emerging for noise reduction include the incorporation of psychoacoustical masking properties of human hearing [Canagarajah, 1991, Canagarajah, 1993, Tsoukalas et al., 1993] and noise reduction in alternative basis expansions, in particular the wavelet domain [Berger et al., 1994] and sub-space representations [Dendrinos et al., 1991, Ephraim and VanTrees, 1993, Ephraim and Van Trees, 1995]. These approaches address various short-comings of existing noise-reduction procedures, and could thus lead to improvements over existing techniques.

4.5.1 Background noise reduction by short-time spectral attenuation

This section deals with a class of techniques known as Short-Time Spectral Attenuation (STSA) [5]. STSA is a single input noise reduction method that basically consists in applying a time-varying attenuation to the short-time spectrum of the noisy signal. STSA techniques are non-parametric and generally need little knowledge of the signal to be processed. They rank among the most popular methods for speech enhancement and their use has been widely predominant for the restoration of musical recordings.

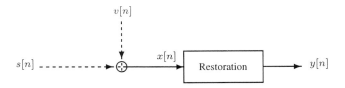

Figure 4.14   Modeled restoration process

General overview.

Hypotheses. Figure 4.14 shows the basic hypotheses common to all short-time spectral attenuation techniques. It is supposed that the original audio signal $s[n]$ has been corrupted by an additive noise signal $v[n]$ uncorrelated with $s[n]$ and that the only observable signal is the degraded signal $x[n]$ [Lim, 1983]. In the field of audio, restoration techniques applicable in such a situation are sometimes referred to as non-complementary [Dolby, 1967] or one-ended [Etter and Moschytz, 1994] to differentiate them from a class of frequently used denoising methods which rely on some pre-processing of the signal prior to the degradation (see [Dolby, 1967]).

The knowledge concerning the noise is usually limited to the facts that it can be considered as stationary and that it is possible to estimate its power spectral density (or quantities that are directly related to it) [Lim and Oppenheim, 1979, Lim, 1986].

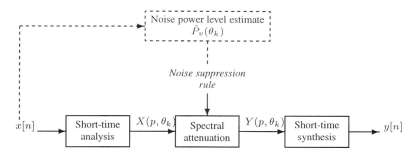

Figure 4.15   Background noise suppression by short-time spectral attenuation

Principle. Figure 4.15 shows the general framework of short-time spectral attenuation: the first step consists in analyzing the signal with a (in general, multirate) filter bank, each channel of the filter-bank is then attenuated (multiplied by a real positive gain, generally smaller than 1), and finally the sub-band signals are put back together to obtain the restored signal. The time-varying gain to be applied in each channel is determined by the so called noise suppression rule [McAulay and Malpass, 1980, Vary, 1985] which usually relies on an estimate of the noise power in each channel (represented by the dotted part of Figure 4.15). The two elements that really characterize a particular STSA technique are the filter-bank characteristics and the suppression rule.

In most STSA techniques the short-time analysis of the signal is performed by use of the Short-Time Fourier Transform (STFT) [Lim and Oppenheim, 1979, Boll, 1991, Ephraim and Malah, 1984, Moorer and Berger, 1986], or with a uniform filter-bank that can be implemented by STFT [Sondhi et al., 1981, Vary, 1985, Lagadec and Pelloni, 1983]. Note that in such cases the two interpretations (multirate filter-bank, and short-time Fourier transform) can be used interchangeably as they are fully equivalent [Crochiere and Rabiner, 1983]. Examples of STSA techniques based on the use of non-uniform filter banks can be found in [Petersen and Boll, 1981, McAulay and Malpass, 1980].

In designing the filter-bank, it is necessary to bear in mind the fact that the sub-band signals will sometimes be strongly modified by the attenuation procedure. As a consequence, while it is of course desirable to obtain a (nearly) perfect reconstruction in the absence of modification, it is also important to avoid effects such as sub-band spectral aliasing which could create distortions in the restored signal [Crochiere and Rabiner, 1983]. With the short-time Fourier transform, satisfying results are obtained with a sub-band sampling rate two or three times higher than the critical-sampling rate (ie. with a 50% to 66% overlap between successive short-time frames) [Cappé, 1991].

Historical considerations. Historically, short-time spectral attenuation was first developed for speech enhancement during the 1970s [Lim and Oppenheim, 1979, Boll, 1991, Sondhi et al., 1981]. The application of STSA to the restoration of audio recordings came afterwards [Lagadec and Pelloni, 1983, Moorer and Berger, 1986, Vaseghi, 1988, Vaseghi and Frayling-Cork, 1992, Valière, 1991, Etter and Moschytz, 1994] with techniques that were generally directly adapted from earlier speech-enhancement techniques.

Prior to works such as [Allen and Rabiner, 1977] and [Crochiere, 1980], there was not necessarily an agreement about the equivalence of the filter-bank and the STFT approaches (see also [Crochiere and Rabiner, 1983]). Traditionally, the filter-bank interpretation is more intuitive for audio engineers [Lagadec and Pelloni, 1983, Moorer and Berger, 1986, Etter and Moschytz, 1994] while the short-time spectrum is typically a speech analysis notion [Lim and Oppenheim, 1979]. Also controversial is the problem of short-time phase: in the STFT interpretation, the short-time attenuation corresponds to a magnitude-only modification of the short-time spectrum. The fact that only the magnitude of the short-time spectrum is processed has been given various interpretations, including an experimental assessment for speech signals in [Wang and Lim, 1982].

The most widespread opinion is that the phase need not be modified because of the properties of the human auditory system [Lim and Oppenheim, 1979]. Strictly speaking however, the assertion that the ear is "insensitive to the phase" was highlighted by psychoacoustic findings only in the case of stationary sounds and for the phase of the Fourier transform [Moore, 1997]. Moreover, it is well known that in the case of STFT, *phase variations* between successive short-time frames can give rise to audible effects (such as frequency modulation) [Vary, 1985].

It should however be emphasized that there is usually no choice but to keep the unmodified phase because of the lack of hypotheses concerning the unknown signal (recall that only the second order statistics of the signal and noise are supposed to be known). This is well known for techniques derived from Wiener filtering (time-domain minimum mean squared error filtering), and a similar result is proved in [Ephraim and Malah, 1984] for a frequency domain criterion (using strong hypotheses concerning the independence of the short-time transform bins). Although other criteria could be used, these results indicate that it may be be difficult to outperform the standard magnitude attenuation paradigm without introducing more elaborate hypotheses concerning the behavior of the signal.

Scope of the method. Until now, STSA techniques have been largely predominant in the field of speech enhancement and appear to have been used almost exclusively for the restoration of musical recordings.

One of the reasons for this wide application of STSA techniques is certainly the fact that they correspond to a non-parametric approach which can be applied to a large class of signals. By contrast, considering that most music recordings contain several simultaneous sound sources, it is unlikely that some of the methods relying on very specific knowledge of the speech signal properties (such as the model-based speech enhancement techniques [Boll, 1991, Ephraim, 1992]) could be generalized for audio restoration.

Another reason for the success of STSA techniques in the field of audio engineering is maybe the fact that they have a very intuitive interpretation: they extend to a large number of sub-bands the principle of well known analog devices used for signal enhancement, such as the noise gate [Moorer and Berger, 1986] (see also [Etter and Moschytz, 1994] for a link with compandors).

Suppression rules. Let $X(p, \theta_k)$ denote the short-time Fourier transform of $x[n]$, where p is the time index, and θ_k the normalized frequency index (θ_k lies between 0 and 1 and takes N discrete values for $k = 1, \ldots, N$, N being the number of sub-bands). Note that the time index p usually refers to a sampling rate lower than the initial signal sampling rate (for the STFT, the down-sampling factor is equal to hop-size between to consecutive short-time frames) [Crochiere and Rabiner, 1983].

The result of the noise suppression rule can always be interpreted as the application of a real gain $G(p, \theta_k)$ to each bin of the short-time transform $X(p, \theta_k)$ of the noisy signal. Usually, this gain corresponds to an 'attenuation', ie. lies between 0 and 1. For most suppression rules, $G(p, \theta_k)$ depends only on the power level of the noisy signal measured at the same bin $|X(p, \theta_k)|^2$ and on an estimate of the power of the noise at the frequency θ_k, $\hat{P}_v(\theta_k) = E\{|V(p, \theta_k)|^2\}$ (which does not depend on the time index p because of the noise stationarity). In the following, the ratio $Q(p, \theta_k) = |X(p, \theta_k)|^2 / \hat{P}_v(\theta_k)$ will be referred to as the relative signal level. Note that since the noise $v[n]$ is un-correlated with the unknown signal $s[n]$, we have

$$E\{Q(p,\theta_k)\} = 1 + \frac{E\{|S(p,\theta_k)|^2\}}{\hat{P}_v(\theta_k)} \qquad (4.18)$$

so that the expected value of the relative signal level is always larger than 1.

Standard examples of noise-suppression rules include the so-called Wiener[6] suppression rule, the power-subtraction (see Figure 4.16), the spectral subtraction [Boll, 1979, Lim and Oppenheim, 1979, McAulay and Malpass, 1980, Vary, 1985], as well as several families of parametric suppression curves [Lim and Oppenheim, 1979, Moorer and Berger, 1986, Etter and Moschytz, 1994].

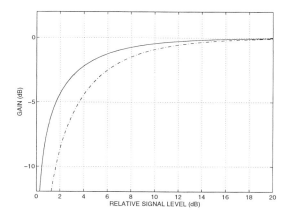

Figure 4.16 Suppression rules characteristics : gain (dB) versus relative signal level (dB). Solid line: Power subtraction; dashed line: Wiener.

All the suppression rules mentioned above share the same general behavior in that $G(p,\theta_k) = 1$ when the relative signal level is high ($Q(p,\theta_k) \gg 1$), and

$$\lim_{Q(p,\theta_k) \to 1} G(p,\theta_k) = 0$$

In many cases, the noise level $\hat{P}_v(\theta_k)$ is artificially over-estimated (multiplied by a factor $\beta > 1$) so that $G(p,\theta_k)$ is null as soon as $Q(p,\theta_k) \leq \beta$ [Lim and Oppenheim, 1979, Moorer and Berger, 1986].

Reference [Boll, 1991] presents a detailed review of suppression rules that are derived from a Bayesian point of view supposing a prior knowledge of the probability distribution of the sub-band signals. These suppression rules are more elaborate in the sense that they generally depend both on the relative signal level (or a quantity directly related to it) and on a characterization of the a priori information (a priori probability of speech presence in [McAulay and Malpass, 1980], a priori signal-to-noise ratio in [Ephraim and Malah, 1984]).

Finally, some suppression rules used for speech enhancement do not require any knowledge of the noise characteristics [Bunnell, 1990, Cheng and O'Shaughnessy, 1991]. These techniques, designed for improving speech intelligibility, can hardly be generalized to the case of audio recordings since they generate non-negligible distortions of the signal spectrum regardless of the noise level.

Evaluation.

'Deterministic' analysis. While it is rather difficult to analyze the results of STSA techniques in a general case, it becomes relatively simple when it is supposed that the unknown input signal is a pure tone, or more generally, a compound of several pure tones with frequencies sufficiently spaced apart. This hypothesis is pertinent since a large proportion of steady instrumental sounds can be efficiently described, both perceptively and analytically, as a sum of slowly modulated pure tones [Deutsch, 1982, Benade, 1976, Hall, 1980].

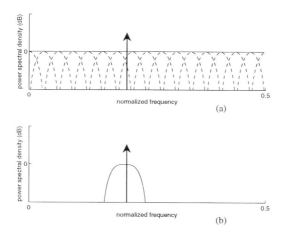

Figure 4.17 Restoration of a sinusoidal signal embedded in white noise (of power 0 dB). (a) The noisy signal (the dotted lines feature the filter bank characteristics); (b) The processed signal.

As pointed out in [Lagadec and Pelloni, 1983], short-time spectral attenuation does not reduce the noise present in the sub-bands that contain signal components.

Figure 4.17 shows an illustration of this fact for a sinusoidal signal embedded in white noise: if the level of the sinusoid is large enough, the channels in which it lies are left unattenuated while the other channels are strongly attenuated. As a consequence the output signal consists of the sinusoidal signal surrounded by a narrow band of filtered noise. Note also that if the sinusoid level is too low, all the channels are strongly attenuated and the signal is completely cancelled.

Cancelling of the signal. For a sinusoidal signal of frequency θ (which is supposed to correspond to the center frequency of one of the filters of the filter-bank), it is easily checked (assuming that the additional noise power spectral density is sufficiently smooth) that Eq. 4.18 becomes

$$E\{Q(p,\theta)\} = 1 + \frac{P_s}{S_v(\theta)W_\theta} \quad (4.19)$$

where P_s is the power of the sinusoid, $S_v(\theta)$ the power spectral density of the noise at frequency θ and W_θ is the bandwidth of the sub-band filter centered around frequency θ (see [Cappé and Laroche, 1995] for a demonstration in the STFT case).

As a consequence, *the level of the signal components that are mistakenly cancelled by the restoration process increases with the bandwidth of the analyzing filter-bank*. Although deceptively simple, this results nonetheless states that the signal enhancement can be made more efficient by sharpening the channel bandwidth as much as allowed by the stationarity hypothesis.

For the STFT case, the bandwidth of the filter-bank is inversely proportional to the duration of the short-time frame and it is shown in [Cappé and Laroche, 1995], using standard results concerning the simultaneous frequency masking phenomenon, that the processing can suppress audible signal components (ie. components that were not masked by the additive noise) if the short-time duration is well below 40 ms.

Audibility of the noise at the output. In the case where the signal component is not cancelled, the processed signal exhibits a band of filtered noise located around the sinusoidal component. It is clear that this phenomenon, if audible, is an important drawback of the method because it makes the remaining nuisance noise correlated with the signal, which was not the case for the original broad-band noise.

It is shown in [Cappé and Laroche, 1995] for the STFT case, that this effect is only perceptible when the frame duration is short (smaller than 20-30 ms)[7].

As for the results mentioned previously concerning signal cancellation, the obtained audibility limit should only be considered as an order of magnitude estimate in real situations since it does not take into account the possible mutual masking between different signal components (a phenomenon which may prevail when the noise level is very low) [Cappé and Laroche, 1995]. These results still support several earlier experimental findings [Moorer and Berger, 1986, Valière, 1991] concerning the influence of the STFT window duration in STSA techniques. In practice, the STFT frame duration should be sufficiently long to avoid creating undesirable modulation effects [Moorer and Berger, 1986] (audible band of noise remaining around signal components). Moreover, for audio signals, the duration of the short-time frame can generally be set advantageously to larger values than those used for speech [Valière, 1991] (because it lowers the limit of signal cancellation).

Transient signals.   The previous results are related to the case of steady portions of musical sounds, however it is well-known that musical recordings also feature many transient parts (note onsets, percussions) that play an important part in the subjective assessment of the signal characteristics [Hall, 1980, Deutsch, 1982].

As with many other techniques which make use of a short-time signal analyzer, it is possible to observe specific signal distortions, generated by the restoration process, which occur when transient signals are present [Johnston and Brandenburg, 1992]. In STSA techniques the distortion manifests itself as a smearing of the signal waveform for low-level signal transients. This phenomenon as well as its perceptive consequences are amplified as the short-time frame duration increases [Valière, 1991, Cappé and Laroche, 1995, Oppenheim and Lim, 1981].

The analysis of such transient effects is made more difficult by the fact that there is no 'prototype' transient signal as simple and as pertinent as the pure tone was for steady sounds. However, the results obtained in a simplistic case (the abrupt onset of a pure tone) seem to indicate that the observed smearing of the transient part of low level signals is mainly due to the modification of the signal spectrum by the suppression rule [Cappé and Laroche, 1995]. This is in contrast with what happens in applications where the magnitude of the short-time spectrum is not drastically modified, such as time-scaling with STFT, where the smearing of transient signals is mostly caused by the phase distortions [Griffin and Lim, 1984b, Oppenheim and Lim, 1981].

As a consequence, methods that exploit the redundancy of the magnitude of the short-time spectra to restore a 'correct' phase spectrum [Griffin and Lim, 1984b, Nawab and Quatieri, 1988b] are not efficient in eliminating the transient distortions caused by STSA.

Consequences of the random nature of the attenuation.   In the previous section we deliberately left apart a major problem: the fact that the attenuation is a random quantity. The randomness of the attenuation comes from the fact that it is (in general) determined as a function of the relative signal level which in turn involves the short-time transform of the noisy signal. This aspect plays a key role in STSA because the relative signal level is estimated by the periodogram (at least in the STFT case) characterized by a very high variance.

172 APPLICATIONS OF DSP TO AUDIO AND ACOUSTICS

A well known result states that the values of the discrete Fourier transform of a stationary random process are normally distributed complex variables when the length of the Fourier transform is large enough (compared to the decay rate of the noise correlation function) [Brillinger, 1981]. This asymptotic normal behavior leads to a Rayleigh distributed magnitude and a uniformly distributed phase (see [McAulay and Malpass, 1980, Ephraim and Malah, 1984] and [Papoulis, 1991]).

Using the normality assumption, it is shown in [Cappé and Laroche, 1995] that the probability density of the relative signal level Q (omitting the two indexes p and θ_k) is

$$f(Q) = e^{-[Q+(\bar{Q}-1)]} I_0\left(2\sqrt{Q(\bar{Q}-1)}\right) \tag{4.20}$$

where $I_o(x)$ denotes the modified Bessel function of order 0, and $\bar{Q}$ denotes the average value of the relative signal level as obtained from Eq. 4.18. The corresponding distributions are shown on figure 4.18 for 6 different average values of the relative signal level.

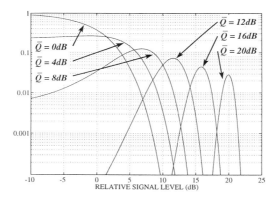

Figure 4.18 Probability density of the relative signal level for different mean values $\bar{Q}$ (from left to right: 0, 4, 8, 12, 16 and 20dB).

What is striking on figure 4.18 is the fact that even for signal components of non-negligible levels (such as $\bar{Q} = 8$dB), the relative signal levels can still take very low values (below 0dB). As a consequence, the use of STSA generates strong random variations of the low-level signal components [Cappé and Laroche, 1995]. Although systematic, these variations are not always heard in practice because they are often perceptively masked either by some other signal components (especially when the noise level is low) or by the fraction of broad band noise that remains after the processing.

The musical noise phenomenon.

What is musical noise?. The other important feature of figure 4.18 is that when only the noise is present (when $\bar{Q} = 1$), the observed relative signal level can still take high values. It is thus practically impossible to separate the noise from the low level signal components on the basis of the relative signal level. As a result, the total cancellation of the noise can only be obtained at the cost of some distortion of the low-level components.

In most STSA techniques, the noise that remains after the processing has a very unnatural disturbing quality, especially in a musical context [Moorer and Berger, 1986]. This phenomenon is generally referred to as musical noise [Ephraim and Malah, 1984] (also as 'musical tones' [Sondhi et al., 1981] or 'residual noise' [Etter and Moschytz, 1994, Vaseghi and Frayling-Cork, 1992]). The musical noise phenomenon is a direct consequence of the fact that the periodogram estimate used for evaluating the relative signal level yields values that are (asymptotically) uncorrelated even for neighboring bins [Brillinger, 1981]. This result, which holds for short-time transform bins belonging to the same analysis frame is complemented by the fact that bins from successive frames will also tends to be uncorrelated for frames which do not overlap in time (again, under the hypothesis of a sufficiently fast decay of the noise autocorrelation function).

Combining these two properties, it is easily seen that STSA transforms the original broad band noise into a signal composed of short-lived tones with randomly distributed frequencies. Moreover, with a 'standard' suppression rule (one that depends only on the relative signal level as measured in the current short-time frame) this phenomenon can only be eliminated by a crude overestimation of the noise level. Using the result of Eq. 4.20 in the case where $\bar{Q} = 1$, it is easily shown that the overestimation needed to make the probability of appearance of musical noise negligible (below 0.1%) is about 9 dB [Cappé, 1991].

Solutions to the musical noise problem. Various empirical modifications of the basic approach have been proposed to overcome this problem. A first possibility consists of taking advantage of the musical noise characteristics: more precisely, the short duration of the musical noise components (typically a few short-time frames) [Boll, 1979, Vaseghi and Frayling-Cork, 1992] and the fact that the appearance of musical noise in one sub-band is independent of that in other sub-bands [Sondhi et al., 1981]. The main shortcoming of this type of approach is that, since they are based on average statistical properties, the musical noise is reduced (ie. its appearance is made less frequent) but not completely eliminated.

Another possibility is to use a smoothed estimate of the relative signal level. Time-smoothing has been considered in [Boll, 1979] and [Etter and Moschytz, 1994], but frequency smoothing can also be used [Canagarajah, 1993, Cappé, 1991]. Limitations of this smoothing approach include the fact that it can generate signal distortion, particularly during transients, when time-smoothing is used. A more elaborate version of the time-smoothing approach aimed at reducing signal distortion is described in [Erwood and Xydeas, 1990].

Finally, an alternative approach consists in concealing the musical noise artifact behind a sufficient level of remaining noise [Moorer and Berger, 1986, Canagarajah, 1993]. One simple way to proceed consists in constraining the computed gain to lie above a preset threshold (which is achieved by the 'noise floor' introduced by Berouti et al. [Berouti et al., 1979]).

The Ephraim and Malah suppression rule. Besides these procedures specifically designed to counter the musical noise artifact, it has been noted that the suppression rules proposed by Ephraim and Malah [Ephraim and Malah, 1983, Ephraim and Malah, 1984, Ephraim and Malah, 1985] do not generate musical noise [Ephraim and Malah, 1984, Valière, 1991, Cappé, 1994]. This is shown in [Cappé, 1994] to be a consequence of the predominance of the time-smoothed signal level (the so called 'a priori signal to noise ratio') over the usual 'instantaneous' relative signal level.

A nice feature of the Ephraim and Malah suppression rule is the 'intelligent' time-smoothing of the relative signal level resulting from the use of an explicit statistical model of the sub-band noise: a strong smoothing when the level is sufficiently low to be compatible with the hypothesis that only the noise is present, and no smoothing otherwise [Cappé, 1994]. Surprisingly, this behavior of the Ephraim and Malah suppression rule is related to the principle adopted in [Erwood and Xydeas, 1990] (which consists in varying the horizon of the time-smoothing depending on the signal level). The Ephraim and Malah suppression rule therefore allows a very 'natural' means (not based on fixed thresholds) of reducing the musical noise artifact without introducing penalizing signal distortions.

When using the Ephraim and Malah suppression rule, it appears that it is still useful to limit the attenuation in order to avoid the reappearance of the musical noise phenomenon at low-levels. In practice, the average attenuation applied to the noisy part can be easily controlled via one of the parameters of the method (see [Cappé, 1994]) in the range from 0dB to approximately -15dB (with lower values the musical noise effect can be audible in some cases). An interesting generalization consists in specifying a frequency dependent average noise reduction in order to take into account the fact that all regions of the noise spectrum do not contribute equally to the loudness sensation [Moore, 1989, Zwicker and Zwicker, 1991].

Current trends and perspectives.

Improving the noise characterization. In many real life applications, the hypothesis that the noise is stationary is unrealistic and it is necessary to track the time-variations of the noise characteristics. For audio restoration, it seems that this aspect can play an important part in the case of old analog disk recordings. Indeed, the noise present on such recordings sounds 'less regular' than the tape hiss heard on analog tapes. It is also common to observe a discrepancy between the noise characteristics measured before and after the recorded part [Cappé, 1991].

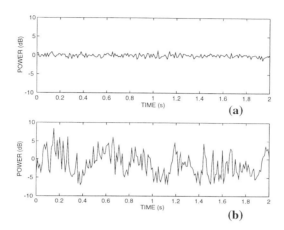

Figure 4.19 Short-time power variations. (a) for a standard analog cassette; (b) for a 78 rpm recording. The signal power is estimated at a 10ms rate and normalized by its average value.

An example of such a behavior can be seen on figure 4.19 which displays the time variations of the short-time power[8] for two noises measured on different audio recording: on a standard analog cassette for part (a), on a 78 rpm record for part (b). The sharp spikes seen on part (b) of figure 4.19 are simply due to the presence of impulsive degradations in the disk noise, which of course is not the case for the tape noise. However, the comparison between the two parts of figure 4.19 shows that the range of the power variations is much more important for the analog disk noise (part [b]) than for the tape noise (part [a]).

It is also interesting to note that the long-term power variations of the disk noise (part [b] of figure 4.19) seem to be related to the disk rotation period (0.77s for a 78 rpm record). This result indicates that the noise present on this particular analog disk is certainly not stationary, but that it could be cyclostationary [Gardner, 1994]. More elaborate tests would be needed to determine if this noise is indeed cyclostationary, and what type of cyclostationarity is actually involved (the simplest model would be an amplitude modulated stationary process) [Gerr and Allen, 1994].

In practice, it is important to emphasize that the various procedures that have been proposed for updating the estimate of the noise characteristics in the context of speech enhancement [Boll, 1979, McAulay and Malpass, 1980, Sondhi et al., 1981, Erwood and Xydeas, 1990] are usually not applicable for audio signals: they rely on the presence of signal pauses that are frequent in natural speech, but not necessarily in musical recordings. The development of noise tracking procedures that are suited for an application to audio signals thus necessitates a more precise knowledge of the noise characteristics in cases where it cannot be considered stationary.

Use of perceptual noise-reduction criteria. Recently, efforts have been devoted to the development of noise suppression strategies based on perceptual criteria [Canagarajah, 1991, Canagarajah, 1993, Tsoukalas et al., 1993, Mourjopoulos et al., 1992]. As of today, the proposed techniques only make use of data concerning the simultaneous masking effect in order to determine the frequency regions where the noise is most audible. A surprising side effect of these techniques is that they notably reduce the musical noise phenomenon [Canagarajah, 1993]. This feature can be attributed to the strong smoothing of the frequency data in the upper frequency range performed in these techniques to simulate the ear's frequency integration properties.

Clearly more work needs to be done to take advantage of other known properties of the auditory system in the context of noise reduction. Interesting clues include the consideration of non-simultaneous masking effects that may be helpful in reducing transient distortions, as well as absolute thresholds of hearing. A troublesome point associated with the use of such perceptual criteria is that they require the knowledge of the listening acoustic intensity [Moore, 1989]. For most applications this requirement cannot be satisfied so that only a worst-case analysis is feasible. However, in cases where the noise reduction is performed directly at the playback level, the adaptation of the noise suppression rule to the effective acoustic intensity of the audio signal is certainly a promising aspect.

Improving the properties of the short-time transform. Another interesting area of research deals with the design of the short-time transform. It is striking to see that while many efforts have been dedicated to the study of advanced suppression rules, little has been done concerning the analysis part of the noise reduction-system.

The first approach that need to be more precisely evaluated is the use of non-uniform filter banks [Petersen and Boll, 1981, Valière, 1991], especially if they are applied in connection with perceptual criteria. Indeed, non-uniform filter banks allow a frequency dependent specification of the time-resolution/bandwidth compromise which could be adapted to the known features of our hearing system. The results of section 4.16 show that a high frequency-resolution is needed anyway, at least in the lower part of the spectrum, to ensure a sufficient separation of sinusoidal signal components from the noise.

A complementary approach is based on the observation that the use of a fixed analysis scheme may be too constraining, which leads to the design of analysis/synthesis structures that are adapted to the local characteristics of the signal. For speech enhancement, various recent works report the successful use of subspace representations in place of the STFT [Dendrinos et al., 1991, Ephraim and VanTrees, 1993, Ephraim and Van Trees, 1995, Jensen et al., 1995]. The subspace representation is still frame-based but it is characterized by a high frequency-resolution (see [Dendrinos et al., 1991, Ephraim and Van Trees, 1995] for the link with damped sinusoidal models). It has however been shown, using stationarity assumptions, that subspace approaches are asymptotically equivalent to STSA techniques for large frame durations [Ephraim and Van Trees, 1995]. For audio restoration, it can thus be expected that both type of methods will yield comparable results. The Adapted Waveform Analysis method described in [Berger et al., 1994] presents a different approach based on a wavelet decomposition of the signal. This promising method basically operates by determining a basis of wavelets [Kronland-Martinet et al., 1987] which is best adapted to the characteristics of the signal.

4.5.2 Discussion

A number of noise reduction methods have been described, with particular emphasis on the short-term spectral methods which have proved the most robust and effective to date. However, it is anticipated that new methodology and rapid increases in readily-available computational power will lead in the future to the use of more sophisticated methods based on realistic signal modelling assumptions and perceptual optimality criteria.

4.6 PITCH VARIATION DEFECTS

A form of degradation commonly encountered in disc, magnetic tape and film sound recordings is an overall pitch variation not present in the original performance. The terms 'wow' and 'flutter' are often used in this context and are somewhat interchangeable, although wow tends to refer to variations over longer time-scales than flutter, which often means a very fast pitch variation sounding like a tremolo effect. This section addresses chiefly the longer term defects, such as those connected variations in gramophone turntable speeds, which we will refer to as wow, although similar principles could be applied to short-term defects.

There are several mechanisms by which wow can occur. One cause is a variation of rotational speed of the recording medium during either recording or playback. A further cause is eccentricity in the recording or playback process for disc and cylinder recordings, for example a hole which is not punched perfectly at the centre of a gramophone disc. Lastly it is possible for magnetic tape and optical film to become unevenly stretched during playback or storage; this too leads to pitch variation in playback. Accounts of wow are given in [Axon and Davies, 1949, Furst, 1946].

In some cases it may be possible to make a physical correction for this defect, as the case of a gramophone disk whose hole is not punched centrally. In most cases, however, no such correction is possible, and signal processing techniques must be used. The only approach currently known to the authors is that of Godsill and Rayner [Godsill, 1993, Godsill and Rayner, 1993b, Godsill, 1994], which is described here.

The physical mechanism by which wow is produced is equivalent to a non-uniform warping of the time axis. If the undistorted time-domain waveform of the gramophone signal is written as $s(t)$ and the time axis is warped by a monotonically increasing function $f_w(t)$ then the distorted signal is given by:

$$x(t) = s(f_w(t)) \tag{4.21}$$

If the time warping function $f_w()$ is known then it is possible to regenerate the undistorted waveform $s(t)$ as

$$s(t) = x(f_w^{-1}(t)). \tag{4.22}$$

A wow restoration system is thus primarily concerned with estimation of the time warping function or equivalently the pitch variation function $p_w(t) = f_w'(t)$. In the discrete signal domain we have discrete observations $x[n] = x(nT)$, where T is the sampling period. If the pitch variation function corresponding to each sampling instant, denoted by $p_w[n]$, is known then it is possible to estimate the undistorted signal using digital resampling operations.

If we have good statistical models for the undistorted audio signal and the process which generates the wow, it may then be possible to estimate the pitch variation $p_w[n]$ from the wow-degraded data $x[n]$. Any of the standard models used for audio signals (see section 4.2) are possible, at least in principle. However, the chosen model must be capable of capturing accurately the pitch variations of the data over the long time-scales necessary for identification of wow. One suitable option is the generalized harmonic model (see section 4.2, equation (4.2)). This represents tonal components in the signal as sinusoids, allowing for a simple interpretation of the wow as a frequency modulation which is common to all components present at a particular time.

Consider a fixed-frequency sinusoidal component $s_i(t) = sin(\omega_{0i} t + \phi_{0i})$ from a musical signal, distorted by a pitch variation function $p_w(t)$. The pitch-distorted component $x_i(t)$ can be written as (see [Godsill, 1993]):

$$\begin{aligned} x_i(t) &= s_i(f_w(t)) \\ &= \sin\left(\omega_{0i} \int_0^t p_w(t) dt + \phi_{0i}\right), \end{aligned} \tag{4.23}$$

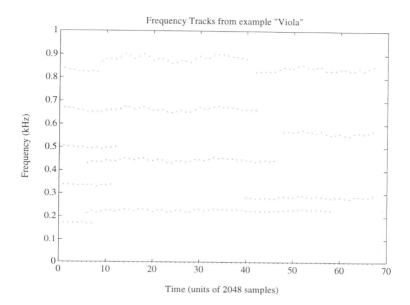

Figure 4.20 Frequency tracks generated for example 'Viola'. (Reprinted with permission from [Godsill, 1994], ©1994, IEEE)

which is a frequency-modulated sine-wave with instantaneous frequency $\omega_{0i} p_w(t)$. The same multiplicative modulation factor $p_w(t)$ will be applied to all frequency components present at one time. Hence we might estimate $p_w[n]$ as that frequency modulation which is common to all sinusoidal components in the music. This principle is the basis of the frequency domain estimation algorithm now described.

4.6.1 Frequency domain estimation

In this procedure it is assumed that musical signals are made up as additive combinations of tones (sinusoids) which represent the fundamental and harmonics of all the musical notes which are playing. Since this is certainly not the case for most non-musical signals, we might expect the method to fail for, say, speech extracts or acoustical noises. Fortunately, it is for musical extracts that pitch variation defects are most critical. The pitch variation process is modelled as a smoothly varying waveform with no sharp discontinuities, which is reasonable for most wow generation mechanisms.

The method proceeds in three stages. The first stage involves estimation of the tonal components using a DFT magnitude-based peak tracking algorithm closely related to

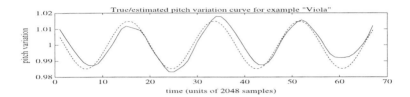

Figure 4.21 Estimated (full line) and true (dotted line) pitch variation curves generated for example 'Viola'. (Reprinted with permission from [Godsill, 1994], ©1994, IEEE)

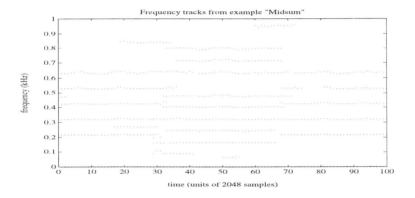

Figure 4.22 Frequency tracks generated for example 'Midsum'. (Reprinted with permission from [Godsill, 1994], ©1994, IEEE)

that described in [McAulay and Quatieri, 1986b] and chapter 9. This pre-processing stage, allowing for individual note starts and finishes, provides a set of time-frequency 'tracks' (see figures 4.20 and 4.22), from which the overall pitch variation is estimated. It is assumed in this procedure that any genuine tonal components in the corrupted signal will have roughly constant frequency for the duration of each DFT block.

The second stage of processing involves extracting smooth pitch variation information from the time-frequency tracks. For the nth block of data there will be P_n frequency estimates corresponding to the P_n tonal components which were being

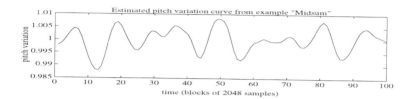

Figure 4.23 Pitch variation curve generated for example 'Midsum'. (Reprinted with permission from [Godsill, 1994], ©1994, IEEE)

tracked at that time. The ith tonal component has a nominal centre frequency f_{0i}, which is assumed to remain fixed over the period of interest, and a measured frequency $f_i[n]$. Variations in $f_i[n]$ are attributed to the pitch variation value $p_w[n]$ and a noise component $v_i[n]$. This noise component is composed both of inaccuracies in the frequency tracking stage and genuine 'performance' pitch deviations (such as vibrato or tremolo) in tonal components. Smooth pitch variations which are common to all tones present may then be attributed to the wow degradation, while other variations (non-smooth or not common to all tones) are rejected as noise, $v_i[n]$. The approach could of course fail during non-tonal ('unvoiced') passages or if note 'slides' dominate the spectrum, and future work might aim to make the whole procedure more robust to this possibility.

Each frequency track has a 'birth' and 'death' index b_i and d_i such that b_i denotes the first DFT block at which f_{0i} is present ('active') and d_i the last (each track is then continuously 'active' between these indices). Frequencies are expressed on a log-frequency scale, as this leads to linear estimates of the pitch curve (see [Godsill, 1993] for comparison with a linear-frequency scale formulation). The model equation for the measured log-frequency tracks $f_{il}[n]$ is then:

$$f_{il}[n] = \left\{ \begin{array}{ll} f_{0il} + p_{wl}[n] + v_{il}[n], & b_i \leq n \leq d_i \\ 0, & \text{otherwise} \end{array} \right\}, \quad 1 \leq i \leq P_{max}, \qquad (4.24)$$

where subscript 'l' denotes the logarithm of the frequency quantities previously defined. P_{max} is the total number of tonal components tracked in the interval of N data blocks. At block n there are P_n active tracks, and the length of the ith track is then given by $N_i = d_i - b_i + 1$.

If the noise terms $v_{il}[n]$ are assumed i.i.d. Gaussian, the likelihood function for the unknown centre frequencies and pitch variation values can be obtained. A singular system of equations results if the Maximum likelihood (ML) (or equivalently least squares) solution is attempted. The solution is regularized by incorporation of the prior information that the pitch variation is a 'smooth' process, through a Bayesian prior probability framework. A second difference-based Gaussian smoothness prior is used, which leads to a linear MAP estimator for the unknowns (see [Godsill, 1993, Godsill, 1994] for full details). The estimate is dependent upon a regularizing parameter which expresses the degree of second difference smoothness expected from the pitch variation process. In [Godsill, 1993, Godsill, 1994] this parameter is determined experimentally from the visual smoothness of results estimated from a small sample of data, but other more rigorous means are available for estimation of such 'hyperparameters' given the computational power (see, e.g. [Rajan, 1994, MacKay, 1992]). Examples of pitch variation curves estimated from synthetic and real pitch degradation are shown in figures 4.21 and 4.23, respectively.

The estimation of pitch variation allows the final re-sampling operation to proceed. Equation (4.22) shows that, in principle, perfect reconstruction of the undegraded signal is possible in the continuous time case, provided the time warping function is known. In the discrete domain the degraded signal $x[n]$ is considered to be a non-uniform re-sampling of the undegraded signal $s[n]$, with sampling instants given by the time-warping function $f_w[n]$. Note, however, that the pitch varies very slowly relative to the sampling rate. Thus, at any given time instant it is possible to approximate the non-uniformly sampled input signal as a uniformly sampled signal with sample rate $1/T' = p_w[n]/T$. The problem is then simplified to one of *sample rate conversion* for which there are well-known techniques (see e.g. [Crochiere and Rabiner, 1983, Rabiner, 1982]). Any re-sampling or interpolation technique which can adjust its sample rate continuously is suitable, and a truncated 'sinc' interpolation is proposed in [Godsill, 1993, Godsill and Rayner, 1993b, Godsill, 1994].

Summary. Informal listening tests indicate that the frequency-based method is capable of a very high quality of restoration in musical extracts which have a strong tonal character. The procedure is, however, sensitive to the quality of frequency tracking and to the constant-frequency harmonic model assumed in pitch estimation. New work in the area might attempt to unify pitch variation estimation and frequency tracking into a single operation, and introduce more robust modelling of musical harmonics.

4.7 REDUCTION OF NON-LINEAR AMPLITUDE DISTORTION

Many examples exist of audio recordings which are subject to non-linear amplitude distortion. Distortion can be caused by a number of different mechanisms such as deficiencies in the original recording system and degradation of the recording through excessive use or poor storage. This section formulates the reduction of non-linear amplitude distortion as a non-linear time series identification and inverse filtering problem. Models for the signal production and distortion process are proposed and techniques for estimating the model parameters are outlined. The section concludes with examples of the distortion reduction process.

An audio recording may be subject to various forms of non-linear distortion, some of which are listed below :

1. Non-linearity in amplifiers or other parts of the system gives rise to intermodulation distortion [Sinclair, 1989].

2. Cross-over distortion in Class B amplifiers [Sinclair, 1989].

3. Tape saturation due to over recording [Sinclair, 1989]: recording at too high a level on to magnetic tape leads to clipping or severe amplitude compression of a signal.

4. Tracing distortion in gramophone recordings [Roys, 1978]: the result of the playback stylus tracing a different path from the recording stylus. This can occur if the playback stylus has an incorrect tip radius.

5. Deformation of grooves in gramophone recordings [Roys, 1978]: the action of the stylus on the record groove can result in both elastic and plastic deformation of the record surface. Elastic deformation is a form of distortion affecting both new and used records; plastic deformation, or record wear, leads to a gradual degradation of the reproduced audio signal.

The approach to distortion reduction is to model the various possible forms of distortion by a non-linear system. Rather than be concerned with the actual mechanics of the distortion process, a structure of non-linear model is chosen which is thought to be flexible enough to simulate the different types of possible distortion.

4.7.1 Distortion Modelling

A general model for the distortion process is shown in figure 4.24 where the input to the nonlinear system is the undistorted audio signal $s[n]$ and the output is the observed distorted signal $x[n]$.

The general problem of distortion reduction is that of identifying the non-linear system and then applying the inverse of the non-linearity to the distorted signal $x[n]$ in order to recover the undistorted signal $s[n]$. Identification of the non-linear system takes two main forms depending on the circumstances. The first is when the physical system which caused the distortion is available for measurement. For example the recording system which produced a distorted recording may be available. Under these circumstances it is possible to apply a known input signal to the system and apply

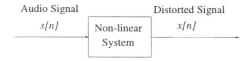

Figure 4.24 Model of the distortion process

system identification techniques in order to determine the non-linear transfer function or apply adaptive techniques to recover the undistorted signal [Preis and Polchlopek, 1984, Schafer et al., 1981, Landau, 1960, Landau and Miranker, 1961]. The second, and much more common, situation is when the only information available is the distorted signal itself. The approach is now to postulate a model for both the undistorted signal and the distortion process. Time series identification techniques must then be used to determine values for the model parameters. This section will concentrate on this situation which might be called *blind identification*.

Choice of a suitable non-linear model to represent the signal and distortion processes is not a straightforward decision since there are many different classes from which to choose.

4.7.2 Non-linear Signal Models

A non-linear time series model transforms an observed signal $x[t]$ into a white noise process $e[t]$, and may be written in discrete form [Priestley, 1988] as:

$$e[t] = F'\{\ldots, x[t-2], x[t-1], x[t], x[t+1], x[t+2], \ldots\}$$

where $F'\{.\}$ is some non-linear function.

Assuming that $F'\{.\}$ is an invertible function this may be expressed as:

$$x[t] = F\{\ldots, e[t-2], e[t-1], e[t], e[t+1], e[t+2], \ldots\} \qquad (4.25)$$

This functional relationship may be expressed in a number of different forms; two of which will be briefly considered.

The Volterra Series. For a time invariant system defined by equation 4.25, it is possible to form a Taylor series expansion of the non-linear function to give [Priestley, 1988]:

$$x[t] = k_0 + \sum_{i_1=-\infty}^{\infty} h_{i_1} e[t-i_1] + \sum_{i_1=-\infty}^{\infty} \sum_{i_2=-\infty}^{\infty} h_{i_1,i_2} e[t-i_1] e[t-i_2] +$$

$$\sum_{i_1=-\infty}^{\infty} \sum_{i_2=-\infty}^{\infty} \sum_{i_3=-\infty}^{\infty} h_{i_1,i_2,i_3} e[t-i_1] e[t-i_2] e[t-i_3] + \cdots \quad (4.26)$$

where the coefficients $k_0, h_{i_1}, h_{i_1,i_2}, \ldots$ are the partial derivatives of the operator F. Note that the summation involving h_{i_1} in the discrete Volterra series corresponds to the normal convolution relationship for a linear system with impulse response $h_{i_1}(n)$. The Volterra Series is a very general class of non-linear model which is capable of modelling a broad spectrum of physical systems. The generality of the model, while making it very versatile, is also its main disadvantage : for successful modelling of an actual system, a very large order of Volterra expansion is often needed, a task which is generally not practical. In view of this, it becomes necessary to consider other representations of non-linear time series.

NARMA Modelling. The NARMA (Non-Linear AutoRegressive Moving Average) model was introduced by Leontaritis and Billings [Leontaritis and Billings, 1985] and defined by:

$$x[n] = f\{x[n-1], \ldots x[n-P_x], e[n-1], \ldots e[n-P_e]\} + e[n]$$

Combining the terms $x[n-1], \ldots x[n-P_x]$ and $e[n-1], \ldots e[n-P_e]$ into a single vector $\mathbf{w}(n)$ and expanding as a Taylor series gives the following representation of a non-linear system [Chen and Billings, 1989].

$$x[n] = a_0 + \sum_{i_1=1}^{P_x+P_e} a_{i_1} w_{i_1}(n) + \sum_{i_1=1}^{P_x+P_e} \sum_{i_2=i_1}^{P_x+P_e} a_{i_1 i_2} w_{i_1}(n) w_{i_2}(n) + \ldots$$

$$\sum_{i_1=1}^{P_x+P_e} \cdots \sum_{i_l=i_{l-1}}^{P_x+P_e} a_{i_1 \ldots i_l} w_{i_1}(n) \ldots w_{i_l}(n) + s(n) \quad (4.27)$$

where:

$$\mathbf{w}(n) = \begin{bmatrix} x[n-1] \\ \vdots \\ x[n-p_x] \\ e[n-1] \\ \vdots \\ e[n-p_s] \end{bmatrix}$$

186 APPLICATIONS OF DSP TO AUDIO AND ACOUSTICS

The advantage of such an expansion is that the model is linear in the unknown parameters a so that many of the linear model identification techniques can also be applied to the above non-linear model. Iterative methods of obtaining the parameter estimates for a given model structure have been developed [Billings and Voon, 1986]. A number of other non-linear signal models are discussed by Priestley [Priestley, 1988] and Tong [Tong, 1990].

4.7.3 Application of Non-linear models to Distortion Reduction

The general Volterra and NARMA models suffer from two problems from the point of view of distortion correction. They are unnecessarily complex and even after identifying the parameters of the model it is still necessary to recover the undistorted signal by some means. In section 4.2 it was noted that audio signals are well-represented by the autoregressive (AR) model defined by equation 4.1:

$$s[n] = \sum_{i=1}^{P} s[n-i]a_i + e[n].$$

Thus a distorted signal may be represented as a linear AR model followed by a non-linear system as shown in figure 4.25.

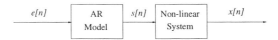

Figure 4.25 Model of the signal and distortion process

Two particular models will be considered for the non-linear system.

Memoryless Non-linearity.   A special case of the Volterra system given by equation 4.26 is:

$$x[n] = h_0 s[n] + h_{00} s^2[n] + \cdots h_{0\ldots 0} s^q[n] + \cdots$$

This is termed a memoryless non-linearity since the output is a function of only the present value of the input $s[n]$. The expression may be regarded as a power series expansion of the non-linear input-output relationship of the non-linearity. In fact this representation is awkward from an analytical point of view and it is more convenient to work in terms of the inverse function. Conditions for invertibility are discussed in Mercer [Mercer, 1993].

$$s[n] = k_1 x[n] + k_2 x^2[n] + \cdots k_q x^q[n] + \cdots$$

$$= \sum_{q=0}^{\infty} k_q x^q[n]$$

An infinite order model is clearly impractical to implement. Hence it is necessary to truncate the series :

$$s[n] = \sum_{q=0}^{Q} k_q x^q[n] \qquad (4.28)$$

A reasonable assumption is that there is negligible distortion for low-level signals, ie $\{x[n] = s[n];$ for $s[n] \approx 0\}$ so that $k_0 = 1$. (Note that this assumption would not be valid for crossover distortion). This model will be referred to in general as the Autoregressive-Memoryless Non-linearity (AR-MNL) model and as the AR(P)-MNL(Q) to denote a AR model of order P and a memoryless non-linearity of order Q.

Note that if the non-linear parameters k_i can be identified then the undistorted signal $\{s[n]\}$ can be recovered from the distorted signal $\{x[n]\}$ by means of equation 4.28.

Non-linearity with Memory. The AR-MNL model is clearly somewhat restrictive in that most distortion mechanisms will involve memory. For example an amplifier with a non-linear output stage will probably have feedback so that the memoryless non-linearity will be included within a feedback loop and the overall system could not be modelled as a memoryless non-linearity. The general NARMA model incorporates memory but its use imposes a number of analytical problems. A special case of the NARMA model is the NAR (Non-linear AutoRegressive) model in which the current output $x[n]$ is a non-linear function of only past values of output and the present input $s[n]$. Under these conditions equation 4.27 becomes:

$$x[n] = + \sum_{i_1=1}^{P_x} \sum_{i_2=i_1}^{P_x} a_{i_1 i_2} x[n-i_1] x[n-i_2] + \ldots$$

$$\sum_{i_1=1}^{P_x} \cdots \sum_{i_l=i_{l-1}}^{P_x} a_{i_1 \ldots i_l} x[n-i_1] \ldots x[n-i_l] + s[n] \qquad (4.29)$$

The linear terms in $x[n-i_1]$ have not been included since they are represented by the linear terms in the AR model. This model will be referred to as Autoregressive Non-linear Autoregressive (AR-NAR) model in general and as AR(P)-NAR(Q) model in which the AR section has order P and only Q of the non-linear terms from equation 4.29 are included. Note that the undistorted signal $\{s[n]\}$ can be recovered from the distorted signal $\{x[n]\}$ by use of equation 4.29 provided that the parameter values can be identified.

4.7.4 Parameter Estimation

In order to recover the undistorted signal it is necessary to estimate the parameter values in equations 4.28 and 4.29. A general description of parameter estimation is given in many texts, e.g. Norton [Norton, 1988, Kay, 1993].

One powerful technique is Maximum Likelihood Estimation (MLE) which requires the derivation of the Joint Conditional Probability Density Function (PDF) of the output sequence $\{x[n]\}$, conditional on the model parameters. The input $\{e[n]\}$ to the system shown in figure 4.25 is assumed to be a white Gaussian noise (WGN) process with zero mean and a variance of σ^2. The probability density of the noise input is:

$$p(e[n]) = \frac{1}{\sqrt{2\pi}\sigma} \exp\left\{-\frac{e^2[n]}{2\sigma^2}\right\}$$

Since $\{e[n]\}$ is a WGN process, samples of the process are independent and the joint probability for a sequence of data $(\{e[n]\}, n = P+1 \text{ to } N)$ is given by :

$$p(e[P+1],\ldots e[N]) = \left[\frac{1}{\sqrt{2\pi}\sigma}\right]^{N-P} \exp\left\{-\frac{1}{2\sigma^2}\sum_{n=P+1}^{N}e^2[n]\right\} \quad (4.30)$$

The terms $\{e[1], e[2]...e[P]\}$ are not included because they cannot be calculated in terms of the observed output $\{x[n]\}$ so that, strictly speaking, the above is a conditional probability but there is little error if the number of observations $N \gg P$.

An expression for the Joint Probability Density Function for the observations $\{x[n]\}$ may be determined by transformations from $\{e[n]\}$ to $\{s[n]\}$ and from $\{s[n]\}$ to $\{x[n]\}$. This gives the likelihood function for the AR-MNL system as :

$$p(x[P+1], x[P+2], \ldots x[N]|\mathbf{a},\mathbf{k},\sigma) =$$

$$\left[\frac{1}{\sqrt{2\pi}\sigma}\right]^{N-P} \left|\prod_{n=P+1}^{N}\{1 + \sum_{q=2}^{Q} qk_q x^{q-1}[n]\}\right| \exp\left\{-\frac{1}{2\sigma^2}\sum_{n=P+1}^{N}e^2[n]\right\} \quad (4.31)$$

where **a** is a vector containing the parameters $a_1 \ldots a_P$ of the AR model and **k** is a vector containing the parameters $k_0 \ldots k_Q$. The noise sequence $\{e[n]\}$ may be expressed in terms of the observed distorted signal $\{x[n]\}$ using equations 4.1 and 4.28

The Likelihood function for the AR-NAR system is:

$$p[x[P+P_x+1], \ldots x[N]|\mathbf{a}, \mathbf{k}, \sigma] = \left[\frac{1}{\sqrt{2\pi}\sigma}\right]^{N-P-P_x} \exp\left\{-\frac{1}{2\sigma^2} \sum_{n=P+P_x+1}^{N} e^2[n]\right\}$$

where **a** is the vector of AR parameters and **k** is a vector containing the parameters a_{i_1}, $a_{i_1 i_2}, \ldots$ of the NAR model. The noise sequence $\{e[n]\}$ may be expressed in terms of the observed distorted signal $\{x[n]\}$ using equations 4.1 and 4.29

The MLE approach involves maximising the Likelihood function with respect to **a**, **k** and σ. The values of **a**, **k** and σ which maximise this equation are the Maximum Likelihood estimates of the model.

Computational aspects. In general there is no analytic solution to maximising the Likelihood equations so that it is necessary to perform a multidimensional optimisation over the unknown model parameters. However before performing the optimisation it is necessary to select a model of appropriate order; too low an order results in a poor system which is unable to correct distortion, too high an order results in an unnecessarily complicated model which imposes a heavy computational burden in determining the optimal parameter values. Model order selection for the memoryless non-linearity is simply a matter of choosing the order of the polynomial expansion in equation 4.28. However the problem is more complex with the NAR model, equation 4.29, since the number of permutations of terms can be extremely large. There is no intuitive means for estimating which non-linear terms should be included and it is necessary to perform the Maximum Likelihood optimisation for each combination of terms in order to find an acceptable system. Such a global search over even a relatively limited subset of the possible model terms is prohibitively expensive and iterative methods have been developed to search the space of model functions to determine an acceptable, although not necessarily optimal, system [Mercer, 1993].

In order to compare the performance of models containing different non-linear terms it is necessary to use a criterion which achieves a compromise between the overly simple model and the overly complex model. One such criterion is the Akaike Information Criterion, AIC(ϕ) (see e.g. Akaike [Akaike, 1974]) given by:

$$AIC(\phi) = -2\log_e \{\text{Maximised Likelihood Function}\} + \phi \times [\text{Number of Parameters}] \quad (4.32)$$

The Akaike Information Criterion is used to select the model which minimises the AIC(ϕ) function for a specified value of ϕ. In the original formulation of the above equation, Akaike used a value of $\phi = 2$ but an alternative selection criterion proposed by Leontaritis and Billings [Leontaritis and Billings, 1987] is based on a value of $\phi = 4$.

4.7.5 Examples

Mercer [Mercer, 1993] presents results for the two models discussed. For the memoryless non-linearity a section of music from a recording of a brass band was passed through the non-linearity defined by:

$$\mathbf{k} = [\ 0.00 \quad 0.30 \quad 0.00 \quad 0.50\ ].$$

An AR model of order 25 was assumed and a non-linearity with $Q \leq 9$ was allowed. Figure 4.26 shows a section of the original, distorted and restored signals.

In order to test the AR-NAR model a section of music was passed through the non-linear system:

$$\begin{aligned} x[n] &= 0.07x[n-1]x[n-4]x[n-6] + 0.05x[n-2]x[n-2]x[n-3] \\ &+ 0.06x[n-3]x[n-6]x[n-8] + 0.06x[n-4]x[n-7]x[n-7] \\ &+ 0.05x[n-8]x[n-9]x[n-9] + s[n] \end{aligned}$$

An AR(30)-NAR(Q) model was fitted to data blocks containing 5000 samples of the distorted data. The non-linear terms allowed in the model were of the form:

$$w(n-i)w(n-j)w(n-k)$$
$$\text{for } i = 1:9,\ j = i:9,\ k = j:9.$$

and a model complexity of $Q \leq 20$ was allowed. Typical results are shown in figure 4.27 which shows a section of the original, distorted and restored signals.

4.7.6 Discussion

The techniques introduced in this section perform well on audio data which have been distorted by the appropriate model. However extensive testing is required to determine whether or not the non-linear models proposed are sufficiently flexible to model real distortion mechanisms.

Further work is required on methods for searching the space of non-linear models of a particular class (eg. AR-NAR) to determine the required model complexity. This may perhaps be best achieved by extending the Maximum Likelihood approach to a full Bayesian posterior probability formulation and using the concept of *model evidence* [Pope and Rayner, 1994] to compare models of different complexity. Some

DIGITAL AUDIO RESTORATION 191

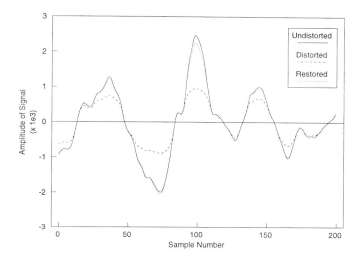

Figure 4.26 Typical section of AR-MNL Restoration

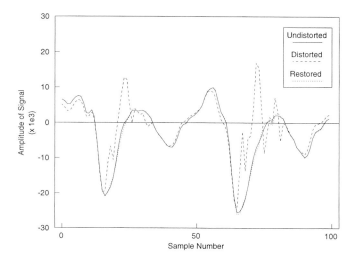

Figure 4.27 Typical section of AR-NAR Restoration

recent work in this field [Troughton and Godsill, 1997] applies Bayesian Markov chain Monte Carlo (MCMC) methods to the problem of non-linear model term selection. It is planned to extend this work in the near future to model selection for the AR-NAR distortion models discussed earlier in this section.

4.8 OTHER AREAS

In addition to the specific areas of restoration considered in previous sections there are many other possibilities which we do not have space here to address in detail. These include processing of stereo signals, processing of multiple copies of mono recordings, frequency range restoration and pitch adjustment.

Where stereo signals are processed, it is clearly possible to treat each channel as a separate mono source, to which many of the above processes could be applied (although correction of pitch variations would need careful synchronization!). However, this is sub-optimal, owing to the significant degree of redundancy and the largely uncorrelated nature of the noise sources between channels. It is likely that a significantly improved performance could be achieved if these factors were utilized by a restoration system. This might be done by modelling cross-channel transfer functions, a difficult process, owing to complex source modelling effects involving room acoustics. Initial investigations have shown some promise, and this may prove to be a useful topic of further research.

A related problem is that of processing multiple copies of the same recording. Once again, the uncorrelated nature of the noise in each copy may lead to an improved restoration, and the signal components will be closely related. In the simplest case, a stereo recording is made from a mono source. Much of the noise in the two channels may well be uncorrelated, in particular small impulsive-type disturbances which affect only one channel of the playback system. Multi-channel processing techniques can then be applied to extraction of the signal from the noisy sources. A Bayesian approach to this problem, which involves simple FIR modelling of cross-channel transfer functions, is described in [Godsill, 1993], while a joint AR-modelling approach is presented in [Hicks and Godsill, 1994]. In the case where sources come from different records, alignment becomes a major consideration. Vaseghi and Rayner [Vaseghi and Rayner, 1988, Vaseghi, 1988, Vaseghi and Rayner, 1989] use an adaptive filtering system for this purpose in a dual-channel de-noising application.

In many cases the frequency response of the recording equipment is highly inadequate. Acoustic recording horns, for example, exhibit unpleasant resonances at mid-range frequencies, while most early recordings have very poor high frequency response. In the case of recording resonances, these may be identified and corrected using a cascaded system model of source and recording apparatus. Such an approach was investigated by Spenser and Rayner [Spenser and Rayner, 1989, Spenser, 1990]. In the case where high frequency response is lacking, a model which can predict high frequency components from low is required, since any low-level high frequency information in the noisy recorded signal is likely to be buried deep in noise. Such a process becomes highly subjective, since different instruments will have different high frequency characteristics. The procedure may thus be regarded more as signal enhancement than restoration.

Pitch adjustment will be required when a source has been played back at a different (constant) speed from that at which it was recorded. This is distinct from wow (see section 4.6) in which pitch varies continuously with time. Correction of this defect can often be made at the analogue playback stage, but digital correction is possible through use of sample-rate conversion technology (see section 4.6). Time-scale modification (see the chapter by Laroche) is not required, since changes of playback speed lead to a corresponding time compression/expansion. We note that correction of this defect will often be a subjective matter, since the original pitch of the recording may not be known exactly (especially in the case of early recordings).

4.9 CONCLUSION AND FUTURE TRENDS

This chapter has attempted to give a broad coverage of the main areas of work in audio restoration. Where a number of different techniques exist, as in the case of click removal or noise reduction, a brief descriptive coverage of all methods is given, with more detailed attention given to a small number of methods which the authors feel to be of historical importance or of potential use in future research. In reviewing existing work we point out areas where further developments and research might give new insight and improved performance.

It should be clear from the text that fast and effective methods are now available for restoration of the major classes of defect (in particular click removal and noise reduction). These will generally run in real-time on readily available DSP hardware, which has allowed for strong commercial exploitation by companies such as CEDAR Audio Ltd. in England and the American-based Sonic Solutions in California. It seems to the authors that the way ahead in audio restoration will be at the high quality end of the market, and in developing new methods which address some of the more complex problems in audio, such as correction of non-linear effects (see section 4.7). In audio processing, particularly for classical music signals, fidelity of results to the original *perceived* sound is of utmost importance. This is much more the case than, say, in speech enhancement applications, where criteria are based on factors such as intelligibility. In order to achieve significant improvements in high quality sound restoration sophisticated algorithms will be required, based on more realistic modelling frameworks. The new models must take into account the physical properties of the noise degradation process as well as the psychoacousical properties of the human auditory system. Such frameworks will typically not give analytic results for restoration, as can be seen even for the statistical click removal work outlined in section 4.3.4, and solutions might, for example, be based on iterative methods such as Expectation-maximize (EM) or Markov chain Monte Carlo (MCMC) which are both powerful and computationally intensive. This is, however, likely to be in accord with continual increases in speed and capacity of computational devices.

To conclude, the range of problems encountered in audio signals from all sources, whether from recorded media or communications and broadcast channels, present challenging statistical estimation problems. Many of these have now been solved successfully, but there is still significant room for improvement in achieving the highest possible levels of quality. It is hoped that the powerful techniques which are now practically available to the signal processing community will lead to new and more effective audio processing in the future.

Notes

1. the 'Packburn' unit achieved masking within a stereo setup by switching between channels

2. With acknowledgement to Mr. B.C. Breton, Scientific Imaging Group, CUED

3. provided that no samples are missing from the first P elements of **s**; otherwise a correction must be made to the data covariance matrix (see [Godsill, 1993])

4. the approximation assumes that the parameter likelihood for the first P data samples is insignificant [Box and Jenkins, 1970]

5. These techniques are also often referred to as 'spectral subtraction'. We will not use this terminology in order to avoid ambiguities between the general principle and the particular technique described in [Boll, 1979], nor will we use the term 'spectral estimation' as quite a number of the STSA techniques are not based on a statistical estimation approach.

6. This suppression rule is derived by analogy with the well-known Wiener filtering formula replacing the power spectral density of the noisy signal by its periodogram estimate.

7. Strictly speaking, this effect could still be perceived for longer window durations when the relative signal level approaches 1. However, it is then perceived more like an erratic fluctuation of the sinusoid level which is hardly distinguishable from the phenomenon to be described in section 4.17.

8. More precisely, the quantity displayed is the signal power estimated from 10ms frames. As the power spectral densities of the two types of noise exhibit a strong peak at the null frequency, the two noises were pre-whitened by use of an all-pole filter [Cappé, 1991]. This pre-processing guarantees that the noise autocorrelation functions decay sufficiently fast to obtain a robust power estimate even with short frame durations [Kay, 1993].

5 DIGITAL AUDIO SYSTEM ARCHITECTURE

Mark Kahrs

Multimedia Group
CAIP Center
Rutgers University
P.O. Box 1390
Piscataway, NJ 08855-1390

kahrs@caip.rutgers.edu

Abstract: Audio signal processing systems have made considerable progress over the past 25 years due to increases in computational speed and memory capacity. These changes can be seen by examining the implementation of increasingly complex algorithms in less and less hardware. In this chapter, we will describe how machines have been designed to implement DSP algorithms. We will also show how progress in integration has resulted in the special purpose chips designed to execute a given algorithm.

5.1 INTRODUCTION

Audio signal processing systems have made considerable progress over the past 25 years due to increases in computational speed and memory capacity. These improvements are a direct result of the ever increasing enhancements in silicon processing technologies. These changes can be demonstrated by examining the implementation of increasingly complex algorithms in less and less hardware. In this chapter, we will describe how sound is digitized, analyzed and synthesized by various means. The chapter proceeds from input to output with a historical bent.

5.2 INPUT/OUTPUT

A DSP system begins at the conversion from the analog input and ends at the conversion from the output of the processing system to the analog output as shown in the figure 5.1:

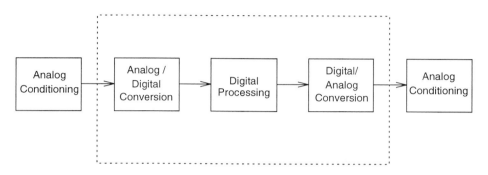

Figure 5.1 DSP system block diagram

Anti-aliasing filters (considered part of "Analog Conditioning") are needed at the input to remove out of band energy that might alias down into baseband. The anti-aliasing filter at the output removes the aliases that result from the sampling theorem.

After the anti-aliasing filter, the analog/digital converter (ADC) *quantizes* the continuous input into discrete levels. ADC technology has shown considerable improvement in recent years due to the development of oversampling and noise-shaping converters. However, a look at the previous technologies [Blesser, 1978] [Blesser and Kates, 1978][Fielder, 1989] will help appreciate the current state-of-the-art.

After digital processing, the output of the system is given to a digital/analog converter (DAC) which converts the discrete levels into continuous voltages or currents. This output must also be filtered with a low pass filter to remove the aliases. Subsequent processing can include further filtering, mixing, or other operations. However, these shall not be discussed further.

5.2.1 Analog/Digital Conversion

Following the discussion in Bennett ([Bennett, 1948]), we define the Signal to Noise Ratio (SNR) for a signal with zero mean and a quantization error with zero mean as follows: first, we assume that the input is a sine wave. Next, we define the root mean square (RMS) value of the input as

$$\sigma_{\text{RMS}} = \frac{\Delta 2^{b-1}}{\sqrt{2}} \qquad (5.1)$$

where Δ is the smallest quantization level and $b-1$ bits are present in the maximum value. The noise energy of such a signal will be the integral of the quantization noise over time:

$$\sigma^2_{\text{noise}} = \frac{\Delta}{\sqrt{12}} \tag{5.2}$$

This will give

$$\sigma_{\text{noise}} = \frac{1}{\Delta}\int_{-\Delta/2}^{+\Delta/2} e^2_{\text{noise}}\, de \tag{5.3}$$

Since SNR is $\frac{\sigma_{\text{RMS}}}{\sigma_{\text{noise}}}$, and deriving the output in decibels, we get

$$SNR(dB) = 10\, \log_{10}\, 2^b\, \sqrt{1.5} \tag{5.4}$$

or, the well known 6 dB per bit.

It is important to remember that this equation depends on the assumption that the quantizer is a fixed point, "mid-tread" converter with sufficient resolution so that the resulting quantization noise (e_{noise}) is white. Furthermore, the input is assumed to be a full scale sinusoidal input. Clearly, few "real world" signals fit this description, however, it suffices for an upper bound. In reality, the RMS energy of the input is quite different due to the wide amplitude probability distribution function of real signals. One must also remember that the auditory system is not flat (see the chapter by Kates) and therefore SNR is at best an upper bound.

Given equation 5.4, we can see how many bits are required for high fidelity. Compact Disks use 16 bits, giving a theoretical SNR of approximately 96 dB; however, this is not as quiet as well constructed analog mixing desks where SNRs of over 120 dB are typically found. An equivalent digital system must therefore be prepared to accommodate fixed point lengths exceeding 20 bits. Recent converters offer 24 bits (but only 112 dB SNR). Floating point converters can provide the same *dynamic range* but with less SNR. We will discuss this shortly, but before it, we will examine the typical fixed point converter.

Fixed Point Converters. The technology of fixed point converters before the introduction of oversampling, is covered amply in a book by Analog Devices [Sheingold, 1986]. Terminology of converter performance are reviewed by Tewksbury, et al. [Tewksbury et al., 1978] Besides oversampling converters (known as "Delta-Sigma" or "Sigma-Delta"), there are four basic types of converters:

- Successive Approximation
- Integration
- Counter (servo)
- Parallel

Audio applications ignore all but the first type; integration converters are too slow to convert a sample in one sampling time. This is also true of counter converters, since it takes 2^b clock cycles to reach the full scale range for b bits. Parallel converters, like the well known "flash" converter, are excessive in their use of silicon area for large b because 2^b comparators are needed as well as a large resistive dividers. Therefore, we will concentrate on successive approximation and oversampling delta-sigma converters.

Successive Approximation. A typical successive approximation converter is shown in figure 5.2.

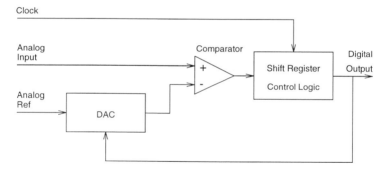

Figure 5.2 Successive Approximation Converter

Under control of a finite state machine, a b-bit shift register is used as the input to a DAC. The output of the DAC is compared with the analog input; if the comparison is negative, then the latest bit was an *overestimation* and therefore the bit should be zero; otherwise the bit is a one. Clearly, the linearity of the DAC effects the overall linearity of the system. Also, the input must not change during the conversion process therefore the addition of a sample and hold must be considered. However, this also introduces other sources of error including droop due to hold capacitor leakage and aperature jitter in the sample and hold clock. However, these errors are further compounded by slew rate limitations in the input and sample and hold amplifiers. These issues were addressed early by Stockham [Stockham, 1972] and were covered extensively by Talambiras [Talambiras, 1976][Talambiras, 1985]. A more recent analysis of jitter effects in oversampling converters was given by Harris [Harris, 1990] who analyzed jitter effects as a form of FM modulation.

Dither. Starting from Robert's pioneering paper [Roberts, 1976], the use of dither in audio was seriously analyzed by Vanderkooy and Lipshitz [Vanderkooy and Lipshitz, 1984]. The basic idea is simple: to whiten the quantization error, a "random" error signal is introduced. While the introduction of noise will make the signal "noisier", it will also decorrelate the quantization error from the input signal (but not totally). Vanderkooy and Lipshitz also propose the use of triangular dither derived from the sum of two uniform random sources [Vanderkooy and Lipshitz, 1989].

Dither can be subtracted out ("subtractive dither") after quantization and thereby whiten the signal. But in most systems this may be either difficult or impossible to do because the dither signal is not available. Therefore "non-subtractive" dither is the most common use of dithering. Yamasaki [Yamasaki, 1983] discusses the use of *large amplitude* dither (as much as 32Δ) in subtractive dither systems.

A typical example of subtractive dither use in an A/D converter can be found in a Teac patent [Nakahashi and Ono, 1990]. Notable features include the use of limiting and overload detectors as well as the ability to control the dither amplitude. A clever example of non-subtractive dither [Frindle, 1995][Frindle, 1992] separates the input signal into two paths. One of the paths is now inverted and then both paths are added to a random noise source. After the DAC, the two analog signals are subtracted; the result is the sum of the conversion errors and twice the input signal.

Oversampling converters. Bennett's [Bennett, 1948] pioneering paper points out that the quantization noise is integrated over a frequency range. For a spectrally flat (i.e., white) signal the noise power is given by the following equation:

$$E_{\text{noise}} = e_{\text{RMS}} \sqrt{2/f_s} = e_{\text{RMS}} \sqrt{2\,T_s} \qquad (5.5)$$

where f_s is the sampling rate and T_s is the sampling period ($1/f_s$)). As the sampling frequency f_s increases, E_{noise} will decrease accordingly. This key fact can be calculated together with the noise power to derive the oversampling factor needed to achieve a given noise floor [Candy and Temes, 1992] as follows:

$$N = \frac{f_{s'}}{2f_s} \qquad (5.6)$$

where $f_{s'}$ is the new sampling frequency. Inserting this into the integral in equation 5.3 will result in noise energy of $\frac{E_{\text{RMS}}}{N}$, so then the SNR will decrease by 3 dB (one half bit) for each doubling of the sampling frequency.

As shown in section 5.2, the resolution is a direct consequence of the degree of oversampling. Audio converters have used oversampling ratios of a much as 128 times leading to an improvement of 21 dB.

Another key development was the introduction of *noise shaping*. The idea is to filter the error output of the quantizer and push the error energy into the higher frequencies. This benefit depends critically on the design of the error filter. The following analysis is after van de Plassche [van de Plassche, 1994]. Suppose that the filter is a simple first-order function $E(z) = |1 - z^{-1}|$. Let $\omega_s = \frac{2\pi f}{f_s}$ where f is the highest frequency component of the input signal. Then, evaluating the amplitude of $E(z)$ gives $2(1 - \cos \omega)$. The shaped noise energy will be the area between D.C. and ω_s, or

$$e_{\text{shaped}}^2 = \int_0^{\omega_s} \epsilon^2 \tag{5.7}$$

where ϵ is the error, in this case, the error function $E(z)$.

So, substituting $E(z)$ into the integral of equation 5.7 gives the noise power of $e_{\text{shaped}}^2 = 2(\omega_s - \sin \omega_s)\epsilon^2$. Without noise shaping, the area will be simply the flat integral, so then $e_{\text{uniform}} = \epsilon^2 \omega_s$.

So, finally, the improvement due to error noise shaping will be the ratio of $\frac{e_{\text{shaped}}}{e_{\text{uniform}}} = \sqrt{2(1 - \sin(\omega_s)/\omega_s)}$.

Candy and Temes [Candy and Temes, 1992] list several criteria for choosing the internal architecture of oversampling converters. These include

- Bit resolution for upsampling factor

- Complexity of the modulator/demodulator

- Stability

- Dithering

Higher order architectures have been successfully used to build converters for digital audio applications [Adams, 1986][Hauser, 1991]. However, the complexity of modulator and demodulator increases with the order. Higher order architectures also exhibit stability problems [van de Plassche, 1994]. Dithering is a remedy to many of the problems of low level signals (see the previous section) and has been used in sigma delta converters as part of the noise shaping loop [Vanderkooy and Lipshitz, 1989].

Today, in 1997, a designer can find oversampling audio converters with a reported resolution of up to 24 bits. In reality, the resulting noise is well below the theoretical maximum (approximately 144 dB).

Besides the elimination of the sample-and-hold, oversampling converters also reduce the complexity of the anti-aliasing filter. Previous anti-aliasing filters required the use of high order active filters (11th order elliptic for example) that resulted in phase

DIGITAL AUDIO SYSTEM ARCHITECTURE   201

distortion [Preis and Bloom, 1984]. The oversampling results in a higher Nyquist frequency and therefore the filter rolloff can be much slower. The typical filter to a Delta-sigma converter is a trivial one pole RC filter. However, note that fidelity problems associated with slew rate limitations in input amplifiers [Talambiras, 1976] [Talambiras, 1985] are *not* eliminated, however.

Floating Point Converters. A floating point converter represents the number as a triple: (S, E, M), where S is the sign, E the exponent and M is the mantissa. The exponent base (B) is a power of two. The mantissa is a rational number of the form P/Q, where $0 \leq P \leq B^{|E|}$ and Q is exactly $B^{|E|}$.

For floating point representations, the variance of the error is

$$\sigma_e^2 = \sigma_x^2 \sigma_\epsilon^2 \tag{5.8}$$

where σ_ϵ is the relative roundoff error. (This assumes that the error is multiplicative).

Therefore, the SNR of a floating point converter is

$$SNR_{\text{floating}} = 10 \log_{10} \left(\frac{\sigma_x^2}{\sigma_x^2 \sigma_\epsilon^2} \right) \tag{5.9}$$

or, simplifying,

$$SNR_{\text{floating}} = 10 \log_{10} \left(\frac{1}{\sigma_\epsilon^2} \right) \tag{5.10}$$

In order to evaluate this equation, we need a value for σ_ϵ. Kontro, et al. [Kontro et al., 1992] used the Probability Distribution Function (PDF) of floating point roundoff error in multiplication:

$$\sigma_\epsilon^2 = \frac{\Delta^2}{8 \log 2} \tag{5.11}$$

where $\Delta = 2^{-b_m}$ and b_m is the number of bits in the mantissa. This gives an SNR_{floating} of $6.26 + 6.02\, b_m$ dB.

One of the early research converters [Kriz, 1976][Kriz, 1975] used floating point for both the DAC and ADC. Kriz used a 12 bit mantissa and a 3 bit exponent (the missing bit is the sign bit) for a theoretical SNR of 78 dB. The DAC is shown in figure 5.3.

Note however, that the amplifier gain must be dynamically changed depending on signal amplitude, which leads to discontinuities. Kriz also points out the slew rate problems in the track and hold. His solution was to use a DC feedback loop with an analog integrator.

Kriz also discusses the construction of a matching 16 bit ADC. He duplicated his integrating solution but used a 8 bit DAC in the DC offset path as shown above.

Fielder [Fielder, 1985] discusses the use of a converter with 12 bit signed magnitude and 6 dB gain steps resulting in a possible dynamic range of 115 dB. An equivalent

202 APPLICATIONS OF DSP TO AUDIO AND ACOUSTICS

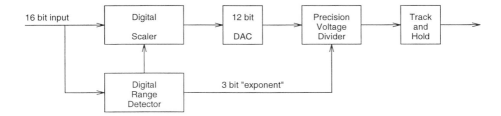

Figure 5.3   16 Bit Floating Point DAC (from [Kriz, 1975])

fixed point converter would have 20 bits! However, as discussed above, the noise performance isn't as good because of the missing codes.

The subsequent development of high quality linear, fixed point converters has largely eliminated floating point converters for audio applications. Note, however, that Yamaha used floating point converters until only recently. This was a direct result of using logarithmic arithmetic internally on their chips (see section 5.14). A relatively recent example of an oversampling Sigma-Delta converter ([Kalliojärvi et al., 1994]) combines oversampling sigma Delta conversion with floating point quantization.

Given the ever increasing number of floating point processors, coupled with the need for better SNR (i.e., over 120 dB) might push the development of matching floating point converters. More work remains to be done.

5.2.2 Sampling clocks

Jitter is the deviation of the sampling instant due to noise and other phenomena. Jitter can be analyzed using a battery of statistical and probabilistic methods [Liu and Stanley, 1965]. In particular, Liu and Stanley analyze the jitter for a storage system where four cases can be considered:

1. input only ("readin")

2. output only ("readout")

3. input and output with identical jitter ("locked")

4. independent input and output jitter

Let us consider the accuracy required for the sampling clock in a DSP system. van de Plassche has the following analysis: If the input is full range and near the Nyquist frequency, then we have the greatest slope. Let us use a simple sinusoid as the input, i.e., $V = A\,sin(\omega t)$. The variation in the output of the converter that depends on the variation in the sampling time instant Δt will be Δv. The slope of the input is:

$$\frac{\Delta A}{\Delta t} = A\omega \cos(\omega t). \tag{5.12}$$

So, then if $\Delta A = \frac{2A}{2^n}$ (peak to peak), then $\Delta t = \frac{2^{-n}}{\omega \cos(\omega t)}$; since, $\omega = 2\pi f$, then

$$\Delta t = \frac{2^{-n}}{\pi f \cos(2\pi f t)} \tag{5.13}$$

And, at $t = 0$, then we have

$$\Delta t_0 = \frac{2^{-n}}{\pi f} \tag{5.14}$$

As an example, consider a CD-audio signal with 16 bits of resolution with a sampling frequency of 44.1 KHz. The time uncertainty (Δt_0) will be 110 picoseconds!

As the digital signal passes through cables and other degrading elements the jitter can and will increase. More attention has been paid to these issues of late [Shelton, 1989][Lidbetter et al., 1988].

5.3 PROCESSING

The history of audio signal processors is directly correlated with the development of silicon technology. As we will see, the first digital synthesizers were constructed from discrete SSI TTL level components. In fact, early synthesizers like the Synclavier used a mixture of digital and analog technology – a prime example being the use of a multiplying DAC for envelope control [Alonso, 1979]. A review of early architecture circa 1975 was given by Allen [Allen, 1975].

Further development of synthesizers was critically dependent on the technology of multiplier implementation. TI's early 2 bit by 4 bit Wallace Tree [Koren, 1993] [Waser and Flynn, 1982] was introduced in 1974. The AMD25S10 was an early improvement: 4 bits by 4 bits. Monolithic Memories' 74S558 multiplier was 8 bits by 8 bits. The major improvement was the 16 by 16 multiplier (the TRW MPY-16). This single chip did more for audio signal processing than almost any other device. It was used for many of the IRCAM machines and also the Alles synthesizer.

Simultaneous improvements in ALU width during the same period can be witnessed by the development of the AMD 2901 4 bit slice [Mick and Brick, 1980]. This flexible ALU building block was very popular in the design of ALUs; combined with a TRW multiplier the resulting machine architectures were flexible and powerful.

The use of Emitter Coupled Logic (ECL) is also possible [Blood, 1980] [Hastings, 1987] but it suffers because of low levels of integration (typically 2 and 4 bit wide parts) as well a high demands for current and fans. In spite of these limitations, the group at Lucasfilm used ECL technology for the ASP [Moorer et al., 1986]. For the era, ECL was the path to high speed computing (an independent example is the Dorado [Clark et al., 1981a]).

Further increases in the density of chips led to the development of the modern signal processor, beginning in 1979 with the Intel i2920. This was followed by the Texas Instruments TMS 320 series (begining with the 32010) that rapidly captured a large segment of the market. The 24 bit width of the Motorola 56000 was welcomed by the digital audio community and found rapid use in many applications. The 56000 also included parallel moves and well designed addressing modes.

In spite of constant improvement in processor technology, there are still applications that are computationally expensive. This will lead to the use of multiple DSP chips in multiprocessors as well as the development of special purpose chips for synthesis. All of these developments will be discussed in the following sections.

5.3.1 Requirements

Before we can discuss the design and architecture of processors for audio DSP tasks, we must discuss the requirements these processors must meet. Then, we will be in a better position to see how the different implementations result in better results. So, we will begin by dividing the requirements into the different tasks. Gordon [Gordon, 1985] has an overview of architecture for computer music circa 1985.

In typical use, a speech algorithm is just as usable for wideband audio with the following caveats in mind: (1) higher sampling rates *decrease* the available processor time per sample (2) higher frequency ranges may mean a greater number of poles and zeros in signal modeling.

Analysis/Synthesis.

Linear Prediction (LPC). LPC is the most popular form of speech coding and synthesis. While it is usually used for speech coding [Markel and Gray, 1976] it can also be used for other time-varying filter models. The LPC analysis process can be divided into two parts: excitation derivation and filter analysis. The filter analysis uses an all-pole model derived from a specific acoustical model. For example, the vocal tract can be modeled as an all pole lattice filter; the lattice filter coefficients are derived from the acoustical vocal tract model (more specifically, the reflection coefficients). LPC filter coefficients can be computed using any number of avenues including the Covariance, Autocorrelation or Lattice methods. These methods all use iterative matrix algebra; the implications for machine architecture are the need for fast array indexing, multiplies and/or vector arithmetic.

The excitation signal depends on the source model (vocal cord models are used in speech for both voice and unvoiced sources). LPC models can be used in other circumstances, for example, the modeling of acoustic musical instruments where the excitation function represents the nonlinear source and the time-varying filter represents the acoustic transmission line of the instrument. When modeling wideband signals, the number of poles can become extremely high (more than 50 poles is not uncommon).

The Phase Vocoder.   The Phase Vocoder [Flanagan and Golden, 1966][Gordon and Strawn, 1985] is a common analysis technique because it provides an extremely flexible method of spectral modification. The phase vocoder models the signal as a bank of equally spaced bandpass filters with magnitude and phase outputs from each band. Portnoff's implementation of the Short Time Fourier Transform (STFT) provides a time-efficient implementation of the Phase Vocoder. The STFT requires a fast implementation of the Fast Fourier Transform (FFT), which typically involves bit addressed arithmetic.

Perceptual Coding.   More recently, coding techniques that take advantage of the masking properties of the inner ear have been developed. These techniques are discussed in Brandenburg's chapter. A typical perceptual model uses frequency domain filtering followed by coding and compression. The receive end (decoder) receives coded and compressed stream and decodes and expands the data. The resulting data is converted back into audio data. Because of the extensive bandwidth reduction of perceptual coding, such algorithms will be finding their way into more and more commercial products including Digital Audio Broadcasting and Digital Video Disk (DVD).

Perceptual coders are an interesting class of DSP algorithm; although it has a signal processing 'core" (typically the DCT (Discrete Cosine Transform)), the algorithm spends most of its time in non-iterative code. For example, the MPEG coder has algorithms that separate noise-components from harmonic-components as well as Huffman coding and bit-rate coding. This code doesn't exhibit the same tight loops DSP processors were designed to handle.

Analysis.   We can divide analysis algorithms into time and frequency domain processes. Certainly, the division between these categories is arbitrary since we can mix them together to solve an audio problem. However, it suffices for our purposes.

Time Domain Analysis.   Perhaps the simplest and most traditional use of a DSP is filtering. DSPs are designed to implement both Finite Impulse Response (FIR) and Infinite Impulse Response (IIR) filters as fast as possible by implementing (a) a single cycle multiply accumulate instruction (b) circular addressing for filter coefficients. These two requirements can be found in *all* modern DSP architectures.

The typical IIR filter is defined by the following equation:

$$y[n] = \sum_{i=0}^{i=N} a[i]x[n-i] - \sum_{j=1}^{j=M} b[j]y[n-j] \tag{5.15}$$

As can be seen, it requires memory (N+M+1 locations) for the tapped delay lines as well as N+M+1 filter coefficients. It also requires a multiplier with a result that is accumulated by the sum. It is important that the accumulator have a "guard band" of sufficient size to avoid overflow during accumulation. The FIR filter is similar except it lacks the feedback terns.

Most audio filters are typically IIR filters because a). They are directly transformed from their analog counterparts via the bilinear transform b). They are faster to compute than the longer FIR version.

IIR filters can be implemented in any number of different ways, including direct form, parallel form, cascaded second order sections and lattices.

It is easy to assume that the filter coefficients in equation 5.15 ($a[i]$ and $b[j]$) are constant. In fact, there are many instances when this is not the case. For example, in real-time audio processing a user moves a slide potentiometer (either physical or possibly on the display); this is digitized and the host processor must change the coefficients in various filters.

The filter coefficients must be updated at a regular interval. If the number of filters coefficients is large (for example, a high order LPC filter) and/or a large number of filters must operate in parallel, then this may interfere with the computation. Moorer called it the "parameter update problem" [Moorer, 1981].

It should also be noted that in audio, many operations are calibrated in decibels. This implies the need for a logarithm (base 10). If possible, such computations should be avoided since the Taylor series calculation method is multiplier intensive. Short cuts, such as direct table lookup are preferable when possible.

Frequency Analysis. The Discrete Fourier Transform (and its fast implementation, the Fast Fourier Transform [Brigham, 1974]) (FFT) as well as its cousin, the Discrete Cosine Transform [Rao and Yip, 1990] (DCT) require block operations, as opposed to single sample inputs. The DFT can be described recursively, with the basis being the 2 point DFT calculated as follows:

$$X[0] = x[0] + x[1] * W_N^0 \tag{5.16}$$
$$X[1] = x[0] + x[1] * W_N^{N/2} \tag{5.17}$$

where $W = e^{-j(2\pi/N)}$. Since $W_N^0 = 1$ and $W_N^{N/2} = -1$, then no multiplications are required, just sign flipping. This is the well known "Butterfly" computation.

Since either the inputs or outputs of the FFT are in bit reversed form, it is useful to have a method to address the data. Although algorithms have been published for bit-reversal [Karp, 1996], the added computation time may not be worth it. Therefore, either table lookup may be used or carries can be propagated from left to right, which produces the needed bit reversal. Since the added feature is just a modification of the carry chain, it is deemed easy enough to implement (leaving aside the issue of the implementation of the addressing mode).

5.3.2 Processing

Some algorithms have a hard time being categorized into one of analysis or synthesis. This includes coders and rate changers. Coders, such as the audio coders discussed in in the chapter by Brandenburg, require the use of a compression technique, such as a Huffman code. Computation of this code at the encoder side is not nearly as simple as its accompanying decoder.

Reverberation, discussed by Gardner in his chapter, points out the need for extremely large memories. Reverberators can be implemented using time-domain filtering with delays or in the frequency domain with convolution. In the time-domain implementation, the delays must be the length of the impulse response. Recall that at the size of the memory will be $M = T_{60}/T_S$. For $T_{60} = 2.0$ seconds and $T_S = 20$ microseconds, $M = 5 \times 10^5$, or a memory address of 19 bits. Therefore, the address space should be larger than 16 bits. The lesson of large number of address bits (and pins!) may seem obvious, but DSPs have used static RAM and therefore the required address space is small to match small wallets.

The frequency domain implementation uses FFTs to convert the incoming window of samples to the frequency domain, multiplies it by the impulse response of the room and then uses the Inverse FFT to get back to the time domain. For large impulses, this generates considerable delay. Gardner discusses how to avoid this in his chapter.

Restoration. With the introduction of the Compact Disc, the restoration of old recordings has become big business. However, the original tape (or disk) masters are often in bad physical condition resulting in pops, clicks and other media maladies. Many of these defects can be overcome with the use of DSP technology. The most famous use of DSP technology was made by Stockham [Stockham et al., 1975] who restored Caruso's recordings from the early part of this century. Stockham used a cepstrum based technique to do blind deconvolution to remove the effect of the recording horn from the recorded master. Computationally, this demanded spectral analysis using the FFT and further processing.

In the past five years, there has been extensive use of statistical methods to recover signals partially obscured by noise (see the chapter by Godsill, Rayner and Cappé).

5.3.3 Synthesis

We can divide synthesis techniques into four basic categories: Additive (Linear), Subtractive, Nonlinear and Physical modeling. Synthesis algorithms depend critically on the implementation of oscillators. For example, in the implementation of Frequency Modulation (F.M.), the output of one oscillator will serve as the input to another. Since the number of real time oscillators depends on the number of simple oscillators, it is important to efficiently and speedily implement the realizations.

Low-noise oscillator synthesis is not trivial however most methods use lookup tables with fractional interpolation. Oscillators can be implemented by (a) table lookup or (b) IIR filters with poles located *exactly* on the unit circle.

Moore [Moore, 1977b] studied the effect of oscillator implementation using lookup tables and found that linear interpolation produces the least distortion and that truncation produces the worst. This result was confirmed by Hartmann [Hartmann, 1987]. Another possibility is to use a recursive (IIR) filter with poles located on the unit circle. This "coupled form" [Tierney et al., 1971] offers a alternate method that avoids using memory space. Frequency resolution requirements were calculated by Snell in a superpipeline oscillator design for dynamic Fourier synthesis [Snell, 1977].

Oscillators also require "control inputs" such as amplitude or frequency parameters. These are often time-varying and so smooth interpolation may be required.

Linear Synthesis. The most popular method of synthesis is so-called "Additive Synthesis", where the output is a sum of oscillators. While it is commonly assumed that the oscillators produce sinusoids (Fourier synthesis), in fact, they can be any waveform. Furthermore, with "static" additive synthesis, a pre-mixed combination of harmonics was stored in the lookup table. Unfortunately, this doesn't permit inharmonic partials. "Dynamic" Fourier synthesis allows the amplitudes and frequencies of the partials to be varied relative to each other. Computationally, it is important to recognize the that updating oscillator coefficients for large numbers of oscillators can be expensive.

Subtractive Synthesis. "Subtractive Synthesis" is the process of filtering a broadband source with a time-varying filter. The most classical example of this is vocal tract synthesis using Linear Prediction Coding (LPC) (see section 5.3.1). This requires a broadband (or noisy) source and, in the case of LPC, an IIR filter with time varying coefficients. The filter coefficients will require interpolation and storage. These seemingly insignificant operations can not be ignored.

Nonlinear synthesis: Frequency Modulation. Frequency Modulation (FM), originally described by Chowning [Chowning, 1973], was patented by Stanford [Chowning, 1977] and later licensed to Yamaha and used in the now famous DX-7 synthesizer. The FM equation

$$y[n] = A \sin(\omega_c t + I \sin(\omega_m t + \phi)) \tag{5.18}$$

requires two oscillators and two amplitude terms. All four of these inputs can be described using envelopes or with constants. The envelopes involve the calculation of either straight lines or exponential curves. The following equation (from the unit generator gen4 from cmusic [Moore, 1990b]) permits both:

$$f(x) = y_1 + (y_2 - y_1)\left(\frac{1 - e^{I(x)\alpha}}{1 - e^\alpha}\right) \tag{5.19}$$

where $x_1 \leq x \leq x_2$ and α is the "transition parameter":

- if $\alpha = 0$, then $f(x) = y_1$, a straight line
- if $\alpha < 0$, then $f(x)$ is exponential
- if $\alpha > 0$, then $f(x)$ is inverse exponential

Lastly, $I(x) = \frac{x - x_1}{x_2 - x_1}$.

FM instruments are made from cascades of FM oscillators where the outputs of several oscillators are mixed together.

Physical Modelling. The last method of synthesis, physical modeling, is the modeling of musical instruments by their simulating their acoustic models. One popular model is the acoustic transmission line (discussed by Smith in his chapter), where a non-linear source drives the transmission line model. Waves are propagated down the transmission line until discontinuities (represented by nodes of impedance mismatches) are found and reflected waves are introduced. The transmission lines can be implemented with lattice filters.

The non-linear sources are founded on differential equations of motion but their simulation is often done by table lookup.

5.3.4 Processors

This section will be presented in a historical fashion from oldest to newest technology. This will demonstrate the effect of technology on the design and implementation of DSP systems and offer a perspective on the effect of underlying technology on architecture of audio signal processing systems.

One of the earliest machines for audio manipulation was GROOVE, an experimental machine built at Bell Laboratories [Mathews and Moore, 1970]. GROOVE was composed of a digital computer controlling an analog synthesizer. The emphasis was on the human interaction, not the analysis or synthesis of sound.

Sequential (MSI scale) machines. The SPS-41 [Knudsen, 1975] was a very early DSP. It had three sections: An ALU, loop control and I/O. The ALU processed *complex* data in the form of two 16 bit words. A multiplier produces the requisite four products forming another complex pair (note that the products must be scaled before storing the result). The loop control has a very fast test: each instruction has four bits of "indirect tests". The four bits address a 16 word memory that permits the testing of 16 bits. The I/O section is basically a DMA controller and resembles the PPUs of the CDC 6000 series [Thornton, 1970].

The Groove machine mentioned above was the inspiration for Moore's "FRMBox" [Moore, 1977a] [Moore, 1985]. It strongly resembles a CPU as shown in figure 5.4. A centralized control unit polls the modules and stores the output from each board in a

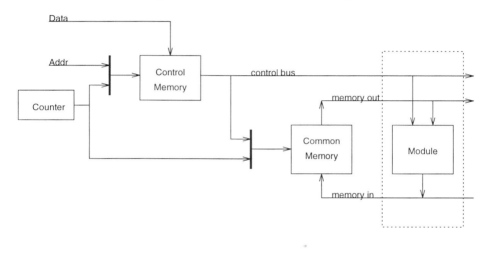

Figure 5.4 Block diagram of Moore's FRMbox

unique location in a centralized common memory. Each board was time multiplexed 32 times, so for an 8 board system there are 256 time slots per sample time. The controller permits multiple sources for a given "virtual" generator via a memory pointer in the control memory. The principal limitation is the size of the bus and the number of virtual units on a given board. However, for a given sampling frequency there are a maximum number of slots per sample time. For example, 256 slots at a 48 KHz sampling frequency is 81 nanosecond per slot. It is possible, of course, to expand the number of slots by adding multiple centralized controllers; but then the issue becomes communication from one controller to another.

In 1973, a small company in San Francisco designed a custom processor from early MSI scale logic. The design was done by Peter Samson [Samson, 1980][Samson,

1985] and was eventually delivered to CCRMA at Stanford in 1975. For its era, it was a large processor, using approximately 2500 integrated circuits and resembling a large green refridgerator. The multipliers were constructed using a modified Wallace Tree [Koren, 1993] and ran at 50 ns per multiply. The sample word width was 20 bits although several data paths were smaller. The overall structure of the machine (known at Stanford as the "Samson Box") is shown in figure 5.5.

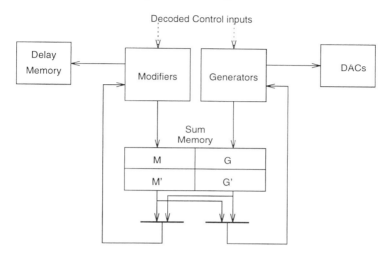

Figure 5.5 Samson Box block diagram

Briefly, each "instruction" is decoded and given to the appropriate part of the machine. The separate fields of each instruction are interpreted by the specific section.

The timing of the machine is quite interesting. A sample time is divided into processing, update and overhead ticks (a tick is 195 ns). There are always 8 overhead ticks since the latency of the pipeline is 8 ticks. So, for a given sample rate, the number of processing cycles is always known and can be divided by the programmer into time spent calculating samples (processing) or updating of coefficients in memory by the host computer (updates). A pause in updating is created via a LINGER command that waits a specific number of ticks.

The architecture of the machine can be divided into three sections: generators, modifiers and delay memory. Each section is pipelined and does *not* share hardware with the other sections. The pipeline timing for generators is detailed below in Table 5.5:

The pipelining for modifiers is more complex: the control for each modifier is different and is therefore not detailed here (see figure 4 of [Samson, 1980] for a sketch

Table 5.1 Pipeline timing for Samson box generators

Tick	Generator	
	Oscillator	Envelope
0	Memory read	
1	Add	
2	Multiply	
3	ROM lookup	Addr Reg read
4	Multiply	ROM lookup
5	Sign negation	Sign negation
6	Multiply envelope & generator	
7	Write into Sum memory	

of the datapath). The "Samson Box" was used extensively for synthesis - not analysis. In part this was due to the unique architecture of the machine and in part from the difficulty of inserting real time data into the machine via the delay memory. Moorer [Moorer, 1981] also points out that the command FIFO often got in the way of time critical updates. Loy [Loy, 1981] mentions the lack of precision in the modifiers (20 bits) and generators (13 bits) sometimes produced audible results. One must remember, however, that the time frame was the mid-1970s and so resulting integration was SSI and MSI scale and word widths were costly in terms of area and wires.

The interconnect of the "Samson Box" was quite novel. It was called "sum memory" and was implemented by parallel loading counters and then writing the result back into RAM. The connection memory acted as a double-buffered multiport memory by dividing the memory into "quadrants" as shown in figure 5.5.

The current generators and modifiers write into the current half while the previous tick's results are available to current processing. Because the output of the sum memory may feed as many as three inputs, it must be time multiplexed over a single tick, which leads to a short time available to the memory.

In 1976 Moorer [Moorer, 1980b] proposed a machine designed for closed-form [Moorer, 1976] summations. With the exception of the amplitude and spectrum envelope generators, it is a conventional microprogrammed machine with a instruction decoder and input and output bus. The envelope generator allows up to 16 different line segments; these are programmed to provide different instrumental timbres.

In 1979, TRW introduced the MPY16, a 16 by 16 bit multiplier. Alles [Alles, 1987] [Alles, 1980] used this new chip to design a 32 voice synthesizer using only 110 chips. Each voice was computed in one microsecond, resulting in a 31.25 KHz sampling rate.

Each slot was further divided into 16 clock cycles (of 64 ns each). At IRCAM, P. diGiugno designed an early machined named the "4A".

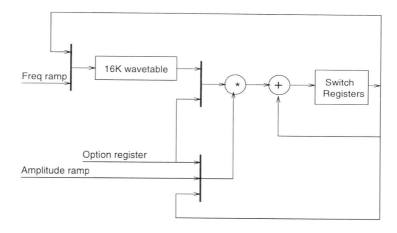

Figure 5.6 diGiugno 4A processor

Note that the start and stop registers are used for creating linear ramps; these ramps are used to control both amplitude and frequency parameters in oscillators. At the endpoint of the ramp, an interrupt is generated on the host (an LSI-11). As Moorer [Moorer, 1981] points out, this can lead to a considerable number of interrupts and delays due to interrupt processing.

H. Alles visited IRCAM and together with P. Di Giugno designed the follow-on to the 4A: the 4B [Alles and di Giugno, 1977]. This machine differed from the 4A in the following ways:

214   APPLICATIONS OF DSP TO AUDIO AND ACOUSTICS

- Expanded interconnect (switch) registers (from 4 to 16)
- Larger wavetable

    This was the direct result of improved technology. Also the wavetable was writable from the host processor.

- New data paths to accommodate FM synthesis

The datapath is shown in figure 5.7.

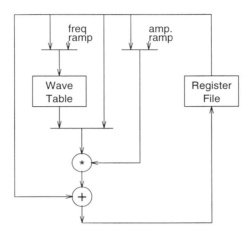

Figure 5.7   IRCAM 4B data path

Alles [Alles, 1977] further expands on the 4B oscillator module by describing the design of a matching filter and reverb module using the TRW multiplier. A separate "switch" module allows for the arbitrary interconnection of the functional units. The switch itself was a small processor with a simple instruction set.

The following problems were identified in the 4B:

- Undersampling the envelopes resulted in audible artifacts
- The interconnect was "hard-wired" and could not be reprogrammed by software
- Since parameter memory and interconnect memory were separate, it was impossible to mix the two

The next generator synthesizer designed by DiGuigno was named the 4C [Moorer et al., 1979]. The 4C represented a technological improvement over the 4B by using larger memories and a variable interconnect. It has five basic functional unit generators: (2) Wavetable oscillators, (2) multiply/accumulators (1) envelope generator, (1) output and (1) timer. Each one of these unit generators is time multiplexed 32 times. A block diagram of the data path of the machine is shown in figure 5.8.

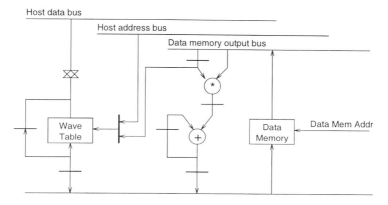

Figure 5.8 IRCAM 4C data path

Redrawn Figure 17.4 (omits some paths) from [Moorer et al., 1979]

Moorer [Moorer, 1981] points out the following shortcomings in the design of the 4C:

- Short table length: Short tables without interpolation result in distortion particularly with stored low frequency sounds

- Lack of time synchronization with updates: Because the 4C clock is different from the host (PDP-11) clock, it is impossible to change parameters sample synchronously.

- Fractional multiplier: There is no way to increase the magnitude of a product except by hacking the output of the multiplier.

The last 4n machine from IRCAM was the 4X [Asta et al., 1980]. The system block diagram of the machine is shown in figure 5.9.

Synthesis units are controlled from the host via an "interface unit" that permits DMA transfers to the private bus (shown here in three buses Data,Ctrl and Addr). The generators can input and output data to and from the data bus under control from the

216   APPLICATIONS OF DSP TO AUDIO AND ACOUSTICS

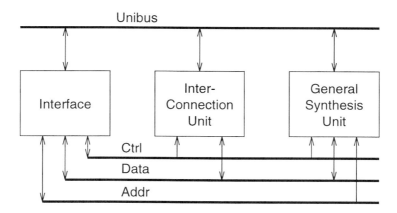

Figure 5.9   IRCAM 4X system block diagram

interface unit. The interconnection unit serves to connect the generator units together via an interconnection memory implemented with queues (FIFOs). It also contains timers for use in controlling amplitude and frequency envelopes. Unlike the previous 4n machines, the 4X included numerous "bypass" paths in the arithmetic pipeline of the synthesis units.

The combination of the AMD 2901 4-bit slice together with the TRW multiplier was very popular. A Sony design was used for a microprogrammable real-time reverberator [Segiguchi et al., 1983]. They used five 2901s together with the TRW multiplier to create a machine with a 170 ns cycle time. A simple 16 bit pipelined machine designed for music synthesis was described by Wallraff [Wallraff, 1987]. The machine was flexible enough to be used in other signal processing tasks. Another example of such a machine was used in an early audio coder [Brandenburg et al., 1982].

The TRW multiplier also found its way into the center of the Sony DAW-1000A editor [Sony, 1986]. A simplified block diagram of the signal processor section is shown in figure 5.10

Note that products can be fed backward; also note that crossfades are found in ROMs which are input to just one side of the multiplier; the inputs including faders and audio inputs are fed to the other side after been converted via logarithmic ROMs.

A very fast CPU customized for the calculation of second order sections was used by Neve [Lidbetter, 1983] (in collaboration with the BBC) in 1979. They used a Time Division Multiplex (TDM) parallel data bus with 16 time slots. At each time slot, the "channel processor" executed 14 microinstructions (at a rate of 100 ns) of a filter

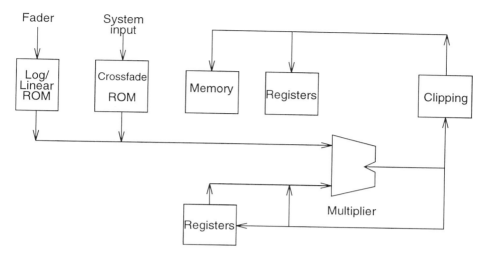

Figure 5.10 Sony DAE-1000 signal processor

program. The TDM output of the channel processor(s) were combined and simply mixed by a separate mixing processor.

The COmputer for Processing Audio Signals (COPAS) [McNally, 1979] was an early microprogrammed mixer. It used a 48 bit microinstruction word executed every 163 ns (about 6 MHz) but could have been improved with a faster multiplier. For a sampling rate of 32 KHz, the machine can execute 191 instructions. This was enough to do 10 biquads (second order sections). It should be noted that the A/D was routed through a microprocessor which offloads the main computational resource.

The Lucasfilm *SoundDroid* [Moorer, 1985b] was a complex system that included a complicated microcoded machine as the arithmetic processor. The first edition of this machine used a horizontally microprogrammed ECL processor called the Audio Signal Processor (or ASP) [Moorer et al., 1986][Moorer, 1983]. The ASP had a 50 ns instruction clock and executed 400 MAcs per 48 KHz sample time by pipelining multiplies and starting a multiply every microcycle. A separate host machine controls the ASPs via a controller over a private bus. The various ASPs have connections to high speed I/O devices whereas the ASP controller takes care of the slower speed DACs and ADCs. Figure 5.11 shows the ALU data path of the ASP (redrawn and simplified from figure 5.3 from [Moorer et al., 1986]). To avoid pipeline "bubbles", the ASP placed branch decisions in the data path by using the result of comparisons to control the output of multiplexers.

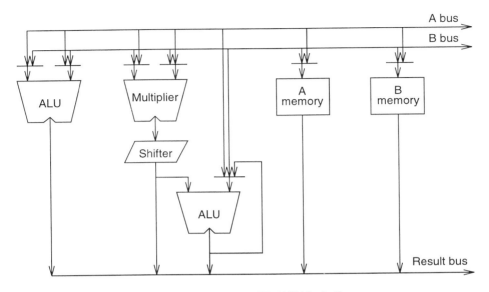

Figure 5.11 Lucasfilm ASP ALU block diagram

The large dynamic RAMs and interconnections to the update and DSP bus are shown in figure 5.12. A direct connection between the disk drives and the signal processor's DRAM banks was provided in Lucasfilm's ASP and SoundDroid processors, enabling real-time multitrack digital audio storage and retrieval for mixing desks.

A special feature of the ASP is the time ordered update queue [Moorer, 1985a] [Moorer, 1980a] shown in figure 5.13 (labeled "255 element queue"). The queue is pipelined and uses pointers implemented in hardware. This queue can be used to update data in either the dynamic RAMs or the static RAM coefficient memories. A more striking application is the use of the update queue to automatically change the microcode on external events. It should be obvious that such a queue is of particular utility when dealing with time stamped musical events. Furthermore, it was equipped with a bypass for passing real-time events (such as the first slope of an envelope upon note-on) to the head of the queue, with associated events (e.g. the subsequent envelope slopes and targets) inserted in time order into the queue.

At this point, the stage is set for the transition from MSI scale logic into LSI and VLSI processors.

Sequential single processors. With the constant improvement of integration technology, it became possible to include a 16 by 16 multiplier as part of the ALU of a

DIGITAL AUDIO SYSTEM ARCHITECTURE   219

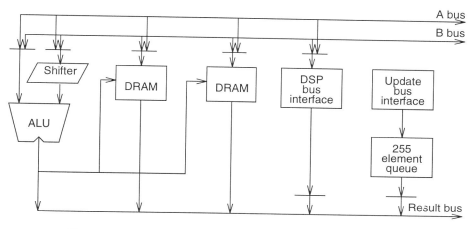

Figure 5.12   Lucasfilm ASP interconnect and memory diagram

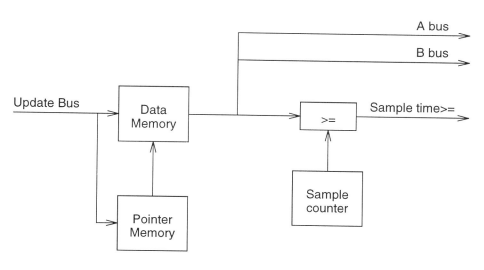

Figure 5.13   Moorer's update queue data path

microprocessor. Together with addressing modes designed for DSP applications (such as bit reversal for FFTs and modular addressing for delay lines and filter coefficients), this produced a powerful combination. Serial input and output was also introduced with DSPs. Serial I/O is quite reasonable for handling audio rates for a small number of channels. Many fixed point processors only have 16 bits of data and address. The limited amount of address bits implies a limited amount of time for reverberation algorithms (at 44.1 KHz sampling rate, 16 bits of address is approximately 1.5 seconds.)

Although other DSPs were utilized before 1982, it was the introduction of the Texas Instruments TI (Texas Instruments),TMS320 series that dramatically changed the environment for DSP algorithm designers. For the first time, an inexpensive, commercially available machine was capable of computing speech and modem algorithms in real-time.

Lee [Lee, 1988][Lee, 1989] surveyed processor architecture circa 1988. He pointed out that principal differences between the arithmetic sections in integer microprocessors and DSPs are:

- more precision
- use of saturation arithmetic
- the ability to accumulate and shift products

The memory organization of DSPs are also different from "ordinary" processors because (1) Memory is typical static RAM and virtual memory support is totally absent (2) Several machines separate data and instruction streams (Harvard Architecture) (at the cost of extra pins). Additionally, modular arithmetic address modes have been added to most processors. This mode finds particular utility in filter coefficient pointers, ring buffer pointers and, with bit reversed addressing, FFTs. One further difference is the use of loop buffers for filtering. Although often called "instruction caches" by the chip manufacturers, they are typically very small (for example, the AT&T DSP-16 has 16 instructions) and furthermore, the buffer is not directly interposed between memory and the processor.

Fixed Point. Texas Instruments introduced the TMS320C10 in 1982. This chip captured a sizable market share due to its simple instruction set, fast multiply accumulate and DSP addressing modes. The TMS320 also features a "Harvard Architecture", which doubles the number of address and data pins but also doubles the bandwidth. The TMS320C10 was followed by the TMS320C20. This chip continues to have considerable market share. The latest edition of this chip is the TMS320C50, which has four times the execution rate of the original C10. There are other members of the family [Lin et al., 1987] that implement various aspects of I/O interfaces and memory configurations.

AT&T introduced a DSP early on called the DSP2 [Boddie et al., 1981]. Although for internal consumption in Western Electric products, this chip led the way to the

DSP-16. The DSP-16 was extremely fast and difficult to program. It featured only two accumulators and 16 registers (two of these were special purpose). The serial I/O also contains a small TDMA section that can be used for multiprocessor communication (see section 5.17).

Motorola introduced the 56000 [Kloker, 1986] in 1986. The 56000 has been used quite successfully in many digital audio projects because

- 24 bits of data width provided room for scaling, higher dynamic range, extra security against limit cycles in recursive filters, better filter coefficient quantization and also additional room for dithering.

- The large accumulator (a 24 bit by 24 bit product + 8 bits of guard = 56 bits of accumulator length) provided a large guard band for filters of high order.

Other positive aspects of the 56000 include memory moves in parallel with arithmetic operations and modular addressing modes [Kloker and Posen, 1988]. The 56000 was used by the NeXT MusicKit [Smith et al., 1989] very effectively.

The Zoran 38000 has an internal data path of 20 bits as well as a 20 bit address bus. The two accumulators have 48 bits. It can perform a Dolby AC-3 [Vernon, 1995] five channel decoder in real time, although the memory space is also limited to one Megaword. It has a small (16 instruction) loop buffer as well as a single instruction repeat. The instruction set has support for block floating point as well as providing simultaneous add and subtract for FFT butterfly computation.

Sony introduced a DSP (the CXD1160) that was quietly listed in the 1992 catalog [Sony, 1992]. It has an astonishingly short instruction memory – only 64 instructions. Likewise, the coefficient and data memories are also 64 locations. The CXD1160 is unusual since it has a DRAM interface directly on chip. It also has a special 40 bit wide serial interface that permits an external host to download instructions, coefficients or data into the chip. This path is used to great effect in the (see section 5.19). The serial I/O is designed to output stereo samples in one sample time; another uncommon feature. The CDX1160 was superceded by the CXD2705 [Hingley, 1994] but was never released to the public.

Other manufacturers have introduced 16 bit fixed point processors. IBM's Mwave is supposed to be for "Multimedia" operations but can only address 32K (15 bits) of data memory. The product register is also only 32 bits (data is assumed to be fractional form of only 15 bits) so constant rescaling is necessary. Perhaps its most noteworthy addition is the wide use of saturation arithmetic in the instruction set and the large number of DMA channels.

Analog Devices also has a 16 bit DSP (the ADSP-2100 series [Roesgen, 1986]) that has found some limited use in audio applications. The 2100 series has limited on chip memory and a limited number of pins (14) for the external memory. Use of a common bus for arithmetic results limits the amount of processor parallelism. However, unlike other 16 bit processors, the ADSP-2100 series can be nicely programmed for multiple precision operations.

The desire to keep many functional units operating in parallel through software scheduling inspired Very Long Instruction Word (VLIW) architectures [Fisher, 1983]. Recently, new VLIW architectures for fixed point signal processing have been released. MPACT [Kalapathy, 1997] (by Chromatic) is a VLIW machine designed for both audio and video. As shown in figure 5.14, it has a large ALU and a fast memory interface.

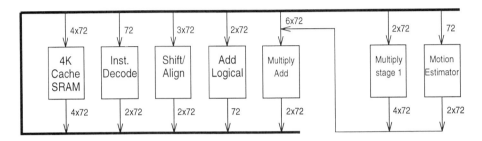

Figure 5.14 MPACT block diagram

Notice how the outputs of the functional units are connected to a huge 792 (72 by 11) wire bus. This acts as a crossbar between outputs and inputs permitting arbitrary connections. The large number of bus wires also permits a very large bus traffic rate reaching one Gigabyte/second in the second generation part.

Floating Point. Integrated floating point units first arrived as separate coprocessors under the direct control of the microprocessor. However, these processors performed arithmetic with numerous sequential operations, resulting in performance too slow for real-time signal processing.

AT&T introduced the first commercially available floating point DSP in 1986 [Boddie et al., 1986]. An important characteristic of the DSP-32 is the exposed four stage pipe. This means scheduling is left up to the compiler or assembler. It was notable, at the time, for its computational speed, but not its ease of use. The DSP-32 ran at 4 million multiply accumulates (MACs) per second. The integer section executed 4 MIPs at peak rate. The processor has double buffered serial DMA so the processor need not be directly involved with stuffing data into I/O registers. Originally fabricated in NMOS, it was recreated in CMOS as the DSP-32C [Fuccio et al., 1988]. The DSP-32C ran at twice the speed of the NMOS version and also increased the address space to 24 bits. Note that the DSP-32 performed floating point arithmetic in an internal format that was not compatible with IEEE-754 [IEEE Computer Society Standards Committee, 1985]. The DSP-32C kept this incompatibility but introduced new instructions to convert back and forth from the internal format to the IEEE 32 bit format. The DSP-32C also introduced interrupts which the DSP-32 did not implement.

Texas Instruments C30/C40/C80

The TMS320C30 [Papamichalis and Simar, 1988] follows the basic architecture of the TMS-320 series. Unlike the DSP-32, it uses pipeline interlocks. Like the DSP-32, it features its own internal format for floating point numbers. Because of the four stage pipeline organization, it can perform a number of operations in parallel. It also features a delayed branch - something of a novelty in DSP processors. The TMS320C40 [Simar et al., 1992] has six parallel bidirectional I/O ports controlled by DMA on top of the basic TMS-320C30 architecture. These ports have been used for multiprocessor communication.

Motorola 96000

Motorola introduced the 96002 [Sohie and Kloker, 1988] as an extension to the existing 56000 architecture. The instruction set is an extension of the 56000 instructions, adding floating point instructions and implementing the IEEE 754 floating point standard directly instead of converting to an internal format (like the DSP-32). It has two parallel ports for multiprocessing.

ADSP-21000

The Analog Devices ADSP-21000 series offers IEEE arithmetic like the 96000 while maintaining the instruction format and addressing modes of the earlier 2100 series. The 21020 processor has a large external memory bus and can process a large I/O rate. The relatively new Analog Devices SHARC (a.k.a. 21060) contains 256K of static memory integrated on the chip along with the processor. The fast internal memory avoids problems with memory stalls, but at the cost of a large die and a hefty price tag. The SHARC also has a large number of pins resulting from the large number of parallel ports (6 nibble wide ports and one byte port). This too increases chip cost but like the TMS320C40 can be used for multiprocessor communication.

Applications. Besides the use of single DSPs as processor adjuncts (as in the NeXT machine [Smith et al., 1989]), the WaveFrame Corporation introduced a modular system [Lindemann, 1987] that uses the notion of a time division multiplexed bus (like the FRMBox). Each slot was 354 ns. A single mixer board [Baudot, 1987] had two memories, one for coefficients and the other for delayed sampled and past outputs. These memories were multiplied together by a 32 by 32 bit multiplier and a 67 bit (3 guard bits) multiplier accumulator. The coefficients can be updated via a one Megabit/second serial link (using 24 bit coefficients, that means one update every 24 microseconds.) The updates are calculated with a DSP chip. However, the DSP chip was not fast enough to do the linear interpolation, so the interpolator was done with a hardware multiplier.

Custom chips. Because of the computational demand of multichannel or multivoice audio processing, custom silicon has been proposed as one way to solve this basically intractable problem. Also, going to custom silicon provides a manufacturer with a not-so-subtle way to hide proprietary algorithms – it's harder to reverse engineer a chip than it is a program for a commercially available processor.

Yamaha. Yamaha has designed numerous custom chips to support its commercial line of music boxes. A number of relevant details can be found in Yamaha's patents. The famous DX-7 has two chips: the first one was an envelope generator; the second one generated the actual samples. The interconnection between these two sections can be found in patents from 1986 and 1988 [Uchiyama and Suzuki, 1986][Uchiyama and Suzuki, 1988]. These patents also describes the use of logarithmic numerical representation to reduce or eliminate multiplication and the use of Time Division Multiplexing (TDM) for multivoice computation. The use of logarithmic representation can be seen in the FM equation (equation 5.18). This is calculated from the inside out as follows from a phase angle $\omega_n t$:

1. Lookup $\omega_m t$ in the *logarithmic* sine table

2. Read the modulation factor $I(t)$ and convert to logarithmic form

3. Add (1) to (2) giving $\log(\sin \omega_m t) + \log(I(t))$

4. Convert back to linear form via an anti-log table giving $I(t) \sin(\omega_m t)$

5. A shift S can be applied, multiplying the output by a power of 2 resulting in $S\ I(t) \sin(\omega_m t)$

6. The carrier is added in, $\omega_c t$ forming $\omega_c t + S\ I(t) \sin(\omega_m t)$

7. This is looked up in the logarithmic sine table in preparation for envelope scaling: $\log(\sin(\omega_c t + S\ I(t) \sin(\omega_m t)))$

8. Finally, the log of the amplitude term $A(t)$ is added to the previous step and looked up in the anti-log table giving $A(t)\ \sin(\omega_c t + S\ I(t) \sin(\omega_m t))$

As remarked earlier, the use of logarithmic arithmetic to avoid multiplication fits in well with floating point converters.

Yamaha has also patented sampling architectures (see Massie's chapter for more information on sample rate conversion and interpolation in samplers). A recent patent [Fujita, 1996] illustrates how fractional addressing from a phase accumulator is used by an interpolation circuit to perform wide range pitch shifting.

Wawrzynek. Wawrzynek [Wawrzynek and Mead, 1985][Wawrzynek, 1986] [Wawrzynek, 1989] proposed the use of "Universal Processing Elements" (UPEs). A UPE implements the following equation (5.20):

$$y = a + (m * b) + (1 - m) * d \qquad (5.20)$$

where a can be used to sum previous results, m is the scaling factor and b and d are arbitrary constants. When $d = 0$, then the UPE computes a sum and a product. When $d \neq 0$, then the UPE computes a linear interpolation between b and d. The UPEs are implemented in serial fashion, as first advocated by Jackson, Kaiser and McDonald [Jackson et al., 1968] and further expounded by Lyon [Lyon, 1981]. The UPEs can be used to implement filters, mixers and any sum of products. Note that tables are lacking, therefore trigonometric functions must be approximated via their Taylor series expansions or via a recursive oscillator (mentioned in section 5.3.3). A second generation chip was proposed by Wawrzynek and von Eicken [Wawrzynek and von Eicken, 1989] that included interprocessor communication however the parameter update bandwidth was severely limited.

E-mu. E-mu Systems (now a subsidiary of Creative Technologies) has designed and patented a number of custom chips for use in their synthesizers and samplers. One example [Rossum, 1992] uses filter coefficient interpolation; the data path provides for dual use of the multiplier; in one use, it is part of the interpolation machinery, in the other path it is used to form the convolutional product and sum. A later example [Rossum, 1994a] uses four times oversampling, pipelining and a pointer arithmetic to implement a basic looping looping (see the chapter by Massie for more information on sampler implementation) There are four memories on-chip: one for the current address, one for the fractional part of the address (as part of the phase interpolator), one for the phase increment and finally, an end point. When the address exceeds the end point, then the memory pointer address is reset. Only one ALU is used on the chip and the output is fed back to the input of the RAMs. It should be noted that this patent also includes the use of logarithmic representations to avoid multiplication: all sounds are stored logarithmically so they can be scaled by the amplitude waveform with a simple addition.

Rossum also proposed the use of a cache memory [Rossum, 1994b] as part of the memory lookup path in a sampler interpolator. Since in many cases, the phase increment is less than one, the cache will be hit on the integer part of the table address, consequently, the memory will be free to use for other voices.

This is illustrated in figure 5.15.

IRIS X-20. The X-20 [Cavaliere et al., 1992] was designed by the IRIS group as a fundamental component of the MARS workstation. It can be considered a VLSI

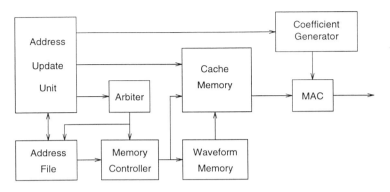

Figure 5.15 Rossum's cached interpolator

follow-on of the 4n series of IRCAM. It had two data memories, one ROM, a 16 by 16 multiplier with a 24 bit output and a 24 bit ALU. It executed from an external fast static memory of 512 instructions. With a clock rate of 20 MHz, it executed at a 40 KHz sample rate.

Sony OXF. Sony designed a custom chip for use in the OXF mixer [Eastty et al., 1995]. It is quite interesting with two different sections: one section devoted to signal processing and the other devoted to interpolation of coefficients. This is shown in figure 5.16.

All signal paths in the processor are 32 bits with a few exceptions. There are four basic sources: one from the vertical bus, one from the horizontal bus, one from DRAM and one fed back from the output of the ALU or multiplier. Note that products must be shifted before they can be accumulated.

The interpolation processor has a double buffered memory that permits updates during processing. Note that interpolation happens on *every* sample, thereby avoiding "zipper noise" due to coefficient quantization. This is an expensive strategy, however it always works.

Ensoniq ESP2. Ensoniq described a custom chip in a patent [Andreas et al., 1996] that included the following features: A single cycle average instruction (add and shift right 1) and a limit instruction (used for checking end points of ramps). It included a special purpose address generator unit (AGEN) that is connected directly to the memory address bus. The AGEN was designed with reverb implementation in mind. Memory is divided into 8 different regions; addresses are wrapped within region boundaries.

DIGITAL AUDIO SYSTEM ARCHITECTURE   227

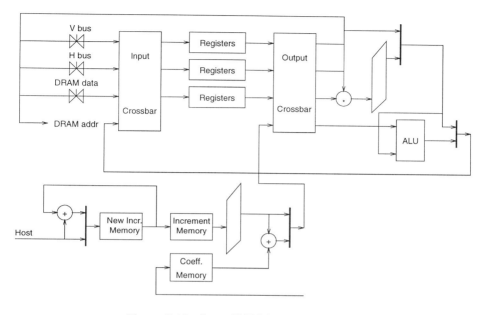

Figure 5.16   Sony OXF DSP block diagram

The AGEN offers the ability to update the base register or address register during the memory cycle.

Parallel machines. Interprocessor communication becomes the problem to solve (for algorithms that won't fit on a single processor, this assumes that the algorithm can be decomposed for multiple processors and intermediate results can be communicated between cooperating processors).

Typical approaches to loosely coupled multiprocessor architectures assume that communication from one processor to another must be dynamic in nature, in particular, that the destination address of one datum can and will change from time to time. Furthermore, such machines also assume that communication is generally uncontrolled and the frequency of communication can vary from sporadic to overwhelming. This is the most general case, but suffers because the interconnection network must accommodate ever changing destination addresses in packets. If the algorithm has *static* addressing patterns, the need for dynamic routing hardware can be totally eliminated, thereby saving cost and complexity.

228   APPLICATIONS OF DSP TO AUDIO AND ACOUSTICS

Cavaliere [Cavaliere, 1991] briefly reviews several parallel DSP architectures for audio processing with an eye toward examining the interconnection strategies and their use for audio algorithms.

Serial Interconnects.   Audio serial rates (768 Kilobits/second per channel at 48 Kilosamples/second) present an opportunity for parallel DSP. By using standard telecommunication time division multiplexing (TDM), it's possible to "trunk" multiple channels as well as use switching techniques.

DSP.* [Kahrs, 1988] (The name is a throwback to Cm* [Gehringer et al., 1987]) was designed around the TDM philosophy exemplified by Moore's (see section 5.3.4). An architectural block diagram of the system is shown in figure 5.17.

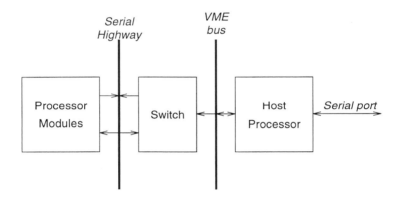

Figure 5.17   DSP.* block diagram

As shown in figure 5.17 processor modules are connected to the serial highway. Each processor module has a unique address on the highway (this address is wired on the card). There are two traces on the backplane per processor: one for the input to the processor (from the switch) and one from the processor (to the switch). All interprocessor communication is via the switch. The switch sits between the host VME bus and the host processor and is programmed by the host processor.

All processors must use the same serial clocks to stay in synchronization with the sample frame clocks. These clocks are buffered by the host interface card and put on the backplane. The card also generates the processor clocks, so the processors can also run in synchronization.

There are two DSP-32s per "processor module". The "master" (with external memory) is connected to the serial lines going to and from the crossbar switch. The "slave" is connected to external I/O devices such as A/D and D/A converters. It is responsible for various I/O operations such as pre and post filtering. The 8 bit parallel data ports of both DSP-32s are indirectly connected to the VME bus. The master DSP-32 can also select which word (load) clocks to use on the transmit side to the switch via the memory mapped I/O; this helps limit the transmission bandwidth (Although serial DMA doesn't directly involve the arithmetic sections of the DSP-32, it does cause wait states to be inserted.)

The time slot interchange switch uses a commercially available telephone switch, the Siemens PEB2040. Besides being able to operate up to 8 megabits per second, it can be programmed to be either a time division, space division or "mixed" time/space division switch.

Unfortunately, the output connections must be fixed depending on the interconnection scheme chosen (time, space or mixed). A "mixed" space/time switch for 16 lines (processors) at 8 Mbits/second requires 32 integrated circuits.

In the worst case, changing a single connection in the switch can take a full frame time. This means that changing the entire topology of the switch is not an action to be taken lightly. However, simple changes can be done relatively rapidly.

Gnusic

The purpose of the *Gnusic* [Kahrs and Killian, 1992] project was to build a music I/O device for an experimental workstation capable of "orchestral synthesis". "Orchestral synthesis" means the synthesis of large number of voices in real time. The basic architecture is shown in figure 5.18.

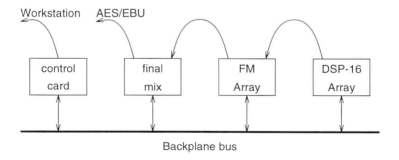

Figure 5.18 Gnusic block diagram

Control and data signals from the 68020 based workstation are bused over to a control card that buffers the signals and places them on the backplane (not illustrated). The instrument cards are plugged into the backplane and controlled by the host processor. Each instrument card has a digital signal processor (an AT&T DSP-16) for mixing the digital outputs as follow: This processor must (a) mix the output from the card "upstream" with the sample generated locally, and (b) perform any effects desired with the leftover time. Such effects might include filtering or feedback control of the on-board oscillators. The basic DSP interface includes a pseudo-dual-ported static program memory for the DSP-16. Because the DSP-16 has a totally synchronous memory interface, memory access must be shared between the 8-bit processor bus and DSP; when the processor sets the DSP to "run", it also prevents itself from accessing the memory.

The "final mix" card is last in the chain. It has the basic ("core") DSP-16 circuitry found on *all* instrument cards. A block diagram is found below:

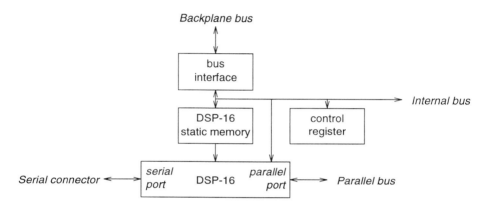

Figure 5.19 Gnusic core block diagram

The host can write into the memory of the DSP-16, but only when the DSP-16 is stopped. The DSP-16 is too fast to allow true dual port access. This is perfectly acceptable since the core program typically doesn't change when the synthesizer is running. The host can also set a control register which contains the run flag and other useful bits.

The DSP-16 can either be a master (called "active" in DSP-16 terminology) or a slave ("passive"). All the DSP-16s *except* the final mix DSP are in passive mode. They are fed clocks from the final mix DSP. This guarantees that all DSPs are on the same output clock. The final mix DSP also provides the left/right clock so that the channels are synchronized as well. The serial data output of the DSP-16 is fed to the serial data input of the next DSP-16 in line. All of the serial I/O is done via flat ribbon cables on the end of the cards.

There are two basic kinds of instrument cards: an FM card and an array of DSP-16s. The DSP-16 array uses the TDM serial bus feature of the DSP-16 and therefore is discussed extensively below. The DSP-16 array card has four DSP-16s and a core DSP-16. The core DSP addresses the 4 satellite DSPs via the 16 bit wide parallel I/O bus of the DSP-16. The core can address any of the satellites as well as address a large external memory specifically designed for reverberation. The serial I/O of the satellites are connected together in a TDM bus using the on chip logic of the DSP-16. The DSP-16 multiprocessor interface permits up to eight processors to be connected on a serial bus.

The data, clock, serial address and sync pins are all bused together. Each processor has its own address; data can be sent from one processor to another in a specific slot. Slots must be reserved (i.e., statically allocated) beforehand as there is no bus contention mechanism. Furthermore, each processor must have a unique time slot "address".

The host has the same interface to the memory of the satellites as it does to the memory of the core DSP-16. It also has a 2K × 8 FIFO attached to the parallel I/O bus of the core for use in parameter passing from the host. Status bits from the FIFO can be used to interrupt the core DSP should the FIFO become too full and risk data overrun.

Sony SDP-1000

The Sony SDP-1000 [Sony, 1989] was an interesting multiprocessor designed around a serial crossbar interconnect. The controlling machine itself can be divided into three sections:

1. A host section featuring a host processor with a graphics processor and video RAM

2. A separate I/O section controlled by a microcontroller including digitizing the trackball and 8 slide potentiometers

3. DSP processing section.

The DSP processing section is shown in figure 5.20. Basically, the 24 serial inputs and 24 serial outputs of the DSPs are connected to a custom crossbar chip. Also included are 8 serial inputs from the outside world and 8 more serial outputs to the outside world which are the final output of the machine.

The crossbar interconnect is under the control of the microprocessor. The DSPs are programmed remotely, via the microprocessor by the processor's serial port (see section 5.13). The processors all run lock-step and are sample synchronous. Note that only four processors have external memory (and only 64K). This severely limits the reverberation time but in fairness, this machine was not designed for effects processing. Also note that coefficient conversion from real time inputs must take place in the host processor and then be converted into serial form and placed in the specific DSP.

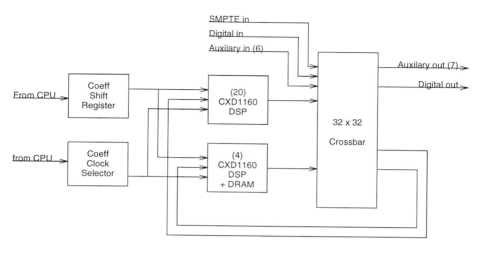

Figure 5.20 Sony SDP-1000 DSP block diagram

Parallel Interconnects. Serial interconnects have the distinct advantage of being easy to connect and easy to multiplex. However, it must be remembered that serial must be converted back to parallel sometime and therefore parallel interconnects can be used if the conversion latency is to be avoided.

The DSP3 [Glinski and Roe, 1994][Segelken et al., 1992] uses a custom interconnect that provides for a four port (NEWS) interconnect. Each board has a 4×4 array of these processors, memories and interconnects. Of the 16 possible outputs at the card edge, eight go to the backplane and the other eight are connected in a toroidal fashion. The backplane has a capacity to handle eight of these boards for a grand total of 128 processors.

The IRCAM Signal Processing Workstation (ISPW) [Lindemann et al., 1991] was designed around a pair of Intel i860s [Intel, 1991]. The i860 was, for its time, an extremely fast and extremely large and expensive chip. It featured a 128 bit internal data bus, a 64 bit instruction bus and internal on chip caching. Its pipeline permitted both scalar and vector operations but the pipeline must be managed explicitly (such as register forwarding and pipe interlocks). The interconnect is a 64 bit by 64 bit crossbar under the control of the host (in this case a NeXT cube). Crossbar interconnects are very flexible (see C.mmp [Wulf et al., 1981]) but can't be expanded. So this makes the two processor 64 bit interconnect a unique design for its time.

Snell [Snell, 1989] also used a crossbar (also influenced by C.mmp) to interconnect 56000s. He connected three 56001s and three DRAM banks to an integrated crossbar

DIGITAL AUDIO SYSTEM ARCHITECTURE   233

with a 7th port connected to a disk drive interface and the 8th port to an expansion port for connection to another crossbar switch for further multiprocessor expansion. Mechanical design limited this expansion technique to several levels, sufficient for interconnection of over a hundred signal processors in a tightly coupled system. He separated the static RAM from the dynamic RAM so that other processors can access the dynamic ram via the crossbar. However, both processors must agree beforehand otherwise a conflict will result. A subsequent design (J. Snell, personal communication, 1997) fixed this problem by integrating FIFOs, a DMA unit and DRAM controller into the crossbar switch.

The Reson8 machine [Barrière et al., 1989] is of many multiprocessors built using Motorola 56000s. Eight Motorola 56000s (see section 5.13) are interconnected on a single shared bus. One of the 56000s is the master, the remaining seven are slaves. The master is also responsible for moving data from one processor to another. It's worth noting that they avoided the use of DRAM because of the added complexity (the limitation in algorithm implementation was considered). Dual-port RAM was used for interprocessor communication, perhaps an influence from their earlier work at Waveframe.

Eastty, et al. [Eastty et al., 1995] describe a very large digital mixer composed of custom chips (see section 5.15 for a description of the custom chip). The interconnect is shown in figure 5.21.

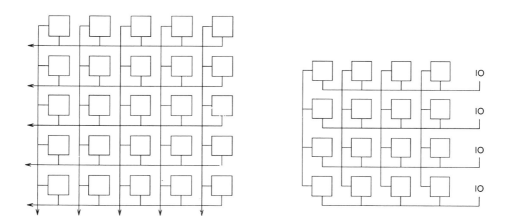

Figure 5.21   Sony's OXF interconnect block diagram

Each board is composed of a 5 by 5 array of processor chips (shown on the left of figure 5.21) connected to the backplane via a horizontal and vertical bus (so each card has 10 buses × 33 bits/bus = 330 signal pins). In turn, the backplane connects the cards in a 4 by 4 matrix where the each bus connects to 3 other cards and an input/output card (shown on the right of figure 5.21).

5.4 CONCLUSION

The processing of audio signals has made great strides since the development of digital synthesizers and their implementation on commodity DSPs. Further improvement will result from better tuning architectures to the specific demands in audio processing, including attention to parameter updates, coefficient calculation and well designed arithmetic. This also includes A/D and D/A converters with improved "sonic" capabilities; the move to 96 KHz sampling rates and longer samples will again push the state of the art in converter design.

This work has been partially supported by ARPA grant DAAL01-93-K-3370. John Snell offered many detailed comments on a draft of this chapter.

6 SIGNAL PROCESSING FOR HEARING AIDS

James M. Kates

AudioLogic
4870 Sterling Drive
Boulder, CO 80301

Jim@audiologic.com

Abstract: This chapter deals with signal processing for hearing aids. The primary goal of a hearing aid is to improve the understanding of speech by an individual with a hearing impairment, although the perception of music and environmental sounds is also a concern. The basic signal-processing system consists of linear filtering followed by amplification, with more sophisticated techniques used to try to compensate for the nature of the hearing impairment and to improve speech intelligibility in noise.20

The chapter starts with a review of auditory physiology and the nature of hearing loss. Linear amplification systems are then discussed along with hearing-aid design objectives and the limitations of conventional technology. Feedback cancellation, which can improve hearing-aid system stability, is presented next. Dynamic-range compression is an important signal-processing approach since the impaired ear has a reduced dynamic range in comparison with the normal ear, and single-channel and multi-channel compression algorithms are described. Noise suppression is also a very important area of research, and several single-microphone approaches are described, including adaptive analog filters, spectral subtraction, and spectral enhancement. Multi-microphone noise-suppression techniques, such as adaptive noise cancellation, are discussed next. Noise can be more effectively suppressed using spatial filtering, and directional microphones and multi-microphone arrays are described. The chapter concludes with a brief summary of the work being done in cochlear implants.

Table 6.1 Hearing thresholds, descriptive terms, and probable handicaps (after Goodman, 1965)

Descriptive Term	Hearing Loss (dB)	Probable Handicap
Normal Limits	-10 to 26	
Mild Loss	27-40	Has difficulty hearing faint or distant speech
Moderate Loss	40-55	Understands conversational speech at a distance of 3-5 feet
Moderately Severe Loss	55-70	Conversation must be loud to be understood and there is great difficulty in group and classroom discussion
Severe Loss	70-90	May hear a loud voice about 1 foot from the ear, may identify environmental noises, may distinguish vowels but not consonants
Profound Loss	> 90	May hear loud sounds, does not rely on hearing as primary channel for communication

6.1 INTRODUCTION

Hearing loss is typically measured as the shift in auditory threshold relative to that of a normal ear for the detection of a pure tone. Hearing loss varies in severity, and the classification of hearing impairment is presented in Table 6.1[Goodman, 1965].

Approximately 7.5 percent of the population has some degree of hearing loss, and about 1.0 percent has a loss that is moderately-severe or greater[Plomp, 1978]. There are approximately 28 million persons in the United States who have some degree of hearing impairment[National Institutes of Health, 1989]. The majority of the hearing-impaired population has mild or moderate hearing losses, and would benefit significantly from improved methods of acoustic amplification. Hearing aids, however, are not as widely used as they might be. Even within the population of hearing-aid users, there is widespread discontent with the quality of hearing-aid amplification [Kochkin, 1992].

One of the most common complaints is that speech is especially difficult to understand in a noisy environment. Pearsons et al. [Pearsons et al., 1976] have shown that in noisy environments encountered in everyday situations, most talkers adjust their voices to maintain a speech-to-noise ratio of 7 to 11 dB. While normal-hearing individuals usually have little difficulty in understanding speech under these conditions, users of hearing aids, or other sensory aids such as cochlear implants, often have great difficulty. In general, the signal-to-noise ratio (SNR) needed by a hearing-impaired person to give speech intelligibility in noise comparable to that for speech in quiet is substantially greater than the corresponding SNR required by a normal-hearing person [Plomp, 1978].

While most commercial hearing aids are still based on analog signal processing strategies, much research involves digital signal processing. This research is motivated by the desire for improved algorithms, especially for dealing with the problem of understanding speech in noise. Cosmetic considerations, however, limit what can be actually implemented in a practical hearing aid. Most users of hearing aids want a device that is invisible to bystanders and thus does not advertise their impairment. As a result, the strongest pressure on manufacturers is to put simple processing into the smallest possible package, rather than develop sophisticated algorithms that require a larger package. Thus practical signal processing, as opposed to research systems, is constrained by the space available for the circuitry and the power available from a single small battery. In order to be accepted in such a market, digital signal-processing systems will have to demonstrate enough performance benefits over their analog counterparts to justify their larger size, shorter battery life, and higher cost.

The emphasis in this chapter is on digital processing algorithms for moderate hearing losses caused by damage to the auditory periphery. Analog processing is also described to give a basis for comparison. The chapter begins with a discussion of peripheral hearing loss, since the behavior of the auditory system motivates hearing-aid algorithm development. Linear hearing aids are then described, followed by the presentation of feedback cancellation to improve the linear system performance. Single-channel and multi-channel compression systems are then presented. Improved speech intelligibility in noise is the next subject, with both single-microphone and multi-microphone noise suppression discussed. The chapter concludes with a brief discussion of cochlear implants and a summary of the hearing-aid material presented.

6.2 HEARING AND HEARING LOSS

The design of a hearing aid should start with a specification of the signal processing objectives. A useful conceptual objective for a peripheral hearing loss is to process the incoming signal so as to give a perfect match between the neural outputs of the impaired ear and those of a reference normal ear. Implementing this ideal system would require access to the complete set of neural fibers in the impaired ear and to a corresponding set of outputs from an accurate simulated normal ear. The simulated neural outputs could then be substituted directly for the neural responses of the impaired ear. In designing

238   APPLICATIONS OF DSP TO AUDIO AND ACOUSTICS

a hearing aid, however, one can only indirectly affect the neural outputs by modifying the acoustic input to the ear. Hearing-aid processing is thus a compromise in which the acoustic input is manipulated to produce improvements in the assumed neural outputs of the impaired ear.

6.2.1 Outer and Middle Ear

The nature of the signal processing, and its potential effectiveness, depends on the characteristics of the auditory system. The ear transforms the incoming acoustic signal into mechanical motion, and this motion ultimately triggers neural pulses that carry the auditory information to the brain. The essential components of the ear are shown in Fig 6.1.

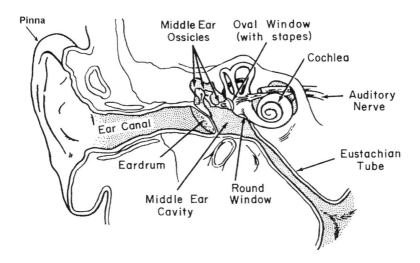

Figure 6.1   Major features of the human auditory system

The ear is divided into three sections, these being the outer, middle, and inner ear. The outer ear consists of the pinna and ear canal. The sound wave enters the pinna and travels through the ear canal to the ear drum (tympanic membrane). The outer ear forms a resonant acoustic system that provides approximately 0-dB gain at the ear drum below 1 kHz, rising to 15-20 dB of gain in the vicinity of 2.5 kHz, and then falling in a complex pattern of resonances at higher frequencies[Shaw, 1974]. The sound energy impinging upon the ear drum is conducted mechanically to the oval window of the cochlea by the three middle ear bones (ossicles). The mechanical transduction in the human middle ear can be roughly approximated by a pressure transformer combined with a second-order high pass filter having a Q of 0.7 and a cutoff frequency of 350 Hz [Lynch et al., 1982][Kates, 1991b].

Problems with the outer or middle ear can lead to a hearing loss even when the inner ear (cochlea) is functioning properly. Such a hearing loss is termed conductive since the sound signal conducted to the inner ear is attenuated. One common pathology, especially in children, is otitis media, in which the middle ear fills with fluid, pus, or adhesions related to infection. Another pathology is otosclerosis, in which the ossicles cease to move freely. Conductive losses are not normally treated with hearing aids since they can usually be corrected medically or surgically.

6.3 INNER EAR

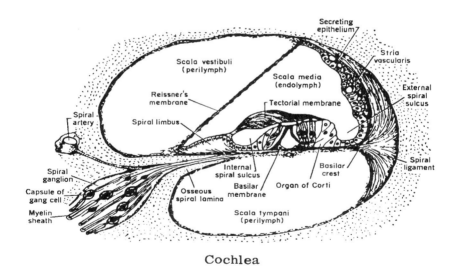

Figure 6.2 Features of the cochlea: transverse cross-section of the cochlea (Reprinted with permission from [Rasmussen, 1943], ©1943, McGraw-Hill)

A transverse cross section through the cochlea is shown in Fig. 6.2. Two fluid-filled spaces, the scala vestibuli and the scala tympani, are separated by the cochlear partition. The cochlear partition is bounded on the top by Reissner's membrane and on the bottom by the basilar membrane, which in turn forms part of the organ of Corti. A more detailed view of the organ of Corti (after Rasmussen[Rasmussen, 1943]) is presented in Fig. 6.3.

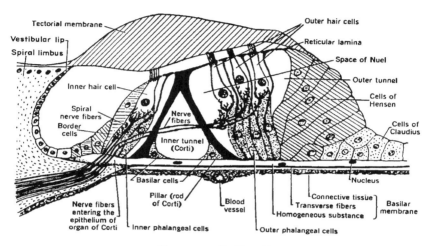

Organ of Corti

Figure 6.3 Features of the cochlea: the organ of Corti (Reprinted with permission from [Rasmussen, 1943], ©1943, McGraw-Hill)

The tectorial membrane rests at the top of the organ of Corti, and the basilar membrane forms the base. Two types of hair cells are found along the basilar membrane. There are three rows of outer hair cells and one row of inner hair cells. The outer hair cells form part of the mechanical system of the cochlear partition, while the inner hair cells provide transduction from mechanical motion into neural firing patterns. There are about 30,000 nerve fibers in the human ear. The vast majority are afferent fibers that conduct the inner hair cell neural pulses towards the brain; approximately 20 fibers are connected to each of the 1,500 inner hair cells. Approximately 1,800 efferent fibers conduct neural pulses from the brain to the outer hair cells[Pickles, 1988].

The organ of Corti forms a highly-tuned resonant system. A set of neural tuning curves for the cat cochlea [Kiang, 1980] is presented in Fig. 6.4.

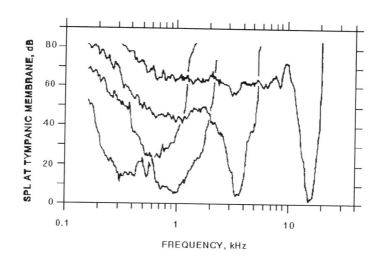

Figure 6.4 Sample tuning curves for single units in the auditory nerve of the cat (Reprinted with permission from [Kiang, 1980]. ©1980 Acoustical Society of America)

A tuning curve is generated by placing an electrode on a single afferent nerve fiber, finding the frequency to which that fiber responds most readily, and then adjusting the stimulus level as the test frequency is varied to maintain the neural firing rate at a level just above threshold. The tip of the tuning curve is the region most sensitive to the excitation, and the tail of the tuning curve is the plateau region starting about one octave below the tip and extending lower in frequency. The ratio of the signal amplitude required to generate a response in the tail region to that required in the region of the tip of the tuning curve is approximately 60 dB. The slopes of the high-frequency portion of the tuning curves are approximately 100-300 dB/octave. The sharpness of the tuning curves, the steepness of the slopes, and the tip-to-tail ratio all decrease at lower characteristic frequencies of the fibers.

An example of what can happen to the tuning curves in a damaged ear is shown in Fig. 6.5[Liberman and Dodds, 1984]. The stereocilia (protruding hairs) of outer and inner hair cells were damaged mechanically in this experiment; the tuning curve for an

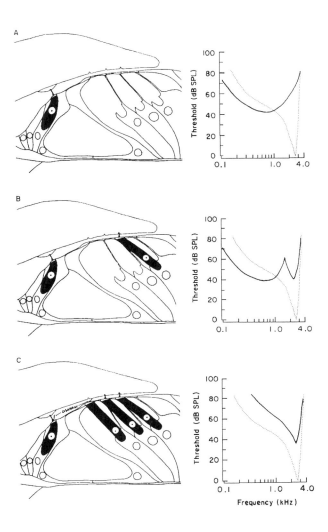

Figure 6.5 Neural tuning curves resulting from damaged hair cells (Reprinted from [Liberman and Dodds, 1984], with kind permission from Elsevier Science – NL, Sara Burgerhartstraat 25, 1055 KV, Amsterdam, The Netherlands)

undamaged cochlea is shown as the dotted line and the tuning curve for the damaged condition is the solid line. In 6.5(A) the outer hair cells have been completely destroyed, resulting in a tuning curve that is much broader and having a response peak shifted to a lower frequency. In 6.5(B) there is partial damage to the outer hair cells, resulting in a "w"-shaped tuning curve having a sharply tuned tip at a greatly reduced sensitivity. In 6.5(C) the inner hair cells have been damaged while the outer hair cells remain mostly intact, resulting in a tuning curve having a nearly-normal shape at all frequencies but which has a much lower sensitivity.

Acoustic trauma and ototoxic drugs usually cause damage to the outer hair cells in the cochlea [Pickles, 1988] similar to that illustrated in Fig 4(A), resulting in a system that is less sharply tuned and which provides much less apparent gain. The auditory filters, which give a high-Q band-pass response in the normal ear, have become much more like low-pass filters, with a resultant reduction in both gain and frequency resolution. The loss of frequency resolution may be related to the excess upward spread of masking[Egan and Hake, 1950] observed in impaired ears [Gagné, 1988], in which low-frequency sounds interfere with perception of simultaneously occurring higher-frequency sounds to a greater than normal degree.

The tuning curves in a healthy ear exhibit compressive gain behavior [Rhode, 1971] [Sellick et al., 1982][Johnstone et al., 1986]. As the signal level increases, the tuning curves become broader and the system exhibits reduced gain in the region of the tip of the tuning curve. The gain in the region of the tail of the tuning curve is essentially unaffected. The compression ratio ranges from about 1.5:1 at low frequencies to about 4:1 at high frequencies, and is about 2.5:1 in the central portion of the speech frequency range [Cooper and Yates, 1994]. In the damaged ear the compression ratio is reduced along with the gain, so the auditory system becomes more linear with increasing hearing loss.

The loss of compression in the damaged ear is a possible cause of the phenomenon of loudness recruitment. Loudness is the perceptual correlate of sound intensity. Loudness recruitment is defined as the unusually rapid growth of loudness with an increase in sound intensity [Moore et al., 1985], and often accompanies sensorineural hearing impairment.

An example of recruitment is presented in Fig. 6.6, for which normal-hearing and hearing-impaired subjects were asked to rate the loudness of narrowband noise on a 50-point scale[Kiessling, 1993]. As the hearing loss increases in severity, the subjects need increasingly intense stimuli to achieve identical estimated loudness scores for sounds near auditory threshold. At high stimulus levels, however, the rated loudness is similar for all degrees of hearing loss. Thus the rate of growth of loudness with increasing stimulus level increases with increasing hearing loss.

In addition to the loss of gain, reduction in compression, and loss of frequency resolution, the impaired ear can also demonstrate a loss of temporal resolution. Gap detection experiments [Fitzgibbons and Wightman, 1982], in which the subjects are

244   APPLICATIONS OF DSP TO AUDIO AND ACOUSTICS

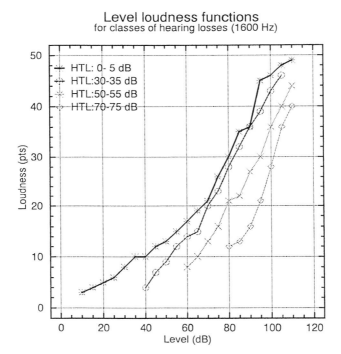

Figure 6.6   Loudness level functions on a 50-point rating scale for different classes of hearing loss.
(Reprinted with permission from [Kiessling, 1993]. ©1993, Canadian Association of Speech-Language Pathologists and Audiologists)

asked to determine if a short pause is present in an otherwise continuous signal, have shown that hearing-impaired listeners require longer gaps for detection in a band-pass filtered noise signal than normal-hearing subjects. However, the difference in performance appears to be closely related to the inability of the hearing-impaired listeners to detect the high-frequency transient portions of the gated test signal [Florentine and Buus, 1984].

As shown in Fig. 6.6, when the presentation levels are corrected for the hearing loss the forward masking results for the hearing-impaired and normal-hearing subjects become nearly identical. In the figure, dB SPL refers to the absolute signal level while dB SL (sensation level) refers to the level above the subject's auditory threshold. Presentation of the stimuli to the normal ear at sensation levels corresponding to those used in the impaired ear results in masking curves that are nearly identical in shape and differ only in the offset used to compensate for auditory threshold.

The effects of sensorineural hearing loss in speech perception are illustrated in Fig. 6.8 to Fig. 6.10 for a simulation of a normal and impaired cochlea [Kates, 1991b] [Kates, 1993a][Kates, 1995]. The time-frequency simulated neural response to the stimulus /da/ at a level of 65 dB SPL is shown for a) a normal ear, b) an ear with a simulated hearing loss obtained by turning off the outer hair cell function in the cochlear model, and c) the stimulus given 30 dB of gain and presented as input to the simulated hearing loss. The speech stimulus is the syllable /da/ digitally generated using a speech synthesizer [Klatt, 1980].

The figure shows the first 25 ms of the neural responses for a simulated normal ear and for an impaired ear in which the outer hair cells have been eliminated. The normal ear of Fig. 6.8 shows regions of synchronized firing activity corresponding to the initial frequencies of each of the three formants (500 Hz, 1.6 kHz, and 2.8 kHz) that give the peaks of the syllable spectrum. In addition, there are high ridges corresponding to the glottal pulses exciting the vocal tract at a fundamental frequency of 120 Hz. Thus the neural firing patterns in the normal ear appear to code the speech formant frequencies both by the region of maximum activity and in the periodic nature of the firing within each of these regions.

The simulated firing pattern for the impaired ear with the outer hair cell function eliminated but with the inner hair cells intact is presented in Fig. 6.9. The complete outer hair cell damage corresponds to a nearly flat hearing loss of about 55-60 dB. The shift in auditory threshold, combined with the auditory filter shapes changing from band-pass to low-pass, has resulted in the first formant dominating the simulated firing behavior. The presence of the second formant can be discerned in a slight broadening of the ridges in the vicinity of 1.6 kHz, while the third formant can not be seen at all. Thus a significant amount of both frequency and temporal information has been lost. Amplifying the input signal for presentation to the impaired ear results in Fig. 6.10. The neural firing rate is substantially increased, but there is little if any information visible beyond that for the unamplified stimulus. Thus amplification can increase the

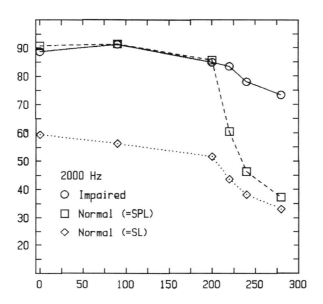

Figure 6.7 Mean results for five subjects with unilateral cochlear impairments (Reprinted with permission from [Glasberg et al., 1987]. ©1987 Acoustical Society of America)

Forward masking, in which a sound can interfere with the perception of sounds that follow it, can also be greater in hearing-impaired subjects. However, this also appears to primarily be a level effect [Glasberg et al., 1987]. Mean results for five subjects with unilateral cochlear impairments showing the threshold for a 10-ms signal as a function of its temporal position relative to a 210-ms masker. Thresholds are plotted as a function of masker-onset to signal-onset delay. The three leftmost points are for simultaneous masking and the three rightmost points are for forward masking. The curves labeled "Normal" are for the subjects' normal ears, while those labeled "Impaired" are for the impaired ears of the same subjects. (after Glasberg et al., 1987)

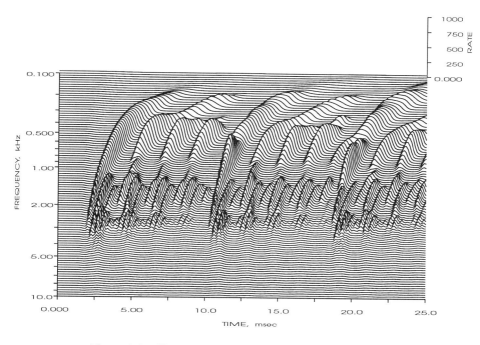

Figure 6.8 Simulated neural response for the normal ear

sensation level above the impaired auditory threshold, but it may not be able to restore the information that has been lost due to the changes in auditory frequency resolution.

6.3.1 Retrocochlear and Central Losses

Hearing loss can also be caused by problems in the auditory pathway carrying the neural signals to the brain, or by problems within the brain itself. Retrocochlear lesions due to tumors in the auditory nerve can cause hearing loss [Green and Huerta, 1994], as can brainstem, cortical, or hemispherical lesions [Musiek and Lamb, 1994]. Furthermore, there is some evidence that the elderly can have increased difficulty in understanding speech even when the auditory periphery exhibits normal or nearly-normal function [Jerger et al., 1989]. Successful signal-processing strategies have not been developed for these central auditory processing deficits, and much more study is needed to characterize the hearing losses and to determine if specialized signal-processing strategies are warranted.

248   APPLICATIONS OF DSP TO AUDIO AND ACOUSTICS

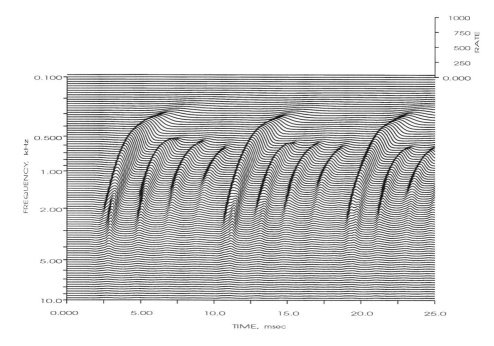

Figure 6.9   Simulated neural response for impaired outer cell function

6.3.2   Summary

The analogy of eyeglasses is often used when discussing hearing aids. As can be seen from the material in this section, however, hearing loss is typically a much more complicated problem than correctable vision. In vision a lens, that is, a passive linear system, provides nearly perfect compensation for the inability of the eye to focus properly at all distances. Hearing loss, on the other hand, involves shifts in auditory threshold, changes in the system input/output gain behavior, and the loss of frequency and temporal resolution. The development of signal processing to compensate for these changes in the impaired ear presents a significant engineering challenge.

6.4   LINEAR AMPLIFICATION

The basic hearing-aid circuit is a linear amplifier, and the simplest hearing aid consists of a microphone, amplifier, and receiver (output transducer). In addition to being commonly prescribed on its own, the linear hearing aid also forms the fundamental building block for more-advanced designs. Thus many of the problems associated with linear amplification will also affect other processing approaches when implemented

SIGNAL PROCESSING FOR HEARING AIDS 249

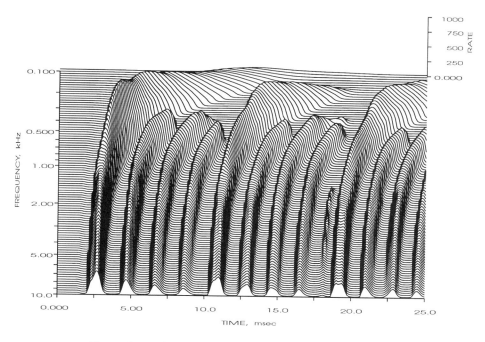

Figure 6.10 Simulated neural response for 30 dB of gain

in practical devices. Conversely, improvements in linear instruments will lead to improvements in all hearing aids.

6.4.1 System Description

A schematic diagram of an in-the-ear (ITE) hearing aid designed to fit within the confines of the pinna and ear canal is shown in Fig. 6.11. Hearing aids are also designed to fit behind the ear (BTE), in a body-worn electronics package, or completely within the ear canal (ITC or CIC). The major external features of the hearing aid are the microphone opening, battery compartment, volume control, and vent opening. The vent is used to provide an unamplified acoustic signal at low frequencies (for individuals having high-frequency hearing losses and who therefore need amplification only at high frequencies), and also provides a more natural frequency response for monitoring the user's own voice. Because of potential feedback problems, discussed in the section on Feedback Cancellation, a vent is not present in all hearing aids.

The microphone is positioned near the top of the hearing-aid faceplate above the battery compartment, and the volume control and the vent are at the bottom. This

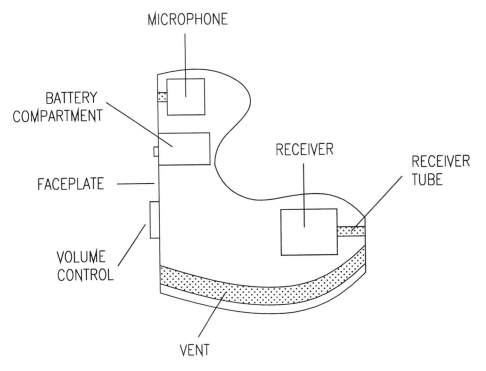

Figure 6.11 Cross-section of an in-the-ear hearing aid

placement maximizes the separation between the microphone and vent opening and helps reduce acoustic feedback problems. Not shown is the circuitry internal to the hearing aid; it is positioned where there is available space since the shell of the hearing aid is molded to fit an impression of the individual ear. The receiver is located in the canal portion of the hearing aid, and the receiver output is conducted into the ear canal via a short tube. The vent runs from the faceplate to the ear canal.

A block diagram of the hearing aid inserted into the ear is presented in Fig. 6.12. The input to the microphone is the sound pressure at the side of the head. The positioning of the hearing aid in the ear canal has destroyed the normal pinna and ear canal resonance at 2.5 kHz. The resultant insertion loss caused by blocking the natural resonance of the outer ear in this frequency region is 15-20 dB, and the corresponding gain should be reintroduced in the frequency response of the electroacoustic system. In addition to the amplified signal path, there is also an unamplified signal path directly through the vent, so the sound pressure in the ear canal is the sum of the amplified and direct signals. Mechanical feedback from the receiver vibrations can excite the microphone

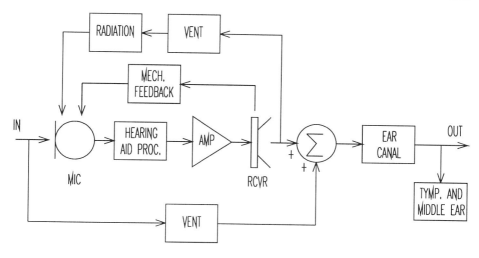

Figure 6.12 Block diagram of an ITE hearing aid inserted into the ear canal

diaphragm in addition to the input sound pressure. Acoustic feedback is also present since the sound pressure generated in the ear canal travels out through acoustic leaks or through the vent, where it is reradiated at the vent opening in the faceplate. The receiver, connected to the ear canal by a short tube, is loaded acoustically by the tube, vent, and ear canal, and the ear canal is terminated by the input impedance of the ear drum. Several simulations have been developed to assist in the design and evaluation of hearing aid acoustics [Egolf et al., 1978][Egolf et al., 1985][Egolf et al., 1986][Kates, 1988][Kates, 1990].

The signal processing in a linear hearing aid consists of frequency-response shaping and amplification. In general, one-pole or two-pole high-pass or low-pass filters are used to shape the frequency response to match the desired response for a given hearing loss. Multi-channel hearing aids are also available that allow the independent adjustment of the gain in each frequency channel. Acoustic modifications to the tubing that connect the output of a BTE instrument to the ear canal can also be used to adjust the hearing-aid frequency response [Killion, 1981][Dillon, 1985].

6.4.2 *Dynamic Range*

The dynamic range of a hearing aid is bounded by noise at low input signal levels and by amplifier saturation at high signal levels. A typical hearing-aid microphone has a noise level of about 20 dB SPL, which is comparable to that of the human ear [Killion, 1976]. The addition of the hearing-aid processing and amplification circuits gives equivalent noise levels of between 25 and 30 dB SPL. More complicated processing, such as a multi-channel filter bank, may generate higher noise levels due to the specific circuit fabrication technology used and the number of circuit components required. The equivalent hearing-aid noise level, after amplification, is therefore about 10 dB higher than that of the normal unaided ear. This noise level tends to limit the maximum gain that a hearing-aid user will select under quiet conditions since, in the absence of the masking provided by intense inputs, a user will reduce the gain in order to reduce the annoyance of the background noise.

At the other extreme, amplifier saturation limits the maximum gain that can be achieved by the hearing aid. A typical hearing aid amplifier clips the signal when the peak input level exceeds about 85 dB SPL. A speech-like signal at an input of 70 dB SPL is therefore amplified cleanly, but a level of 80 dB SPL causes large amounts of distortion [Preves and Newton, 1989]. Speech input at 65 to 70 dB SPL is typical of normal conversational levels [Pearsons et al., 1976];[Cornelisse et al., 1991], but the spectra of individual speech sounds can be as much as 15 dB higher when monitoring the talker's own voice at the ear canal [Medwetsky and Boothroyd, 1991]. Thus the typical hearing aid amplifier does not have enough headroom to guarantee that the user's own voice will be amplified without distortion.

The available hearing-aid dynamic range is thus about 55 dB from the noise floor to the saturation threshold. Selecting an amplifier with more gain, and turning down the volume control, will raise the saturation threshold, but will also raise the noise level by a similar amount. Thus a typical hearing aid, due to the compromises made in battery size and circuit design, can only handle half the dynamic range of a normal ear. Some progress is being made, however, since the development of class-D hearing-aid amplifiers [Carlson, 1988] provides 10 to 20 dB more output at saturation than does a class-A amplifier having comparable gain [Fortune and Preves, 1992]. The small class-B amplifiers that are becoming available in hearing aids also greatly reduce the problems associated with amplifier saturation[Cole, 1993].

6.4.3 Distortion

Amplifier saturation most often takes the form of symmetric peak clipping (S. Armstrong, personal communication, 1989). If a single sinusoid is input to the hearing aid, the clipping will generate harmonic distortion, and for two or more simultaneous sinusoids, intermodulation (IM) distortion will also result. The amount of distortion influences judgments made about hearing-aid quality. Fortune and Preves [Fortune and Preves, 1992], for example, found that reduced coherence in the hearing-aid output signal was related to a lower hearing-aid amplifier saturation level and a lower loudness discomfort level (LDL). LDL is the maximum level at which an individual is willing to listen to speech for an extended period of time. This result suggests that hearing-aid users will select reduced gain in order to reduce the distortion. In another study, a large majority of hearing-aid users indicated that good sound quality was the most important property of hearing aids, with clarity being the most important sound-quality factor [Hagerman and Gabrielsson, 1984]. Thus reduced distortion would be expected to lead to greater user comfort and satisfaction, and could lead to improved speech intelligibility at high sound levels.

6.4.4 Bandwidth

The bandwidth of a hearing aid should be wide enough for good speech intelligibility and accurate reproduction of other sounds of interest to the user. French and Steinberg [French and Steinberg, 1947] determined that a frequency range of 250-7000 Hz gave full speech intelligibility for normal-hearing subjects, and more recent studies [Pavlovic, 1987] extend this range to 200-8000 Hz for nonsense syllables or continuous discourse. For music, a frequency range of 60-8000 Hz reproduced over an experimental hearing aid was found to compare favorably with a wide-range loudspeaker system, again using normal-hearing listeners as subjects [Killion, 1988]. Thus a reasonable objective for a hearing aid is a 60-8000 Hz bandwidth.

Most hearing aids have adequate low-frequency but inadequate high-frequency response for optimal speech intelligibility, with the high-frequency response typically decreasing rapidly above 4-6 kHz. Increasing the high-frequency gain and bandwidth in laboratory systems generally yields improved speech intelligibility [Skinner, 1980]. However, the benefits of increased bandwidth will accrue in hearing aids only if the amplifier can cope with the increased power demands without undue distortion and if the system would remain stable in the presence of increased levels of acoustic and mechanical feedback. Thus increasing the hearing-aid bandwidth, while desirable, must wait for other problems to first be solved.

6.5 FEEDBACK CANCELLATION

Mechanical and acoustic feedback limits the maximum gain that can be achieved in most hearing aids and also degrades the system frequency response. System instability caused by feedback is sometimes audible as a continuous high-frequency tone or whistle emanating from the hearing aid. One would also expect distortion to be increased in an instrument close to the onset of instability since the feedback oscillations will use up most of the available amplifier headroom. Mechanical vibrations from the receiver in a high-power hearing aid can be reduced by combining the outputs of two receivers mounted back-to-back so as to cancel the net mechanical moment; as much as 10 dB additional gain can be achieved before the onset of oscillation when this is done. But in most instruments, venting the BTE earmold or ITE shell establishes an acoustic feedback path that limits the maximum possible gain to about 40 dB[Kates, 1988] or even less for large vents. Acoustic feedback problems are most severe at high frequencies since this is where a typical hearing aid has the highest gain. The design criterion for effective feedback suppression would be to usefully increase maximum gain while preserving speech information and environmental awareness.

The traditional procedure for increasing the stability of the hearing aid is to reduce the gain at high frequencies[Ammitzboll, 1987]. Controlling feedback by modifying the system frequency response, however, means that the desired high-frequency response of the instrument must be sacrificed in order to maintain stability. Phase shifting and notch filters have also been tried[Egolf, 1982], but have not proven to be very effective. A more effective technique is feedback cancellation, in which the feedback signal is estimated and subtracted from the microphone input. Simulations and digital prototypes of feedback cancellation systems [Bustamante et al., 1989][Engebretson et al., 1990][Kates, 1991a] [Dyrlund and Bisgaard, 1991][Engebretson and French-St.George, 1993][French-St.George et al., 1993] indicate that increases in gain of between 6 and 17 dB can be achieved before the onset of oscillation with no loss of high-frequency response. In laboratory tests of a wearable digital hearing aid [French-St.George et al., 1993], a group of hearing-impaired subjects used an additional 4 dB of gain when the adaptive noise cancellation was engaged and showed significantly better speech recognition in quiet and in a background of speech babble. Field trials of a practical adaptive feedback-cancellation system built into a BTE hearing aid have shown increases of 8-10 dB in the gain used by severely-impaired subjects[Bisgaard, 1993].

An example of a feedback-cancellation system is shown in figure 6.13. This system is typical of the majority that have been proposed in that a noise signal is injected to probe the feedback path [Kates, 1991a][Engebretson and French-St.George, 1993] [Bisgaard, 1993]. The characteristics of the feedback path are determined by cross-correlating the noise signal p(n) with the error signal e(n); this measurement includes the amplitude and phase effects of the receiver, the acoustic and mechanical feedback paths, and the microphone response. The error signal e(n) is minimized by a least-mean-squares (LMS) adaptive weight update algorithm[Widrow et al., 1975].

In some systems, the noise is continuously injected at a low level [Engebretson and French-St.George, 1993][Bisgaard, 1993], and the LMS weight update also proceeds on a continuous basis. This approach results in a reduced SNR for the user due to the presence of the injected probe noise. In addition, the ability of the system to cancel the feedback may be limited due to the presence of the speech signal while the system is adapting[Kates, 1991a][Maxwell and Zurek, 1995]. Better estimation of the feedback path can occur if the processed hearing-aid signal g(n) is disconnected during a short time interval (50 ms) while the adaptation occurs[Kates, 1991a]; for stationary conditions up to 7 dB of additional feedback cancellation is observed as compared to a continuously adapting system, but this approach can have difficulty in tracking a changing acoustic environment[Maxwell and Zurek, 1995] since the adaptive weights are updated only when a decision algorithm ascertains the need. Another approach is to restrict the adaptation to those time intervals where the speech input is reduced

SIGNAL PROCESSING FOR HEARING AIDS   255

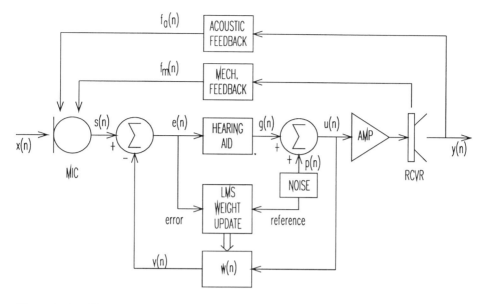

Figure 6.13   Block diagram of a hearing aid incorporating signal processing for feedback cancellation

in intensity; this technique also yields a feedback cancellation advantage over the continuously adapting system, although the quality of the speech can be reduced due to the bursts of injected noise[Maxwell and Zurek, 1995].

6.6   COMPRESSION AMPLIFICATION

Dynamic-range compression, also termed automatic gain control (AGC, hearing aid) in hearing aids, is used for two different purposes. The first, and most prevalent use in hearing aids, is as a limiter to prevent overloading of the amplifier circuits or the user's ear when an intense sound occurs. The second use, sometimes termed recruitment compensation, is to match the dynamic range of speech and environmental sounds to the restricted dynamic range of the hearing-impaired listener. These two uses imply different and even contradictory criteria for setting the compression parameters.

The most common form of compression in hearing aids is a single-channel system used to reduce amplifier overload. For this application, a rapid attack time is desired so as to give a rapid response to a sudden intense sound, a high compression ratio is desired to limit the maximum signal level, and a high compression threshold is desired so as not to limit sounds that could otherwise be amplified without distortion.

For recruitment compensation, on the other hand, longer release times are desired to minimize any deleterious effects of the compression on the speech envelope[Plomp, 1988][Boothroyd et al., 1988] or in modulating the background noise[Cole, 1993] [Neuman et al., 1995]. Low compression ratios are often chosen to match the dynamic range of the impaired ear to that of a normal ear, and a low compression threshold is often chosen so that speech sounds at any level of presentation can be perceived above the impaired auditory threshold [Waldhauer and Villchur, 1988][Killion, 1993].

6.6.1 Single-Channel Compression

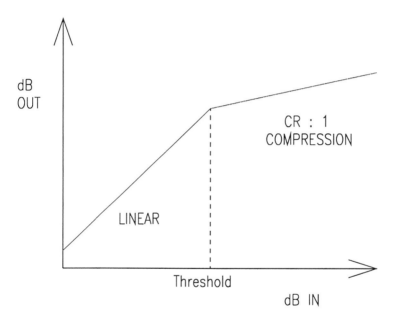

Figure 6.14 Input/output relationship for a typical hearing-aid compression amplifier

The steady-state input/output relationship for a hearing-aid compressor is shown in Fig. 6.14. For input signal levels below the compression threshold, the system is linear. Above the compression threshold, the gain is reduced so that the output increases by 1/CR dB for each dB increase in the input where CR is the compression ratio.

A block diagram of a typical AGC hearing aid is shown in Fig. 6.15 [Cole, 1993]; the system is assumed to be operating as a high-level limiter. The hearing-aid designer has several options as to where the volume control is to be placed. Each option, as shown by the results for locations A, B, and C in the figure, gives a different family of input/output curves as the volume control is adjusted by the user. Control point

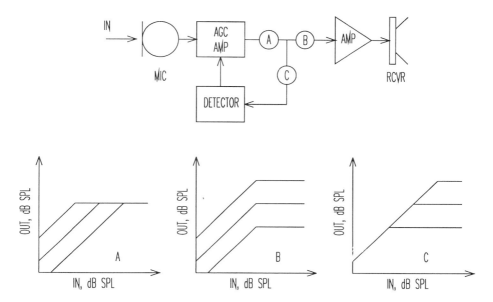

Figure 6.15 Block diagram of a hearing aid having feedback compression Input/output functions for attenuation at points A, B, and C are shown along the bottom (after Cole, 1993) (Reprinted with permission from [Cole, 1993]. ©1993, Canadian Association of Speech-Language Pathologists and Audiologists)

B gives input AGC, in which the gain and maximum output level are simultaneously adjusted by the volume control but where the input-referred compression threshold is unaffected. A separate trimmer adjustment is normally provided for the compression threshold. Control point A for the volume control gives an example of output AGC, in which the volume control simultaneously adjusts the gain and compression threshold, and a separate trimmer is used to set the maximum output level. Another option is feedforward compression, in which the detector is driven directly by the microphone signal; a delay in the amplified signal relative to the control signal can then be used to reduce the attack overshoot in the compression circuit[Verschuure and Dreschler, 1993].

The choice of optimum compression parameters to maximize speech intelligibility or speech quality is still open to contention. Rapid attack time constants (less than 5 ms) are accepted in the industry and also by researchers to prevent transients from saturating the output power amplifier. Arguments for fast release times (less than 20 ms), also termed syllabic compression, are based on considerations of the syllabic variations in speech and the desire to amplify soft speech sounds on a nearly instantaneous basis

[Villchur, 1973]. Arguments for long release times are based on the desire to preserve the envelope structure of speech (Plomp, 1988) and to avoid annoying fluctuations in the perceived level of background noise ("dropouts", "breathing", or "pumping") that can be caused by the rapidly changing gain of a compression amplifier[Cole, 1993].

The concept behind syllabic compression, that different speech sounds need different amounts of amplification, has lead to experiments in modifying the consonant-vowel (CV) ratio in speech[Gordon-Salant, 1986][Gordon-Salant, 1987][Montgomery and Edge, 1988] [Kennedy et al., 1996]. In the CV-ratio enhancement experiments of Kennedy et al., for example, the amplitude of consonants relative to that of vowels in vowel-consonant syllables was adjusted by hand for each syllable, with the modified syllables then presented to hearing-impaired subjects at a comfortable level. The results indicated that the recognition of some consonants was substantially improved given a higher relative level of presentation, but that the recognition of other consonants was unaffected or even decreased. The degree of consonant amplification that produced the maximum recognition scores varied significantly as a function of consonant type, vowel environment, and audiogram shape. Large individual differences among the subjects were also observed.

Recently, the effect of release time on perceived speech quality was investigated [Neuman et al., 1995]. Three compression ratios, these being 1.5:1, 2:1, and 3:1, were used in a digitally simulated hearing aid in combination with release times of 60, 200, and 1000 ms. The attack time was 5 ms and the input compression threshold was set to to be 20 dB below the RMS level of the speech. Twenty listeners with sensorineural hearing loss were asked to give paired-comparison judgments of speech quality for speech in different background noises. For each judgment, the subjects were allowed to toggle between speech processed through two different systems until they reached a decision. The results indicated a statistically significant interaction between the release time and the noise level. There was a significant preference for the longer release times as the background noise level was increased. No significant preference among the release times was observed for the lowest noise level. Significant individual differences from the mean preferences were observed, however, indicating that individual adjustment of hearing-aid release times may lead to greater user satisfaction than using a single pre-set release time.

The effect of compression ratio on speech quality was also investigated under similar experimental conditions[Neuman et al., 1994]. Compression ratios ranging from 1.5:1 to 10:1 were investigated, along with linear amplification. The release time was held constant at 200 ms. Analysis of the paired-comparison data revealed that the twenty hearing-impaired subjects showed a significant preference for the lower compression ratios, with ratios in excess of 3:1 preferred least often. The majority of listeners preferred a linear hearing aid when the noise level was high. Quality ratings(A. Neuman, personal communication, 1995) indicated that the simulated hearing aids with the lower compression ratios were judged to be more pleasant and to have less background noise. Listeners with a reduced dynamic range (less than 30 dB) selected compression at a significantly greater rate than listeners with a larger (greater than 30 dB) dynamic range.

The experimental results cited above for compression ratio are consistent with considerations from auditory physiology. In a healthy cochlea, the active mechanism of the outer hair cells provides about 50-60 dB of gain for a sinusoid at auditory threshold [Kiang, 1980]. Increasing the signal level results in a reduction of gain and a broadening of the auditory filters[Johnstone et al., 1986], until at high levels the gain is reduced to about 0-10 dB. In a cochlea with extensive outer hair-cell damage, the filter shape and gain is similar at all input levels to that of the healthy cochlea at high levels[Harrison et al., 1981]. As an approximation for speech frequencies, assume that in the healthy cochlea an input of 0 dB SPL gets 60 dB of gain, while an input of 100 dB SPL gets 0 dB of gain, giving a compression ratio of 2.5:1 . A severely-impaired cochlea, on the other hand, has 0 dB of gain at all input levels resulting in a linear system. One could therefore argue that the highest compression ratio needed for wide dynamic-range compression in a hearing aid, corresponding to complete outer hair-cell damage, is 2.5:1, and that lesser amounts of damage would require correspondingly lower compression ratios.

Total outer hair-cell damage results in a threshold shift of no more than 60 dB since that is the maximum amount of gain provided by the cochlear mechanics. Hearing losses greater than 60 dB must therefore be accompanied by damage to the neural transduction mechanism of the inner hair cells, and the Liberman and Dodds [Liberman and Dodds, 1984] data presented in Fig 4 indicates that inner hair-cell damage results in a threshold shift but no apparent change in the mechanical behavior of the cochlea. Thus outer hair-cell damage, in this model of hearing loss, causes a loss of sensitivity combined with a reduction in compression ratio, while inner hair-cell damage causes a linear shift in sensitivity. Thus the family of hearing-aid input/output curves would be as indicated in Fig. 6.16, where the compression ratio is increased as the hearing loss increases up to 60 dB of loss, after which the compression ratio remains constant and the gain is increased.

The choice of compression ratio and attack and release times will also effect the distortion of the hearing aid, especially for wide-dynamic-range compression where the signal is nearly always above the compression threshold. The distortion in a simulated compression hearing aid was investigated by Kates [Kates, 1993b] for a hearing aid having an idealized flat frequency response from 100 to 6000 Hz. An input of speech-shaped noise at 70 dB SPL was used, and the distortion metric was the signal-to-distortion ratio (SDR) at 1000 Hz computed from the unbiased coherence function [Kates, 1992]. The results for a compression ratio of 2:1 with a compression threshold of 50 dB SPL show that the distortion decreases as the attack and release times are increased; the SDR is approximately 30 dB for any combination of ANSI [ANSI, 1987] attack time greater than 2 ms and release time greater than 50 ms.

260   APPLICATIONS OF DSP TO AUDIO AND ACOUSTICS

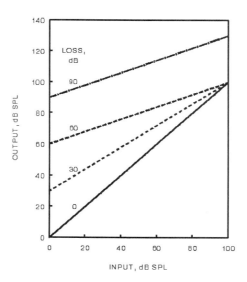

Figure 6.16   Compression amplifier input/output curves derived from a simplified model of hearing loss.

The ANSI [ANSI, 1987] attack time is defined as the length of time it takes for the overshoot at an upward jump from 55 to 80 dB SPL in the signal level to decay to within 2 dB of steady state, and the release time is defined as the length of time it takes for the signal to recover to a value 2 dB below steady state after the test signal returns to the 55-dB SPL level. Thus the distortion would not be expected to reduce speech intelligibility or significantly effect speech quality in quiet for time constants within this range [Kates and Kozma-Spytek, 1994]. Increasing the compression ratio to 8:1 reduces the SDR by about 5 dB, which again would not be expected to substantially affect speech intelligibility or quality.

6.6.2   Two-Channel Compression

Several different systems have been proposed for two-channel dynamic-range compression. The most common approach is independent operation of the two channels [Villchur, 1973]. Commercial products incorporating adjustable gains, compression ratios, and an adjustable crossover frequency between the two channels have been introduced by several manufacturers [Hodgson and Lade, 1988][Johnson and Schnier, 1988] [Pluvinage and Benson, 1988]. Systems offering compression limiting, as opposed to wide-dynamic-range compression, are also available[Branderbit, 1991]. These systems are programmable to allow tailoring of the response to the individual hearing loss, and use digital control of an analog signal path.

Variations in the compression system are also available. Compression in the low-frequency channel, combined with a linear high-frequency channel, has been used in an attempt to reduce the upward spread of masking that can be caused by intense low-frequency noise [Ono et al., 1983][Kates, 1986]. Such systems can also be implemented using a high-pass filter having a filter cutoff frequency controlled by the estimated low-frequency signal level; their performance is discussed in the section on single-microphone noise suppression. Alternatively, compression in the high-frequency channel combined with a linear low-frequency channel has been proposed to compensate for recruitment in a high-frequency hearing loss and is also commercially available[Killion, 1993].

There is some evidence that two-channel compression can offer small improvements in speech intelligibility over a linear or single-channel compression system[Villchur, 1973][Moore, 1987]. A two-channel system investigated by Moore [Lawrence et al., 1983][Moore, 1987] used a two-stage compression system comprising a front-end compressor having a 5-ms attack time and a 300-ms release time, followed by two-channel syllabic compression. The front-end AGC served to keep the speech within a relatively narrow dynamic range, while the syllabic compressors were designed to modify the speech spectrum within the confines of the nearly-constant average level. Results for an experiment involving hearing-impaired listeners in which noise level was kept at a constant level and the speech level adjusted to maintain constant percent correct speech recognition indicated that the two-channel compression system allowed the subjects to listen at a 2-3 dB worse SNR while maintaining performance comparable to that of a matched linear hearing aid or single-channel compression system.

6.6.3 Multi-Channel Compression

Multi-channel compression systems divide the speech spectrum into several frequency bands, and provide a compression amplifier for each band. The compression may be independent in each of the bands, or the compression control signals and/or gains may be cross-linked. Independent syllabic compression has not been found to offer any consistent advantage over linear amplification [Braida et al., 1979][Lippmann et al., 1981][Walker et al., 1984]. One problem in multi-channel compression systems has been the unwanted phase and amplitude interactions that can occur in the filters used for frequency analysis/synthesis [Walker et al., 1984] and which can give unwanted peaks or notches in the system frequency response as the gains change in each channel.

Linear-phase digital filters can remove the problem of filter interactions, but speech intelligibility results for multi-channel digital systems are no more encouraging than those for their analog counterparts [Kollmeier et al., 1993].

A multi-channel compression system using longer time constants has also been investigated[van Dijkhuizen et al., 1991]. This slow-acting compression system gave significantly better speech recognition than single-channel compression for noisy speech when the noise contained an intense narrow-band component. However, adjustments to the amplified spectrum had very little effect on intelligibility when the spectrum of the noise matched that of the speech[van Dijkhuizen et al., 1987][van Dijkhuizen et al., 1989]. One can therefore conclude that the success of this system is based primarily on the reduction of the spread of masking caused by the intense narrow-band noise.

More-complicated multi-channel compression systems involve linking the compression control signals and/or gains across channels. These systems are intended to dynamically modify the spectrum so as to maximize the amount of the speech signal that is placed within the residual hearing region of a hearing-impaired listener. One proposed system adjusts the relative amplitudes of the principle components of the short-time spectrum[Bustamante and Braida, 1987], while another varies the coefficients of a polynomial series fit to the short-time spectrum[Levitt and Neuman, 1991]. These systems improve the intelligibility of low-level speech, but most of the benefit comes from increasing the magnitude of the low-order coefficients, an effect that is equivalent to single-channel compression. In general, magnification of the higher-order spectral fluctuations does not appear to lead to an improvement in intelligibility [Haggard et al., 1987][Stone and Moore, 1992].

Compression systems have also been developed on the principal of matching the estimated loudness in the impaired ear to that of a normal ear [Yund et al., 1987] [Dillier et al., 1993][Kollmeier et al., 1993]. In the Dillier et al. system[Dillier et al., 1993], the loudness of the smoothed short-time spectrum is determined in each of eight frequency bands, and gains in each band are selected to give the same band loudness in the impaired ear as would occur in a normal ear. The overall loudness of the summation of the eight bands is then estimated, and a correction factor is generated to prevent the composite signal from becoming overly loud. Results in comparison with conventional hearing aids for hearing-impaired subjects show that the greatest improvement in intelligibility occurs for low-level consonants, and that the benefit of the processing is substantially reduced in noise.

Kollmeier et al.[Kollmeier et al., 1993] use a similar approach in computing the gain in each frequency band to create in the impaired ear the loudness level that would occur in a normal ear. The overall loudness of the modified signal is not computed, but spread of masking is incorporated into the procedure. The level that is used to compute the gain in a given band is the maximum of the signal level in that band or the masking patterns from the adjacent bands. This system produced a significant increase in the rated quality of the processed speech in comparison with a linear frequency-shaping system, but the speech intelligibility was not significantly improved.

In general, the design goal for a hearing aid intended for peripheral hearing loss is to process the incoming sound so as to give the best possible match between the neural firing patterns in the impaired ear to those in a reference normal ear. Kates [Kates, 1993d] has proposed a system based on a minimal mean-squared error match between models of normal and impaired hearing. The incoming signal is divided into a set of frequency bands. The compression ratio in each band is computed from the hearing loss using the procedure outlined in Fig 13. This system differs from the loudness-based systems described above in that the compression control signal is the maximum spectral level measured over a region extending one octave below to one-half octave above each band center frequency. This form of compression control mimics aspects of two-tone suppression[Sachs and Kiang, 1968] as observed psychophysically in human subjects [Duifhuis, 1980], in which the total neural firing rate at the place most sensitive to a given frequency can be reduced by the addition of a second tone at a different frequency. The practical result of the compression rule is that intense sounds, such as the formants in vowels, control the gain in the surrounding frequency region. As the signal spectrum changes, the control signals and regions shift in frequency; in most cases, there are two or three frequency regions, with the gain in each region governed by the most-intense spectral peak within the region. This system has not been tested with hearing-impaired subjects.

6.7 SINGLE-MICROPHONE NOISE SUPPRESSION

Improving speech intelligibility in noise has long been an important objective in hearing-aid design. In cases where the interference is concentrated in time (e.g. clicks) or frequency (e.g. pure tones) intelligibility can in fact be improved; clipping the signal in the former or using a notch filter in the latter case will reduce the noise level by a much greater amount than the speech[Weiss and Aschkenasy, 1975]. A much more difficult problem is to improve speech intelligibility in the presence of broadband noise. The single-microphone techniques that have been developed are based on the assumption that an improvement in SNR will yield a corresponding improvement in intelligibility, but this has not been found to be true in practice.

6.7.1 Adaptive Analog Filters

Adaptive filters for reducing the low-frequency output of a hearing aid in the presence of noise have been available for several years [Ono et al., 1983][Kates, 1986]. While some instruments have been based on a two-channel approach having compression in the low-frequency channel in order to limit the amplification of intense low-frequency noise, it is more common to find a system using a high-pass (low-cut) filter having a slope of 6 or 12 dB/octave and having an automatically adjustable cutoff frequency.

264   APPLICATIONS OF DSP TO AUDIO AND ACOUSTICS

The adjustable high-pass filter reduces the low-frequency bandwidth of the hearing aid as the estimated low-frequency noise level increases. Tests of such systems have demonstrated that there is no net improvement in intelligibility when the volume control is kept in a fixed position [Kuk et al., 1989][Fabry and Tasell, 1990], but some improvement has been reported when the subjects are free to adjust the volume [Sigelman and Preves, 1987]. This dependence on volume setting suggests that the major effect on intelligibility is actually a result of a reduction in distortion; the reduced gain at low frequencies allows for increased amplification at high frequencies before the amplifier saturates[Fabry, 1991].

An alternative to using SNR as the criterion for noise-suppression effectiveness is to maximize the Articulation Index (AI) computed for the system. The AI is based on a weighted sum of the steady-state SNR values in frequency bands from 200 to 6000 Hz [French and Steinberg, 1947][Kryter, 1962] and includes spread of masking in the frequency domain. Reducing the gain in any one critical band will not affect the signal-to-noise ratio (SNR) in that band, but may still increase the AI if the noise in that band, at the original level, was intense enough to mask speech in a nearby band. Masking effects extend primarily upward in frequency [Egan and Hake, 1950]. Thus, if the masking effects of the noise in a given band on sounds in a higher-frequency band exceed the noise level in the higher-frequency band, the gain in that band should be reduced, and the gain reduction should only be enough to make the out-of-band masking and in-band noise approximately equal in the high-frequency band[Kates, 1989]. Filtering in excess of the amount needed to maximize the AI will not improve speech intelligibility[Fabry and Tasell, 1990]. This argument is consistent with the results of Dillon and Lovegrove [Dillon and Lovegrove, 1993], who found some benefit for analog noise-suppression filters when the noise was concentrated at low frequencies, but no net benefit for speech babble. It is also consistent with the findings of Kuk et al.[Kuk et al., 1989] that subjects preferred an adaptive filter with a 6-dB/oct slope to one with a 12-dB/oct slope in daily use since the 6-dB/octave slope is adequate to remove the frequency-domain masking for most everyday noises [Kates, 1993b].

6.7.2 Spectral Subtraction

In spectral subtraction, an estimate of the average noise magnitude spectrum is subtracted from the short-time speech-plus-noise spectrum to give an improved estimate of the speech signal[Boll, 1979]. Since the magnitude spectrum of the noise can not be estimated accurately when the speech is present, it is approximated as the average noise magnitude spectrum observed during non-speech intervals. Spectral subtraction is illustrated in Fig. 6.17.

The magnitude and phase of the incoming signal are computed on a block-by-block basis using the FFT. The noise magnitude spectrum is calculated as a running average

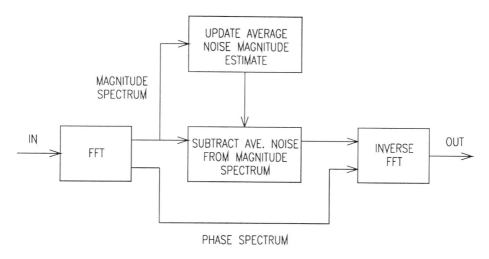

Figure 6.17 Block diagram of a spectral-subtraction noise-reduction system.

of those signal blocks determined to be primarily noise alone. The average noise magnitude spectrum, or a modified estimate, is then subtracted from the magnitude spectrum of the incoming signal; negative differences are set to zero. The modified signal magnitude spectrum is then recombined with the unaltered phase spectrum, and the output signal generated via an inverse FFT.

A variant on spectral subtraction is the INTEL technique [Weiss et al., 1975], in which the square root of the magnitude spectrum is computed and the rooted spectrum is then further transformed via a second FFT. Processing similar to that described above is then performed in this pseudo-cepstral domain. The estimate of the speech amplitude function in this domain is transformed back to the magnitude spectral domain and squared to remove the effect of rooting the spectrum.

Speech intelligibility and quality have been tested for the INTEL method. For the intelligibility tests, isolated speech syllables were presented in a background of white noise, speech-shaped noise, or cafeteria babble, and the noise level was adjusted to give an average score of 50-percent correct. Although the INTEL processing resulted in an 8-dB improvement in the measured stimulus SNR, speech recognition performance for the processed and unprocessed stimuli did not differ significantly. Despite the fact that overall speech intelligibility did not improve with the processing, subjects expressed a preference for the sound of the processed speech[Neuman et al., 1985].

A perceptual analog of spectrum subtraction is the REDMASK processing technique [Neuman and Schwander, 1987]. In this technique, the gain of the hearing aid is adjusted so that the background noise is just inaudible at all frequencies. This gain adjustment is accomplished by measuring the threshold of detection of noise in one-third octave bands, and then adjusting the frequency-gain characteristic so that the noise spectrum lies just below the auditory threshold. The experimental evaluation showed an improvement in sound quality for approximately half of the hearing-impaired subjects tested, but this improvement in quality was accompanied by a reduction in speech intelligibility. It was also found that intelligibility could be improved by raising the noise spectrum to a suprathreshold level, thereby providing more gain to the speech signal; however, this was accompanied by a reduction in sound quality.

6.7.3 Spectral Enhancement

Several techniques have been proposed for modifying the shape of the short-time spectrum so as to emphasize those portions deemed to be important for speech perception or to reduce the amplitude of those portions assumed to be noise. One approach is adaptive comb filtering[Lim et al., 1978]. In this method, the fundamental frequency of voiced speech sounds is estimated, and a comb filter is then constructed to pass signal power in the regions of the pitch harmonics and to suppress power in the valleys in between. Experimental results with normal-hearing subjects, however, have shown no significant improvement in intelligibility with this type of system [Perlmutter et al., 1977][Lim et al., 1978].

Another approach is to construct an optimal filter for improving the speech SNR. A multi-channel filter bank or equivalent FFT system is used to provide the frequency analysis, and the signal and noise powers are estimated in each frequency band. The gain in each band is then adjusted based on the signal and noise power estimates; the system is adaptive since the power estimates fluctuate with changes in the signal or noise characteristics. Various rules have been used for implementing the filter gains, including Wiener filter, power subtraction, magnitude subtraction, and maximum-likelihood envelope estimation [McAulay and Malpass, 1980][Doblinger, 1982] [Vary, 1983][Ephraim and Malah, 1984] While improvements in measured SNR of up to 20 dB have been reported [Vary, 1983], no improvement in speech intelligibility has been observed[Sandy and Parker, 1982].

Instead of trying to remove the noise, one can try instead to enhance the speech. The general approach that has been used is to increase the spectral contrast of the signal short-time spectrum by preserving or increasing the amplitude of frequency regions containing spectral peaks while reducing the amplitude of regions containing valleys. Techniques include squaring and then normalizing the spectral magnitude [Boers, 1980], increasing the spectral magnitude in pre-selected spectral regions while reducing it in others[Bunnell, 1990], filtering the spectral envelope to increase the higher rates of fluctuation [Simpson et al., 1990][Stone and Moore, 1992][Baer et al., 1993], and using sinusoidal modeling of the speech to remove the less-intense spectral components while preserving the peaks[Kates, 1994].

In general, spectral enhancement has not yielded any substantial improvement in speech intelligibility. The systems that filter the spectral envelope [Simpson et al., 1990][Stone and Moore, 1992] have demonstrated intelligibility improvements corresponding to changes in SNR of less than 1 dB, but these results are confounded by the fact that for some speech sounds the processing can increase the amplitude of the second or third formant relative to that of the first formant, so the improvement in performance may be due to the small change in spectral tilt rather than to the change in spectral contrast. To test this hypothesis, Baer et al. [Baer et al., 1993] repeated the experiment correcting for the spectral tilt, and still found a small performance advantage for the processed speech. The Bunnell [Bunnell, 1990] system, which also increased the relative amplitude of the second-formant peaks relative to those of the first formant, also showed a small improvement in stop-consonant recognition for the increased spectral contrast condition. In a system that reduced the gain in the spectral valleys without amplifying the level of the peaks [Kates, 1994], there was no improvement in intelligibility. In fact, Kates [Kates, 1994] found that reducing the number of sinusoids used to represent the speech reduced speech intelligibility in a manner similar to increasing the background noise level, suggesting that the valleys and sidelobes in the speech spectrum convey useful speech information.

6.8 MULTI-MICROPHONE NOISE SUPPRESSION

In many situations, the desired signal comes from a single well-defined source, such as a person seated across a table, while the noise is generated by a large number of sources located throughout the area, such as other diners in a restaurant. Under these conditions the speech and the noise tend to have the same spectral distribution, but the spatial distributions differ. The spatial separation of the speech and the noise can be exploited to reduce the noise level without any deleterious effects on the speech. Furthermore, unlike the situation for single-microphone noise-suppression techniques, the improvements in SNR measured with directional microphones and microphone arrays give corresponding improvements in speech intelligibility.

6.8.1 Directional Microphone Elements

A directional microphone will improve the SNR by maintaining high gain in the direction of the desired source and reduced gain for sources coming from other directions. An ideal cardioid response will improve the SNR by 4.8 dB when compared with an omnidirectional microphone for an on-axis sound source and an isotropic (diffuse) noise field [Olson, 1957]. Measurements of an actual directional microphone mounted on the head, however, indicate that the advantage is only about 2.5 dB in comparison with an omnidirectional hearing-aid microphone in a diffuse noise field [Soede et al., 1993a]. Larger benefits can be obtained under more constrained conditions; a relative improvement of 3-4 dB for the directional microphone was found when a sound source was positioned in front and a noise source behind the head in a reverberant room [Hawkins and Yacullo, 1984]. Even though these improvements are relatively small, directional microphones are the only practical hearing-aid technique that has consistently demonstrated benefit in enhancing speech intelligibility in noise.

6.8.2 Two-Microphone Adaptive Noise Cancellation

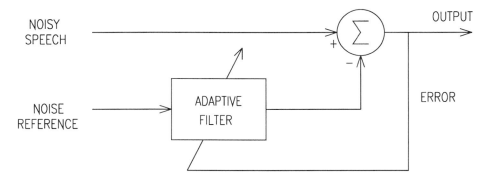

Figure 6.18 Block diagram of an adaptive noise-cancellation system.

Adaptive noise cancellation [Widrow et al., 1975] is illustrated in the block diagram of Fig. 6.18. The system uses one microphone to acquire the noisy speech signal and a second microphone to acquire a signal that is predominantly noise. This latter signal is termed the noise reference. The noise reference is processed by an adaptive filter to match the noise corrupting the speech as closely as possible; the least mean square (LMS) error criterion is often used. The filtered noise sequence is then subtracted from the noisy speech to cancel its noise component and thus improve the SNR. The maximum improvement that can be realized is limited by the noise-to-speech ratio at the reference microphone. Under favorable conditions, adaptive noise cancellation can give improvements in SNR in excess of 30 dB [Chabries et al., 1987][Weiss, 1987], and improvements in speech intelligibility of 35 to 38 percent have been reported [Brey et al., 1987]. Thus there is a high correlation between the improvements in SNR and improvements in speech intelligibility.

The performance of a head-mounted two-microphone adaptive noise-cancellation system was investigated by Weiss [Weiss, 1987] and Schwander and Levitt [Schwander and Levitt, 1987]. In this system, an omnidirectional microphone was used for the speech signal and a rear-facing hypercardioid microphone mounted directly above the speech microphone was used for the noise reference. In a room having a reverberation time of 0.4 sec, this system improved the speech recognition score to 74 percent from 34 percent correct for the unprocessed condition for normal-hearing listeners given a single interfering noise source. Moderate amounts of head movement reduced the recognition score to 62 percent correct, and increasing the number of active noise sources to more than one also reduced the effectiveness of the processing.

6.8.3 Arrays with Time-Invariant Weights

Greater improvements in the SNR and speech intelligibility require arrays that combine the outputs of several microphones. The simplest multi-microphone processing approach is delay-and-sum beamforming. The benefit of delay-and-sum microphone arrays of the sort that can be built into an eyeglass frame, for example, is an improvement of 5-10 dB in SNR, with the greatest improvement at higher frequencies [Soede et al., 1993a][Soede et al., 1993b]. The improvement in the speech reception threshold (SRT), the SNR at which half the words in a list are correctly identified, was found to be 7 dB. These arrays used five cardioid microphone elements uniformly spaced over 10 cm. Furthermore, the performance of both broadside arrays (across the front of the eyeglasses) and endfire arrays (along the temple) did not appear to be affected by the head to any great extent.

The performance of delay-and-sum beamforming can be bettered by using superdirective array processing [Cox et al., 1986] to give the optimum improvement in SNR for a stationary noise field. Simulation studies for a spherically isotropic noise field [Stadler and Rabinowitz, 1993][Kates, 1993c], using endfire array configurations similar to that used by Soede et al. [Soede et al., 1993a][Soede et al., 1993b], show that a superdirective array will give a SNR about 5 dB better than that obtained for delay-and-sum beamforming using the same set of microphones. For the broadside array orientation, the superdirective weights average less than 1-dB better SNR than delay-and-sum beamforming. When using omnidirectional microphones, a null at 180 deg can be imposed on the endfire array beam pattern with only a 0.3 dB penalty in SNR [Kates, 1993c], and increasing the number of microphones in the array from five to seven without increasing the overall length improves the SNR by only about 0.3 dB [Stadler and Rabinowitz, 1993].

6.8.4 Two-Microphone Adaptive Arrays

Optimal performance in a wide variety of listening environments may require more flexibility than can be obtained with an array using a data-independent set of weights. Optimal, or nearly optimal, performance in a noise field that is not known a priori or one which is nonstationary can be achieved with adaptive array processing. The simplest adaptive array geometry uses two microphones. An example of a two-microphone Griffiths-Jim adaptive array [Griffiths and Jim, 1982] is shown in figure 6.19 [Greenberg and Zurek, 1992]. A version of this array using directional microphone elements has also been implemented [Kompis and Dillier, 1994].

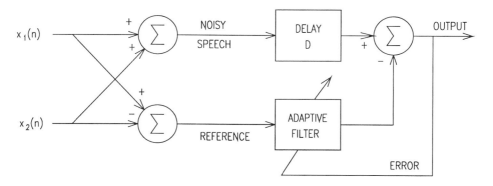

Figure 6.19 Block diagram of an adaptive two-microphone array.

The operation of the two-microphone Griffiths-Jim beamformer [Greenberg and Zurek, 1992] is related to adaptive noise cancellation, described in a previous section, with a pre-processor that forms the sum and difference of the microphone signals. The sum signal forms the noisy speech signal input, and the difference signal forms the noise reference. The performance of this array is governed by the amount of reverberation, the ratio of the amplitude of the desired speech signal to that of the noise source (target-to-jammer ratio, or TJR), the accuracy with which the assumed direction of the speech signal matches the actual alignment of the array, and the length of the adaptive filter.

This system was tested mounted on the KEMAR anthropometric manikin [Burkhard and Sachs, 1975] with the target straight ahead and the jammer at 45 deg. The adaptive filter had 169 taps, and the delay D was set to half the filter length. The system showed improvements in SNR of approximately 30 dB in anechoic conditions when the array was correctly aligned and the TJR approached negative infinity (i.e. no speech signal). Increasing the TJR to 20 dB reduced the improvement in SNR to about 24 dB for a broadside array and about 3 dB for an endfire array configuration. In a moderately reverberant room, the improvement in SNR was about 10 dB at all TJR values for the target 0.8 m from the array, and was reduced to about 2 dB when the target was 2.6 m away. Simulation results show that the array performance degrades as the relative amount of reverberation increases, asymptotically approaching the performance of a two-microphone delay-and-sum beamforming array. Furthermore, at high TJR's, increasing the filter length from 100 to 1000 taps leads to signal cancellation in the simulated reverberant environment. Misalignment of the array gave small but consistent reductions in performance under all test and simulation conditions.

These results indicate that array performance in reverberation is an important concern for hearing-aid applications. Modifications to the Griffiths-Jim array can improve the performance in the presence of reflections. Removing or reducing the delay of D samples shown in the signal path of Fig. 6.19 can greatly reduce the deleterious effects of reflections (Hoffman, 1994). Adding a processing constraint to inhibit adaptation at a positive TJR (in dB) also improves performance[Greenberg, 1994]. Measurements of a system having both modifications[Greenberg, 1994] show that improvements in SNR of 20-30 dB occur at a direct-to-reverberant ratio of 20 dB, independent of the TJR. However, when the reverberant power exceeds that of the direct signal, the system performance degrades to about a 3 dB improvement in SNR, which is what would be obtained from delay-and-sum beamforming.

6.8.5 Multi-Microphone Adaptive Arrays

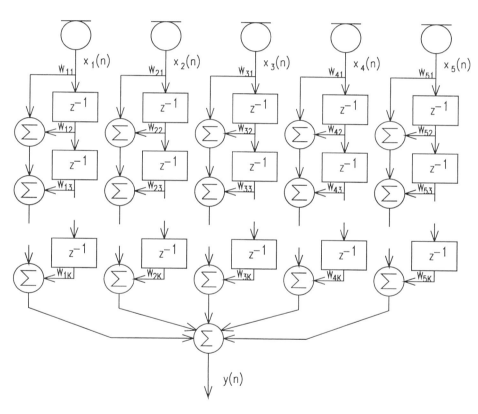

Figure 6.20 Block diagram of a time-domain five-microphone adaptive array.

The signal processing for a general time-domain five-microphone adaptive array is shown in the block diagram of Fig. 6.20. Each microphone is attached to a tapped delay line, and the complete set of weights is adapted to optimize the selected design criterion. Such algorithms commonly use a minimum mean-squared error criterion [Monzingo and Miller, 1980], with the consequence that the correlation matrix used in computing the sensor weights includes the signal as well as the noise. For this type of processing, a perturbed signal wavefront, as occurs when a strong highly-correlated reflection is present or when there are random displacements in the microphone locations, can lead to the suppression of the desired signal [Seligson, 1970][McDonough, 1972] [Cox, 1973]. Signal suppression has been observed in arrays designed for hearing-aid applications, with the suppression caused by strong reflections [Greenberg and Zurek, 1992] or by a missteered array [Hoffman et al., 1994].

Prevention of signal suppression can be achieved by applying a constraint to the adaptive algorithm. It can be shown that suppression of the desired signal is accompanied by an increase in the magnitude of the adaptive weight vector [Cox, 1973]. Thus constraining the magnitude of the weight vector to be less than or equal to a pre-set limit guarantees that the desired signal will never be eliminated by the processing; this approach gives rise to the scaled projection algorithm[Cox et al., 1987]. The scaled projection algorithm has been found to be effective in hearing-aid applications [Hoffman et al., 1994].

Hoffman et al.[Hoffman et al., 1994] simulated head-mounted microphone arrays having from three to seven microphones, with eight- or sixteen-tap filters at each microphone used to implement a Griffiths-Jim beamformer incorporating the scaled projection algorithm. The speech source was in front of the head, and a single noise source was at a 45-deg angle. Reverberant environments were simulated giving direct-to-reverberant power ratios of infinity, 6.9, 1.3, and -4.0 dB. The results show that, for the single noise source, increasing the number of microphones in the array has no significant effect on the array performance, while the sixteen-tap filters performed slightly better than the eight-tap filters. The improvement in SNR produced by the array, however, was strongly affected by the direct-to-reverberant ratio, with the array benefit going from about 17 dB at the ratio of infinity to about 3 dB at the ratio of -4.0 dB. The scaled projection algorithm was adjusted so that under conditions of misalignment the adaptive array performance was reduced to that of delay-and-sum beamforming.

The Hoffman et al. [Hoffman et al., 1994] results show that even though the scaled projection algorithm prevents suppression of the desired signal, it does not give optimal array performance in the presence of the correlated interference that often occurs in reverberant environments. An approach that has been taken to deal with correlated interference is to modify the signal-plus-noise correlation matrix. One technique is to form a correlation matrix of reduced rank by averaging the correlation matrices obtained from subsets of the sensor array [Evans et al., 1981][Shan et al., 1985][Takao et al., 1986]. However, for the short array having a small number of microphones that would be appropriate for hearing-aid applications, the reduced-rank correlation matrix would greatly limit the number of nulls that could be generated in the array response pattern. An alternative technique is to average the values along each diagonal of the full-rank signal-plus-noise correlation matrix [Godara and Gray, 1989][Godara, 1990][Godara, 1991][Godara, 1992] thus forcing it to have a Toeplitz structure. This structured correlation matrix has been shown to have performance for correlated interference that is nearly identical to that for uncorrelated interference given an array many wavelengths long [Godara, 1990][Godara, 1991].

A further consideration in the selection of the adaptive processing algorithm is the rate of convergence and the computational burden. A frequency-domain implementation of the adaptive processing generally offers faster convergence than a time-domain version due to reduced eigenvalue spread in the correlation matrices [Narayan et al., 1983][Chen and Fang, 1992]. However, given M microphones in the array and a sampling rate of T samples per second, the frequency-domain processing requires approximately MT transforms per second independent of the transform block size. In order to reduce the computational burden, a block frequency-domain implementation incorporating a causality constraint on the adaptive weights can be chosen for the processing [Ferrara, 1980][Clark et al., 1981b][Clark et al., 1983]. The block frequency-domain processing preserves the advantageous properties of the reduced correlation matrix eigenvalue spread [Mansour and Gray, 1982] while reducing the required number of transforms; only T transforms per second are needed when there are five microphones in the array and a 32-point FFT is used.

An example of a five-microphone array using block frequency-domain processing is shown in Fig. 6.21. The incoming signals at each microphone are read into a buffer, and an FFT is used to transform the signal at each microphone into the frequency domain. The set of weights for each FFT bin is adaptively updated, and the weighted sum at each frequency is then formed. An inverse transform then returns the signal to the time domain.

6.8.6 Performance Comparison in a Real Room

Most of the microphone array results have been obtained from computer simulations. However, the acoustic field in a real room is far more complex than anything that is produced by a simple simulation. A recent experiment[Kates and Weiss, 1996] compared several time-invariant and adaptive array-processing algorithms for an array placed in a large office containing bookshelves, filing cabinets, several desks, chairs, and tables. An end-fire array, consisting of five microphones at a uniform 2.5-cm spacing to give a 10-cm array, was built using Knowles EK-3033 omnidirectional hearing-aid microphones. The microphone signals were acquired at a 10-kHz sampling rate in each channel using an A/D converter having simultaneous sample-and-hold circuits. The desired signal was a sentence from a loudspeaker in front of the array, and the interference was multi-talker speech babble from a loudspeaker positioned to

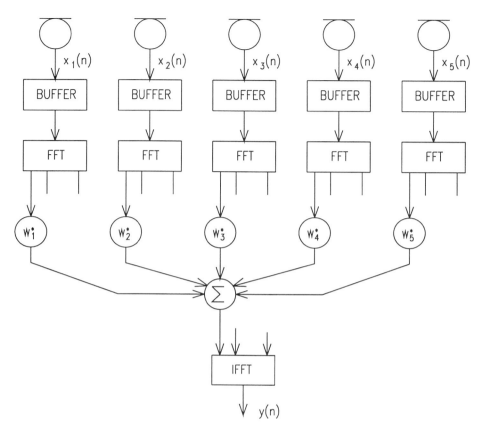

Figure 6.21 Block diagram of a frequency-domain five-microphone adaptive array.

one side or behind the array. Three array positions in the room were investigated: the array mounted on a stand near the middle of the room, the array on a short desk stand with the speech loudspeaker at the other end of the desk, and the array mounted above one ear of the KEMAR manikin with the manikin near the middle of the room.

Five signal-processing algorithms were investigated in frequency-domain implementations. The equivalent time-domain filter length was 16 taps at the 10-kHz sampling rate. The three algorithms using time-invariant coefficients were delay-and-sum beamforming, an oversteered superdirective array [Cox et al., 1986] in which the time delays used in delay-and-sum beamforming are greater than those corresponding to the inter-element sound propagation times, and superdirective processing optimized at each frequency[Kates, 1993c] with a causality constraint imposed on the weights

[Clark et al., 1981b][Clark et al., 1983]. The two adaptive algorithms were a frequency-domain version of the scaled projection algorithm [Cox et al., 1987], and the same algorithm but using the composite structured correlation matrix in which the structured correlation matrix [Godara, 1990] was used at low SNR values, gradually shifting to the superdirective system at high SNR values.

The results for a single source of interference in the room were an average SNR improvement of 5.6 dB for the delay-and-sum beamforming, 7.1 dB for the oversteered array, and 9.8 dB for the optimum superdirective array. The SNR improvement of the scaled projection algorithm was 11.3 dB at a TJR of -10 dB, reducing to 8.9 dB at a TJR of +10 dB. The composite structured correlation matrix yielded an improvement in SNR of 10.0 dB at all TJR values. The greatest improvement in SNR was observed for the desk-top array position, giving an average of 10.1 dB, and the least for the array mounted on KEMAR, where an average of 8.4 dB was obtained. An analysis of variance (ANOVA) showed that there was no significant difference between the optimum superdirective array and either adaptive algorithm at a 0-dB TJR.

These results indicate that a short microphone array can be very effective in a real room, yielding an improvement in SNR of approximately 10 dB. Good performance was obtained for all of the array positions, including being positioned on a desk top and above the ear of KEMAR. The optimum superdirective array worked as well as the adaptive arrays under almost all test conditions, while the delay-and-sum and oversteered arrays were noticeably inferior. The results for the optimum superdirective array suggest that the complexity of an adaptive system may not be needed for a short hearing-aid microphone array, but that simple analog systems, such as the oversteered array, will not perform as well as an optimum digital system.

6.9 COCHLEAR IMPLANTS

Cochlear implants have become a viable option for individuals with profound sensorineural hearing loss who obtain negligible benefit from hearing aids. A summary of the principles and auditory performance for cochlear implants already exists (Working Group on Communication Aids for the Hearing-Impaired, 1991), and readers are referred to that paper[Working Group on Communication Aids for the Hearing-Impaired, 1991] for background. One cochlear implant system, the Nucleus 22-channel device, has received premarket approval from the FDA. Two additional systems, the Inneraid and the Clarion, are currently undergoing clinical trials. As of June 1993, there were about 3340 implant users in the United States (A. Boothroyd, personal communication, 1994).

Two areas of signal processing research in cochlear implants are coding strategies for multi-electrode excitation and the development of noise-suppression systems. One of the problems in cochlear implants is that there is a large spread of the electrical stimulation within the cochlea. Because of this, simultaneous electrical pulses at spatially separated electrodes interact strongly, and the general strategy has therefore been to excite the cochlea with sequential pulse trains where only one electrode is stimulated at a time. As an example, the Nucleus cochlear implant, which is the most widely used of the devices, originally used an electrode excitation scheme in which the estimated speech second formant was used to select the electrode that was excited and the estimated pitch controlled the rate of excitation[Tong et al., 1979][Patrick and Clark, 1991].

Recent research has indicated that a system in which the overall structure of the speech spectrum is encoded, rather than a small set of features, leads to improved speech intelligibility [McDermott et al., 1992][Wilson et al., 1993][Dillier et al., 1993]. The Wilson et al.[Wilson et al., 1993] continuously interleaved sampling (CIS) scheme, for example, uses five or six channels, with the amplitude of the biphasic pulses encoding the magnitude of the envelope within each channel. The channels are strobed sequentially at the maximum rate possible within the processing system, thereby avoiding any overlap between pulses, and giving about 800 pulses/sec in each channel. The newer Nucleus system[McDermott et al., 1992] uses 16 channels, with electrodes corresponding to the six highest peaks after equalization being excited. Again, an asynchronous sequential set of pulses is generated at the maximum rate allowed by the system. Other strategies, in which speech feature extraction is used for voiced speech and the CIS approach used for unvoiced speech, or in which the lowest-frequency electrode is excited with pitch information and the remainder using the CIS approach, are also being investigated[Dillier et al., 1993].

Noise suppression has also been explored for cochlear implants. A common complaint of users of cochlear implants is that performance deteriorates rapidly with increasing levels of background noise. However, the performance of the original feature-extraction form of the Nucleus processing can be improved in broadband noise by pre-processing the signal with the INTEL method of spectral subtraction [Weiss et al., 1975]. While the estimates for the speech formants and pitch are reasonably accurate in quiet, they become increasingly inaccurate in the presence of background noise[Weiss and Neuman, 1993]. The addition of the INTEL processing resulted in a reduction of 4-5 dB in the SNR needed for the accurate identification of phonemes by implant users[Hochberg et al., 1992].

6.10 CONCLUSIONS

The human auditory system is marvelously complicated. The healthy cochlea has extremely sharp frequency analysis and high gain at low signal levels. The system is nonlinear, with reduced gain and broader filter bandwidths as the signal level increases, and the presence of more than a single tone leads to gain interactions and suppression effects in the system behavior. Hearing impairment involves a loss of gain that causes a shift in auditory threshold, a linearization of the system input/output behavior that results in loudness recruitment, and a loss of frequency and temporal resolution that leads to larger than normal amounts of masking in the frequency and time domains.

Most signal-processing strategies concentrate on a single characteristic of the cochlear damage. Linear amplification is intended to overcome the loss in auditory sensitivity for normal conversational speech levels. Wide dynamic-range compression amplification is intended to compensate for loudness recruitment and the reduced dynamic range in the impaired ear. Noise suppression algorithms are intended to compensate for the loss of frequency and temporal resolution in the impaired ear. Improved performance may result from something as simple as reducing the distortion in an analog amplifier or the sensitivity to feedback in a linear hearing aid, or may require complex models of the nonlinear interactions that occur in the cochlea.

Engineering solutions depend on having a well-defined problem. When the problem and the criteria to be met for its solution can be clearly stated, a solution is often found. Examples of such well-defined problems include reduced amplifier distortion, feedback cancellation, and directional microphone arrays for noise suppression. But where the problem definition is nebulous and success is based on perceptual criteria, a solution may be difficult to find. Examples of these difficult problems include recruitment compensation for complex signals and single-microphone noise suppression. As a further consideration, the loss of frequency resolution in the impaired ear may well mean that total compensation for hearing loss is not possible in a conventional hearing aid. Improved signal processing for hearing aids is thus a deceptively difficult engineering problem.

Acknowledgements

The preparation of this chapter was supported by NIDCD under Grants 2-P01-DC00178 and 1-R01-DC01915-01A1. The author greatly appreciates the comments of Mark Weiss and Arlene Neuman on draft versions of this chapter. The conversion to LaTeX was done by the editor, Mark Kahrs.

7 TIME AND PITCH SCALE MODIFICATION OF AUDIO SIGNALS

Jean Laroche

Joint E-mu/Creative Technology Center
1600 Green Hills Road
POB 660015
Scotts Valley, CA 95067

jeanl@emu.com

Abstract: The independent control over the time evolution and the pitch contour of audio signals has always stood very high on the wish-list of the audio community because it breaks the traditional tie between pitch and rate of playback. Rudimentary solutions to that problem date back to the analog days, but the recent advent of fast digital audio processors has made it possible for better methods to make their way into consumer products, with applications in karaoke, phone answering-systems, post-production audio/video synchronization, language learning, to name only a few. This chapter describes several popular methods for the time-scale, pitch-scale and formant-scale modification of audio signals, and summarizes their respective strengths and weaknesses as well as their relative costs in terms of computation power.

7.1 INTRODUCTION

In many situations, one needs to be able to control *in an independent way* both the time-evolution and the pitch of audio signals.

- Controlling and modifying the time-evolution of a signal is referred to as time-scale modification: the aim is to slow down or speed up a given signal, possibly in a time-varying manner, without altering the signal's spectral content (and in particular its pitch when the signal is periodic).

- Controlling and modifying the pitch of a signal is referred to as pitch-scale modification: the aim is to modify the pitch of the signal, possibly in a time-varying manner, without altering the signal's time-evolution (and in particular, its duration).

Situations where independent control on time and pitch is required include the following:

Synthesis by Sampling: Synthesizers based on the sampling technique typically hold a dictionary of pre-recorded sound units (e.g., musical sounds or speech segments) and generate a continuous output sound by splicing together the segments with a pitch and duration corresponding to the desired melody. Because there can only be a limited number of sound segments stored in the dictionary, one cannot afford sampling all possible pitches and durations, hence the need for independent time-scale and pitch-scale control. Such systems include sampling machines (see Massie's chapter) and text-to-speech synthesizers based on the concatenation of acoustical units [Allen, 1991], which scan a written text as an input, typically a computer ascii file, and 'read' it aloud.

Post-synchronization: Synchronizing sound and image is required when a soundtrack has been prepared independently from the image it is suppose to accompany. By modifying the time evolution of the sound track, one is able to re-synchronize sound and image. A typical example is dialogue post-synchronization in the movie industry.

Data compression: Time-scale modification has also been studied for the purpose of data compression for communications or storage [Makhoul and El-Jaroudi, 1986]. The basic idea consisted of shrinking the signal, transmitting it, and expanding it after reception. It was found however, that only a limited amount of data reduction could be obtained using this method.

Reading for the blind: For visually impaired people, listening to speech recordings can be the only practical alternative to reading. However, one can read at a must faster rate than one can speak, so 'reading by listening' is a much slower process than sight reading. Time-scale modification makes it possible to increase this listening rate, and to scan an audio recording as one can do for a written text.

Foreign language learning: Learning a foreign language can be significantly facilitated by listening to foreign speakers with an artificially slow rate of elocution which can be made faster as the student's comprehension improves. This, again, is a task for time-scaling systems.

Computer interface: Speech-based computer interfaces suffer from the same limitations as encountered in 'reading by listening'. The pace of the interaction is controlled by the machine and not by the user. Techniques for time-scale modifications can be used to overcome the 'time bottleneck' often associated with voice interfaces.

Post-production Sound Editing: In the context of sound recording, the ability to correct the pitch of an off-key musical note can help salvage a take that would otherwise be unusable. Multi-track hard-disk recording machines often offer such capabilities.

Musical Composition: Finally, music composers working with pre-recorded material find it interesting to be given an independent control over time and pitch. In this context, time and pitch-scale modification systems are used as composition tools and the 'quality' of the modified signal is an important issue.

Solutions for time or pitch scaling of audio signals can be tracked back to the 1950's with Fairbanks, Everitt and Jaeger's modified tape recorder [Fairbanks et al., 1954]. The machine, described in more detail in section 7.4, achieved time compression/expansion by constantly discarding/repeating portions of signal, a mechanism which was to be called the 'sampling or splicing technique'. This method inspired a number of *time-domain techniques* (i.e., techniques based on the time-domain representation of the signal) whose algorithmic simplicity made them suitable to real-time implementation. The rotating head method was transposed in the digital domain by Lee [Lee, 1972], with the magnetic tape replaced by a 'circular' memory buffer, and the record and playback heads replaced by read/write memory pointers. The method was further improved by Scott and Gerber [Scott and Gerber, 1972], Malah [Malah, 1979] and others [Roucos and Wilgus, 1985, Moulines and Charpentier, 1990] for speech but also for music [Dattorro, 1987, Roehrig, 1990, Laroche, 1993, Truax, 1994].

An alternative solution to the problem of time or pitch scaling appeared in the 1960's with the use of the short-time Fourier transform [Schroeder et al., 1967]. A frequency-domain representation of the signal was used, obtained by either the short-time Fourier transform or by a filter bank. Later on, improved frequency-domain techniques for time or frequency modification of speech were proposed by Portnoff [Portnoff, 1981] and Seneff [Seneff, 1982] and applied to music by Moorer [Moorer, 1978]. Since then, several frequency-domain time or pitch scaling techniques have been studied, most being slight variations of the original short-time Fourier transform scheme [Dolson, 1986, Dembo and Malah, 1988].

Finally, a third class of time or pitch scaling methods appeared in the 1980's with the use of 'signal models'. The signal to be modified is first modeled, then the model parameters are modified to achieve the desired time or frequency transformation. Such parametric models include linear prediction models [Makhoul, 1975, Griffin and Lim, 1988], in which the signal is the output of a time-varying filter fed with an excitation signal, sinusoidal models where the signal is represented by a sum of sinusoids with time-varying parameters [Marques and Almeida, 1989, Quatieri and McAulay, 1986, George and Smith, 1992] possibly embedded in additive noise [Serra and Smith, 1990, Laroche et al., 1993a], or 'granular models' in which the signal is modeled as the succession of 'grains' of sound [Poirot et al., 1988, Depalle, 1991, Arfib, 1991, Arfib and Delprat, 1993, Jones and Parks, 1988]. See Massie's chapter for more detail on signal models.

This chapter is organized as follows: By reference to a signal model, time-scale and pitch-scale modifications are defined in the first part. The second part presents frequency-domain techniques while the third part describes time-domain techniques. In the fourth part, the limitations of time-domain and frequency-domain methods are discussed along with improvements proposed in the last few years.

7.2 NOTATIONS AND DEFINITIONS

7.2.1 An underlying sinusoidal model for signals

Coming up with a rigorous definition of time-scaling or pitch-scaling is not easy because time and frequency characteristics of a signal, being related by the Fourier transform, are not independent. However, the task is greatly facilitated by referring to a parametric model of audio signals, even when this model is not used explicitly for analysis/synthesis purposes. Perhaps the simplest and most efficient model in our context is the quasi-stationary sinusoidal model, introduced practically simultaneously by Almeida, Silva and Marques [Almeida and Silva, 1984a, Marques and Almeida, 1987] and McAulay and Quatieri [McAulay and Quatieri, 1986b] (see the chapter by Quatieri and McAulay for more detail). In this model, the signal is represented as a sum of sinusoids whose instantaneous frequency $\omega_i(t)$ and amplitude $A_i(t)$ vary slowly with time. This can be written as:

$$x(t) = \sum_{i=1}^{I(t)} A_i(t) \exp(j\phi_i(t)) \quad \text{with} \quad \phi_i(t) = \int_{-\infty}^{t} \omega_i(\tau) d\tau \qquad (7.1)$$

in which $\phi_i(t)$ and $\omega_i(t)$ are called the instantaneous phase and frequency of the ith sinusoid. By reference to this underlying model, a proper definition of time-scale and pitch-scale operations can be given.

7.2.2 A definition of time-scale and pitch-scale modification

Time-scale modification:. The object of time-scale modification is to alter the signal's apparent time-evolution without affecting its spectral content. Defining an arbitrary time-scale modification amounts to specifying a mapping between the time *in the original signal* and the time *in the modified signal*. This mapping $t \to t' = T(t)$ is referred to as the *the time warping function*; in the following, t refers to the time in the original signal, t' to the time in the modified signal. It is often convenient to use an integral definition of T:

$$t \to t' = T(t) = \int_0^t \beta(\tau) d\tau \qquad (7.2)$$

in which $\beta(\tau) > 0$ is the time-varying *time-modification rate*[1]. For a constant time-modification rate $\beta(\tau) = \beta$, the time warping function is linear $t' = T(t) = \beta t$. The case $\beta > 1$ corresponds to slowing down the signal by means of time-scale expansion, while $\beta < 1$ corresponds to speeding it up by means of time-scale compression. For time-varying time-modification rates, the function $t' = T(t)$ is non-linear. Note that $\beta(t)$ is implicitly assumed to be a regular and 'slowly' varying function of time, i.e., its bandwidth is several orders of magnitude smaller than the effective bandwidth of the signal to be modified.

In the sinusoidal model above, an ideal time-scaled signal corresponding to the signal described by Eq. (7.1) would be

$$x'(t') = \sum_{i=1}^{I(T^{-1}(t'))} A_i(T^{-1}(t')) \exp(j\phi_i'(t')) \qquad (7.3)$$

with

$$\phi_i'(t') = \int_{-\infty}^{t'} \omega_i\left(T^{-1}(\tau)\right) d\tau \qquad (7.4)$$

Equation 7.3 states that the instantaneous amplitude of the ith sinusoid at time t' in the time-scaled signal corresponds to the instantaneous amplitude in the original signal at time $t = T^{-1}(t')$. Similarly, by differentiating $\phi_i'(t')$ with respect to t' one can verify that the instantaneous frequency of the ith sinusoid at time t' in the time-scaled signal corresponds to the instantaneous frequency in the original signal at time $t = T^{-1}(t')$. As a result, the time-evolution of the signal is modified, but its frequency content remains unchanged.

Pitch-scale modification:. The object of pitch-scale modifications is to alter the frequency content of a signal without affecting its time evolution. Defining an arbitrary pitch-scale modification amounts to specifying a (possibly time-varying) pitch-modification factor $\alpha(t) > 0$. As above, $\alpha(t)$ is implicitly assumed to be a regular and 'slowly' varying function of time.

By reference to our sinusoidal model, the ideal pitch-scaled signal corresponding to the signal described by Eq. (7.1) would be:

$$x'(t') = \sum_{i=1}^{I(t')} A_i(t') \exp(j\phi_i'(t')) \quad \text{with} \quad \phi_i'(t') = \int_{-\infty}^{t'} \alpha(\tau) \omega_i(\tau) d\tau \qquad (7.5)$$

The above equation indicates that the sinusoids in the modified signal at time t' have the same amplitude as in the original signal at time $t = t'$, but their instantaneous frequency are multiplied by a factor $\alpha(t')$, as can be seen by differentiating $\phi'(t')$ with respect to t'. As a result, the time-evolution of the original signal is not modified but its frequency content is scaled by the pitch-modification factor. When the signal is periodic (as can be the case in speech and music), the fundamental frequency of the modified signal is that of the original signal multiplied by the factor $\alpha(t)$.

Combined modification:. It is possible to combine time-scale and pitch-scale modifications. Given a time warping function $T(t)$ and a pitch-modification factor $t \to \alpha(t)$, the ideal modified signal corresponding to the signal described in Eq. (7.1) is given by

$$x'(t') = \sum_{i=1}^{I(T^{-1}(t'))} A_i(T^{-1}(t')) \exp(j\phi'_i(t')) \quad \text{with}$$

$$\phi'_i(t') = \int_{-\infty}^{t'} \alpha\left(T^{-1}(\tau)\right) \omega_i\left(T^{-1}(\tau)\right) d\tau$$

Duality:. As might be expected, there is a duality between time-scale and pitch-scale modifications: Starting for example from the ideal time-scale modification of Eq. (7.3),

$$x'(t') = \sum_{i=1}^{I(T^{-1}(t'))} A_i(T^{-1}(t')) \exp\left(j \int_{-\infty}^{t'} \omega_i\left(T^{-1}(\tau)\right) d\tau\right)$$

The modified signal $x'(t')$ can be time warped arbitrarily by a time warping function $T(t)$ in order to obtain a signal $y(t)$

$$y(t') \stackrel{\triangle}{=} x'(T(t')) = \sum_{i=1}^{I(t')} A_i(t') \exp\left(j \int_{-\infty}^{T(t')} \omega_i\left(T^{-1}(\tau)\right) d\tau\right)$$

and a change of variable $\tau \to \tau' = T^{-1}(\tau)$ in the integral yields:

$$y(t') = \sum_{i=1}^{I(t')} A_i(t') \exp\left(j \int_{-\infty}^{t'} \left(\frac{dT}{d\tau'}\right) \omega_i(\tau') d\tau'\right)$$

This corresponds to the ideal pitch-scaling operation described by Eq. (7.5) with

$$\alpha(\tau) = \frac{dT}{d\tau}$$

This result means that a pitch-scale modification specified by $\alpha(t)$ can be achieved by first performing a time-scale modification whose factor is given by $\alpha(t) = dT/dt$

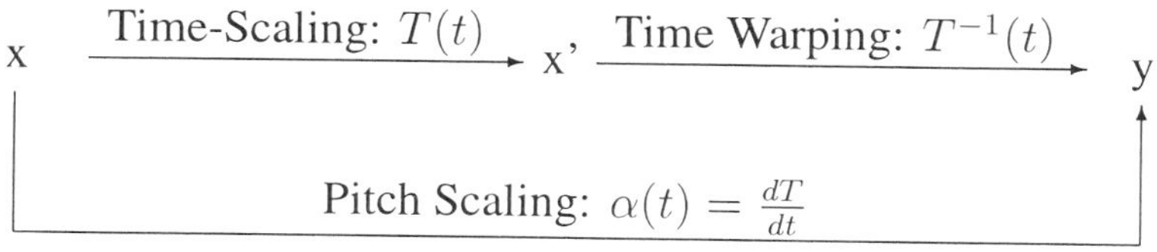

TIME AND PITCH SCALE MODIFICATION OF AUDIO SIGNALS   285

$$x \xrightarrow{\text{Time-Scaling: } T(t)} x' \xrightarrow{\text{Time Warping: } T^{-1}(t)} y$$

$$\text{Pitch Scaling: } \alpha(t) = \frac{dT}{dt}$$

Figure 7.1 Duality between Time-scaling and Pitch-scaling operations. The pitch-scaled signal y can be obtained by a time-scale modification followed by a simple time warping. Conversely, the time-scaled signal x' can be obtained from the pitch-scale modified signal y by a simple time warping.

then time warping the result by the function $T^{-1}(t)$, as summarized in the following figure. Note that the time warping operation above is a mere scaling of the time-axis defined by the equation $y(t) = x'(T(t))$ which in effect modifies both the duration and the pitch of x'. This time warping gives the signal $y(t)$ the original duration of $x(t)$. For discrete signals (i.e., when times t are integer multiples of a sampling period ΔT) time warping is a process known as resampling (see section 8.4). Similarly, any time-scaling modification can be achieved by a pitch-scaling modification followed by a time warping operation. In Fig. 7.1, the signal x' can be obtained from y by a time warping operation $x'(t') = y(T^{-1}(t'))$.

7.3 FREQUENCY-DOMAIN TECHNIQUES

Frequency-domain time or pitch scaling techniques make use of a frequency-domain description of the signal, usually (but non necessarily) obtained by the Fourier transform. Because time-scale and pitch-scale modifications have been defined with reference to a signal model (the sinusoidal decomposition of Eq. (7.1)), any time or pitch scaling technique is by essence parametric. However, frequency-domain time or pitch scaling techniques can be classified according to whether they make *explicit* or *implicit* use of the signal parameters. Methods making explicit use of the signal parameters require a preliminary signal analysis stage during which the parameters are estimated. Such methods can be based on purely sinusoidal models, mixed sinusoid/noise models, or granular models. Methods making implicit use of the signal parameters do not require this analysis stage (or require a much simplified analysis stage). Such methods are usually base on the short-time Fourier transform.

7.3.1 Methods based on the short-time Fourier transform

The reader can refer to chapter 9.2.2 for an alternative presentation of the short-time Fourier transform, in the context of sinusoidal modeling.

Definitions and assumptions. In the following, all signals will be supposed to be discrete signals.

The short-time Fourier transform. The short-time Fourier transform method has been used for signal analysis, synthesis and modifications for many years and its applications are numerous. In this section, we give a brief overview of the technique and of the main results. Interested readers can refer to [Crochiere, 1980, Crochiere and Rabiner, 1983, Allen, 1982, Nawab and Quatieri, 1988b] for mode detailed analyses of its theory and implementation. The short-time Fourier transform can be viewed as an alternate way of representing a signal in a joint time and frequency domain. The basic idea consists of performing a Fourier transform on a limited portion of the signal, then shifting to another portion of the signal and repeating the operation. The signal is then described by the values of the Fourier transforms obtained at the different locations. When the values of the Fourier transforms are expressed in polar coordinates, the short-time Fourier transform is alternately called the "phase vocoder".

Analysis: In standard applications, the short-time Fourier transform analysis is performed at a constant rate: the analysis time-instants t_a^u are regularly spaced, i.e. $t_a^u = uR$ where R is a fixed integer increment which controls the analysis rate. However, in pitch-scale and time-scale modifications, it is usually easier to use regularly spaced synthesis time-instants, and possibly non-uniform analysis time-instants. In the so-called band-pass convention, the short-time Fourier transform $X(t_a^u, \Omega_k)$ is defined by:

$$X(t_a^u, \Omega_k) = \sum_{n=-\infty}^{\infty} h(n) x(t_a^u + n) \exp(-j\Omega_k n) \quad (7.6)$$

in which $h(n)$ is the analysis window and $\Omega_k = \frac{2\pi k}{N}$. Note that the Fourier transform used here is the *discrete Fourier transform*, since both time and frequencies are discrete. The Fourier transform is calculated on N points, N being usually longer than the length T of the analysis window $h(n)$.

Synthesis: Given an arbitrary sequence of synthesis short-time Fourier transforms $Y(t_s^u, \Omega_k)$, there is in general no time-domain signal $y(n)$ of which $Y(t_s^u, \Omega_k)$ is the short-time Fourier transform: the stream of short-time Fourier transforms of a given signal must satisfy strong consistency conditions since the Fourier transforms usually correspond to *overlapping* short-time signals (these conditions are given for example in [Portnoff, 1980]). Consequently, many methods exist to obtain approximate $y(n)$ from $Y(t_s^u, \Omega_k)$. The most general reconstruction formula is:

$$y(n) = \sum_{u=-\infty}^{\infty} w(n - t_s^u) \frac{1}{N} \sum_{k=0}^{N-1} Y(t_s^u, \Omega_k) \exp(j\Omega_k n - t_s^u) \quad (7.7)$$

in which $w(n)$ is called the synthesis window.

Perfect reconstruction: One can show that the short-time Fourier transform yields perfect reconstruction in the absence of modification (i.e. a synthesis signal $y(n)$ exactly similar to the original $x(n)$ when $t_s^u = t_a^u$ and $Y(t_s^u, \Omega_k) = X(t_a^u, \Omega_k)$) if

$$\sum_{u=-\infty}^{\infty} w(n - t_a^u) h(t_a^u - n) = 1 \qquad \forall n \tag{7.8}$$

Short-time Fourier transform of a sinusoidal signal. When the signal corresponds to the model of Eq. (7.1), the short-time Fourier transform can be expressed in terms of the model parameters. Substituting Eq. (7.1) into Eq. (7.6) yields

$$X(t_a^u, \Omega_k) = \sum_{n=-\infty}^{\infty} h(n) \left(\sum_{i=1}^{I(t_a^u+n)} A_i(t_a^u + n) \exp(\phi_i(t_a^u + n)) \right) \exp(-j\Omega_k n) \tag{7.9}$$

Now we will assume that the analysis window $h(n)$ is sufficiently short so that the instantaneous frequencies and amplitudes of the sinusoids can be assumed constant over the duration of h. As a result, we have

$$\phi_i(t_a^u + n) = \phi_i(t_a^u) + n\omega_i(t_a^u)$$

and the short-time Fourier transform becomes, after straightforward manipulations

$$X(t_a^u, \Omega_k) = \sum_{i=1}^{I(t_a^u)} A_i(t_a^u) \exp(j\phi_i(t_a^u)) H(\Omega_k - \omega_i(t_a^u)) \tag{7.10}$$

where $H(\omega)$ is the Fourier transform of the analysis window $h(n)$. Equation (7.10) shows that the short-time Fourier transform of the sinusoidal signal is the sum of $I(t_a^u)$ images of $H(\omega)$, translated by $\omega_i(t_a^u)$ and weighted by $A_i(t_a^u) \exp(j\phi_i(t_a^u))$. We will further assume that $h(n)$ is real, symmetric around $n = 0$, so that $H(\omega)$ is real symmetric, and that the cutoff frequency ω_h of the analysis window $h(n)$ is less than the spacing between two successive sinusoids. This means that the shifted versions of $H(\omega)$ do not overlap, and Eq. (7.10) simplifies to

$$X(t_a^u, \Omega_k) = \begin{cases} A_i(t_a^u) \exp(j\phi_i(t_a^u)) H(\Omega_k - \omega_i(t_a^u)) & \text{if } |\Omega_k - \omega_i(t_a^u)| \leq \omega_h \\ 0 & \text{otherwise} \end{cases} \tag{7.11}$$

Eq. (7.11) shows that the short-time Fourier transform gives access to the instantaneous amplitude $A_i(t_a^u)$, and the instantaneous phase $\phi_i(t_a^u)$ of the sinusoid i which falls into

Fourier channel k. The phase is known up to a multiple of 2π (since only $\exp(j\phi_i(t_a^u))$ is known). Time-scale modifications also require the knowledge of the instantaneous frequency $\omega_i(t_a^u)$. $\omega_i(t_a^u)$ can also be estimated from successive short-time Fourier transforms: for a given value of k, computing the backward difference of the short-time Fourier transform phase yields

$$\Delta\phi = \phi_i(t_a^{u+1}) - \phi_i(t_a^u) + 2m\pi = (t_a^{u+1} - t_a^u)\omega_i(t_a^u) + 2m\pi \qquad (7.12)$$

in which we have assumed that the instantaneous frequency $\omega_i(t_a^u)$ remained constant over the duration $t_a^{u+1} - t_a^u$. The term $2m\pi$ which comes from the fact that only the principal determination of the phase is known (e.g., as given by an four-quadrant inverse tangent) is estimated the following way. Denoting $R(u) = t_a^{u+1} - t_a^u$, we have (since the ith sinusoid falls within the kth channel)

$$|(\omega_i(t_a^u) - \Omega_k)R(u)| < \omega_h R(u)$$

in which ω_h is the bandwidth of the analysis window. If $R(u)$ is such that $\omega_h R(u) < \pi$ we have

$$|\Delta\phi - \Omega_k R(u) - 2m\pi| < \pi \qquad (7.13)$$

and there is only one integer m that satisfies the latter inequality. Once m is determined (by adding or subtracting multiples of 2π until the preceding inequality is satisfied - a process known as phase unwrapping), the instantaneous frequency can be obtained by Eq. (7.12), yielding

$$\omega_i(t_a^u) = \Omega_k + \frac{1}{R(u)}\left(\Delta\phi - \Omega_k R(u) - 2m\pi\right) \qquad (7.14)$$

Time scaling. Because the phase-vocoder (the short-time Fourier transform) gives access to the implicit sinusoidal model parameters, the ideal time-scale operation described by Eq. (7.3) can be implemented in the same framework. Synthesis time-instants t_s^u are usually set at a regular interval $t_s^{u+1} - t_s^u = R$. From the series of synthesis time-instants t_s^u, analysis time-instants t_a^u are calculated according to the desired time warping function $t_a^u = T^{-1}(t_s^u)$. The short-time Fourier transform of the time-scaled signal is then:

$$\begin{aligned} Y(t_s^u, \Omega_k) &= \left|X(T^{-1}(t_s^u), \Omega_k)\right| \exp(j\hat{\phi}_k(t_s^u)) \quad \text{with} \\ \hat{\phi}_k(t_s^u) &= \hat{\phi}_k(t_s^{u-1}) + \left(t_s^u - t_s^{u-1}\right)\lambda_k(T^{-1}(t_s^u)) \end{aligned} \qquad (7.15)$$

in which $\lambda_k(T^{-1}(t_s^u))$ is the instantaneous frequency calculated in channel k, as given by Eq. (7.14). As previously, $\lambda_k(t_a^u)$ is supposed to be constant over the duration

$t_a^u - t_a^{u-1}$. It is easy to verify that this short-time Fourier transform corresponds to the ideal time-scaled signal. According to the preceding equation, the modulus of the modified short-time Fourier transform in a given channel at time t_s^u is the same as that of the original short-time Fourier transform at time $t_a^u = T^{-1}(t_s^u)$, and its phase is calculated so that the instantaneous frequency of any sinusoid in the modified signal at time t_s^u is the same as in the original signal at time $T^{-1}(t_s^u)$. In other words, the phase runs freely and only its derivative with respect to time is controlled. The modified short-time Fourier transform $Y(t_s^u, \Omega_k)$ can then be used in Eq. (7.7) to obtain the time-scaled signal. The complete algorithm can be summarized as follows:

1. Set the initial instantaneous phases $\hat{\phi}(0, \Omega_k) = \arg(X(0, \Omega_k))$.

2. Set the next synthesis instant $t_s^{u+1} = t_s^u + R$ and calculate the next analysis instant $t_a^{u+1} = T^{-1}(t_s^{u+1})$

3. Compute the short-time Fourier transform at next analysis time-instant t_a^{u+1} and calculate the instantaneous frequency in each channel according to Eq. (7.14).

4. Calculate the instantaneous phase $\hat{\phi}_k(t_s^{u+1})$ according to Eq. (7.15).

5. Reconstruct the time-scaled short-time Fourier transform at time t_s^{u+1} according to Eq. (7.15).

6. Calculate the $(u+1)$-th short-time modified signal by use of the synthesis formula Eq. (7.7) and return to step 2.

Note that if an analysis time-instant t_a^u is not an integer, it can be rounded to the nearest integer prior to the calculation of the instantaneous frequency, provided that the corrected value of $R(u-1) = t_a^u - t_a^{u-1}$ is used in Eq. (7.14). It is easy to show that for a constant-amplitude, constant-frequency sinusoid, the procedure above outputs a perfect time-modified sinusoid[2] provided

$$\sum_{u=-\infty}^{\infty} w(n - t_s^u) h(t_s^u - n) = 1 \qquad \forall n \qquad (7.16)$$

which is similar to the standard condition of perfect reconstruction, Eq. (7.8). For an output overlap factor of 75%, a possible choice for $w(n)$ and $h(n)$ is Hanning windows.

Pitch-scaling. There are several ways of using the phase-vocoder for pitch-scaling operations.

290   APPLICATIONS OF DSP TO AUDIO AND ACOUSTICS

Using a bank of sinusoidal oscillators. The Fourier analysis described above gives access to the sinusoidal parameters of the signal: time-varying sinusoidal amplitudes $A_i(t_a^u)$ (Eq. (7.11)) and instantaneous frequencies $\omega_i(t_a^u)$ Eq. (7.14). A simple means of performing pitch-scaling consists of resynthesizing the signal after multiplying the instantaneous frequencies by the factor $\alpha(t)$:

$$y(t') = \sum_{i=1}^{I(t')} A_i(t') \exp(j\hat{\phi}_i(t')) \quad \text{with} \quad \hat{\phi}(t) = \int_{-\infty}^{t} \alpha(\tau)\omega_i(\tau)d\tau \quad (7.17)$$

This solution has the advantage of being very simple, but the drawback of being expensive in terms of calculations, even when tabulated sinusoids are used in Eq. (7.17). In fact, additive synthesis (of which Eq. (7.17) is an example) can be implemented at a much lower cost by use of the Fourier transform [Rodet and Depalle, 1992]. This last remark is a strong motivation for using the following alternative:

Using time-scaling and resampling. Pitch-scaling can be performed by using the duality between time and pitch-scaling operations, as discussed in section 7.2.2. Pitch-scaling is implemented in two steps:

1. The original signal $x(t)$ is first time-scaled by a factor $T(t)$ such that $\alpha(t) = \frac{dT}{dt}$ by use of the technique described above.

2. The resulting signal $x'(t')$ is then *resampled*, yielding

$$y(t) = x'(T(t))$$

As was previously shown, $y(t)$ is then the pitch-scaled version of the original signal $x(t)$.

Resampling can be achieved in several ways, either in the time or in the frequency domain. Because $x'(t')$ is only known at integer values of t', resampling amounts to interpolating (since in general, $T(t)$ is not an integer). As is well known [Oppenheim and Schafer, 1989], for constant modification rates ($\alpha(t) = \alpha$ or $T(t) = \alpha t$), the ideal interpolation is the so-called band-limited interpolation in which y is obtained by convolving x' with a sinc function:

$$y(t) = \sum_{i=-\infty}^{\infty} x'(i) \frac{\sin \pi(\mu t - i)}{\pi(\mu t - i)}$$

in which $\mu = \min(\alpha, 1/\alpha)$. This convolution can be very expensive in terms of calculation, especially for non-rational values of α. For rational α, multirate implementations can be used [Crochiere and Rabiner, 1983], with a significant reduction of the computational cost. However, for time-varying pitch-scaling factors $\alpha(t)$, multirate resampling can no longer be used. See section 8.4 for alternative ways of performing sampling-rate conversion in the context of sampling synthesizers.

The short-time Fourier transform provides another way of performing resampling with possibly non rational, time-varying factors. The major advantage of this technique is that it can be combined with the time-scaling in a single step to minimize the amount of calculations required. Frequency domain resampling is achieved the following way:

1. A short-time Fourier transform analysis is first carried out on the signal to be resampled, at regularly spaced analysis time-instants $t_a^u = Ru$.

2. The short-time spectra are scaled to account for the modification of the sampling frequency.

$$Y(t_s^u, \Omega) = X\left(t_a^u, \frac{\Omega}{\alpha(t_a^u)}\right) \qquad (7.18)$$

Note that when $\alpha(t_a^u) > 1$, the upper part of the original spectrum is discarded, and when $\alpha(t_a^u) < 1$, the upper part of the modified spectrum is null.

3. The resampled time-domain signal is obtained by use of the synthesis formula Eq. (7.7), noticing that because the signal is resampled, the synthesis time-instants t_s^u differ from the analysis time-instants, and are now given by

$$t_s^u = \int_{-\infty}^{t_a^u} \alpha(\tau) d\tau$$

For a *constant modification rate* $\alpha(t) = \alpha$, it can be shown that the above procedure leads to perfect resampling provided the analysis window and the synthesis window verify

$$\sum_{u=-\infty}^{\infty} w(n - t_s^u) h(t_s^u - n\alpha) = 1 \qquad \forall n$$

As with time-domain resampling, the scaling in step 2 above is in fact an interpolation since the short-time spectra are known only at discrete frequencies $\Omega_k = \frac{2\pi k}{N}$. Ideally, band limited interpolation should be used here. In practice however, a mere linear interpolation is used:

$$Y(t_s^u, \Omega_k) = \mu X(t_a^u, \Omega_{k'}) + (1-\mu) X(t_a^u, \Omega_{k'+1}) \quad \text{with} \qquad (7.19)$$

$$k' = \lceil \frac{k}{\alpha(t_a^u)} \rceil \quad \text{and} \quad \mu = \frac{k}{\alpha(t_a^u)} - k' \qquad (7.20)$$

in which $\lceil x \rceil$ is the integer immediately below x. Using linear interpolation in place of band limited interpolation generates *time-aliasing*, an artifact that can be minimized by using an FFT size N much larger than $\max(T, T/\alpha)$, T being the length of the analysis window.

Frequency domain resampling is attractive in the present context because the time-scaling and the resampling operations can be performed in a single phase-vocoder analysis, thus reducing the computational cost. The interested reader may consult [Moulines and Laroche, 1995] for a complete description of the algorithm.

Choice of the analysis parameters. We can now summarize the various constraints introduced so far on the length T of the analysis window, its cutoff frequency ω_h, and the analysis rate R:

- For the short-time Fourier analysis to resolve the sinusoids, the cutoff frequency of the analysis window must satisfy $\omega_h < \min_i \Delta\omega_i$, i.e. be less than the spacing between two successive sinusoids.

- The duration T of the analysis window must be small enough so the amplitudes and instantaneous frequencies of the sinusoids can be considered constant within the analysis window.

- To make phase unwrapping possible, the cutoff frequency and the analysis rate must satisfy $\omega_h R < \pi$ (see Eq. (7.13)).

For standard analysis windows (e.g. Hanning, Hamming,) the cutoff frequency is inversely proportional to the window length, $\omega_h \approx 4\pi/T$. The first condition implies that $T > 4\Delta\omega_{min}$: *The window must be longer than 4 times the period corresponding to the interval between the closest frequencies.* The last constraint above implies $R < T/4$, i.e. successive analysis windows must have a minimum overlap of 75%. The larger the cutoff frequency, the larger the minimum overlap between successive analysis windows.

Puckette in [Puckette, 1995] proposes an alternate way of computing the phases and the amplitudes of the short-time Fourier transform at the synthesis instants, replacing the calculation of the arc tangent and the phase-unwrapping stage by another Fourier transform. Essentially, in Eq. 7.15 the phase increment can also be estimated if the phase $\phi_k(\tilde{t}_a^u)$ of the input signal at time $\tilde{t}_a^u = t_a^u + t_s^u - t_s^{u-1}$ is known:

$$\hat{\phi}_k(t_s^u) = \hat{\phi}_k(t_s^{u-1}) + \phi_k(\tilde{t}_a^u) - \phi_k(t_a^u) \qquad (7.21)$$

$\phi_k(\tilde{t}_a^u)$ can be obtained by calculating an additional Fourier transform at the analysis time-instant $\tilde{t}_a^u$. Moreover, since phases are only added or subtracted, all the operations above can be done by mere complex multiplications and divisions, which are far less costly than trigonometric functions. Since the parameter λ_k is no longer needed, phase unwrapping is no longer required. In many cases, the additional Fourier transform ends up being less costly than the computationally expensive arc tangent and the phase-unwrapping.

7.3.2 Methods based on a signal model

Chapter 9 presents several models that can be used to parametrize audio signals, and possibly modify them. In that context, time or pitch scale modifications are implemented by adequately modifying the model parameters before the resynthesis stage.

7.4 TIME-DOMAIN TECHNIQUES

Historically, the first techniques designed to achieve independent control over pitch or duration were carried out in the time domain: Fairbanks, Everitt and Jaeger's modified tape recorder [Fairbanks et al., 1954] probably is the first known automatic time-domain system for speech transposition. By contrast with frequency-domain methods, time-domain techniques for time or pitch scale modification manipulate short-duration time-segments extracted from the original signal, a mechanism usually called 'sampling' or 'splicing'; As a result, they tend to require much fewer calculations and lend themselves quite well to real-time implementations.

7.4.1 Principle

The basic idea consists of decomposing the signal into successive segments of relatively short duration (of the order of 10 to 40 ms). Time-scale compression (respectively, expansion) is achieved by discarding (respectively, repeating) some of the segments, while leaving the others unchanged, and by copying them back in the output signal as shown in Fig. 7.2. As was the case for frequency-domain techniques, pitch-scale

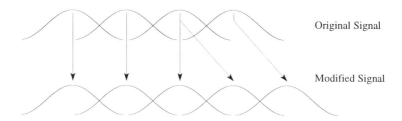

Figure 7.2 Time stretching in the time-domain. Segments excised from the original signal are copied back into the modified signal, with possible repetition (e.g., the third segment in this example).

modifications can be obtained by combining time-scaling and resampling. For this scheme to work properly, one must make sure that no discontinuity appears at time-instants where segments are joined together. This is the reason why the segments are usually overlapped and multiplied by weighting windows.

Most time-domain methods are based on this simple idea, and differ only in the choice of the segment durations, splicing times and weighting windows. They can be classified according to whether they make use of pitch information or not.

7.4.2 Pitch independent methods

Fairbanks, Everitt and Jaeger's modified tape recorder [Fairbanks et al., 1954] is the simplest and oldest example of a pitch-independent time-domain time or pitch scaling system. Although Pitch independent methods are no longer used in practice, they offer a simple illustration of the basic principles underlying most of the more recent algorithms.

The analog origin. Fairbanks, Everitt and Jaeger used a tape recorder equipped with four playback heads attached to a rotating cylinder. The signal is recorded via a fixed head on the moving magnetic tape, then read by the moving playback heads. Depending on the direction and speed of rotation of the cylinder, the signal is read faster or slower than it has been recorded, hence the pitch modification. To simplify the discussion, assume that only 2 playback heads are used, and that the tape is in contact with half the cylinder's perimeter as shown in Fig. 7.3. When the cylinder

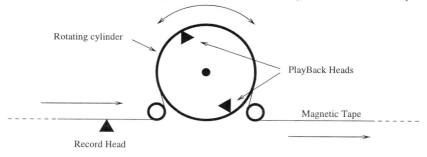

Figure 7.3 A modified tape recorder for analog time-scale or pitch-scale modification. The rotating cylinder is fitted with two playback heads (triangles on the figure), while the record head is fixed (triangle on the left)

is rotating contrary to the tape motion, the speed of the tape relative to the playback head is higher than the recording speed, and therefore the pitch is raised (this is the resampling stage mentioned above). When the first playback head leaves the tape (point A), the second one comes into contact with it at point B, and therefore the signal recorded on the tape between points A and B is repeated (this is the time-scaling stage). The duration of the repeated segment is constant (depending only on the tape transport speed and on the cylinder diameter), and the splicing points are regularly spaced (for a constant pitch-transposition factor) as shown schematically in Fig. 7.4. Because the heads gradually leave (or come into contact with) the tape at points A and B, their output signals gradually fade in or out: adding the outputs of the two playback heads guarantees continuity. The pitch can be lowered by making the cylinder rotate in the same direction as the tape motion. In that case, when the first playback head leaves the tape (point B), the signal recorded on the tape between points A and B will not be read by any playback head (i.e., it is discarded). Time-scaling is achieved by first

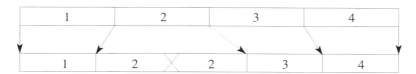

Figure 7.4 Pitch modification with the sampling technique. The segments are resampled (here to a lower sampling rate) and possibly repeated (as segment 2), giving the output signal the same duration as the input signal.

recording the signal on the tape, then playing back the tape with a slower speed (for time expansion) or a higher speed (for time-compression), using the rotative playback heads to compensate for the resulting modification of pitch.

Digital counterpart. The modified tape recorder suffered from problems associated with the use of high-speed rotating mechanical components (in particular, the low-level playback signals had to pass through sliprings, a potential source of problems). These mechanical limitations were easily overcome by the digital implementation of this technique which appeared in the beginning of the 1970s [Lee, 1972]. The idea is to replace the tape by a circular memory register (a memory region addressed modulo its length), to replace the record head by an input address pointer and the two playback heads by two output address pointers pointing to different locations in the memory. The original signal is written into the memory via the input address pointer which is incremented by 1 every sampling period. The output address pointers are incremented by α every sampling period, so the samples are read at a different rate than they were recorded. When the current output pointer meets with the input pointer, the corresponding output samples are faded out while the samples read by the other output pointer fade in, avoiding any discontinuity. Note that for non-integer modification rate α, the output address pointers are incremented by non-integer values, and the output must be obtained by interpolation. The 'digital modified tape recorder' functions exactly as its analog counterpart, without suffering from the usual limitations of analog/mechanical systems.

296 APPLICATIONS OF DSP TO AUDIO AND ACOUSTICS

Input-time/Output-time characteristic. It is interesting to study more closely the relation between the time in the input signal and the time in the output signal. Suppose a time-scale modification of constant-ratio $\beta > 1$ is performed by use of the sampling method. As shown above, the output signal is obtained by periodically repeating segments of the input signal, while leaving the other segments unchanged. The relation between the elapsed time in the original signal t and the elapsed time in the modified signal $\hat{t}'$ is shown in Fig. 7.5. Ideally, one would have $t' = \beta t$, however

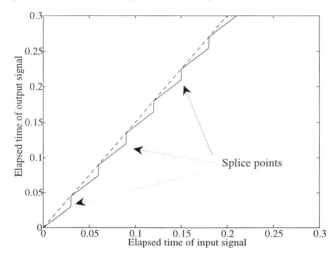

Figure 7.5 Output elapsed time versus input elapsed time in the sampling method for Time-stretching ($\beta > 1$). Dashed line: ideal curve, solid line: approximate curve.

as seen in the figure, $\hat{t}' \neq \beta t$. For segments that are not repeated (and therefore are just copied back into the output signal), the output time increases at the same rate as in the input signal (straight lines with a slope 1 in Fig. 7.5). When a segment is repeated however, the output time increases while the input time remains the same (vertical lines in Fig. 7.5). Thus the ideal curve $t' = \beta t$ (the dashed line of slope β) is approximated by successive segments of slope unity, separated by vertical jumps. The height of the vertical jumps is equal to the fixed length t_b of the repeated segments (half the duration of the circular buffer, or in the analog case, the duration corresponding to half the perimeter of the cylinder). As a result, the number of repeated segments per input second is approximately

$$N_s = \frac{|\beta - 1|}{t_b} \tag{7.22}$$

The problem of tempo. Consider again a time-scale modification of constant factor β. Because the ideal straight line of slope β in Fig. 7.5 is approximated by successive segments of unity slope, the regularity of 'tempo' in the processed signal is altered: consider regularly spaced instants in the original signal (e.g., the ticking of a metronome of period P). The corresponding ticks in the output signal are no longer regularly distributed, although in average they are separated by a duration βP: the time-scaled metronome limps! The maximum discrepancy between the ideal output-time and the actual output-time is precisely the height of the vertical segment t_b. For speech signals, fairly large irregularities of tempo (up to ± 60 ms) can be accepted without any impact on the naturalness of the modified speech. For music however, the regularity of tempo is an extremely important issue and large time-discrepancies cannot be accepted. As a consequence, t_b should not be allowed to excess a maximum value t_b^{max} which, depending on the kind of music to be processed, can be as small as 10 ms.

Splicing artifacts. The time-scale/pitch-scale modification system described above requires very few calculations, and lends itself very well to real-time implementation. Unfortunately, it is prone to artifacts because no precaution is taken at the splicing points, other than to guarantee continuity. Assume a sinusoidal input signal, if the length of the segments repeated or discarded is not equal to the period of the sinusoid, then an artifact will be generated, as shown in Fig. 7.6. Although the waveform is

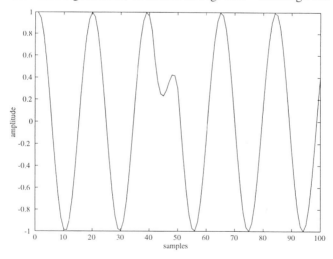

Figure 7.6 Time-scale modification of a sinusoid.

'continuous', the local aperiodicity generates a soft "plop" in the output signal. Such plops appear periodically in the output signal at a frequency which increases with the modification rate as shown by Eq. (7.22). The result is a quite audible degradation of the signal, discrete enough for speech signals but unacceptable for musical signals. A standard way of reducing this degradation consists of using longer cross-fades (50 ms and above) at the splicing points. Although this does make the splicing artifacts much less conspicuous, it does not eliminate them, and has the side-effect of introducing phasing or chorusing (resulting from cross-fading delayed signals). Clearly, one way to improve this system consists of making use of the pitch information, when available.

7.4.3 Periodicity-driven methods

For strictly periodic signals, the splicing method functions perfectly provided the duration of the repeated or discarded segments is equal to a multiple of the period. This is still true to a large extend for nearly periodic signals such as voiced speech and many musical sounds. A number of methods have been proposed, similar in principle to the splicing method, but in which an actual estimate of the pitch or some measure of 'waveform similarity' are used to optimize the splicing points and durations. The method proposed by Scott and Gerber in 1972 [Scott and Gerber, 1972] for speech used an estimate of the pitch obtained by a laryngograph (a device placed on both sides of the thyroid cartilage, which detects the closure of vocal folds via an impedance measurement). The pitch however can be extracted from the signal itself, thus avoiding the use of a specific device, as was proposed by Malah [Malah, 1979] and others [Cox et al., 1983]. Other methods presented below have also been proposed, in which the pitch is not explicitly determined, but the waveform is inspected for self similarity so the length of the segments can be adjusted accordingly.

Methods based on waveform similarity. The SOLA (Synchronized OverLap Add) method originally proposed by Roucos and Wilgus [Roucos and Wilgus, 1985] and its many variations [Verhelst and Roelands, 1993, Wayman and Wilson, 1988, Suzuki and Misaki, 1992, Laroche, 1993, Hardam, 1990] are all based on the following principle. The idea consists of adjusting the length of the repeated/discarded segment so the overlapped parts (e.g., the beginning and the end of the second segment in Fig. 7.4) are 'maximally similar', so the kind of artifact shown in Fig. 7.6 is avoided. When the signal is quasi-periodic, the optimal duration is a multiple of the quasi-period since in that case, the overlapping parts are nearly similar (separated by an integer number of periods). Note that the idea is very similar to what is done in sampling machines to adapt the loop length to the signal (see section 8.2.4). Many methods can be used to measure the similarity between the overlapping parts, of which the normalized cross-correlation $c_c(t_a^u, k)$ and the average magnitude difference function (AMDF) $c_a(t_a^u, k)$

are the most standard.

$$c_c(t_a^u, k) = \frac{\sum_{i=1}^{N_c} x(t_a^u + i) x(t_a^u + i + k)}{\left[\sum_{i=1}^{N_c} x^2(t_a^u + i)\right]^{1/2} \left[\sum_{i=1}^{N_c} x^2(t_a^u + i + k)\right]^{1/2}} \quad \text{and} \quad (7.23)$$

$$c_a(t_a^u, k) = \frac{1}{N_c} \sum_{i=1}^{N_c} |x(t_a^u + i) - x(t_a^u + i + k)| \quad (7.24)$$

in which N_c controls the duration over which 'similarity' is estimated. At time t_a^u, the optimal segment duration is the value of k at which $c_c(t_a^u, k)$ is maximal or at which $c_a(t_a^u, k)$ is minimal. The calculation of the normalized cross-correlation is somewhat costly, but can be simplified for example by down-sampling the signal or by making use of a fast Fourier Transform [Laroche, 1993][3]. By contrast, the average magnitude difference function requires fewer calculations, and its minimum can be found without calculating the sum over all values of i in Eq. (7.24) for all values of k. However, the average magnitude difference function is more sensitive to noise.

Because the duration of the repeated segments no longer is a fixed parameter, the optimized algorithm now requires an upper bound t_b^{max} for acceptable time-discrepancies t_b. In most of the methods based on waveform similarity, the measure of similarity ($c_c(t_a^u, k)$ or $c_a(t_a^u, k)$) is evaluated at regularly spaced time-instants $t_a^u = Ru$. Only those values of k are tested such that repeating or discarding a segment of length k keeps the time-discrepancy below its limit t_b^{max}. In other words, in Fig. 7.5 vertical jumps are tested every R samples of input signal with heights k such that the time discrepancy lies within acceptable limits: $|\hat{t'} + k - \beta t| < t_b^{max}$. When $|\hat{t'} - \beta t| < t_b^{max}$, the value of k that maximizes the measure of similarity is trivially $k = 0$ and no splicing is performed. When $|\hat{t'} - \beta t| > t_b^{max}$ then $k = 0$ is not an acceptable value (because the time-discrepancy is already too large) and a splicing operation must be performed. The method in [Laroche, 1993] suggests a simplification in that the measure of similarity is calculated only when necessary (i.e., when $|\hat{t'} - \beta t| = t_b^{max}$). Fig. 7.7 shows the relation between the input elapsed time and the output elapsed time. The vertical lines no longer have a fixed duration and splicing occurs whenever the time-discrepancy between the ideal curve and the approximated curve reaches the limit t_b^{max} (i.e., splicing no longer occurs at a regular rate). Standard values for the maximum splice length range from 10 ms to 60 ms depending on the source material. The duration of the cross-fade can be short (less than 5 ms) or longer (over 60 ms) depending on how periodic the signal is.

The PSOLA method. The PSOLA (Pitch Synchronous OverLap-Add) method [Moulines and Charpentier, 1990] was designed mainly for the modification of speech signals. For time-scale modifications, the method is a slight variation of the technique described above, in which the length of the repeated/discarded segments is adjusted

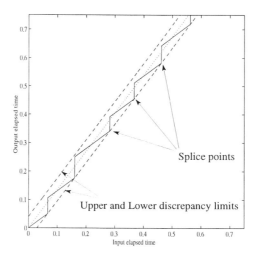

Figure 7.7 Output elapsed time versus input elapsed time in the optimized sampling method for Time-stretching ($\beta > 1$). Dashed line: ideal curve, solid line: approximate curve.

according to the local value of the pitch given by a preliminary pitch estimation. For pitch-scale modifications however, the method has the advantage of not modifying the location and bandwidth of the formants, by contrast with techniques based on the combination of time-scale modification and resampling which merely perform a local scaling of the frequency axis, therefore shifting the location of the formants as well as the fundamental frequency. As is well known, even a small change in the formant location or bandwidth can considerably alter the naturalness of the modified speech, a result that makes the PSOLA method very attractive for pitch-scale modifications of speech. Pitch-scale modifications that do not affect the bandwidth and location of the formants are usually called formant-preserving pitch-modifications.

The PSOLA method is based on the assumption that the local value of the pitch is known (for segments exhibiting periodicity) as well as the locations of *glottal pulses*. A standard, simple speech production model assumes that the speech signal is obtained by filtering a periodic series of glottal pulses by a time-varying resonant filter [Markel and Gray, 1976]. The resonant filter models the acoustic propagation in the vocal tract, while the glottal pulses model the peaks of pressure resulting from the rapid closure of the vocal folds. Based on these assumptions, the basic idea behind PSOLA pitch-scale modifications consists of extracting short-time segments of signal centered around the successive glottal pulses and adding them together at a different rate, as shown in Fig. 7.8. The short-time signals are extracted at a pitch-synchronous rate (denoted

$P(t)$ on the figure), by use of a weighting window. By contrast with the preceding

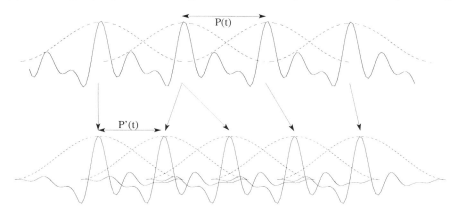

Figure 7.8 Pitch-scale modification with the PSOLA method. The short-time segments extracted from the original signal (top) are overlap/added at a different rate in the modified signal. Here, the pitch is raised ($P'(t) < P(t)$), and segment 2 is repeated to compensate for the modification of the duration.

method (see Fig. 7.4) the short-time signals *are not resampled*, but are overlap/added at a modified rate $P'(t) = P(t)/\alpha(t)$, and repeated (or discarded) to compensate for the corresponding modification of duration. It is possible to describe theoretically the modifications performed by the PSOLA method, and therefore to assess to what extend speech formants are unaltered by the pitch-modification. The interested reader can refer to [Moulines and Laroche, 1995] or [Bristow-Johnson, 1995] for a thorough analysis of the algorithm. It is interesting, however, to understand intuitively what is going on: Each short-time signal can be thought of as the convolution of a glottal pulse by the impulse response of the resonant filter. Because the short-time signals are not resampled but merely copied back, the locations and bandwidths of the formants are not altered: the impulse response of the resonant filter is unchanged. However, the periodicity of the glottal pulses is modified, which corresponds to an alteration of the pitch. By contrast, in the standard splicing method the short-time signal would be resampled (see Fig. 7.4) before being copied back thereby causing a modification of the impulse response of the resonant filter. One of the main limitation of the PSOLA method comes from the assumption that the pitch contour of the signal is known: obtaining the pitch contour usually requires a preliminary analysis stage, which makes the real-time implementation of the PSOLA method difficult. Also, estimating the location of the glottal pulses can be difficult. In order to alleviate this difficulty, the analysis time-instants can be set at a pitch synchronous rate, regardless of the true location of the glottal pulses, at the cost of a slight decrease in the otherwise excellent quality of the modified signal.

7.5 FORMANT MODIFICATION

Some of the methods described above for pitch-scale modifications can also be used to control the location of the formants independently from the pitch. This can be used to make a voice more feminine or masculine (by altering the location of the formants), or to "munchkinize" a voice, an effect made popular by the movie "Wizard of Oz".

7.5.1 Time-domain techniques

A nice byproduct of the PSOLA pitch-scale modification technique (see section 7.7) is that formant-scaling can be done at a fairly low cost. The idea consists of recognizing that the information about the location of the formants lies in the short-term signal segments. As explained above, because the original PSOLA technique merely copies back the short-term segments without modifying them, the original signal formants are preserved. But if the short-term segments are resampled prior to overlap-adding, the formants will be modified accordingly. For example, to raise the formants by a constant factor $\gamma > 1$, without otherwise modifying the pitch or the duration of the signal, one would resample each short-term segments at a sampling rate lower than the original by a factor γ, thereby making them *of shorter duration*, before overlap-adding them together at a rate identical to the original rate $P'(t) = P(t)$. Because the overlap rate has not been modified, the pitch and the time evolution of the signal remain the same, but the formants have been shifted by the resampling operation. The resampling operation should ideally be band-limited, and care must be taken to preserve the position of the middle of each short-term signal during resampling, prior to overlap-adding. The PSOLA formant-modification technique has the advantage of a low computational cost, but relies on the assumption that the signal is periodic, and that its pitch is known. The technique breaks-down when any of these assumptions is violated. Also, the technique only allows for linear scaling of the formants.

7.5.2 Frequency-domain techniques

Because they give access to the spectral representation of the signal, frequency-domain techniques are well suited for formant modification. The first step in frequency-domain formant modification techniques consists of obtaining a estimation of the spectral envelope. Based of the short-time representation of the signal, it is possible to derive a spectral envelope function using a variety of different techniques. If the pitch of the signal is available, the short-time Fourier spectrum is searched for local maxima located around harmonic frequencies, then an envelope can be obtained by joining the local maxima with linear segments, or using more elaborate cepstrum techniques [Cappé et al., 1995]. If the pitch is not available, the problem of reliably determining the spectral envelope is more difficult because harmonic spectral peaks have to be sorted from non pitch-related peaks, a task very similar to pitch-estimation. Once the spectral envelope at a given analysis time t_a^u has been estimated (we'll denote it $E(t_a^u, \Omega_k)$), formant modification is obtained by modifying the modulus of the short-time Fourier transform prior to resynthesis Eq. (7.7). For example, raising the formants by a constant factor $\gamma > 1$ is done the following way:

$$Y(t_s^u, \Omega_k) = X(t_a^u, \Omega_k) E(t_a^u, \Omega_k/\gamma) / E(t_a^u, \Omega_k) \qquad (7.25)$$

where $X(t_a^u, \Omega_k)$ is the short-time Fourier transform of the original signal, and $Y(t_s^u, \Omega_k)$ is the short-time Fourier transform of the formant-modified signal. Because Ω_k/γ does not necessarily correspond to a discrete frequency, the spectral envelope needs to be interpolated, which can be done linearly in a dB scale. Of course, Eq. (7.25) can be integrated with time-scale or pitch-scale modifications for combined effects. The advantage of frequency-domain techniques for formant-modification becomes obvious when the original signal is not periodic (mixture of voices, polyphonic musical signal), in which case time-domain techniques break down entirely. In addition, unlike PSOLA, frequency-domain techniques allow for non-linear formant modifications (the factor γ in Eq. (7.25) can be made a function of the frequency Ω_k). At the time this book went to press, a few commercial implementations of formant-modification techniques were available, in sound editing software, stand-alone effect boxes and hard-disk recorders.

7.6 DISCUSSION

The preceding sections introduced various time or pitch scaling methods performing in the time-domain or in the frequency-domain. In this section, we will summarize their performance, and point out limitations and problems often encountered when using them.

7.6.1 Generic problems associated with time or pitch scaling

The time-scale/pitch-scale modification methods presented so far have been used successfully in the domains of speech or music processing. However, a number of problems are almost systematically encountered in practice.

Reverberation and shape invariance. One problem often associated with the use of time-scale or pitch-scale modifications, pointed out in [Portnoff, 1981] is commonly called the reverberation, chorusing or phasiness effect (chorusing refers to the subjective sensation that several persons are speaking/playing at the same time, as in a chorus). For moderate to large modification factors (say, above 1.5 or under 0.7), and especially for time expansion, the modified signal tends to be reverberated or chorused. This artifact is encountered mainly with frequency-domain techniques, but is also present in time-domain techniques, to a much lesser degree.

In the case of *non-sinusoidal signals* (signals that are better represented by noise) the problem is a simple consequence of the inadequacy of the underlying model. Time-expanded noise generally tends to acquire a definite 'buzzy', sinusoidal quality due to the inadequate sinusoidal representation used in the phase-vocoder, or in time-domain techniques to the fact that segments are repeated, thus introducing undesirable long-term correlation. In frequency-domain techniques, this phenomenon can be significantly reduced by increasing the size of the Fourier transform. More generally, the problem can be solved by use of methods based on a mixed sinusoidal/noise representation of the signal [Griffin and Lim, 1988, Serra and Smith, 1990, Poirot et al., 1988, Laroche et al., 1993b]. See chapter 9.5 for a description of a sinusoid/noise model.

In the case of *quasi-periodic sinusoidal signals*, the 'buzziness' can often be linked to the fact that the phase coherence between sinusoidal components is not preserved. Shape invariant modification techniques for quasi-periodic signals are an attempt to tackle this problem. As explained in 9.4.2, quasi-periodic signals such as speech voiced segments or sounds of musical instruments can be thought of as sinusoidal signals whose frequencies are multiples of a common fundamental $\omega_0(\tau)$, but with additional, slowly varying phases $\theta_i(t)$:

$$x(t) = \sum_{i=1}^{I(t)} A_i(t) \exp(j\phi_i(t) + \theta_i(t)) \qquad (7.26)$$

where

$$\phi_i(t) = \int_{-\infty}^{t} i\omega_0(\tau) d\tau \qquad (7.27)$$

Although common knowledge has it that fixed phase relations do not influence the perception of timbre [Zwicker and Fastl, 1990, Zwicker, 1982], this is not true for time-varying phases: disturbing phase relations is known to introduce buzziness or reverberation in the modified signal [McAulay and Quatieri, 1986b]. In Eq. (7.26), phase relations are controlled through the terms $\theta_i(t)$. Therefore, an ideal shape-invariant time-scale modification would be

$$x'(t') = \sum_{i=1}^{I(T^{-1}(t'))} A_i(T^{-1}(t')) \exp(j\phi_i'(t') + \theta_i(T^{-1}(t'))) \qquad (7.28)$$

where

$$\phi_i'(t') = \int_{-\infty}^{t'} i\omega_0\left(T^{-1}(\tau)\right) d\tau \qquad (7.29)$$

in which the slowly varying phase θ_i at time t' in the modified signal corresponds to that in the original signal at time $t = T^{-1}(t')$. This guarantees that the phase relations in the modified signal at time t' correspond to that in the original signal at time $t = T^{-1}(t')$.

Pitch-driven time-domain modification systems by construction are immune to such problems since no sinusoidal decomposition is performed (i.e., no phase is extracted). This is the reason why time-domain techniques are known to produce high-quality modifications for quasi-periodic signals. By contrast, frequency domain techniques and, more specifically, methods based on the phase vocoder very often exhibit problems connected to shape-invariance because the sinusoidal phases in the modified signal are allowed to run free (i.e. only their derivatives are controlled as in in Eq. (7.15)): specific phase-relations that existed in the original signal are destroyed in the modified signal. Fig. 7.9 shows the time-domain evolution of an original speech signal (top) and its time-scaled version (bottom). The time-scale modification was carried out by use of a time-domain pitch-driven technique, and the modification factor was 1.5. As is clear in the picture, the phase relations between the harmonics in the original signal are preserved in the modified signal (the shapes of the two signals in the time-domain are similar).

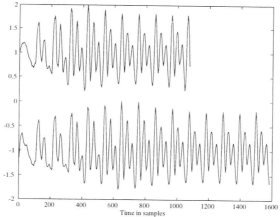

Figure 7.9 Time-domain representation of an original speech signal (top), and of its time-stretched version (bottom), showing shape invariance. Time-domain modification technique.

By contrast, Fig. 7.10 shows the modified signal obtained by use of the standard phase-vocoder time-scaling technique. Clearly, the shapes of the signals are quite different, illustrating the lack of shape invariance. The standard phase-vocoder technique described in section 7.3 cannot ensure shape invariance because *the signal is not assumed to be quasi-periodic* and the time-scale modification is at best that described by

APPLICATIONS OF DSP TO AUDIO AND ACOUSTICS

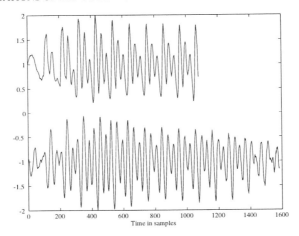

Figure 7.10 Time-domain representation of an original speech signal (top), and of its time-stretched version (bottom), showing the loss of shape-invariance. Phase-vocoder modification technique.

Eq. (7.3), which differs from Eq. (7.28). Without the assumption of quasi-periodicity, shape invariance cannot be obtained. The original method proposed by Portnoff in [Portnoff, 1981] makes use of pitch information to evaluate the slowly varying phase component $\theta(t)$ in Eq. (7.26) and modify it according to Eq. (7.28). However, the phases in the modified signal are still obtained by a discrete integration (a cumulative sum as in Eq. (7.15)) and phase errors inevitably accumulate, thereby altering the phase relations in the original signal. A modification of Portnoff's technique has been proposed in [Sylvestre and Kabal, 1992] to solve this problem: contrary to the standard phase-vocoder technique (Eq. (7.15)), the phase in each channel at synthesis instant t_s^u is not obtained through a cumulated sum. Rather, phases are reset at each synthesis time-instant t_s^u to theirs values in the original signal at time t_a^u. As a result, the phase continuity between two successive short-time synthesis signals is no longer guaranteed. To recover phase continuity, a fixed phase offset is added to each channel, and the remaining phase discontinuity is exactly cancelled by slightly modifying the instantaneous frequency in each short-time Fourier transform channel. The modified signal is then obtained by concatenating the short-time signals, rather than overlap-adding them. This algorithm guarantees some degree of shape invariance because the phase relations between the pitch-harmonics in the vicinity of a given synthesis time-instant are the same as in the original signal in the vicinity of the corresponding analysis time-instant, up to a linear phase-shift. Some parametric modification methods have also been modified to ensure shape invariance in the case of quasi-periodic signal [Quatieri and McAulay, 1989, Quatieri and McAulay, 1992]. See section 9.4.2 for the description of a shape-invariant parametric modification technique. For non quasi-periodic as well as quasi-periodic signals, the reverberation has also been linked to the lack of phase-synchronization between the short-time Fourier transform channels *around each sinusoidal component*. As indicated by Eq. (7.11), the phases in the short-time Fourier transform channels around a given sinusoidal component are not independent. For example, for a Hanning window without zero-padding, the term $H(\Omega_k - \omega_i(t_a^u))$ for successive values of k around the maximum exhibits π phase shifts. In the standard implementation of the phase vocoder, this phase coherence is not guaranteed at the synthesis stage, and phase misalignment between adjacent channels generates beatings perceived as chorusing or reverberation. Puckette in [Puckette, 1995] suggests an inexpensive solution for ensuring phase-coherence of adjacent short-time Fourier transform channels. The phase of any given channel is obtained from the sum of the short-time Fourier transform values at the channel and its two neighbors with alternating signs (reflecting the sign alternation of the Fourier transform of the Hanning window). This "phase locking" solution does not require any explicit phase calculation and is therefore inexpensive, but provides in some cases a significant decrease in the perceived reverberation.

Transient smearing. Another important issue in techniques involving the phase vocoder is the reconstruction process. As explained above, the modified short-time Fourier transform does not necessarily correspond to any existing time-domain signal. The phase modification inherent to time or pitch-scaling does not necessarily preserve the phase coherence that originally existed in successive original short-time spectra[4]. This problem inherent to the use of the phase-vocoder has been connected to undesirable reverberation and smearing effects in other contexts (e.g., coding [Johnston and Brandenburg, 1992]). The phenomenon becomes more and more conspicuous as the size of the Fourier transform increases. To avoid this, it was proposed [Hayes et al., 1980, Nawab et al., 1983, Griffin and Lim, 1984a] that the phase information in the short-time spectra be discarded, and that the modified short-time Fourier transform be reconstructed from the knowledge of its magnitude only, using the large data redundancy in the short-time Fourier transform to make up for the loss of information. This idea, originally proposed in the field of image-processing, leads to iterative algorithms whose convergence has been proved in some cases [Griffin and Lim, 1984a]. However, it has been remarked that the *global* minimum is not always reached. These iterative reconstruction methods have been applied to the problem of time-scale modification [Nawab et al., 1983, Griffin and Lim, 1984a, Roucos and Wilgus, 1985] and have been shown to improve significantly the quality of the modified speech signal. In particular, the reverberation/chorusing effect is significantly diminished. However, the convergence is usually quite slow [Roucos and Wilgus, 1985], and the algorithms extremely time-consuming. Transient smearing has also been observed in parametric methods. See section 9.4.3 for the description of techniques used to overcome this problem.

7.6.2 Time-domain vs frequency-domain techniques

The pitch-driven time-domain techniques presented in section 7.4 are relatively immune to reverberation/chorusing problems, at least for moderate modification factors and for quasi-periodic signals. Because of the underlying assumption of quasi-periodicity, these methods perform poorly with non-periodic signals, except for small modification factors. Standard difficult cases include noisy signals (breathy voice, wind instruments) which tend to acquire a buzzy quality upon modification, or complex multi-pitch signals (as in music) where splicing operations are more difficult to conceal. Also, a commonly encountered problem is that of *transient doubling* in which, due to a splicing operation taking place in the vicinity of a transient, that transient ends up being repeated or discarded in the modified signal. Advanced time-domain techniques attempt to detect and gracefully handle transients, but this can be done only when the modification factor is small (say less than 20%). Time-domain methods still prove extremely useful for their simplicity and their fairly good performance for small modification factors [Laroche, 1993]. Non-parametric frequency-domain techniques that do not rely on the hypothesis of periodicity but rather on the assumption that the signal is sinusoidal tend to perform much better in all situations when time-domain methods fail. In particular, time or pitch-scale modifications by large factors cannot be carried out by time-domain methods and usually require the use of the more elaborate frequency-domain techniques. However, as was mentioned above they are prone to artifacts (reverberation, buzziness, transient smearing) and a compromise has to be found for the size of the Fourier transform: large sizes improve the results in the case of noisy signals, but make transient smearing worse. On the other hand, the Fourier transform size cannot be made smaller than a limit, as explained in section 7.3.1. Finally, the solutions proposed to eliminate these artifacts are either excessively costly and complex or only marginally efficient. Parametric techniques tend to outperform non-parametric methods when the adequation between the signal to be modified and the underlying model is good. When this is not the case however, the methods break down and the results are unreliable. Parametric techniques usually are more costly in terms of computations, because they require an explicit preliminary analysis stage for the estimation of the model parameters.

Conclusion. The development of low-complexity time-domain methods for time-scale or pitch-scale modifications has already made it possible to incorporate such systems in consumer products such as telephone answering systems, effect boxes and semi-professional CD players. They could easily be implemented in record or playback devices such as DAT or CD players, offering the user additional control over the play-back. Time-scale or pitch-scale modification techniques have also become standard tools in editing softwares for professional or home-studio post-production or broadcast studios, but are also quite common in sound-editing softwares. As of today, time-domain techniques seem to predominate over frequency-domain methods, due to the heavier computational cost of the latter. This tendency might be reversed soon, following the steady increase of the computation power available in standard microprocessors.

Notes

1. $\beta(\tau) > 0$ guarantees that $T(t)$ is never decreasing and therefore that $T^{-1}(t')$ exists.

2. i.e., a sinusoid with the same amplitude, the same initial phase and the same frequency as the original sinusoid.

3. Note that the division in Eq. (7.23) needs not be calculated since only the value k that maximizes $c_c(t_a^u, k)$, not the actual value, of $c_c(t_a^u, k)$ is needed.

4. Notice that while the preceding paragraph mentioned phase relations *between harmonics* in connection with shape invariance, this paragraph addresses the problem of phase coherence *between successive analysis windows*.

8 WAVETABLE SAMPLING SYNTHESIS

Dana C. Massie

Joint E-mu/Creative Technology Center
1600 Green Hills Road
POB 660015
Scotts Valley, CA 95067

dana@emu.com

Abstract: Sampling Wavetable Synthesis ("sampling") is possibly the most commercially popular music synthesis technique in use today (1997). The techniques used in sampling include the traditional playback of digitized audio waveforms from RAM wavetables, combined with sample rate conversion to provide pitch shifting. Sampling evolved from traditional computer music wavetable synthesis techniques, where the wavetable size simply grew to include an entire musical note. Extensions to simple wavetable playback include looping of waveforms, enveloping of waveforms, and filtering of waveforms to provide for improved expressivity, i.e., spectral and time structure variation of the perfomed notes. A simple comparison is given between band limited sample rate conversion for pitch shifting, linear interpolation, and traditional computer music phase increment oscillator design.

8.1 BACKGROUND AND INTRODUCTION

Electronic methods for creating musical sounds have been used at least since the late 1890's, with Thaddius Cahill's Telharmonium, which used multi-ton alternators to generate organ like sounds intended to be sold to listeners over telephone lines. But digital generation of musical sounds only dates from the late 1950's. Max Mathews of Bell Labs led a group of researchers who pioneered the use of the digital computer to generate musical sounds. At that time, all of the sounds generated were created out of real-time. Work on non real-time synthesis of musical sounds with digital computers continued through the 1960's and the 1970's, mostly at university research labs. During this period, the analog electronic music synthesizer was developed and flourished. Don Buchla and Robert Moog, from opposite sides of the USA are generally credited with independently inventing the analog voltage controlled patchable analog music synthesizer in the early 1960's.

The analog electronic music synthesizer used modular elements such as oscillators, filters (including low pass, high pass, band pass, and notch), multipliers (both 2-quadrant, and 4-quadrant), and adders (known as "mixers"), all interconnected with telephone style patch cords. Programming analog music synthesizers consisted of establishing an interconnection, and then laboriously adjusting the module parameters by trial and error to produce a musically useful sound. Since the modules drifted with temperature changes, and parameters were hard to store, sounds were rarely reproducible from one day to the next. Still, these machines opened up the musical world to a new class of timbres, which permanently changed music production.

The analog synthesizer flourished at about the same time that analog computing matured. There were many interesting parallels between analog computing techniques and analog music synthesizer principles. While analog computing developed a bit earlier than analog electronic music, probably there were few direct influences from analog computer design on analog electronic music synthesizer designers. Instead, the parallels probably represented parallel evolution of two technologies with similar constraints.

Analog computing used modular computing elements such as adders, multipliers, integrators, piecewise non-linear function generators, and input/output devices such as precision potentiometers, oscilloscopes, and paper strip chart recorders. While the precision of analog computers rarely exceeded 1.0performance could exceed that of digital computers up until the 1980's, for many applications.

Hybrid analog computers with digital computers as control elements largely replaced pure analog computers in the 1970's. Digital controlled analog music synthesizers replaced pure analog synthesizers by the early 1980's. At that time, pure digital music synthesis in real-time still required very exotic technology, so using digital elements for control allowed rapid recall of parameters and voices.

8.1.1 Transition to Digital

Digital computing methods have of course nearly replaced analog computing in one field after another. The advantages of reliability and programmability combined with the economics of Very Large Scale Integration of digital computing technologies have almost completely displaced analog technologies.

8.1.2 Flourishing of Digital Synthesis Methods

Dozens of digital synthesis techniques have been developed and used in the past 40 years. A very short list of techniques include:

- Additive (Sums of amplitude modulated sinusoids)
- FM – Frequency Modulation [Chowning, 1973]
- Waveshaping [LeBrun, 1979]
- Granular Synthesis (Chant, Vosim) [Rodet et al., 1989][Kaegi et al., 1978]
- Switched Wavetables (PPG, Korg Wavestation, Ensoniq VFX, MicroWave, others)
- Spectral Modelling Synthesis [Serra and Smith, 1990]
- Source-filter Synthesis (Subtractive synthesis, Karplus-Strong Plucked string algorithm [Karplus and Strong, 1983])
- Physical Modeling families
- LASynthesis (Linear Arithmetic) (Roland D-50)

These methods (and many others) are described quite well in the exhaustive work by Roads [Roads, 1996].

Each of these synthesis methods has been explored to varying degrees by numerous researchers, university composers, and commercial implementors, but today, sampling synthesis has come to largely dominate the entire commercial synthesis industry. The largest number of music synthesis instruments sold today (1997) are based on sampling synthesis technology. In multi-media markets, personal computer sound cards based on using FM synthesis still dominate in numbers of units installed, but the fastest growth rates in sales for personal computer sound card technologies are for wavetable synthesis cards, which is just the name used in the multi-media industry for sampling synthesis.

With so many other synthesis techniques available, why has sampling become so dominant? An analogy in the visual arts is the comparison between photography and painting. In the visual arts, photography has become dominant in the sense that many more photographs are taken than pictures painted. But in music, one could argue that sampling synthesis has become more dominant in music than photography has in graphic arts. Drawing and painting and other means for synthetic image generation are very widespread, where pure synthesis for music is rapidly becoming unusual.

Simplicity is probably the primary factor in the dominance of both sampling and photography. Sampling is simple to implement, but it is not simpler than all other synthesis methods. However, it appears to be simpler and more efficient to produce sound libraries for samplers than for any other synthesis method. With sampling, producing libraries of sounds can be quickly accomplished with minimal training, where with most synthesis techniques, the time needed to produce a sound library is often an order of magnitude greater, and such a task usually takes considerably more training. Here again, the analogy with the visual arts is strong. A painter typically needs much more training than a photographer. True, a professional photographer may have a great deal of training, but amateurs can create photographs with much less training that it would take to produce a painting of similar detail.

8.1.3 Metrics: The Sampling - Synthesis Continuum

Metrics for music synthesis techniques in the past have been dominated by simple tabulations of CPU costs or brief subjective evaluations of sound quality. Recently, efforts have been made to apply metrics to many aspects of synthesis techniques [Jaffe, 1995]. While it may be difficult to exactly define metrics for some of the subjective elements of synthesis techniques, it is still very helpful to try to apply metrics for evaluating different methods of synthesis. Two of the most important metrics could be accuracy and expressivity.

Expressivity. Here we define expressivity as the variation of the spectrum and time evolution of a signal for musical purposes. That variation is usually considered to have two components, a deterministic component and a random component. The deterministic element of expressivity is the change in spectrum and time evolution controlled by the user during performance. For example, hitting a piano key harder makes the note louder and brighter (more high frequency content). The random component is the change from note to note that is not possible to control by the musician. Two piano notes played in succession, for example, are never identical no matter how hard the musician attempts to create duplicate notes. While the successive waveforms will always be identified as a piano note, careful examination shows that the waveform details are different from note to note, and that the differences are perceivable.

Accuracy. We can describe accuracy in one sense as the fidelity of reproduction of a given musical instrument sound. This fidelity can even be given objective measurements, such as percentage distortion. Some musicians have argued that accuracy of reproduction of existing musical instruments should not be the only goal of music synthesis, so we might expand our definition to include measures appropriate for novel musical instrument sounds for which there is no reference point for objective measures of accuracy. In this case, perhaps some measure of the acoustical or perceptual sophistication of a sonic event can be devised, but for this discussion, we only consider imitative synthesis. Many advocates of imitative synthesis suggest that the imitative process can be simply a benchmark for the quality possible in a synthesis method, and that if acoustical instruments can be synthesized with acceptable quality, then novel instruments should then be possible by extrapolation.

8.1.4 Sampling vs. Synthesis

Synthesis can have arbitrarily high expressivity, since the sound is generated from fully parametric descriptions of a synthesis process. However, synthesis typically has very poor accuracy. FM, a powerful and popular synthesis method, is notoriously unsuccessful in synthesizing the sound of an acoustic piano. But FM has very simple methods for varying the brightness, harmonicity, and other parameters of a signal under musical control.

The goal of many researchers and engineers in synthesis is to move synthesis technology towards greater accuracy. The goal of many in sampling technology is to move sampling towards greater expressivity. Physical modeling is an example of a synthesis method that is specifically oriented towards more accurate yet extremely expressive synthesis of natural musical instruments. The goal of adding more and more post-processing technology to sampling instruments is to add the ability to mold and shape the spectrum and time structure of sampled musical events, in order to produce a more expressive yet still accurate re-creation of natural musical instruments.

Also, sampling and synthesis both can produce novel instrument sounds that have not been heard before, breaking away from traditional acoustical models.

Another analogy can be drawn between music and speech. In speech technology, expressivity is analogous to fluency, or the naturalness of the pitch contours ("prosody") and articulation as a function of time for speech. Accuracy is another term for quality in speech reproduction or synthesis. In speech, synthesis by rule is capable of very good fluency, but still is plagued by poor quality. In contrast, waveform encoding is capable of arbitrarily high quality, but is very weak at constructing arbitrary sentences with good fluency, since the ability to modify the encoded waveform is weak.

Improving the expressivity of sampling or the accuracy of synthesis inevitably increases implementation costs. In figure 8.1, costs are shown as a single axis, but it is worthwhile to consider costs as having two sets of components; hardware costs and labor costs. Often, hardware cost analyses are restricted to the costs for the sound engine calculations only, and ignore the costs for the control stream. In sampling, memory costs often outweigh the costs for the sound engine. In additive synthesis, memory costs are reduced at the expense of greater sound engine cost, and a much greater control engine cost. This is shown in figure 8.2. The most overlooked cost metrics are labor costs (See figure 8.3). One of the hardest lessons many algorithm designers learn is how difficult it can be to program a sound using a given synthesis

316   APPLICATIONS OF DSP TO AUDIO AND ACOUSTICS

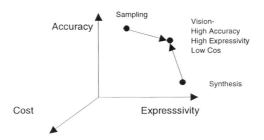

Figure 8.1   Expressivity vs. Accuracy - Sampling has high accuracy but weak expressivity, where Synthesis has high expressivity and weak accuracy. Both technologies are evolving towards an ideal with both high accuracy and expressivity, but at a higher cost for implementation.

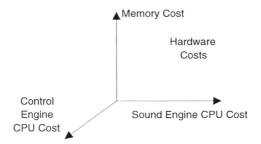

Figure 8.2   Tradeoffs between sound engine costs, memory costs, and control engine costs are often ignored in synthesis algorithms, where researchers tend to focus only on simple measures of multiply-add rates within the sound engine. Typically, increased sound engine complexity can decrease memory cost, but at an increase in control engine complexity. Conversely, sampling uses trades off increased memory costs at a reduction of sound engine cost.

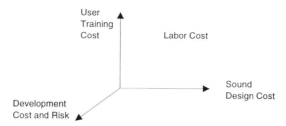

Figure 8.3 The labor costs for synthesis techniques are also often overlooked. The complexity and expense of developing sounds for various synthesis techniques varies widely. User learning difficulty can thwart acceptance of a promising new synthesis technique, and development costs and risks can stiffle a company's desire to commercialize a new technique. These hidden costs often dominate the success or failure of a synthesis technique in the market place, rather than the expressive power of the technique.

technique. FM synthesis, while extremely popular due to its low hardware costs and rich expressivity, is famous for its difficulty of programming. Sampling on the other hand is extremely efficient for sound designers to produce instrument data sets for. While still requiring a fair amount of technical skill, it is very straight forward to produce a data set for a sampler once a representative set of the desired instrument sounds is recorded.

Physical modeling is another significant example of a promising new synthesis technology that has had a slow introduction partly due to the large amount of technical skill needed to develop instrument sounds. Future synthesis techniques may well be limited more by the expense of sound development rather than the costs for implementation.

Learning how to perform a synthesis technique is also a major hidden cost. Physical modeling offers greater expressivity, but this in turn requires greater skill and learning investment on the part of the composer. In some cases, this is a burden that working musicians are hesitant to take on, especially with the rapid evolution of technology making users unsure how long before their equipment purchases will become obsolete.

Finally, development efforts and risks have a big impact on whether a company will undertake a commercial project of a given synthesis technique. While there are dozens of promising synthesis techniques waiting in the research community for new hardware to host, companies are usually only willing to invest in techniques that show the least risk for development.

8.2 WAVETABLE SAMPLING SYNTHESIS

In a typical full featured sampling synthesizer, several capabilities are combined:

- Playback of digitized musical instrument events
- Entire note recorded, not just a single period
- Pitch shifting technologies
- Looping (usually) of more than one period of sustain
- Multi-sampling
- Enveloping
- Filtering
- Amplitude variations as a function of velocity
- Mixing or summation of channels
- Multiplexed wavetables

8.2.1 Playback of digitized musical instrument events.

The basic operation of a sampling synthesizer is to playback digitized recordings of entire musical instrument notes under musical control. Playback of a note can be triggered by depressing a key on a musical keyboard, or from some other controller, or from a computer. The simplest samplers are only capable of reproducing one note at a time, while more sophisticated samplers can produce polyphonic (multi-note), multi-timbral (multi-instrument) performances.

Of course, digitized recordings of musical performances had been used long before sampling synthesis was invented. With sampling, each recording is of a musical instrument playing a single note. A performance is constructed by triggering the playback of sequences of notes which can be overlapped and added ("mixed", in audio engineering terms). Digitized recordings of musical performances involve making a recording of performers playing complete musical compositions, not individual notes.

8.2.2 Entire note - not single period

Sampling playback oscillators are similar to earlier table lookup oscillators in that waveform data are stored in a waveform memory and then output at some rate determined by the desired frequency. An important distinction between a simple table lookup oscillator and a sampling oscillator is that a simple oscillator has only one period of a static waveform stored in it, while a sampling playback oscillator typically has an entire note stored.

The Allen Organ Company built an organ using a digital oscillator that stored a single period of a pipe organ sound. At the time, this was a major technical feat, which required a considerable investment in computer technology. RMI built an electronic organ that allowed the user to enter in waveform sample values into a short wavetable.

A major qualitative change occurs when more than a single period is stored. It is not clear exactly how many periods are needed before this qualitative change occurs, but when a complete musical note is digitized, including the entire onset of the acoustical event, the result is perceptually much different from creating a perfectly periodic waveform. Many instruments have highly distinctive transients at the start of a note: the hammer striking the strings of a piano, the chiff of wind noise at the start of a flute note, or the pick snapping off of a guitar string. So, it is essential to store the attack (or onset) of the musical note to make the instrument reliably recognizable. Also, it is often essential to retain enough of the body or sustain of the note to prevent rigid periodicity. A piano, for example, has multiple strings that vibrate at slightly different frequencies to create motion in the sound.

Effect of storing attacks. Storing the attack of a musical instrument is needed because musical instruments are identified to a large extent by their onset characteristics [Winckel, 1967]. To illustrate, when the onset of one musical instrument is grafted onto the sustain or steady-state portion of a second instrument, listeners usually identify the instrument based on the attack, not the sustain. Synthesizing the attack of an instrument accurately with other music synthesis methods is very difficult.

Loops. As described below in more detail, the sustain portion of a sampled sound is generated by looping or repeating a small segment. By making this segment long enough, the sound produced can seem non-stationary or "animated". This synthesis of animation is another important element in avoiding the objectionable qualities of a single period loop.

8.2.3 Pitch Shifting Technologies

Simple sampling playback synthesizers only use a sample address counter which increments by one each sample period and reads out each sample of the waveform successively to reproduce the original sound at the playback sample rate, as shown in figure 8.4. While playback of sampled waveforms at their original pitch is widely used in toys, telephone answering machines, telephone information services, etc.; the ability to change the oscillator playback frequency greatly widens the usefulness of sampling synthesis. There are several methods used for pitch shifting wave forms in samplers which we can divide into synchronous and non-synchronous categories.

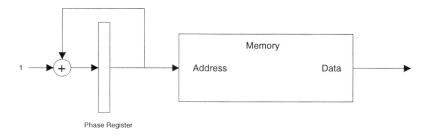

Figure 8.4 The most rudimentary form of sampling replays a waveform at its original pitch from samples stored in memory. The sample address simply increments by one sample per output cycle, and each word in memory is read out to feed a digital to analog converter, reproducing the original waveform.

Asynchronous Pitch Shifting. Asynchronous pitch shifting, the simplest pitch shifting method, simply changes the clock rate of each output digital to analog converter (DAC) to vary the pitch. Each channel requires a separate DAC. Each DAC has its own clock whose rate is determined by the requested frequency for that channel. When a DAC clock occurs, the DAC issues a request to a memory controller that supplies a waveform sample to the DAC. The earliest samplers had a separate memory for each DAC.

This method is considered asynchronous because each output DAC runs at a different clock rate in order to generate different pitches.

Disadvantages of asynchronous pitch shifting include the need for a single DAC per channel, system cost which increases with channel count, and the inability to digitally mix multiple channels for further digital post processing such as reverberation. Also, each channel requires an analog re-construction filter that tracks the playback sample rate of that channel. Output re-construction filters have a constant cut-off frequency in traditional digital signal reconstruction where sample rates are constant. When the sample rate varies, which typically happens every note in music synthesis, the re-construction filter should also change its cutoff frequency. High order analog low pass filters with a variable cut-off frequency are expensive and difficult to design. Switched-capacitor IC tracking filters have been popular in this role, but typically have limited signal to noise ratio.

Advantages of asynchronous pitch shifting include easy circuit design and no pitch shifting artifacts, as long as the analog tracking filter is of high quality.

Numerous commercial instruments were built in the early 1980's that used asynchronous pitch shifting, including the Fairlight Computer Music Instrument [Roads, 1996], Kurzweil 250 [Byrd and Yavelow, 1986], the E-mu Emulator and Emulator 2 [Massie, 1985], and the New England Digital Synclavier.

Synchronous Pitch Shifting. Synchronous pitch shifting techniques are methods to change the pitch of wavetable playback data through sample rate conversion algorithms. This makes it easier to read wavetable memory in regular time slots and also allows the digital summation or mixing of multiple output channels into a single digital stream for further post-processing.

It is difficult to integrate multiple DACs into a single chip, but integrating the pitch shifting circuitry onto a single chip has been economical since the middle 1980's. Only a single DAC is then required for output, since the data can be mixed in the digital domain.

Reverberation and other digital signal processing effects are very popular, and are not practical in sampling synthesizers without digital mixing of the multiple sample channels.

These factors motivated a complete shift in sampler design from asynchronous to synchronous pitch shifting techniques in the middle 1980's.

Sample rate conversion. All of the synchronous pitch shifting techniques essentially involve sample rate conversion techniques. The theory and practice of sample rate conversion has received extensive coverage in many excellent texts and articles, but it is illuminating to compare the computer music perspective with the traditional sample rate conversion literature. Insights from the sample rate conversion literature provide insights to the computer music perspective, and vice versa.

Frequency Scaling. Pitch shifting can be defined as a simple frequency scaling operation. For example, if we have a sine wave at 440 Hz, and we need to transpose or shift the pitch of the sine wave to 220 Hz, we can view this as simply scaling the frequency of the signal by $1/2$. We define the Fourier transform of the continuous time signal $x(t)$ as $x(\omega)$. The frequency scaled signal would simply be, $2x(2\omega)$ and the corresponding time domain signal is just $x(t/2)$. For a discrete time signal $x[n]$, we cannot use this simple relation, because we cannot define $x[an]$ with a being non-integer. Instead, we need to define an interpolation or sample rate conversion process.

Note that frequency scaling also scales the time domain features of a signal as well. This simply means that the signal is stretched in time when it is compressed in frequency. This operation is the same as playing back a tape recording at a different speed. If a tape recording is played back at half of its original speed, frequencies are scaled down by one octave, and the recording also takes twice as long to playback.

Some authors prefer to reserve the term "pitch shifting" for scaling frequency without scaling the time domain features of a signal (see Laroche's chapter) Signal processing methods exist that allow independent scaling of time and frequency domain features, but these techniques are far costlier than simple sample rate conversion. Also, time and pitch scaling methods are themselves imperfect, and introduce artifacts.

Finally, the term pitch shifting is well entrenched to mean simple sample rate conversion within the sampler engineering community.

The perceptual consequence of time-domain scaling is that attacks (onsets) or other time domain features of a musical event are either compressed (if the event is pitch shifted up) so that they occur faster, or they are elongated (if the event is pitch shifted down). This is noticeable in some cases, depending on the instrument and how far the note is pitch shifted. In recordings of signals that have a significant time domain features, as in certain percussion sounds such as shakers, the time domain re-scaling can be quite objectionable. For small amounts of pitch shifting, the elongation or compression is not objectionable, so the simple method of sample rate conversion is quite successful and is widely used.

To surmount the difficulties of the time domain re-scaling that pitch shifting introduces, techniques such as multi-sampling are used as described below.

Formant Re-scaling. For signals with discernible formant structure, such as speech, an unnatural side effect of frequency scaling is that the formants themselves will be re-scaled. A singer or speaker changes the pitch of their voice by changing the frequency of the glottal pulse, but the overall size of the vocal tract does not change. The vocal tract changes shape with different vowels, of course, but the vocal tract size is an important cue in speaker identification.

When the formant is re-scaled, the new formant is equivalent to a formant from a person with a different vocal tract size. Scaling frequencies down is equivalent to increasing the size of the speaker's vocal tract and vice-versa. This scaling of the formant spectrum is clearly noticeable with only a few semitones of pitch shift. Workers in sampling playback technology have called this artifact "munchkin-ization" or "chipmunk effect" after movie and television sound effects where voices were played back at a different speed from their original recording speed to produce an altered formant spectrum intentionally.

Formant re-scaling artifacts can also be somewhat circumvented by using multi-sampling as described below.

Sample Rate Conversion Techniques. The simplest form of sample rate conversion is called either drop sample tuning or zero order hold interpolator. This technique is the basis for the table lookup phase increment oscillator, well known in computer music [Moore, 1990a].

The basic element of a table lookup oscillator is a table or wavetable, that is, an array of memory locations that store the sampled values of the waveform to be generated.

Once the table is generated, the waveform is read out using a simple algorithm. The pre-computed values of the waveform are stored in a table denoted `WaveTable`, where `WaveTable[n]` refers to the value stored at location n of the table.

We define a variable called Phase which represents the current offset into the waveform, which has both an integer and a fractional part. The integer part of the Phase variable is denoted IntegerPart(Phase).

The oscillator output samples $x[n]$ are generated for each output sample index n;

$$x[n] = \text{WaveTable[IntegerPart(Phase)]}; \quad (8.1)$$
$$\text{Phase} = \text{Phase} + \text{PhaseIncrement}; \quad (8.2)$$

With PhaseIncrement = 1.0, each sample for the wavetable is read out in turn, so the waveform is played back at its original sampling rate. With PhaseIncrement = 0.5, the waveform is reproduced one octave lower in pitch. Each sample is repeated once. With PhaseIncrement = 2.0, the waveform is pitch shifted up by one octave, and every other sample is skipped, effectively decimating the waveform by 2.

We can look at an equivalent hardware block diagram. Here we have a wavetable being addressed by what is essentially a counter whose rate is changed to vary the pitch. The term "drop sample tuning" refers to the fact that samples are either dropped (skipped) or repeated to change the frequency of the oscillator. The phase increment is added to the current value of the phase register every sample, and the integer part of the phase is used as an address to lookup a sample in waveform memory to output to a DAC.

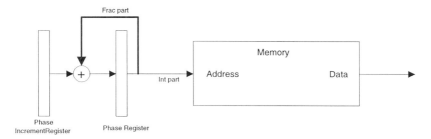

Figure 8.5 "Drop Sample Tuning" table lookup sampling playback oscillator. The phase Increment Register adds an increment to the current phase, which has a fractional part and an integer part. The integer part is used to address a wavetable memory, and the fractional part is used to maintain tuning accuracy.

The frequency of the waveform produced is simply the frequency of the original waveform scaled by the phase increment. For example, assuming that the waveform stored in WaveTable is a sine wave at 440 Hz, and PhaseIncrement is 0.5, then the frequency of the reproduced waveform is simply $440 * 0.5 = 220$ Hz.

Drop sample tuning can introduce significant artifacts from changing the pitch of a waveform. This method originated in the design of early computer music oscillators where the waveform stored has many samples per period, from 256 to 2048 or more. With a large number of samples per period, the signal is essentially highly oversampled, and artifacts are minimized [Moore, 1977b].

Linear Interpolation Table Lookup Oscillators. Significant improvements in the signal to noise are obtained by using linear interpolation to change sample rate. Adding linear interpolation changes the calculation of the table lookup as follows:

$$x[n] = (\text{WaveTable[IntegerPart(Phase)]} * (1 - \text{FractionalPart(Phase)}) + \\ (\text{WaveTable[IntegerPart(Phase)} + 1] * \text{FractionalPart(Phase)}) \quad (8.3)$$

Most sampling synthesizers today are implemented using this basic two point interpolation table lookup oscillator design. The difference between the use of the table lookup oscillator in computer music and in sampling is the degree of over-sampling of the waveforms. An over-sampled signal is one whose highest frequency component is much less than $1/2$ of the sample rate, or Nyquist frequency. A signal that is not over-sampled, or alternatively, which is "critically sampled", has its highest frequency component very near the Nyquist frequency.

With recorded instrument waveforms, it is not practical to store as many as 256 samples per period for most waveforms, as done in sine wave table lookup oscillators. This would correspond to a very high sampling rate. At 44100 Hz sampling rate, waveforms two octaves above "A-440" have about 50 samples per period. To maintain 256 samples per period would require a sampling rate of over 200 kHz, which is impractical due to the expense of the memory required.

A brief review of sample rate conversion is helpful, even though this topic has been covered in detail elsewhere. The reader is referred to numerous references [Crochiere and Rabiner, 1983, Vaidyanathan, 1993, Smith and Gossett, 1984] for more detailed reviews of sample rate conversion.

To summarize, the classical method for analyzing sample rate conversion is to assume that the sample rate conversion factor is or can be approximated by a rational number $\frac{L}{M}$. Then the sample rate conversion can be viewed as a three stage process, up-sample by an integer factor L, filter by $h[n]$, and down-sample by M. Up-sampling by L inserts $L - 1$ zero valued samples in between the existing samples of $x[n]$ and decimating by M retains only each M-th sample. This approach is an analytical tool rather than an actual implementation strategy, and it allows the comparison of different sample rate conversion methods in a similar framework.

The filter $h[n]$ would be an ideal low pass filter with a cutoff frequency $\omega \leq$ min $\left(\frac{\pi}{L}, \frac{\pi}{M}\right)$. In other words, if the desired conversion ratio $\frac{L}{M}$ is greater than one, i.e., we are increasing the sample rate (or decreasing the pitch) of the sound, then the cutoff frequency of the filter $h[n]$ is simply $\frac{\pi}{L}$. If the desired conversion ratio $\frac{L}{M}$ is less

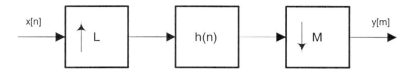

Figure 8.6 Classical sample rate conversion chain; up-sampling by L, filtering, and down-sampling by M

than one, then we are decreasing the sample rate or increasing the pitch of the sound, and the cutoff frequency of the filter $h[n]$ is simply $\frac{\pi}{M}$. Informally, we tend to call increasing the pitch of a sound "decimation" since the sample rate is being reduced, and decreasing the pitch "interpolation" since the sample rate is increased. When a signal is decimated, the cutoff of the filter must be below the original Nyquist rate $\frac{\pi}{M}$, while with interpolation, the cutoff can be at the original Nyquist rate $\frac{\pi}{L}$. Example: $\frac{L}{M} = \frac{2}{1}$ implies a pitch decrease of one octave, so: cutoff $= \min\left(\frac{\pi}{2}, \frac{\pi}{1}\right) = \frac{\pi}{2}$. The drop sample tuning approach can be viewed in this framework by choosing a zero order hold for $h[n]$. The zero order hold is defined in this case for discrete time sequences as $h[n+m] = x[n]$, $m = 0$ to $L-1$. This just means that the sample $x[n]$ is repeated $L-1$ times, rather than having $L-1$ zeros inserted in between samples. L is determined by the number of fractional bits in the phase accumulator. The frequency response of the zero-order hold is just the "digital Sinc" function [Oppenheim and Willsky, 1983]

$$H(e^{j\omega}) = \frac{1}{L}\left[\frac{\sin\frac{\omega L}{2}}{\sin\frac{\omega}{2}}\right] \quad (8.4)$$

For $L = 4$, i.e., 2 fractional bits in the phase accumulator, we have the following frequency response: There is 3.9 dB of attenuation at the original passband edge frequency (0.25π) and the peak sidelobe attenuation is only about -12 dB, allowing a considerable amount of energy to alias into the passband. There are zeros at multiples of the original sampling rate (in this case, at 0.5π), which means that images of signals very near 0 Hz frequency will be well suppressed.

Linear interpolation has been studied with this framework as well (see [Crochiere and Rabiner, 1983]). The equivalent filter $h[n]$ for linear interpolation is called a first order hold, and is simply a triangular window. The frequency response for the triangular linear interpolation filter is the "digital" Sinc2

$$H(e^{j\omega}) = \frac{1}{L}\left[\frac{\sin\frac{\omega L}{2}}{\sin\frac{\omega}{2}}\right]^2 \quad (8.5)$$

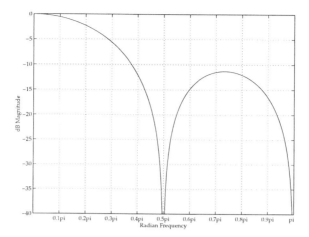

Figure 8.7 Digital Sinc function – the frequency response for a zero order hold interpolator sample rate converter with $L = 4$, which puts the original Nyquist frequency at 0.25π. We can see rolloff in the passband of about -3.9 dB and very poor rejection of images outside of the passband, which result in artifacts perceived as pitch shifting distortion.

shown in figure 8.8. Again, we see roll-off in the passband. A linear interpolator has almost 8 dB of attenuation at the edge of the passband (0.25π in this case), but its peak stopband attenuation is now down to -24 dB. There are still zeros near multiples of the sampling rate, which means that for signals that are highly over-sampled, a linear interpolator performs very well since images of low frequency signals will fall near multiples of the sampling rate.

Today (in 1997), most commercial sampling playback implementations use only two point linear interpolation. This is described as "two point" because only two input samples are involved in the interpolation calculation. While this method works reasonably well, implementors have to be careful not to use waveforms with significant high frequency content (energy above $\frac{\pi}{4}$), or aliasing distortion will be noticeable.

By using more than two points, higher quality sample rate conversion filters can be implemented. The traditional sample rate conversion literature usually describes techniques, such as polyphase filters, that are appropriate for sample rate conversion by a fixed ratio. In sampling synthesis applications, of course, the conversion ratio is usually time varying, so the polyphase techniques are not the most suitable. Smith and Gosset[Smith and Gosset, 1984] showed a method for sample rate conversion that is more appropriate for arbitrary sample rate conversion ratios.

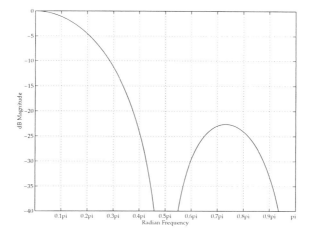

Figure 8.8 Frequency response of at linear interpolation sample rate converter with $L = 4$ showing better stopband rejection around 0.5π (or at the original Nyquist rate) but increased rolloff in the passband below 0.25π of almost -7 dB.

In the Smith and Gosset interpolator design, each output sample is formed from an inner product of the stored signal $x[n]$ and a set of coefficients from the interpolating filter.

The implementation is different from the standard polyphase filters of Crochiere and Rabiner et al. In the polyphase filter approach, each sub-filter is composed of coefficients which are reordered from a prototype filter. In the Smith and Gosset approach, the coefficients of the prototype filter are stored non- reordered, and each output sample is formed from an inner product of input samples and coefficients which are selected from the filter table at run time. The indexing equation is very similar to the equation that Crochiere and Rabiner give for re-ordering the prototype filter to generate their polyphase filter designs.

The set of coefficients which are chosen is determined at run time, rather than being fixed as in traditional polyphase filter design. Also, Smith and Gosset describe a method for interpolating between coefficients to reduce the size of the table holding the filter coefficients. Performing this interpolation between coefficients adds additional incentive to keep the coefficients in their original non-reordered form to simplify the coefficient interpolation calculation.

If a large filter coefficient table is practical, then the interpolation of coefficients that Smith and Gosset describe is not needed, and an approach closer to Crochiere and Rabiner becomes more preferable. Basically, that approach is to create a prototype filter table for a large value of L and stored the coefficients in a re-ordered form as is done with standard polyphase filters. Then at run time, the subphase of the filter needed is chosen based on the fractional part of the phase accumulator. This differs from standard polyphase filters only in that a standard polyphase filter sample rate converter will use all of its subphases, since the design assumes that the conversion ratio is fixed. Here, the subphases are selected every output sample, by the fractional part of the phase register, from a large set of possible sub phases available.

The Crochiere and Rabiner equation for interpolation/decimation by a fixed factor of L/M is summarized here. The output sample sequence $y[m]$ is formed by the convolution:

$$y[m] = \sum_{n=-W/2}^{W/2} g_m(n) x \left(\lfloor \frac{mM}{L} \rfloor - n \right) \qquad (8.6)$$

where the filter table $g_m(n) = h(nL + (mM \bmod L))$, for all m and all n, and L/M is the sample rate conversion ratio. Note that M/L is the phase increment in the computer music oscillator. m refers to output time. n refers to input time. The symbol $\lfloor x \rfloor$ means greatest integer smaller than x. $\lfloor \frac{mM}{L} \rfloor$ thus refers to the integer part of $\frac{mM}{L}$. $\frac{mM}{L}$ is exactly equivalent to the phase register of the computer music phase accumulator oscillator at output time m. $g_m(n)$ are the filter coefficients of the prototype filter $h[n]$ reordered into L separate subfilters. In sampling implementations, M is set to one for the purposes of re-ordering the filter coefficient table. The selection of filter subphases at run time, indexed by the fractional phase register. In block diagram form, the algorithm has some of the same elements as the traditional linear interpolation table lookup oscillator. The fractional part of the phase register is now used to select one of the filter sub phases (one of g_m). W values of the WaveTable are read out (indexed by the Filter Counter), and the output sample is simply the inner product of the filter coefficient set (the polyphase sub filter) and the W samples from the wavetable. L sets of filter coefficient vectors are stored in memory.

Instead of reading out two samples from the wavetable memory in the case of linear interpolation, W samples are read out. A "Filter Counter" is shown which performs this indexing. The base address in the wavetable where the signal vector is read out is provided by the integer part of the phase register, as in the case with linear interpolation.

When the phase increment M/L is greater than one, and the original signal is being reduced in sample rate, classical sample rate techniques require that the cutoff frequency of the prototype filter change to $\frac{\pi}{M}$. One approach is to time- scale the polyphase subfilters, but this increases the computation rate, which is undesirable in a typical VLSI implementation. Another approach is to switch filter tables. In practice, may sampler implementations only have one filter table, and pitch shifting up is restricted to be less than one octave. With this restriction, it is usually OK to use only one filter table with its cutoff equal to $\frac{\pi}{L}$. The artifacts resulting from this compromise are usually acceptable.

WAVETABLE SAMPLING SYNTHESIS 329

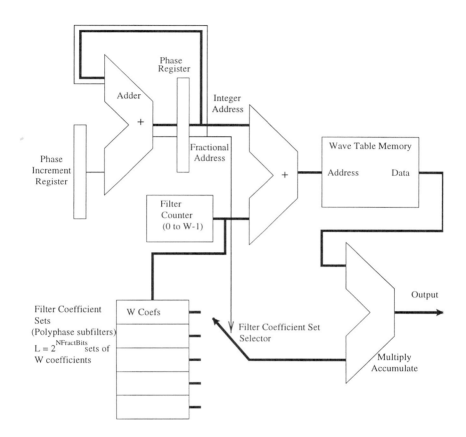

Figure 8.9 A sampling playback oscillator using high order interpolation. Every output sample is a vector dot product of W input samples and one of the filter coefficient vectors, stored re-ordered from the original prototype filter. The fractional phase address selects the filter coefficient vector used.

Re-ordering the filter coefficients into the polyphase components reduces the cost of indexing the filter coefficients. Memory costs have declined to the point where storing a large number of filter coefficients is very practical on many signal processors, and also on general purpose host processors as well. For many implementations, the memory is less precious than CPU cycles, further motivating the use of polyphase filter strategies.

How Many Fractional Phase Register Bits are Needed. The choice of how many bits to make the phase register is an important issue in computer music design. While other authors have covered this issue in relation to traditional sine wave oscillators, there are some subtle differences in the design of sample playback oscillators. Here, the fractional part of the phase register essentially determines how much pitch resolution is available, while the integer part determines how many octaves up the waveform can be transposed (pitch shifted).

Denoting $\alpha > 0$ the pitch shift ratio, and assuming that the loop buffer contains one period, the frequency F_{out} of the output sine wave is simply $F_{out} = \alpha F_{loop}$ where F_{loop} denotes the frequency at which the loop samples are output. From this, we derive that relative variations of α and F_{out} are equal:

$$\frac{\Delta F_{out}}{F_{out}} = \frac{\Delta \alpha}{\alpha} \tag{8.7}$$

The smallest available variation of the pitch shift ratio α is given by the number N_f of bits used to represent its fractional part. More specifically,

$$\Delta \alpha = 2^{-N_f} \tag{8.8}$$

It is usually assumed that people can hear pitch tuning errors of about one cent, which is 1% of a semi-tone. A semitone is a ratio of $2^{1/12}$, so a ratio of one cent would be $2^{1/1200}$. For the variation ΔF_{out} to be smaller than 1 cent, one must have

$$\frac{\Delta F_{out}}{F_{out}} < 2^{1/1200} - 1 \tag{8.9}$$

and combining the three equations above leads to

$$2^{-N_f} < \alpha(2^{1/1200} - 1) \tag{8.10}$$

Clearly, the constraint on N_f is more stringent as α becomes small: tuning errors will be more audible in downward pitch shifting than in upward pitch shifting. Unless a limit is imposed on the required amount of downward pitch shifting, an arbitrary large number of bits must be used to represent α. Denoting N_{oct} the maximum number of octaves one wishes to pitch-shift *down*, we always have $\alpha > 2^{-N_{oct}}$ and equation (8.10) now reads

$$2^{-N_f} < 2^{-N_{oct}}(2^{1/1200} - 1) \tag{8.11}$$

or
$$N_f > N_{oct} - \log_2(2^{1/1200} - 1) \approx N_{oct} + 10.76 \quad (8.12)$$

This tells us that in order to maintain one cent of accuracy, we need eleven bits more than the maximum number of octaves of downward pitch shift. Typically, implementors do not pitch shift by a large amount. 12 to 16 fractional bits is fairly typical in practice.

8.2.4 Looping of sustain

Another practical limitation of sampling involves long notes. A violinist can play a note for as long as necessary to meet a musical requirement, but a sampler needs to have a recording of the entire note event in order to play for a required duration.

To solve this problem, the concept of looping was developed. Looping is the process where after the onset of a musical note has transpired, a section of the steady state portion of the note is simply repeated over and over. This technique is called looping, after a similar technique used with analog tape recordings where a segment of tape was cut and spliced literally in a loop, allowing playback of a segment to repeat over and over. This allows a short segment of a waveform to be used to substitute for an arbitrary length of the steady state portion of a musical instrument signal. A

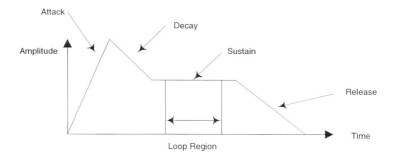

Figure 8.10   The traditional ADSR amplitude envelope used in electronic and computer music, with the looping region shown.

number of empirical methods have been employed to make the process of identifying start and end points for a loop that do not produce clicks or other artifacts [Massie, 1986].

A simple City Block or L1 metric has been found to be fairly successful in comparing candidate splice points. One simply computes

$$score(m,n) = \sum_{0<i<W} |X_{n+i} - X_{m+i}|, 0 < i < W \quad (8.13)$$

where n is the start point for a loop, m is the end point, and W is the window length. This function satisfies the definitions of a metric, and a score of zero indicates an identical waveform segment[Ross et al., 1974]. As scores increase, the closeness of match decreases.

Since the L1 metric only involves differences, absolute values, and addition, it is inexpensive to compute with fixed point arithmetic, but searching by varying both the loop start and end (n and m) means that the test would be applied at $n*m$ points, which drives the cost up for this test. Many people further constrain the candidate loop points to zero crossings, and then apply this test only to those zero crossings. This tends to work very well and dramatically reduces the cost of the test. While having a zero crossing is neither necessary nor sufficient to ensure a good loop, it seems to be a good starting point for searches.

The search region can further be constrained if a the candidate loop start point is moved far enough away from the start of the note to find a relatively stable segment of the signal.

Window sizes have been picked empirically, but the window size seems to determine the frequency spectrum of the loop splice "click". If the window size is made small, then sub-optimal loops tend to have a large low frequency thump. If the window size is made large, remaining artifacts will be high frequency clicks.

Good choices for window sizes tend to be about one period of a 500 Hz to 100 0Hz waveform, which seems to be a good compromise between low and high frequency artifacts.

Another perspective of looping is to consider the waveform as a sum of sinusoids. Each sinusoid at the loop start must have the same amplitude and phase as the sinusoid at the loop end in order to avoid splice artifact. If any component sinusoid does not line up in amplitude and phase, then there will be a click at the frequency of the sinusoid. The click should have the spectrum of a Sinc function translated to the frequency of the sinusoid that does not line up.

Backwards-Forwards loops. A clever trick for looping has been used a few commercial samplers, and is known as back-forwards looping. Few samplers use this technology now, but it still is a useful and interesting method. The technique simply is to advance through the waveform loop segment forwards in time, and when the end point of the segment is reached, reverse the read pointer and move backwards through the loop segment. When the beginning of the segment is reached, then the read pointer direction is reversed again, and the loop body is read out again forwards.

This technique immediately ensures first order continuity, since the read pointer does not jump to an arbitrary sample point at the loop boundary, but immediately continues reading at an adjacent sample. However, this does not guarantee a perfect loop.

WAVETABLE SAMPLING SYNTHESIS   333

Interestingly, the requirement for an non-clicking backwards-forwards loop is even symmetry around the loop points. A signal x[n] is referred to as "even" if $x[n] = x[-n]$. A signal is odd if $x[-n] = -x[n]$. A cosine wave is even around the origin and a sine wave has odd symmetry around the origin.

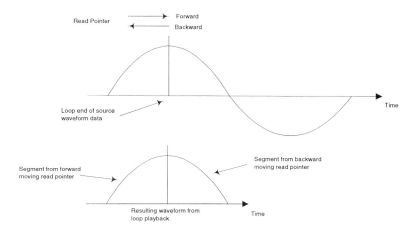

Figure 8.11   Backwards forwards loop at a loop point with even symmetry.

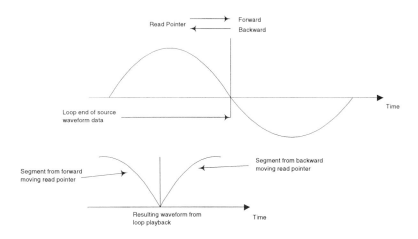

Figure 8.12   Backwards forwards loop at a loop point with odd symmetry.

Loop body. Although loop splice points can be found with reasonable ease, the resulting loop usually is still unsatisfactory unless the body of the loop is stationary enough. A single cycle loop will produce a perfectly stationary spectrum, with no amplitude variations, but it will also sound quite static. In most cases, sampling instruments use many cycles for a loop which then produce the impression of movement or animation. Strict definitions for animation or strict criteria for determining if a sequence of waveforms will seem animated have not been developed. Instead, sound designers use trial and error to find waveform segments that are long enough to sound animated but as short as possible to save memory.

Research into the acoustics of musical instruments has revealed considerable evidence that aperiodicity and noise play an important role in the sound quality of a musical instrument. This research reinforces the justifications for using more than one period for looping in sampling. Since the loop is actually a periodic waveform, the number of samples in that loop of course determines the number of spectral components that can be present in the spectrum, and their frequency spacing. N samples are completely specified by $N/2$ complex Fourier components. At $44100\,Hz$ sample rate, for a 256 sample loop, the spacing between frequencies would be $44100/256 = 172\,Hz$. Noise or other aperiodic components would be forced to fall on one of these bins. The longer the loop, the closer that spectral components can become, and the more aperiodic they can become. A truly aperiodic waveform would need an infinite loop, but our perception mechanism can be fooled into perceiving aperiodicity with a much shorter loop.

Crossfade looping. In many cases, a perfect loop is difficult or impossible to find. It can be helpful to perform what is called a crossfade loop. Here the loop data are modified to produce an acceptable loop with a signal processing rather than a signal analysis operation.

The basic principle of a crossfade loop is to average data from the beginning and the end of a loop. If the waveform data at the each end of a loop are replaced with the average of the data from the both the beginning and the end of the loop, the splice point will be inaudible. This is a simple operation; denote the loop start as $x[s]$ and the loop end as $x[e]$, where s denotes the loop start offset and e denotes the loop end offset, then we have the modified waveform data

$$x'[s+n] = x'[e+n] = (x[s+n] + x[e+n])/2, n = -W \text{ to } W \qquad (8.14)$$

where W is the window size for the average.

While the sampler is playing data back from within the loop itself, there will be no click at the loop splice point. Of course, as the sampler read pointer crosses the transition between unmodified waveform data into the modified (averaged) data, there potentially will be a discontinuity. So the second element of crossfade looping is the "crossfade". The crossfade is simply a weighting function to fade in the averaged waveform data and to fade out the original data.

We take the simplistic crossfade equation and window it:

$$x'[s+n] = (x[s+n] * (1 - A[n]) + A[n] * (x[s+n] + x[s+n]))/2, n = -W \text{ to } W \tag{8.15}$$

$A[n]$ can be any number of window shapes, but the two most commonly used are a triangular window, and a $\cos^2$ window. In both cases, the window should sum to one when overlapped with itself by $W/2$ samples, and should be unity at the center when the waveform segments that are overlapped correlate highly. If the overlapped waveform segments have a very low correlation, then it is preferable to use an equal power crossfade, where the sum of the windows is 1.414. This strategy tends to keep the RMS energy close to constant across the cross fade.

We do essentially the same with the data at the end of the loop

$$x'[e+n] = (x[e+n] * (1 - A[n]) + A[n] * (x[s+n] + x[e+n]))/2, n = -W \text{ to } W \tag{8.16}$$

Since the window $A[n]$ is equal to unity at its center, the waveform data at the start and end of the loop are identical, but the transition from the original data into the averaged data is smooth, preventing any discontinuities.

Appropriate sizes for the window length have been typically found empirically; usually they are on the order of tens of milliseconds or longer.

Backwards Forwards Crossfade Looping. A backwards forwards loop at any arbitrary point can also be created (or improved) by modifying the loop data. Here the crossfade is performed not between the data at the beginning and end of the loop, but from before and after the loop points.

All sequences can be decomposed into even and odd components. The even part of a sequence (about $n = 0$) is simply $even(x[n]) = x[n] + x[-n]$. The odd part of the sequence will cancel out when we perform this addition. The odd part of a sequence (about $n = 0$) is $odd(x[n]) = x[n] - x[-n]$.

The objective of a backwards forward crossfade is to modify the data at the loop so that it is a purely even function around the loop point. Thus we simply replace the data at the loop point with the even part of the function, and fade back into the original data going away from the loop point itself.

$$x[n+e] = x[-n+e] = x[n+e] * A[n+e] + even(x[n+e]) * (1 - A[n+e]) \tag{8.17}$$

$$x[n+e] = x[-n+e] = x[n+e] * A[n+e] + (x[n+e] + x[-n+e]) * (1 - A[n+e]) \tag{8.18}$$

where $-W < n < W$ and $A[n]$ is a triangular window function as described above for crossfade looping, where

$$A[0] = 1.0$$
$$A[W] = A[-W] = 0$$
$$e = \text{loop point}$$

The process of a backwards forwards crossfade loop does produce artifacts if the signal has significant anti-symmetric (odd) components around the loop point, so it is best performed on a loop point that has already been screened for a close fit to being symmetric.

In general, backwards forwards looping has fallen out of favor in the sampler design community, probably because it does not offer enough advantages to be worth the small extra effort to implement.

Relation of Crossfade Looping to Time and Pitch Scaling. The operation of crossfade looping is essentially the same operation that is performed in pitch or time scaling schemes as described in the chapter by Laroche except that in sampling the operation is typically performed once, off line, and with pitch or time scaling, the operation is performed in real time continuously, in order to scale pitch independently from time.

Crossfade looping is not only helpful in producing an acceptable loop splice point, but it can also help to smooth out variations within the loop body. Jupiter Systems has introduced commercial implementations of even more elaborate crossfade looping schemes to help smooth out the loop body of signals for which simple crossfade looping is not satisfactory [Collins, 1993]. In one of the approaches followed by Jupiter Systems, many copies of the loop are time-shifted, faded in and out and summed to the loop body. This has been effective at looping such signals as non-stationary orchestral string sections.

Perception of periodicity in loops. While a loop is clearly a periodic signal, listeners will not perceive a loop as being periodic under certain conditions, for example, if the loop is long enough. A single cycle loop will immediately be noticed as periodic. This perception is extremely striking, in fact. The contrast is so strong between a recorded musical transient and a single cycle loop of the same instrument sound, that it is somewhat surprising how tolerant musicians have been of electronic instruments (including electronic organs) that relied on purely periodic waveforms such as sawtooth waves, square waves, and even waveforms with complex harmonic content but with a static spectrum.

An example of musician's coping strategies was the Leslie Rotating Speaker system used on Hammond organs to literally add motion to an otherwise static timbre by rotating speakers.

Schemes to introduce instabilities and irregularities into purely periodic waveforms have tended to be less successful than simply using a loop with enough periods of a natural waveform to seem non-stationary. For many instruments, a loop of a less than

250 milliseconds can produce the perception of a non-stationary or animated sound. For a sound with a pitch of 440 Hz, at 44100 Hz sampling rate, this corresponds to more than 100 periods of the waveform.

Even with a very long loop of perhaps several seconds, when the loop repeats enough times, listeners will in time become aware of the periodicity. It seems that our perception mechanism evolved to sense change, and any purely repetitive event quickly becomes filtered out or uninteresting. In typical musical applications, loops of sampling instruments are not repeated very many times. Instead, the note is ended and another note is started before the loop becomes noticeable. It is rare for musical notes to last more than a second or two. The Mellotron was an keyboard analog musical instrument popular in the 1970's which had a segment of tape for each key with a recording of a musical instrument note on each tape segment. The Mellotron did not have the capability for looping. Instead, its tape segments lasted for 9 seconds, and users rarely would ever get to the end of a tape in normal musical uses. (Since there was a separate tape for each note, there was no problem with time compression.)

8.2.5 Multi-sampling

A single recording of a musical instrument can sometimes be pitch shifted over the entire range of a musical instrument and still be useful, but often the artifacts of the pitch shifting process become too noticeable to allow pitch shifting over such a large range. Thus, the process of "multi-sampling" was invented to allow mapping of more than one recording of an instrument to the controlling keyboard. For example, it seems

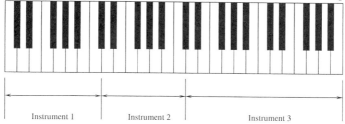

Figure 8.13 Multi-sampling is the practice of mapping or assigning individual sounds to play only over a small region of the musical keyboard controller.

to be satisfactory to record one note per octave for a piano to be able to reproduce a piano without drastic artifacts. Horn sounds or human voice sounds, on the other hand, require several recordings per octave to avoid sounding unnatural.

With the ability to map arbitrary sounds to arbitrary regions of the keyboard, sound designers went on to create unusual mappings, where individual keys controlled individual sound effects for film and television sound effects production. Samplers have become a common tool for sound effects production the film and video industry.

8.2.6 Enveloping

One of the easiest ways to modify recorded musical signals is to change the amplitude envelope of the signal. This is accomplished easily by multiplying the signal by a time-varying envelope function. The onset or attack of a musical event is probably the most important element in the identification of the instrument, and the actual attack contours of many acoustical instruments are quite complex. Producing a synthetic version of a complex envelope is usually quite difficult, but the complex envelope is preserved in a simple recording. This means that samplers do not usually attempt to produce complex envelope functions. Instead, simple modifications of the pre-recorded envelope can be accomplished using a traditional electronic music envelope generator.

Some applications for modifying the envelope of a musical sound include producing a slower attack onset for a piano. This produces the effect of a "bowed piano". This can be a very musically useful sound. Other examples include ramping the decay of a sound to zero faster than the sound normally does. This can be used to increase the effective damping of a drum sound, for example.

Without much more sophisticated schemes, enveloping of musical instruments cannot make the attack of a sound faster, however. If a violin attack takes 50 milliseconds, it is not easy to decrease this to 25 milliseconds, for example. But simple schemes have been introduced to get around this limitation. Changing the sample start point is an effective method to shorten the attack time. While it may seem to be a invalid operation, simply advancing the start pointer of the attack time to start playback of the musical event to some few milliseconds into the musical note, works well for varying the attack time of an instrument. The scheme can produce clicks in some instrument sounds, but usually the instrument has a great deal of noise in its initial transient which will mask the clicks produced by advancing the start pointer. Also, it is easy to generate a short taper in the onset by using an amplitude envelope generator. The envelope generator can eliminate the click caused by changing the start point of a sound.

8.2.7 Filtering

Time domain filtering is a popular and effective technique for further modifying the timbre of a wavetable played back in a sampler. Many commercial samplers today have one to four pole filters per channel to filter the signal being reproduced. Typically the filter is a low pass filter with adjustable Q or resonance, in the case where the filter has more than one pole. Applications for using the filters include altering the effective brightness as a function of keyboard velocity. Also, sweeping the filter cutoff frequency using an envelope generator creates a popular synthesizer type sound that is similar to early electronic music synthesizers.

There are instruments that have used higher order filtering schemes to allow more complex spectral modifications. The E-mu Morpheus sound module uses 14th order

ARMA filters to modify waveforms. The filters allow interpolation of stored filter coefficient sets under user control to modify the spectrum of the recorded signals.

8.2.8 Amplitude variations as a function of velocity

Musicians are today used to having control over the loudness of the sounds produced by their instruments. The piano is actually short for piano-forte which means soft-loud. The piano was an improvement over the harpsichord, which did not have any significant control over loudness.

It is rather easy today to provide control over the amplitude of a signal produced by a sampler, simply by scaling the signal as a function of the initial key velocity of the played note. This simple dimension is one of the most important forms of expressivity.

8.2.9 Mixing or summation of channels

Today musicians take for granted that their instruments are polyphonic. Early analog instruments were monophonic, that is, they only could produce one note at a time like a clarinet, which can only produce one note at a time. Musicians used to pianos expected to have arbitrary polyphony available since with a piano, a player could conceivably press up to all 88 keys simultaneously.

Today, (1997) samplers and other keyboard musical instruments typically can produce from 32 to 64 notes or channels simultaneously, although instruments are available that can generate 128 independent channels. Many of these instruments have more than one output D-A converter, but having more than 8 output D-A converters is rare.

All of these channels need to be added together (mixed, in audio engineering terms) before being routed to the D-A converter. Summation is of course a simple process with digital circuits, but a subtle point about this output summation is the choice of scaling rule. When several channels of digital audio signals are summed together, word growth occurs.

When two signals are added together, the worst case word growth in two's complement number representation would be one bit. To prevent the possibility of overflow, N channels added together would be scaled by $1/N$, which corresponds to using an L1 Norm scaling rule [Oppenheim and Schafer, 1989].

This scaling rule is considered too conservative in most cases. An L2 scaling rule is usually more successful, where N channels would be scaled by $1/\sqrt{N}$. Typically, when channels are summed together (mixed) the signals are uncorrelated. Summing two uncorrelated signals will only increase the RMS value of the signal by 3 dB, while the peak value can increase by 6 dB.

In most cases, the loudness of the result of mixing two signals corresponds to the RMS value of the sum, and not the peak value, so scaling by the L2 norm is appropriate. Leaving headroom for the resulting signal's crest factor (the ratio of the peak to the

RMS value of the signal) and scaling by the square root of the number of channels is usually a very effective rule.

Musical signals in a sampling playback synthesizer rarely are closely correlated. When many notes are all started at exactly the same time, clipping can sometimes occur, but this is a very rare event. When clipping does occur, it usually only happens for a few samples or at most a few periods of the waveform. Quite often, this small amount of clipping cannot be perceived, or if it is perceived, it is often preferred to having no clipping! The clipping distortion can increase the harmonic content of the signal and the apparent loudness of the signal, and so some musicians often will deliberately introduce some small amount of this distortion, even though it is "digital distortion" which is considered a "harsh" non-linearity.

Interestingly, most implementors of samplers have not explicitly formulated any headroom rule; instead, they have set the amount of headroom needed empirically, and often the empirical choices correspond closely to a L2 rule, although they are usually a bit less conservative.

8.2.10 Multiplexed wavetables

Samplers can use waveform memories that are either multiplexed or non- multiplexed. A multiplexed wavetable is simply a waveform memory that is accessed more than once per sampling period to allow many notes to share the same waveform memory, while non-multiplexed waveform memory dedicates a separate memory for each playback channel. A simplistic analysis of memory access times suggests that a non-multiplexed wavetable is wasteful because sampling period is about 20 microseconds while the time for a single access for random access memory (even in the late 1970's, when samplers first appeared) is around 100 nanoseconds, many times faster. But there are many difficulties in multiplexed wavetable sampler design, especially related to pitch shifting issues, so the earliest samplers used a single wavetable per playback channel. Non- multiplexed wavetables allowed a very simple digital circuit for both pitch shifting and for connection to D-A converters. The Fairlight Computer Music Instrument, released about 1979, was probably the most prominent instrument that used non-multiplexed wavetables, but many instruments were designed and sold for many years after this using non-multiplexed memories.

John Snell's article on digital oscillator design is an excellent reference on digital oscillators that describes multiplexing quite well [Snell, 1977].

In 1981, the E-mu Emulator sampler was introduced which used a waveform memory multiplexed to allow 8 channels or voices to be produced in real time from one 128 Kbytes RAM bank.

A single large wavetable memory offered a number of advantages. Notably, more flexibility is available to allocate memory buffers for waveforms. A large memory array of 128 Kbytes allowed much more flexibility than eight separate 16 Kbytes memories.

Today, synchronous outputs have become the standard in sampler design.

8.3 CONCLUSION

Today, in 1997, sampling wavetable synthesis dominates commercial musical instrument synthesis, in spite of concerted effort on the part of many researchers and companies to replace it with some form of parametric synthesis.

The future of sampling synthesis is probably as hard to predict as it has ever been, but some probable trends are:

- Cost will decrease.

- Data compression may become more common.

- More methods for signal modification yielding more expressivity will become common, but only at the expense of more computation cost.

- More overlap between hard disk recording, editing, and synthesis will occur.

- Generalized sampling schemes will become more common, in the guise of Analysis/Resynthesis schemes.

The author appreciates the generous contributions and support from Dr. Jean Laroche, Scott Wedge, Byron Sheppard, Dave Rossum, Julius O. Smith, Michelle Massie, and the patience of the editor, Mark Kahrs.

9 AUDIO SIGNAL PROCESSING BASED ON SINUSOIDAL ANALYSIS/SYNTHESIS

T.F. Quatieri and R.J. McAulay

Lincoln Laboratory
Massachusetts Institute of Technology
Lexington, MA 02173
{tfq,rjm}@sst.ll.mit.edu

Abstract: Based on a sinusoidal model, an analysis/synthesis technique is developed that characterizes audio signals, such as speech and music, in terms of the amplitudes, frequencies, and phases of the component sine waves. These parameters are estimated by applying a peak-picking algorithm to the short-time Fourier transform of the input waveform. Rapid changes in the highly resolved spectral components are tracked by using a frequency-matching algorithm and the concept of "birth" and "death" of the underlying sine waves. For a given frequency track, a cubic phase function is applied to the sine-wave generator, whose output is amplitude-modulated and added to sines for other frequency tracks. The resulting synthesized signal preserves the general waveform shape and is nearly perceptually indistinguishable from the original, thus providing the basis for a variety of applications including signal modification, sound splicing, morphing and extrapolation, and estimation of sound characteristics such as vibrato. Although this sine-wave analysis/synthesis is applicable to arbitrary signals, tailoring the system to a specific sound class can improve performance. A source/filter phase model is introduced within the sine-wave representation to improve signal modification, as in time-scale and pitch change and dynamic range compression, by attaining phase coherence where sine-wave phase relations are preserved or controlled. A similar method of achieving phase coherence is also applied in revisiting the classical phase vocoder to improve modification of certain signal classes. A second refinement of the sine-wave analysis/synthesis invokes an additive deterministic/stochastic representation of sounds consisting of simultaneous harmonic and aharmonic contributions. A method of frequency tracking is given for the separation of these components, and is used in a number of applications. The sine-wave model is also extended to two additively combined signals for the separation of simultaneous talkers or music duets. Finally, the use of sine-wave analysis/synthesis in providing insight for FM synthesis is described, and remaining challenges, such as an improved sine-wave representation of rapid attacks and other transient events, are presented.

9.1 INTRODUCTION

The representation of signals by a sum of amplitude-frequency modulated sine waves and analysis/synthesis techniques based on this representation have become essential tools in music and speech sound processing. Common objectives include duplication of natural sounds, creation of new and enhanced sounds through modification and splicing, and separation of components of a complex sound. The physical generation of music signals is in part similar to the generation of speech signals, and thus it is not surprising that sinusoidal-based processing useful in one area is useful to the other.

In certain wind instruments, for example, a vibrating reed excites the instrument's cavity; while in speech the vibrating vocal cords excite the vocal tract. Moreover, this "source/filter" representation is made up of common signal classes in the two domains, all of which can be represented approximately by a sum of amplitude- and frequency-modulated sine waves. Quasi-periodic signals, as from steady speech vowels and sustained musical notes, consist of a finite sum of *harmonic* sine waves with slowly-time-varying amplitudes and frequencies; while noise-like signals, as from speech fricatives and musical turbulence, have no clear harmonic structure. Transients, as from speech plosives and musical attacks and decays, may be neither harmonic nor noise-like, consisting of short acoustic events that occur prior, during, or after steady regions. Noise-like and transient sounds although *aharmonic*, nevertheless, can be represented approximately by a sum of sine waves, but generally without coherent phase structure. A typical sound is often a mixture of these components whose relative weights, timing, and duration can be key to accurate modeling.

The example given in Figure (9.1), the waveform and spectrogram of an acoustic signal from a trumpet, illustrates these sound classes. Quasi-periodic sounds occur during sustained note segments, while transients occur at note attacks and decays, the latter seen to be harmonic-dependent. There is also an often-present noise-like component, due in part to turbulent air jet flow at the mouth piece, contributing a "breathiness" to the sound. In addition, the example shows sine-wave amplitude and frequency modulation which, in the music context, are referred to, respectively, as tremolo and vibrato. This modulation of the quasi-periodic portion of notes, the harmonic-dependent note attack and decay, and noise-like components, all contribute to the distinguishing character of the sound.

AUDIO SIGNAL PROCESSING BASED ON SINUSOIDAL ANALYSIS/SYNTHESIS

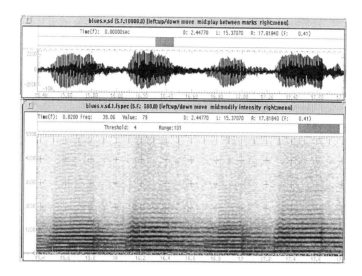

Figure 9.1 Signal (upper) and spectrogram (lower) from a trumpet.

The purpose of this chapter is to describe the principles of signal analysis/synthesis based on a sine-wave representation and to describe its many speech and music applications. As stated, an important feature of the sinusoidal representation is that the aforementioned sound components can be expressed approximately by a sum of amplitude- and frequency-modulated sine waves. Moreover these sound components, as well as the "source" and "filter" contribution to their sine-wave representation, are separable by means of a sine-wave-based decomposition. This separability property is essential in applying sinusoidal analysis/synthesis in a number of areas.

Section 5.2 of this chapter gives a brief description of an early filter bank-based approach to sine-wave analysis/synthesis referred to as the phase vocoder [Flanagan and Golden, 1966], and shows how the phase vocoder motivates the more general approach to sine-wave analysis/synthesis [McAulay and Quatieri, 1986b] which is the primary focus of the chapter. The phase vocoder is also described in chapter 7 in the context of other filter bank-based approaches to estimating parameters of a sine-wave model. In section 5.3, a baseline sine-wave analysis/synthesis system, based on frequency tracking through the short-time Fourier transform and interpolation of resulting sine-wave amplitude and phase samples, is described. In this section, refinements and some applications of this system, including signal modification, splicing, and estimation of vibrato, are presented. This section ends with an overview of time-frequency resolution considerations for sine-wave analysis. Although this basic sine-wave analysis/synthesis is applicable to arbitrary signals, tailoring the system to a specific class can improve performance. In section 5.4, a source/filter phase model for quasi-periodic signals is introduced within the sine-wave representation. This model is particularly important for signal modifications such as time-scale and pitch modification and dynamic range compression where *phase coherence*, i.e., preserving certain phase relations among sine-waves, is essential. A similar phase representation is then used in revisiting the phase vocoder to show how phase coherence can be introduced to improve modification of various signal classes. In section 5.5, an *additive* model of deterministic and stochastic components is introduced within the sine-wave representation. This two-component model is particularly important for representing sounds from speech and musical instruments with simultaneous harmonic and aharmonic contributions. Section 6 then describes a sine-wave-based approach to separating two signal "voices" that are additively combined; the technique is applied to separation of two simultaneous talkers, as well as musical duets. Section 7 reviews an approach for generating a sum of sine waves by modulating the frequency of a single sine wave. This technique, referred to as FM Synthesis [Chowning, 1973], gives the potential of a compact representation of a harmonic complex and has been the basis of many electronic music synthesizers. The sine-wave analysis/synthesis of this chapter may provide insight for further refinements of the FM synthesis approach. Finally in section 8, the chapter is summarized, applications not covered within the chapter are briefly discussed, and some of the many fascinating unsolved problems are highlighted.

9.2 FILTER BANK ANALYSIS/SYNTHESIS

Early approaches to music analysis relied on a running Fourier transform to measure sine-wave amplitude and frequency trajectories. This technique evolved into a filter-bank-based processor and ultimately to signal analysis/synthesis referred to as the phase vocoder [Flanagan and Golden, 1966]. This section describes the history of the phase vocoder, its principles, and limitations that motivate sinusoidal analysis/synthesis. Other formulations and refinements of the phase vocoder are given in chapter 7.

9.2.1 Additive Synthesis

An early approach to music processing, referred to as *additive synthesis* [Moorer, 1977], used the sinusoidal model of a quasi-periodic music note

$$s(n) = \sum_{k=1}^{L} A_k \cos(k\omega_o n + \phi_k) \tag{9.1}$$

where A_l and ϕ_k represent the amplitude and phase of each sine-wave component associated with the kth harmonic $k\omega_o$, and L is the number of sine waves. In this model the amplitude A_k, fundamental frequency ω_o, and phase offset ϕ_k are slowly varying. The changing ω_o accounts for pitch movement, while the change in ϕ_k accounts for

AUDIO SIGNAL PROCESSING BASED ON SINUSOIDAL ANALYSIS/SYNTHESIS   347

time-varying deviations in frequency from each harmonic[1]. The derivative of the phase function $\theta_k(n) = k\omega_o n + \phi_k$ approximates the slowly-changing frequency of each harmonic, and is given by $\omega_k(n) = k\omega_o + \dot{\phi}_k$. Equation (9.1) serves not only as a model, but also as a means for sound synthesis. In synthesis, the control functions $A_k(n)$ and $\omega_k(n)$, initially were set manually based on knowledge of the musical note; the absolute phase offset ϕ_k in $\theta_k(n)$ was not used.

One of the first attempts to estimate the control functions relies on a running discrete-Fourier transform (DFT (Discrete Fourier Transform)) [Moorer, 1977]. Assuming the presence of one periodic note in a measurement $x(n)$, the DFT length is set equal to the waveform's pitch period N. The real and imaginary components are then given by

$$c_k(n) = \sum_{r=n}^{n+N-1} x(r)\cos(rk\omega_o)$$

$$d_k(n) = \sum_{r=n}^{n+N-1} x(r)\sin(rk\omega_o) \qquad (9.2)$$

where $\omega_o = 2\pi/N$ and from which one obtains the estimates of the slowly time-varying amplitude and phase of each harmonic

$$\hat{a}_k(n) = \sqrt{c_k^2(n) + d_k^2(n)}$$

$$\hat{\theta}_k(n) = \arctan[d_k(n)/c_k(n)] \qquad (9.3)$$

ane where the frequency of each harmonic is given approximately by the derivative of the unwrapped version of $\hat{\theta}_k(n)$[2].

A limitation of this method is that the pitch period must be known exactly to obtain reliable estimates. The running DFT can be viewed as a filter bank where each filter is a cosine modulated version of a prototype filter given by a rectangular window of length N over the interval $0 \leq n < N$. Based on this interpretation, an improvement in sine-wave parameter estimation can be made by generalizing the window shape as described in the following section.

9.2.2 Phase Vocoder

An analysis/synthesis system based on a filter bank representation of the signal can be derived from the time-dependent short-time Fourier transform (STFT) [Nawab and Quatieri, 1988a]

$$X(n,\omega) = \sum_m w(n-m)x(m)e^{-j\omega m} \qquad (9.4)$$

By changing the variable $n - m$ to m, Equation (9.4) becomes

$$X(n,\omega) = e^{-j\omega n}\sum_m x(n-m)w(m)e^{j\omega m}$$

348   APPLICATIONS OF DSP TO AUDIO AND ACOUSTICS

$$= e^{-j\omega n}(x(n) * [w(n)e^{j\omega n}]) \quad (9.5)$$

Equation (9.5) can be viewed as first a modulation of the window to frequency ω, thus producing a bandpass filter $w(n)e^{j\omega n}$, followed by a filtering of $x(n)$ through this bandpass filter. The output is then demodulated back down to baseband. The temporal output of the filter bank can be interpreted as discrete sine waves that are both amplitude- and phase-modulated by the time-dependent Fourier transform.

In expanding on this latter interpretation, consider a sequence $x(n)$ passed through a discrete bank of filters $h_k(n)$ where each filter is given by a modulated version of a baseband prototype filter $h(n) = w(n)$, i.e.,

$$h_k(n) = w(n)e^{j(2\pi/R)kn} \quad (9.6)$$

where $h(n)$ is assumed causal and lies over a duration $0 \leq n < S$, and $2\pi/R$ is the frequency spacing between bandpass filters, R being the number of filters. The output of each filter can be written as

$$\begin{aligned} y_k(n) &= x(n) * h_k(n) \\ &= x(n) * [w(n)e^{j\omega_k n}] \\ &= e^{j\omega_k n} X(n, \omega_k) \end{aligned} \quad (9.7)$$

which is Equation (9.5) without the final demodulation, evaluated at discrete frequency samples $\omega_k = (2\pi/R)k$ that can be thought of as center frequencies for each of the R "channels".

Since each filter response $h_k(n)$ in Equation (9.7) is complex, each filter output $y_k(n)$ is complex so that the temporal envelope $a_k(n)$ and phase $\phi_k(n)$ of the output of the kth channel is given by

$$\begin{aligned} a_k(n) &= |y_k(n)| \\ \phi_k(n) &= \tan^{-1}(\operatorname{Im}[y_k(n)]/\operatorname{Re}[y_k(n)]) \end{aligned} \quad (9.8)$$

Thus the output of each filter can be viewed as an amplitude- and phase-modulated complex sine wave

$$y_k(n) = a_k(n)e^{j\phi_k(n)} \quad (9.9)$$

and reconstruction of the signal[3] can be viewed as a sum of complex exponentials

$$x(n) = \sum_k a_k(n)e^{j\phi_k(n)} \quad (9.10)$$

where the amplitude and phase components are given by Equation (9.8) (see Figure (9.2)). The resulting analysis/synthesis structure is referred to as the *phase vocoder* [Flanagan and Golden, 1966].

AUDIO SIGNAL PROCESSING BASED ON SINUSOIDAL ANALYSIS/SYNTHESIS

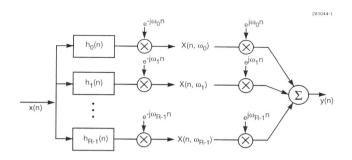

Figure 9.2 Phase vocoder based on filter bank analysis/synthesis.

The amplitudes and phases in Equation (9.10) can correspond to physically meaningful parameters for quasi-periodic signals typical of speech and music [Flanagan and Golden, 1966, Rabiner and Schafer, 1978a]. In order to see this, the STFT is first written as

$$X(n, \omega_k) = |X(n, \omega_k)| e^{j\theta(n,\omega_k)} \qquad (9.11)$$

where ω_k is the center frequency of the kth channel. Then, from Equations (9.7) and (9.11), the output of the kth channel is expressed as

$$y_k(n) = |X(n, \omega_k)| e^{j[\omega_k n + \theta_k(n,\omega_k)]} \qquad (9.12)$$

Consider now two filters that are symmetric about π so that $\omega_{R-k} = 2\pi - \omega_k$ where $\omega_k = 2\pi k/R$ and assume for simplicity that R is even. Then it is straightforward to show that

$$X(n, \omega_k) = X^*(n, \omega_{R-k}) \qquad (9.13)$$

From Equations (9.12) and (9.13), the sum of two symmetric channels k and $R-k$ can be written as

$$\hat{y}_k(n) = 2|X(n, \omega_k)| \cos(\omega_k n + \theta_k(n, \omega_k)) \qquad (9.14)$$

which can be interpreted as a real sine wave which is amplitude- and phase-modulated by the STFT, the "carrier" of the later being the kth filter's center frequency.

Consider now a sine-wave input of frequency ω_o, $x(n) = A_o \cos(\omega_o n + \theta_o)$, that passes through the kth channel filter without distortion, as illustrated in Figure (9.3). Then it can be shown that for the kth channel

$$\begin{aligned} |X(n, \omega_k)| &= A_o/2 \\ \theta(n, \omega_k) &= (\omega_o - \omega_k)n + \theta_o \end{aligned} \qquad (9.15)$$

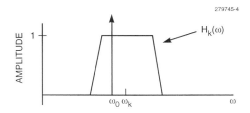

Figure 9.3 Passage of single sine wave through one bandpass filter.

and thus the output of the kth (real) channel is given by

$$\hat{y}_k(n) = A_o \cos[\omega_o n + \theta_o] \qquad (9.16)$$

A similar analysis can be made for quasi-periodic signals which consist of a sum of sine waves with slowly-varying amplitude and instantaneous frequency each of which is assumed to pass through a single filter.

The phase vocoder has been useful in a number of applications[4]. In time-scale modification, for example, the goal is to maintain the perceptual quality of the original signal while changing its apparent rate of "articulation". In performing time-scale modification with the phase vocoder, the instantaneous frequency and amplitude of each channel are interpolated or decimated to a new time scale[5]. In one scenario, the phase of each filter output in Equation (9.9) is first unwrapped, and the channel amplitude and unwrapped phase are then time scaled. With time-scale modification by a factor ρ, the modified filter output is given by

$$\tilde{y}_k(n) = \tilde{a}_k(n) \cos[\rho \tilde{\phi}_k(n)] \qquad (9.17)$$

where $\tilde{a}_k(n)$ and $\tilde{\phi}_k(n)$ are the interpolated/decimated amplitude and phase functions, respectively. The modified phase is scaled by ρ to maintain the original frequency trajectory, i.e., phase derivative, of each filter output.

9.2.3 Motivation for a Sine-Wave Analysis/Synthesis

In spite of the many successes of the phase vocoder, numerous problems have limited its use. In the applications of time-scale modification and compression, for example, it is assumed that only one sine wave enters each bandpass filter within the filter bank. When more than one sine wave enters a bandpass filter, the meaning of the input sine-wave amplitude and phase envelope is lost. A particular sine wave also may not be adequately estimated when it falls between two adjacent filters of the filter bank. In addition, sine waves with rapidly-varying frequency due to large vibrato or fast pitch change are difficult to track. A result of using a fixed filter bank is that the frequency of an input sine wave cannot be measured outside the bandwidth of each bandpass filter. Although these measurement problems may be resolved by appropriate combining of adjacent filter bank outputs, such solutions are likely to be quite cumbersome [Serra, 1989]. Yet another problem is that of phase dispersion. In time-scale modification, for example, the integration of the phase derivative and scaling of the unwrapped phase results in a loss of the original phase relation among sine waves, thus giving an objectionable "reverberant" quality characteristic of this method. Finally, the phase vocoder was formulated in the context of discrete sine waves and hence was not designed for the representation of noise components of a sound.

A number of refinements of the phase vocoder have addressed these problems [Dolson, 1986, Portnoff, 1981, Malah and Flanagan, 1981, Malah, 1979]. For example, the assumption that only one sine wave passes through each filter motivates a filter bank with filter spacing equal to the fundamental frequency, thus allowing one harmonic to pass through each filter [Malah, 1979]. An alternative is to oversample in frequency with the hope that only one harmonic passes through each filter. One approach to prevent phase dispersion is to use an overlap-add rendition of the synthesis with windows of length such that the overlap is always in phase [Malah and Flanagan, 1981]. Another refinement of the phase vocoder was developed by Portnoff who represented each sine-wave component by a source and filter contribution [Portnoff, 1981]. Portnoff also provided a rigorous analysis of the stochastic properties of the phase vocoder to a noise-like input. An extension of the Portnoff phase vocoder that attempts to avoid phase dispersion is reviewed in chapter 7.

The analysis stage of the original phase vocoder and its refinements views sine-wave components as outputs of a bank of uniformly-spaced bandpass filters. Rather than relying on a filter bank to extract the underlying sine-wave parameters, an alternate approach is to explicitly model and estimate time-varying parameters of sine-wave components by way of spectral peaks in the short-time Fourier transform [McAulay and Quatieri, 1986b]. It will be shown that this new approach lends itself to sine-wave tracking through frequency matching, phase coherence through a source/filter phase model, and estimation of a stochastic component by use of an additive model of deterministic and stochastic signal components. As a consequence, the resulting sine-wave analysis/synthesis scheme resolves many of the problems encountered by the phase vocoder, and provides a useful framework for a large range of speech and music signal processing applications.

9.3 SINUSOIDAL-BASED ANALYSIS/SYNTHESIS

In this section it is shown that a large class of acoustical waveforms including speech, music, biological, and mechanical impact sounds can be represented in terms of estimated amplitudes, frequencies and phases of a sum of time-varying sine waves. There are many signal processing problems for which such a representation is useful, including time-scale and pitch modification, sound splicing, interpolation, and extrapolation. A number of refinements to the baseline analysis, accounting for closely spaced frequencies and rapid frequency modulation, are also described. The approach of this section explicitly estimates the amplitudes, frequencies, and phases of a sine-wave model, using peaks in the STFT magnitude, in contrast to the phase vocoder and the methods described in chapter 7 that rely on filterbank outputs for parameter estimation.

9.3.1 Model

The motivation for the sine-wave representation is that the waveform, when perfectly periodic, can be represented by a Fourier series decomposition in which each harmonic component of this decomposition corresponds to a single sine wave. More generally, the sine waves in the model will be aharmonic as when periodicity is not exact or turbulence and transients are present, and is given by

$$s(n) = \sum_l A_l(n) \cos[\theta_l(n)]$$

where $A_l(n)$ is a time-varying envelope for each component and with phase

$$\theta_l(n) = \int_0^{nT} \omega_l(\tau) d\tau + \phi_l \tag{9.18}$$

where $\omega_l(n)$ is the *instantaneous frequency* which will also be referred to as the *frequency track* of the the $k\text{th}$ sine wave.

In this section, the model in Equation (9.18) is used to develop an analysis/synthesis system which will serve to test the accuracy of the sine-wave representation for audio signals. In the analysis stage, the amplitudes, frequencies, and phases of the model are estimated, while in the synthesis stage these parameter estimates are first matched and then interpolated to allow for continuous evolution of the parameters on successive frames. This sine-wave analysis/synthesis system forms the basis for the remainder of the chapter.

9.3.2 Estimation of Model Parameters

The problem in analysis/synthesis is to take a waveform, extract parameters that represent a quasi-stationary portion of that waveform, and use those parameters or modified versions of them to reconstruct an approximation that is "as close as possible" to a desired signal. Furthermore, it is desirable to have a robust parameter extraction algorithm since the signal in many cases is contaminated by additive acoustic noise. The general identification problem in which the signal is to be represented by multiple sine waves is a difficult one to solve analytically. In an early approach to estimation by Hedelin [Hedelin, 1981], the sine-wave amplitudes and frequencies were tracked using Kalman filtering techniques, and each sine-wave phase is defined as the integral of the associated instantaneous frequency. The approach taken here is heuristic and is based on the observation that when the waveform is perfectly periodic, the sine-wave parameters correspond to the harmonic samples of the short-time Fourier transform (STFT). In this case, the model in Equation (9.18) reduces to

$$s(n) = \sum_{l=1}^{L} A_l \cos(nl\omega_0 + \phi_l) \qquad (9.19)$$

in which the sine-wave frequencies are multiples of the fundamental frequency ω_0 and the corresponding amplitudes and phases are given by the harmonic samples of the STFT. If the STFT of $s(n)$ is given by

$$Y(\omega) = \sum_{n=-N/2}^{N/2} s(n) \exp(-jn\omega) \qquad (9.20)$$

then Fourier analysis gives the amplitude estimates as $A_l = |Y(l\omega_0)|$ and the phase estimates as $\phi_l = \angle Y(l\omega_0)$. Moreover, the magnitude of the STFT (i.e., the periodogram) will have peaks at multiples of ω_0. When the speech is not perfectly voiced, the periodogram will still have a multiplicity of peaks but at frequencies that are not necessarily harmonic and these can be used to identify an underlying sine-wave structure. In this case the sine-wave frequencies are simply taken to be the frequencies at which the slope of the periodogram changes from positive to negative and the amplitudes and phases are obtained by evaluating the STFT at the chosen frequencies [McAulay and Quatieri, 1986b].

The above analysis implicitly assumes that the STFT is computed using a rectangular window. Since its poor sidelobe structure will compromise the performance of the estimator, the Hamming window was used in all experiments. While this resulted in a very good sidelobe structure, it did so at the expense of broadening the mainlobes of the periodogram estimator. Therefore, in order to maintain the resolution properties that were needed to justify using the peaks of the periodogram, the window width is made at least two and one-half times the average pitch period. During aharmonic frames, the window is held fixed at the value obtained on the preceding harmonic frame[6].

Once the width of the analysis window for a particular frame has been specified, the pitch-adaptive Hamming window $w(n)$ is computed and normalized according to

$$\sum_{n=-N/2}^{N/2} w(n) = 1 \qquad (9.21)$$

so that the periodogram peak will yield the amplitude of an underlying sine wave. Then the STFT of the Hamming-windowed input is taken using the DFT. Peaks in the periodogram are obtained by finding all values which are greater than their two nearest neighbors, and peaks below a specified threshold from the maximum peak (about 80 dB) are eliminated. The location of the selected peaks give the sine-wave frequencies and the peak values give the sine-wave amplitudes. The sine-wave phases ϕ_l are computed from the real and imaginary components of the STFT evaluated at ω_l.

It should be noted that the placement of the analysis window $w(n)$ relative to the time origin is important for computing the phases. Typically in frame-sequential processing the window $w(n)$ lies in the interval $0 \leq n \leq N$ and is symmetric about $N/2$, a placement which gives the measured phase a linear term equal to $-\omega N/2$. Since N is on the order of 100–400 discrete time samples, any error in the estimated frequencies results in a large random phase error and consequent hoarseness in the reconstruction. An error of one DFT sample, for example, results in a $\frac{2\pi}{M}\frac{N}{2}$ phase error (where M is the DFT length) which could be on the order of π. To improve the robustness of the phase estimate the Hamming window is placed symmetric relative to the origin defined as the center of the current analysis frame; hence the window takes on values over the interval $-N/2 \leq n \leq N/2$.

The approximations leading to the above periodogram estimator were based on the quasi-harmonic waveform assumption; the estimator can also be used with sustained sine waves that are not necessarily quasi-harmonic. Nowhere however have the properties of aharmonic noise-like signals been taken into account. To do this in a way that results in uncorrelated amplitude samples requires use of the Karhunen-Loève expansion for noise-like signals [Van Trees, 1968]. Such an analysis shows that a sinusoidal representation is valid provided the frequencies are "close enough" such that the ensemble power spectral density changes slowly over consecutive frequencies. If the window width is constrained to be at least 20 ms wide then, "on the average," there will be a set of periodogram peaks that will be approximately 100 Hz apart, and this should provide a sufficiently dense sampling to satisfy the necessary constraints[7]. The properties of aharmonic transient sounds have also not been addressed. Here the justification for the use of the periodogram estimator is more empirical, based on the observation that peak-picking the STFT magnitude captures most of the spectral energy so that, together with the corresponding STFT phase, the short-time waveform character is approximately preserved. This interpretation will become more clear when one sees in the following sections that sine-wave synthesis is roughly equivalent to an overlap-and-adding of triangularly weighted short-time segments derived from the STFT peaks.

The above analysis provides a heuristic justification for the representation of the waveform in terms of the amplitudes, frequencies, and phases of a set of sine waves that applies to one analysis frame. As the signal evolves from frame to frame, different sets of these parameters will be obtained. The next problem to address then is the association of amplitudes, frequencies, and phases measured on one frame with those that are obtained on a successive frame in order to define sets of sine waves that will be continuously evolving in time.

9.3.3 Frame-to-Frame Peak Matching

If the number of peaks were constant from frame to frame, the problem of matching the parameters estimated on one frame with those on a successive frame would simply require a frequency-ordered assignment of peaks. In practice, however, the locations of the peaks will change as the pitch changes, and there will be rapid changes in both the location and the number of peaks corresponding to rapidly varying signal regions, such as at harmonic to noise-like transitions. In order to account for such rapid movements in the spectral peaks, the concept of "birth" and "death" of sinusoidal components is introduced. The problem of matching spectral peaks in some "optimal" sense while allowing for this birth-death process, is generally a difficult problem. One method, which has proved to be successful is to define sine-wave tracks for frequencies that are successively "nearest-neighbors". The matching procedure is made dynamic by allowing for tracks to begin at any frame (a "birth") and to terminate at any frame (a "death"), events which are determined when successive frequencies do not fall within some "matching interval". The algorithm, although straightforward, is a rather tedious exercise in rule-based programming [McAulay and Quatieri, 1986b].

An illustration of the matching algorithm showing how the birth-death procedure accounts for rapidly varying peak locations is shown in Figure 9.4 for a speech waveform. The figure demonstrates the ability of the tracker to adapt quickly through transitory speech behavior such as voiced/unvoiced transitions and mixed voiced/unvoiced regions.

9.3.4 Synthesis

As a result of the frequency-matching algorithm described in the previous section, all of the parameters measured for an arbitrary frame k have been associated with a corresponding set of parameters for frame $k+1$. Letting $(A_l^k, \omega_l^k, \theta_l^k)$ and $(A_l^{k+1}, \omega_l^{k+1}, \theta_l^{k+1})$ denote the successive sets of parameters for the lth frequency track, then an obvious solution to the amplitude interpolation problem is to take

$$\hat{A}^k(n) = A^k + (A^{k+1} - A^k)(n/T) \qquad (9.22)$$

where $n = 0, 1, \cdots, T-1$ is the time sample into the kth frame. (The track subscript "l" has been omitted for convenience.)

Unfortunately, such a simple approach cannot be used to interpolate the frequency and phase because the measured phases θ^k and θ^{k+1} are obtained modulo 2π. Hence, phase unwrapping must be performed to ensure that the frequency tracks are "maximally smooth" across frame boundaries. The first step in solving this problem is to

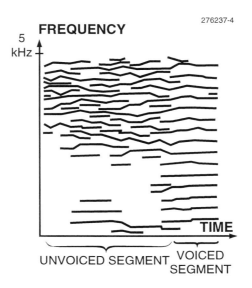

Figure 9.4 Sine-wave tracking based on frequency-matching algorithm. An unvoiced/voiced transition is illustrated. (Reprinted with permission from [McAulay and Quatieri, 1986b], ©1986, IEEE)

postulate a phase interpolation function that is a cubic polynomial[8], namely,

$$\hat{\theta}(t) = \zeta + \gamma t + \alpha t^2 + \beta t^3 + 2\pi M \tag{9.23}$$

where the term $2\pi M$, M an integer, is used to account for the phase unwrapping. It is convenient to treat the phase function as though it were a function of a continuous time variable t, with $t = 0$ corresponding to the center of frame k and $t = T$ corresponding to the center of frame $k + 1$. Since the derivative of the phase is the frequency, it is necessary that the cubic phase function and its derivative equal the measured phase and frequency measured at the frame k. Therefore, Equation (9.23) reduces to

$$\hat{\theta}(t) = \theta^k + \omega^k t + \alpha t^2 + \beta t^3 + 2\pi M \tag{9.24}$$

Since only the principal value of the phase can be measured, provision must also be made for unwrapping the phase subject to the constraint that the cubic phase function and its derivative equal the measured frequency and phase at frame $k + 1$. An explicit solution can be obtained for interpolation and phase unwrapping by invoking an additional constraint requiring that the unwrapped cubic phase function be "maximally smooth". The problem then reduces to finding that multiple of 2π that leads to the "smoothest" phase interpolation function while meeting the constraints on the frequency and phase at frame $k + 1$. It can be shown that these constraints are met for values of α and β that satisfy the relations [McAulay and Quatieri, 1986b]

$$\begin{bmatrix} \alpha(M) \\ \beta(M) \end{bmatrix} = \begin{bmatrix} \frac{3}{T^2} & \frac{-1}{T} \\ \frac{-2}{T^3} & \frac{1}{T^2} \end{bmatrix} \begin{bmatrix} \theta^{k+1} - \theta^k - \omega^k T + 2\pi M \\ \omega^{k+1} - \omega^k \end{bmatrix} \tag{9.25}$$

The phase unwrapping parameter M is then chosen to make the unwrapped phase "maximally smooth" [McAulay and Quatieri, 1986b].

Letting $\hat{\theta}_l(t)$ denote the unwrapped phase function for the lth track, then the final synthetic waveform for the kth frame will be given by

$$\hat{s}(n) = \sum_{l=1}^{L} \hat{A}_l(n) \cos[\hat{\theta}_l(n)] \tag{9.26}$$

where $\hat{A}_l(n)$ is given by Eq. (9.22), $\hat{\theta}_l(n)$ is the sampled data version of Eq. (9.24), and L is the number of sine waves[9].

This completes the theoretical basis for the new sinusoidal analysis/synthesis system. Although extremely simple in concept, the detailed analysis led to the introduction of the birth-death frequency tracker and the cubic interpolation phase unwrapping procedure chosen to ensure smooth transitions from frame-to-frame. The degree to which these new procedures result in signal synthesis of high-quality will be discussed in the next section.

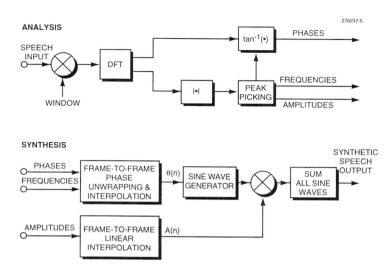

Figure 9.5 Block diagram of baseline sinusoidal analysis/synthesis. (Reprinted with permission from [McAulay and Quatieri, 1986b], ©1986, IEEE)

9.3.5 Experimental Results

A block diagram description of the analysis/synthesis system is given in Figure 9.5. A non-real-time floating-point simulation was developed in order to determine the effectiveness of the proposed approach. The signals processed in the simulations were low-pass-filtered at 5 kHz, digitized at 10 kHz, and analyzed at 2-10 ms frame intervals with a 1024-point FFT. In order to attain the time and frequency resolution required to reconstruct a large variety of signals, the duration of the analysis window $w(n)$, the number of sine-wave peaks N and the frame interval Q are adapted to the signal type. For quasi-periodic signals such as voiced speech and steady-state waveforms from wind instrumentals, using a pitch-adaptive Hamming window, having a width which was two and one-half times the average pitch, was found to be sufficient for accurate peak estimation. In reconstructing sharp attacks, on the other hand, short (and fixed) window durations and frames are used. The maximum number of peaks that are used in synthesis was set to a fixed number (≈ 80) and, if excessive peaks were obtained, only the largest peaks were used.

Although the sinusoidal model was originally designed in the speech context, it can represent almost any waveform. Furthermore, it was found that the reconstruction does not break down in the presence of interfering background. Successful reconstruction was obtained of multi-speaker waveforms, complex musical pieces, and biologic signals such as bird and whale sounds. Other signals tested include complex acoustic

AUDIO SIGNAL PROCESSING BASED ON SINUSOIDAL ANALYSIS/SYNTHESIS 359

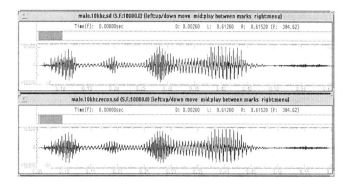

Figure 9.6 Reconstruction (lower) of speech waveform (upper) using sinusoidal analysis/synthesis.

signals from mechanical impacts as, for example, from a bouncing can, a slamming book, and a closing stapler. These signals were selected to have a variety of time envelopes, spectral resonances, and attack and decay dynamics. In addition, numerous background signals, both synthetic and real comprising random signals (e.g., synthetic colored noise or an ocean squall) and AM-FM tonal interference (e.g., a blaring siren) were tested. The synthesized waveforms were essentially perceptually indistinguishable from the originals with little modification of background.

An example of sine-wave analysis/synthesis of a speech waveform is shown in Figure 9.6, which compares the waveform for the original speech and the reconstructed speech during a number of unvoiced/voiced speech transitions. The fidelity of the reconstruction suggests that the quasi-stationarity conditions seem to be satisfactorily met and that the use of the parametric model based on the amplitudes, frequencies, and phases of a set of sine-wave components appears to be justifiable for both voiced and unvoiced speech. To illustrate the generality of the approach, an example of the reconstruction of a waveform from a trumpet is shown in Figure 9.7; while Figure 9.8 shows the reconstruction of a complex sound from a closing stapler. In each case, the analysis parameters were tailored to the signal, and the reconstruction was both visually and aurally nearly imperceptible from the original; small discrepancies are found primarily at transitions and nonstationary regions where temporal resolution is limited due to the analysis window extent. Illustrations depicting the performance of the system in the face of the interfering backgrounds are provided in [McAulay and Quatieri, 1985, Quatieri et al., 1994a].

Although high-quality analysis/synthesis of speech has been demonstrated using amplitudes, frequencies, and phases at the spectral peaks of the high-resolution STFT, it is often argued that the ear is insensitive to phase, a proposition that forms much

360   APPLICATIONS OF DSP TO AUDIO AND ACOUSTICS

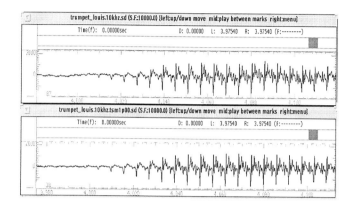

Figure 9.7   Reconstruction (lower) of trumpet waveform (upper) using sinusoidal analysis/synthesis.

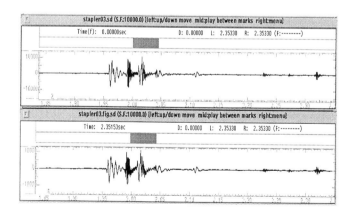

Figure 9.8   Reconstruction (lower) of waveform from a closing stapler (upper) using sinusoidal analysis/synthesis.

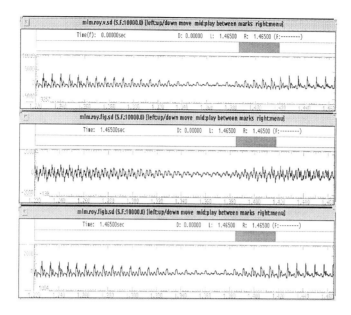

Figure 9.9 Magnitude-only reconstruction of speech (middle) is compared against original (upper) and reconstruction with the true phase estimate (bottom).

of the work in narrowband speech coders. The question arises whether or not the phase measurements are essential to sine-wave synthesis. An attempt to explore this question was made by performing "magnitude-only" reconstruction by replacing the cubic phase tracks in Equation (9.23) by a phase that was simply the integral of the instantaneous frequency. One way to do this is to make the instantaneous frequency be the linear interpolation of the frequencies measured at the frame boundaries and then perform the integration. Alternately one can simply use the quadratic frequency derived from the cubic phase via initiating the cubic phase offset at zero upon the birth of a track. While the resulting speech was very intelligible and free of artifacts, in both cases it was perceived as being different in quality from the original speech and the differences were more pronounced for low-pitched speakers (i.e. pitch $<\sim 125Hz$). An example of a waveform synthesized by the magnitude-only system (case two above) is given in Figure 9.9b. Compared to the original speech shown in Figure 9.9a and the reconstruction with the true phase estimate shown in Figure 9.9c, the synthetic speech is quite different because of the failure to maintain the true sine-wave phases. In these cases the synthetic speech was "hollow", "reverberant" and "mechanical". When the magnitude-only system was used to synthesize noisy speech, the synthetic noise took on a tonal quality that was unnatural and annoying.

For a large class of signals, therefore, the sine-wave analysis/synthesis is nearly a perceptual identity system; and the signals are expressed in terms of a functional model describing the behavior of each of its sine-wave components. The sine-wave representation therefore provides an appropriate framework for developing signal modification and enhancement techniques based on transforming each of the functional descriptors.

9.3.6 Applications of the Baseline System

Time-Scale Modification. In time-scale modification, the magnitude, frequency, and phase of the sine-wave components are modified to expand the time scale of a signal without changing its frequency characteristic. Consider a time-scale modification by a factor β. By time-warping the sine-wave frequency tracks, i.e., $\omega_l(\beta t) = \dot{\theta}_l(\beta t)$, the instantaneous frequency locations are preserved while modifying their rate of change in time [Quatieri and McAulay, 1986]. Since $d/dt[\theta_l(t\beta)/\beta] = \omega_l(\beta t)$, this modification can be represented by

$$\tilde{s}(t) = \sum_{k=1}^{N} A_l(\beta t) \cos[\theta_l(\beta t)/\beta] \quad (9.27)$$

where the amplitude functions are also time-warped. In the baseline analysis/synthesis system illustrated in Figure 9.5, the analysis and synthesis frame intervals are Q samples. In contrast, in the implementation of an analysis/synthesis system based on the model in Equation (9.27), the synthesis interval is mapped to $Q' = \rho Q$ samples. Q' is constrained to an integer value since the synthesis frame requires an integer number of discrete samples. The modified cubic phase and linear amplitude functions, derived for each sine-wave component, are then sampled over this longer frame interval. This modification technique has been successful in time-scaling a large class of speech, music, biologic, and mechanical impact signals [Quatieri and McAulay, 1986, Quatieri and McAulay, 1992, Quatieri et al., 1994b]. Nevertheless, a problem arises in the inability of the system to maintain the original sine-wave phase relations through $\theta_l(\beta t)/\beta$, and thus some signals suffers from the reverberance typical of other modification systems, as well as the "magnitude-only" reconstruction described in section 3.5. An approach to preserve *phase coherence*, and thus improve quality, imparts a source/filter phase model on the sine-wave components and is described in section 4. In spite of this drawback, the technique of Equation (9.27) remains the most general. Similar approaches, using the baseline sine-wave analysis/synthesis, have been used for frequency transformations, including frequency compression and pitch modification [Quatieri and McAulay, 1986].

Sound Splicing, Interpolation, and Extrapolation. Other applications of the baseline system are sound splicing, interpolation, and extrapolation. Sound splicing is sometimes used in music signal synthesis. Many instrumental sounds, for example, have a noisy attack and periodic steady-state and decay portions. Therefore, except for the attack, synthesis of quasi-periodic signals captures most of the sound characteristic. A realistic attack can be obtained by splicing the original attack into the synthesized sound which can be performed with an extension of the sinusoidal analysis/synthesis [Serra, 1989, Serra and Smith, 1989]. This is not possible with most synthesis techniques because the synthesized waveform does not adequately blend short-time phase at the transition. In sine-wave analysis, however, the amplitude, frequency and phase of every frequency is tracked; the amplitudes and phases of the original waveform at the attack and during steady-state can be matched and interpolated using the linear and cubic phase interpolators of section 3.4. It may also be desired to create hybrid sounds or to transform one sound to another. Splicing the attack of a flute with the sustained portion of a clarinet, for example, tests the relative importance of the temporal components in characterizing the sound. This is performed by matching the phases of the two synthesized sounds at a splice point.

In sound interpolation, in contrast to splicing temporal segments of a sound, entire frequency tracks are blended together. In music synthesis, the functional form for amplitude, frequency, and phase gives a natural means for moving from one instrument into another. For example, a cello note can be slowly (or rapidly) interpolated into the note of a French horn. A new frequency track is created as the interpolation of tracks $\omega_1(n)$ and $\omega_2(n)$ from the two instruments, represented by $\omega(n) = (N - n)\omega_1(n)/N + n\omega_2(n)/N$, over a time interval $[0, N]$. A similar operation is performed on the corresponding amplitude functions. Extensions of this basic idea have been used to interpolate passages of very different features, including pitch, vibrato and tremolo [McMillen, 1994, Tellman et al., 1995]. Such time-varying blend of different signals can also be performed in the framework of the phase vocoder [Moorer, 1977].

Finally sine-wave analysis/synthesis is also suitable for extrapolation of missing data [Maher, 1994]. Situations occur, for example, where a data segment is missing from a digital data stream. Sine-wave analysis/synthesis can be used to extrapolate the data across the gap. In particular, the measured sine-wave amplitude and phase are interpolated using the linear amplitude and cubic phase polynomial interpolators, respectively. In this way, the slow variation of the amplitude and phase function are exploited, in contrast with rapid waveform oscillations.

Tracking Vibrato. For quasi-periodic waveforms with time-varying pitch, each harmonic frequency varies synchronously[10]. The sinusoidal analysis has also been useful in tracking such harmonic frequency and amplitude modulation in speech and music and can have a clear advantage over the phase vocoder that requires the modulated frequency to reside in a single channel. The presence of vibrato in the analyzed tone may cause unwanted "cross-talk" between the bandpass filters of the phase vocoder; i.e., a partial may appear in the passband of two or more analysis filters during one vibrato cycle[11]. Sine-wave analysis, on the other hand, was found by Maher and

Beachamp [Maher and Beauchamp, 1990] to be an improvement over fixed filter bank methods for the analysis of vibrato since it is possible to track changing frequencies thereby avoiding the inter-band cross-talk problem.

An interesting application is the analysis of the singing voice where vibrato is crucial for the richness and naturalness of the sound. Here in the case of vibrato, it appears that the resonant character of the vocal tract remains approximately fixed while the excitation from the vocal folds changes frequency in some quasi-sinusoidal manner. The output spectrum of the singing voice will show frequency modulation due to the activity of the vocal folds and amplitude modulation, i.e., tremolo, due to the source partials being swept back and forth through the vocal tract resonances [McAdams, 1984].

Maher and Beauchamp have used sine-wave analysis/synthesis to trace out sine-wave amplitudes and frequencies and investigate their importance in maintaining natural vibrato. Inclusion of vibrato in sine-wave synthesis induces spectral modulation resulting in a substantial improvement over examples having constant spectra. The importance of spectral modulation due to vibrato was investigated by resynthesizing tones with measured amplitude fluctuations for each partial, but with constant partial frequencies replacing the measured frequency oscillations; and also with measured frequency oscillations, but with constant partial amplitudes. It was also observed that the phase relationship between the time-varying fundamental frequency and the amplitude fluctuation of an individual partial can be used to identify the position of that partial relative to a vocal tract resonance through tracing the resonance shape by frequency modulation[12] [Maher and Beauchamp, 1990].

9.3.7 Time-Frequency Resolution

For some audio signal processing applications, it is important that the sine-wave analysis parameters represent the actual signal components. Although a wide variety of sounds have been successfully analyzed and synthesized based on the sinusoidal representation, constraints on the analysis window and assumptions of signal stationarity do not allow accurate estimation of the underlying components for some signal classes. For example, with sine-wave analysis of signals with closely-spaced frequencies (e.g., a waveform from a piano or bell) it is difficult to achieve adequate temporal resolution with a window selected for adequate frequency resolution; while for signals with very rapid modulation or sharp attacks (e.g., a waveform from a violin or symbol), it is difficult to attain adequate frequency resolution with a window selected for adequate temporal resolution. In Figures 9.6 and 9.7, for example, the window duration was set to obtain adequate spectral resolution; a 25ms analysis window and a 10ms frame were used. In some cases these parameter setting can result in temporal smearing of signal components of very short duration or with sharp attacks, and may be perceived as a mild dulling of the sound. In Figure 9.8, on the other hand, the parameters were selected for good temporal resolution; a 7ms analysis window and a 2 ms frame were used. Such a short-duration window may prevent accurate representation of low frequencies and closely-spaced sine waves.

Prony's method has had some success in improving frequency resolution of closely-spaced sines in complex music signals over short analysis windows [McClellan, 1988, Laroche, 1989]. In one form of Prony's method, the signal is modeled over the analysis window as a sum of damped sine waves

$$x(n) = \sum_{l=1}^{L} a_l e^{\alpha_l n} \cos(2\pi f_l n + \phi_l) \qquad (9.28)$$

whose frequencies f_l, amplitudes a_l, phases ϕ_l, and damping factors α_l are estimated by various formulations of least-squared error minimization with respect to the original signal. In one two-step approach [Laroche, 1989], the frequencies and damping factors are first estimated, followed by the calculation of amplitudes and phases. These parameters are estimated over each analysis frame, frequency-matched over successive frames using the algorithm of section 3.3, and then allowed to evolve in time by polynomial interpolation. To improve estimation of a large number of sinusoids, the signal is filtered into subbands and damped sine-wave parameters are estimated separately within each band. This analysis/synthesis method has been applied to a variety of musical sounds (e.g., glockenspiel, marimba, bell, gong, piano, bass, vibraphone, tam-tam) over a 16kHz bandwidth [Laroche, 1989]. For signals with very closely-spaced frequencies, the Prony method can provide improved parameter estimation over peak-picking the STFT magnitude under a constrained window. The full analysis/synthesis however generally performs worse on long-duration complex signals [Laroche, 1989, Laroche, 1994]. One reason for the lower performance is that the "noise" and attack portion of such signals are not accurately represented because the signal does not fit the damped sine-wave model. This occurs particularly when the signal is noisy (e.g., for flute or violin signals). To account for this signal type, Laroche [Laroche, 1989] suggests generalizing the deterministic model in Equation (9.28) with a colored noise component obtained as the output of an autoregressive (all-pole) filter with a white-noise input. In section 5 of this chapter, a similar approach is described in the sine-wave context. Refinements of sine-wave parameter estimation using other variations of Prony's method have also been investigated [Laroche, 1994, Therrien et al., 1994, Victory, 1993].

One approach to address the nonstationary nature of sine-wave parameters over a constrained analysis window relies on a time-varying amplitude and frequency model. A specific model assumes a linear evolution of frequency over the analysis window[13]. With a Gaussian analysis window (this selection includes constant and exponential as special cases), Marques and Almeida [Marques and Almeida, 1989] has shown that the windowed signal can be written in complex form as

$$s(t) = \sum_{1}^{N} A_l o_l(t - t_o)$$

with

$$o_l(t) = e^{\mu k t^2} e^{\lambda k t} e^{j(\Delta_l t^2 + \omega_l t)} \quad (9.29)$$

where the center frequency for each basis is ω_l, the frequency slope is $2\Delta_l$, and the Gaussian envelope is characterized by μ_l and λ_l. The Fourier transform of $s(t)$ in Equation (9.29) is given by

$$S(\omega) = \sum_{k=1}^{N} A_l O_l(\omega) \quad (9.30)$$

where for a Gaussian window $O(\omega)$ can be evaluated analytically and also takes on a Gaussian function. This convenient form allows for estimation of the unknown parameters by iterative least-squared error minimization using a log spectral error. To relieve the multiple sine estimation problem, estimation is performed one sinusoid at a time, successively subtracting each estimate from the signal's spectrum[14]. Improvement in segmental signal-to-noise ratio was observed in speech signals whose pitch varies very rapidly.

The problem of tracking frequency variation is particularly severe for high-frequency sine waves. Since for periodic waveforms each harmonic frequency is an integer multiple of the fundamental frequency, higher frequencies will experience greater variation than low frequencies. Thus to obtain equivalent time resolution along each frequency trajectory, one would need to decrease the window duration with increasing frequency[15]. This high-frequency variation can be so great that the signal spectrum can appear noise-like in high-frequency regions, thus reducing the efficacy of spectral peak peaking. A preprocessing approach to address this problem was introduced by Ramalho [Ramalho, 1994]. In this approach, the waveform is temporally warped according to an evolving pitch estimate, resulting in a nearly monotone synthesis. A fixed analysis window is selected for a desired frequency resolution; dewarping the frequency estimate yields the desired sine-wave frequency trajectory.

9.4 SOURCE/FILTER PHASE MODEL

For signals represented approximately by the output of a linear system driven by periodic pulse or noise excitations (e.g., human speech or woodwind instruments), the sine-wave model of the previous section can be refined by imposing a source/filter representation on the sine waves components. Within this framework, the notion of *phase coherence* [Quatieri and McAulay, 1989] is introduced, becoming the basis for a number of applications, including time-scale modification and dynamic range compression. The section ends with revisiting the phase vocoder in which phase coherence is used for time-scale expansion of a class of short-duration aharmonic signals.

9.4.1 Model

In speech and certain music production models [Rabiner and Schafer, 1978a], the waveform $s(t)$ is assumed to be the output of passing a vibratory excitation waveform $e(t)$ through a linear system $h(t)$ representing the characteristics of the vocal tract or music chamber. With periodic excitation, it is assumed for simplicity that the excitation pulse shape as well as the system impulse response is part of the response $h(t)$. Usually the excitation function is represented as a periodic pulse train during harmonic sounds, where the spacing between consecutive pulses corresponds to the "pitch" of the sound, and is represented as a noise-like signal during aharmonic sounds. Alternately, the binary harmonic/aharmonic excitation model is replaced by a sum of sine waves of the form [McAulay and Quatieri, 1986b, Quatieri and McAulay, 1986, Quatieri and McAulay, 1992]

$$e(t) = \sum_{k=1}^{L} a_k(t) \cos[\Omega_k(t)]$$

where for the kth sine wave, the excitation phase $\Omega_k(t)$ is the integral of the time-varying frequency $\omega_k(t)$

$$\Omega_k(t) = \int_0^t \omega_k(\sigma) d\sigma + \phi_k \quad (9.31)$$

where ϕ_k is a fixed phase offset to account for the fact that the sine waves will generally not be in phase. L represents the number of sine waves at time t and $a_k(t)$ is the time-varying amplitude associated with each sine wave. Since the system impulse response is also time-varying, the system transfer function (i.e., the Fourier transform of $h(t)$) can be written in terms of its time-varying amplitude $M(\omega;t)$ and phase $\psi(\omega;t)$ as

$$H(\omega;t) = M(\omega;t) \exp[j\psi(\omega;t)] \quad (9.32)$$

The system amplitude and phase along each frequency trajectory $\omega_k(t)$ are then given by

$$\begin{aligned} M_k(t) &= M[\omega_k(t);t] \\ \psi_k(t) &= \psi[\omega_k(t);t] \end{aligned} \quad (9.33)$$

Passing the excitation Equation (9.31) through the linear time-varying system Equation (9.32) results in the sinusoidal representation for the waveform

$$s(t) = \sum_{k=1}^{N} A_k(t) \cos[\theta_k(t)]$$

where

$$A_k(t) = a_k(t) M_k(t)$$

and

$$\theta_k(t) = \Omega_k(t) + \psi_k(t) \qquad (9.34)$$

represent the amplitude and phase of each sine-wave component along the frequency trajectory $\omega_k(t)$. The accuracy of this representation is subject to the caveat that the parameters are slowly-varying relative to the duration of the system impulse response.

In developing a source/filter phase model, the excitation phase representation in Equation (9.31) is simplified by introducing a parameter representing the pitch pulse *onset time*. In the context of the sine-wave model, a pitch pulse occurs when all of the sine waves add coherently (i.e., are in phase). The excitation waveform is modeled as

$$e(t) = \sum_{k=1}^{N} a_k(t) \cos[(t - t_o)\omega_k] \qquad (9.35)$$

where t_o is the onset time of the pitch pulse and where the excitation frequency ω_k is assumed constant over the duration of the analysis window. Comparison of Equation (9.31) with Equation (9.35) shows that the excitation phase $\Omega_k(t)$ is linear with respect to frequency. With this representation of the excitation, the excitation phase can be written in terms of the onset time t_o as

$$\Omega_k(t) = (t - t_o)\omega_k \qquad (9.36)$$

According to Equation (9.34), the system phase for each sine-wave frequency is given by the phase residual obtained when the linear excitation phase $(t - t_o)\omega_k$ is subtracted from the composite phase $\theta_k(t)$ which consists of both excitation and system components.

$$\psi_k(t) = \theta_k(t) - (t - t_o)\omega_k \qquad (9.37)$$

Similarly, an amplitude decomposition can be made through the amplitude function in Equation (9.34) when the system function $M(\omega)$ is know or estimated.

9.4.2 Phase Coherence in Signal Modification

Time-Scale Modification. A simplified linear model of the generation of speech and certain music signal predicts that a time-scaled modified waveform takes on the appearance of the original except for a change in time scale. This section develops a time-scale modification system that preserves this shape invariance property for quasi-periodic signals, sometimes referred to as *phase coherence* [Quatieri and McAulay, 1989, Quatieri and McAulay, 1992]. A similar approach can be applied to pitch modification [Quatieri and McAulay, 1992].

Excitation/System Model: For a uniform change in the time scale, the time t_0 corresponding to the original articulation rate is mapped to the transformed time t'_0 through the mapping

$$t'_0 = \rho t_0 \qquad (9.38)$$

The case $\rho < 1$ corresponds to slowing down the rate of articulation by means of a time-scale expansion, while the case $\rho > 1$ corresponds to speeding up the rate of articulation by means of a time-scale compression. Events which take place at a time t'_0 according to the new time scale will have occurred at $\rho^{-1} t'_0$ in the original time scale.

In an idealized sine-wave model for time-scale modification, the "events" which are modified are the amplitudes and phases of the system and excitation components of each underlying sine wave. The rate of change of these events are functions of how fast the system moves and how fast the excitation characteristics change. In this simplified model, a change in the rate at which the system moves corresponds to a time scaling of the amplitude $M(\omega;t)$ and the phase $\psi(\omega;t)$. The excitation parameters must be modified so that frequency trajectories are stretched and compressed while maintaining pitch. While the excitation amplitudes $a_k(t)$ can be time scaled, a simple time scaling of the excitation phase $\Omega_k(t)$ will alter pitch. Alternatively, the transformation given by $\Omega_k(\beta t)/\beta$ maintains the pitch but results in waveform dispersion, as in the baseline sine-wave modification system of Equation (9.27), because the phase relation between sine waves is continuously being altered. A different approach to modeling the modification of the excitation phase function, which provides *phase coherence*, relies on the representation of the excitation in terms of pitch pulse locations, i.e., *onset times*, introduced in the previous section. In time-scale modification, the excitation onset times extend over longer or shorter time durations relative to the original time scale. This representation of the time-scaled modified excitation function is a primary difference from time-scale modification using the baseline system of Equation (9.26), described in section 3. The model for time-scale modification is illustrated in Figure 9.10.

Equations (9.31)-(9.38) form the basis for a mathematical model for time-scale modification. To develop the model for the modified excitation function, suppose that the pitch period $P(t)$ is time-scaled according to the parameter ρ. Then the time-scaled pitch period is given by

$$\tilde{P}(t') = P(t'\rho^{-1}) \qquad (9.39)$$

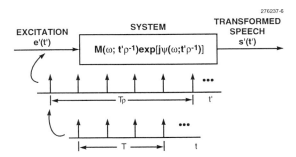

Figure 9.10 Onset-time model for time-scale modification. (Reprinted with permission from [Quatieri and McAulay, 1992], ©1992, IEEE)

from which a set of new onset times can be determined. The model of the modified excitation function is then given by

$$\tilde{e}(t') = \sum_{k=1}^{N} \tilde{a}_k(t') \cos[\tilde{\Omega}_k(t')] \qquad (9.40a)$$

where

$$\tilde{\Omega}_k(t') = (t'\rho^{-1} - t'_o)\omega_k \qquad (9.40b)$$

and where t'_o is the modified onset time. The excitation amplitude in the new time scale is the time-scaled version of the original excitation amplitude function $a_k(t)$ and is given by

$$\tilde{a}_k(t') = a_k(t'\rho^{-1}) \qquad (9.40c)$$

The system function in the new time scale is a time-scaled version of the original system function so that the magnitude and phase are given by

$$\begin{aligned} \tilde{M}_k(t') &= M_k(t'\rho^{-1}) \\ \tilde{\psi}_k(t') &= \psi_k(t'\rho^{-1}) \end{aligned} \qquad (9.41)$$

where $M_k(t)$ and $\psi_k(t)$ are given in Equations (9.33), (9.33). The model of the time-scaled waveform is then completed as

$$\tilde{s}(t) = \sum_{k=1}^{N} \tilde{A}_k(t') \cos[\tilde{\theta}_k(t')]$$

where

$$\begin{aligned} \tilde{A}_k(t'_k) &= \tilde{a}_k(t')\tilde{M}_k(t') \\ \tilde{\theta}_k(t') &= \tilde{\Omega}_k(t') + \tilde{\psi}_k(t') \end{aligned} \qquad (9.42)$$

which represent the amplitude and phase of each sine-wave component.

The time-scale modification model was developed for the harmonic case where approximate periodicity is assumed. There exist many transient aharmonic sounds such as voiced stops /b/, /d/, and /t/ and unvoiced stops /p/, /t/, and /k/ in speech, and attacks and decays in music which violate this assumption. In these cases a change in the rate of articulation may not be desired and so an adaptive rate change might be invoked [Quatieri and McAulay, 1992]. A change in the rate of articulation may be desired, however, in noise-like aharmonic sounds such as unvoiced fricatives (/s/) and voiced fricatives (/z/) in speech and turbulence in musical instruments (flute). In these cases the spectral and phase characteristics of the original waveform, and therefore the "noise-like" character of the sound, are roughly preserved in an analysis/synthesis system based on the rate-change model in Equation (9.42), as long as the synthesis interval is 10 ms or less to guarantee sufficient decorrelation of sine waves from frame-to-frame, and as long as the analysis window is 20 ms or more to guarantee approximate decorrelation in frequency of adjacent sine waves [McAulay and Quatieri, 1986b, Quatieri and McAulay, 1992].

For time-scale expansion this noise-like property however is only approximate since some slight tonality is sometimes perceived due to the determinism introduced by temporal stretching of the sine-wave amplitude and phase. In this case and when the 10ms synthesis frame condition is not feasible due to a very slow time scale (ρ greater than 2 with an analysis frame no less than 5 ms), then frequency and phase dithering models can be used to satisfy the decorrelation requirements [Quatieri and McAulay, 1992, Macon and Clements,]. One approach for reducing tonality is to add a random phase to the system phase in Equation (9.42) in only those spectral regions considered "noise-like" (more generally aharmonic)[16]. For the kth frequency track, the phase model is expressed as

$$\tilde{\theta}_k(t') = \tilde{\Omega}_k(t') + \tilde{\psi}_k(t') + b_k(\omega_c)\varphi(t;\omega_c) \qquad (9.43a)$$

where $b_k(\omega_c)$ is a binary weighting function which takes on a value of unity for a frequency track declared "aharmonic" and a value of zero for a "harmonic" track

$$\begin{aligned} b_k(\omega_c) &= 1 \ if \ \omega_k > \omega_c \\ &= 0 \ if \ \omega_k \leq \omega_c \end{aligned} \qquad (9.43b)$$

where ω_k are sine-wave frequencies estimated on each frame and $\varphi(t)$ is a phase trajectory derived from interpolating over each frame, and differently for each sine wave, random phase values selected from a uniformly distributed random variable on $[-\pi, \pi]$. The cutoff frequency ω_c is the harmonic/aharmonic cutoff for each frame and varies with a "degree of harmonicity" measure V_h; i.e.,

$$\omega_c = V_h B \qquad (9.44)$$

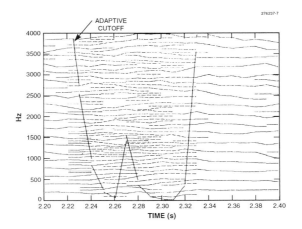

Figure 9.11 Transitional properties of frequency tracks with adaptive cutoff. Solid lines denote voiced tracks, while dashed lines denote unvoiced tracks. (Reprinted with permission from [Quatieri and McAulay, 1991], ©1991, IEEE)

over a bandwidth B and where the harmonicity measure V_h is obtained from a sine-wave-based pitch estimator [McAulay and Quatieri, 1990]. Figure 9.11 illustrates an example of track designations in a speech voiced/unvoiced transition. Alternate modification schemes for dealing with these two sound classes are discussed in section 5.5 and in chapter 7.

Analysis/Synthesis: With estimates of excitation and system sine-wave amplitudes and phases at the center of the new time-scaled synthesis frame, the synthesis procedure becomes identical to that of the baseline system of section 3.6. The goal then is to obtain estimates of the amplitudes, $\tilde{A}_k(t)$, and phases, $\tilde{\theta}_k(t)$, in Equation (9.42) at the center of the synthesis frame of duration $Q' = \rho Q$ where Q is the analysis frame interval as defined in section 3.5. Since in the time-scale modification model, the system and excitation amplitudes are simply time scaled, from Equations (9.40a) and (9.42) the composite amplitude need not be separated and therefore the required amplitude can be obtained from the sine-wave amplitudes measured on each frame m by spectral peak-picking.

$$\tilde{A}_k(m) = a_k(m) M_k(m) \tag{9.45}$$

where for convenient reference will be made to the mth analysis frame. The system and excitation phases, however, must be separated from the measured phases since the components of the composite phase $\tilde{\theta}_k(t)$ in Equation (9.42) are manipulated in different ways.

To estimate the required system phase, the excitation phase, which is estimated relative to the analysis frame, is subtracted from the measured phase samples. The first step in estimating the excitation phase is to obtain the onset time with respect to the center of the mth frame, denoted in discrete time by $n_o(m)$. At this point the phase model is dependent on the measured onset time which can introduce phase jitter, rendering the synthetic modified speech "rough" unless it is estimated with considerable accuracy[17]. Since the function of the onset time is to bring the sine waves into phase at times corresponding to the sequence of pitch pulses, it is possible to achieve the same effect simply by keeping track of successive onset times generated by a succession of pitch periods. If $n_0(m-1)$ is the onset time for frame $m-1$ and if $P(m)$ is the pitch period estimated for frame m, then a succession of onset times can be specified by

$$q_0(m;j) = n_0(m-1) + jP(m) \quad j = 1,2,3... \tag{9.46}$$

If $q_0(m;J)$ is the onset time closest to the center of the frame m, then the onset time for frame m, is defined by

$$n_0(m) = q_0(m;J) = n_0(m-1) + JP(m) \tag{9.47}$$

An example of a typical sequence of onset times is shown in Figure 9.12a. Implied in the figure is that in general there can be more than one onset time per analysis frame. Although any one of the onset times can be used, in the face of computational errors due to discrete Fourier transform (DFT) quantization effects, it is best to choose the onset time which is nearest the center of the frame, since then the resulting phase errors will be minimized. This procedure determines a relative onset time, which is in contrast to finding the absolute onset time which is the actual time at which the excitation pulses occur [McAulay and Quatieri, 1986a]. Since the onset time is obtained from the pitch period, which is derived from a fractional pitch estimate [McAulay and Quatieri, 1990], very accurate relative onset times are obtained.

The excitation phase is then given by

$$\Omega_k(m) = -[m - n_o(m)]\omega_k(m) \tag{9.48}$$

Finally, an estimate of the system phase at the measured frequencies, is computed by subtracting the estimate of the excitation phase $\Omega_k(m)$ from the measured phase at the sine-wave frequencies.

$$\tilde{\psi}_k(m) = \theta_k(m) - \Omega_k(m) \tag{9.49}$$

When the excitation phase, derived from the relative onset time, is subtracted, some residual linear phase will be present in the system phase estimate. This linear phase residual is consistent over successive frames and therefore does not pose a problem to the reconstruction since the ear is not sensitive to a linear phase shift.

The remaining step is to compute the excitation phase relative to the new synthesis interval of Q' samples. As illustrated in Figure 9.12b, the pitch periods are accumulated

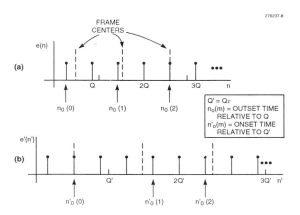

Figure 9.12 Estimation of onset times for time-scale modification: (a) Onset times for system phase; (b) Onset times for excitation phase. (Reprinted with permission from [Quatieri and McAulay, 1992], ©1992, IEEE)

until a pulse closest to the center of the mth synthesis frame is achieved. The location of this pulse is the onset time with respect to the new synthesis frame and can be written as

$$n'_0(m) = n'_0(m-1) + J'P(m) \tag{9.50}$$

where J' corresponds to the first pulse closest to the center of the synthesis frame of duration Q'. The phase of the modified excitation $\tilde{e}(n')$, at the center of the mth synthesis frame, is then given by

$$\Omega'_k(m) = -[m - n'_o(m)]\omega_k(m) \tag{9.51}$$

Finally, in the synthesizer the sine-wave amplitudes over two consecutive frames, $\tilde{A}_k(m-1)$ and $\tilde{A}_k(m)$, are linearly interpolated over the frame interval Q'. The phase components are summed and the resulting sine-wave phases, $\tilde{\theta}_k(m-1)$ and $\tilde{\theta}_k(m)$ are interpolated across the duration Q' using the cubic polynomial interpolator[18]. A block diagram of the complete analysis/synthesis system is given in Figure 9.13. An important feature of the sine-wave-based modification system is its straightforward extension to time-varying rate change, details of which are beyond the scope of this chapter [Quatieri and McAulay, 1992]. As a consequence, the corresponding analysis/synthesis system can be made to adapt to the events in the waveform (e.g., harmonic/aharmonic), which may better emulate signal generation mechanisms as discussed in the previous section. One way to achieve this adaptivity is through the measure of "harmonicity" V_h [Quatieri and McAulay, 1992]. In addition, the rate change can be controlled in a time-varying fashion independent of this adaptivity or superimposed upon it.

AUDIO SIGNAL PROCESSING BASED ON SINUSOIDAL ANALYSIS/SYNTHESIS

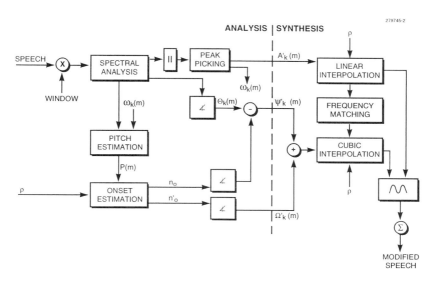

Figure 9.13 Analysis/synthesis for time-scale modification. (Reprinted with permission from [Quatieri and McAulay, 1992], ©1992, IEEE)

An example of time-scale expansion of a waveform from a trumpet is shown in Figures 9.14 where the time scale has been expanded by a factor of two and has been made to adapt according to the harmonicity measure; i.e., the rate change is given by $\rho = 1$ for $V_h \leq 0.5$ and by $\rho = [1 - V_h] + 2V_h$ for $V_h > 0.5$. The analysis frame interval was set at 5ms in these experiments so that the synthesis frame interval will be no greater than 10 ms. In this example, it is interesting to observe that, although the waveform shape is preserved, the harmonic bandwidth has decreased, as seen in the superimposed spectra. This narrowing of the bandwidth results from reduction in the vibrato rate (see Figure 9.1) by the time-scale expansion. An example of time-scale modification of speech is shown in Figure 9.15 where in this case the rate change is controlled to oscillate between a compression and expansion of about a factor of two. The results show that details of the temporal structure of the original waveforms have been maintained in the reconstructions; phase dispersion, characteristic of the original baseline sine-wave analysis/synthesis and phase vocoder, does not occur. The reconstructions are generally of high quality, maintain the naturalness of the original, and are free of artifacts. Interfering backgrounds, including typewriter and engine sounds, were also reproduced at faster and slower speeds. Although the phase model is pitch-driven, this remarkable property of robustness is likely due to the use of the original sine-wave amplitudes, frequencies, and phases in the synthesis rather, than forcing a harmonic structure onto the waveform.

376   APPLICATIONS OF DSP TO AUDIO AND ACOUSTICS

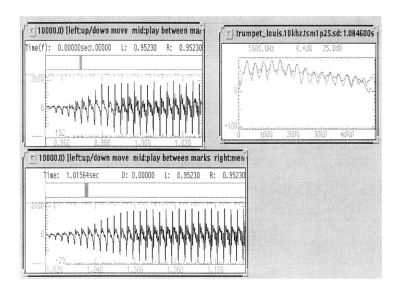

Figure 9.14   Example of time-scale modification of trumpet waveform: Original (upper left); Expansion of $\rho = 1.25$ (lower); Superimposed spectra (upper right). Top spectrum is original and lower spectrum is modified.

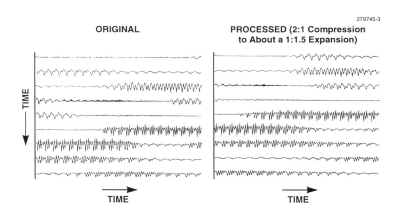

Figure 9.15   Example of time-varying time-scale modification of speech waveform. (Reprinted with permission from [Quatieri and McAulay, 1992], ©1992, IEEE)

Dynamic Range Compression. In the next application, dynamic range compression, phase coherence is again required to control sine-wave phase relations. Here however the original phase relation is not preserved, but rather intentionally modified to yield a response with a minimum peak-to-rms value relying on a phase design technique derived in a radar signal context.

Key-Fowle-Haggarty Phase Design: In the radar signal design problem, the signal is given as the output of a transmit filter whose input consists of impulses. The spectral magnitude of the filter's transfer function is specified and its phase is chosen so that the response over its duration is flat. The resulting response is an FM pulse-compressed chirp signal. This design allows the waveform to have maximum average power given a peak-power limit on the radar transmitter.

The basic unit of the radar waveform is the impulse response $h(n)$ of the transmit filter. It is expedient to view this response in the time domain as an FM chirp signal with envelope $a(n)$ and phase $\phi(n)$

$$h(n) = a(n)cos[\phi(n)] \quad 0 \leq n < L \tag{9.52a}$$

which has a Fourier transform $H(\omega)$ with magnitude $M(\omega)$ and phase $\psi(\omega)$

$$H(\omega) = M(\omega)\exp[\psi(\omega)] \tag{9.52b}$$

By exploiting the analytic signal representation of $h(n)$, Key, Fowle, and Haggarty [Key et al., 1959] have shown that, under a large time-bandwidth product constraint, specifying the two amplitude components, $a(n)$ and $M(\omega)$, in Equation (4.22) is sufficient to determine approximately the remaining two phase components. How large the time-bandwidth product must be for these relations to hold accurately depends on the shape of the functions $a(n)$ and $M(\omega)$ [Cook and Bernfeld, 1967, Fowle, 1967].

Ideally, for minimum peak-to-rms ratio in the radar signal, the time envelope $a(n)$ should be flat over the duration L of the impulse response. With this and the additional constraint that the spectral magnitude is specified (a flat magnitude is usually used in the radar signal design problem), Key, Fowle, and Hagarty's (KFH) general relation among the envelope and phase components of $h(n)$ and its Fourier transform $H(\omega)$ reduces to an expression for the unknown phase $\psi(\omega)$ as

$$\psi(\omega) = L \int_0^\omega \int_0^\beta \hat{M}^2(\alpha) d\alpha d\beta \tag{9.52a}$$

where "hat" indicates that the magnitude has been normalized by it's energy, i.e.,

$$\hat{M}^2(\omega) = M^2(\omega) / \int_0^\pi M^2(\alpha) d\alpha \tag{9.52b}$$

and where π represents the signal bandwidth in the discrete-time signal representation [Oppenheim and Schafer, 1975]. The accuracy of the approximation in Equation (9.52) increases with increasing time-bandwidth product [Quatieri and McAulay, 1991].

378   APPLICATIONS OF DSP TO AUDIO AND ACOUSTICS

Equations (9.52) shows that the resulting phase $\psi(\omega)$ depends only on the normalized spectral magnitude $\hat{M}(\omega)$ and the impulse response duration L. It can be shown that the envelope level of the resulting waveform can be determined, with the application of appropriate energy constraints, from the unnormalized spectrum and duration [Quatieri and McAulay, 1991]. Specifically, if the envelope of $h(n)$ is constant over its duration L and zero elsewhere, the envelope constant has the value

$$A = [\frac{1}{2\pi L} \int_0^\pi M^2(\omega) d\omega]^{1/2}, \quad 0 \leq nL \qquad (9.53)$$

The amplitude and phase relation in Equations (9.52) and (9.53) will be used to develop the sine-wave-based approach to peak-to-rms reduction.

Waveform Dispersion: In the above discussion, in the radar signal design context, the spectral magnitude was assumed known and a phase characteristic was estimated from the magnitude. Alternately, a filter impulse response with some arbitrary magnitude or phase might be given and the objective is to disperse the impulse response to be maximally flat over some desired duration L. This requires first removing the phase of the filter and then replacing it with a phase characteristic from the KFH calculation. This problem is similar to that required to optimally disperse a speech or music waveform. The goal is to transform the system impulse response which has some arbitrary spectral magnitude and phase into an FM chirp response which is flat over the duration of a pitch period.

A "zero-phase" version of the sine-wave system has been developed for removing the natural dispersion in the waveform during harmonic segments. Use of the system phase during aharmonic (noise-like) regions does not change the preprocessor's effectiveness in reducing the peak-to-rms ratio since these regions contributes negligibly to this measure. Moreover, the preservation of as much of the original waveform as possible helps to preserve the original quality. The sine-wave system first separates the excitation and system phase components, from the composite phase of the sine waves that make up the waveform, as was done in for time-scale modification. The system component is then removed and a zero-phase synthesis system is produced. The new KFH phase then replaces the natural phase dispersion to produce the dispersed waveform.

Applying the KFH phase to dispersion requires the estimation of the spectral magnitude $M(\omega; m)$ of the system impulse response and the pitch period of the excitation $P(m)$. The duration of the synthetic impulse response is set close to the pitch period $P(m)$ so that the resulting waveform is as "dense" as possible. The sine-wave analysis produces estimates of the spectral and pitch characteristics. The synthetic system phase derived using the KFH solution, denoted by $\psi_{kfh}(\omega; m)$, is given by

$$\psi_{kfh}(\omega; m) = \mu P(m) \int_0^\omega \int_0^\beta \hat{M}^2(\alpha; m) d\alpha d\beta \qquad (9.54)$$

where "kfh" denotes the KFH phase, where "hat" denotes that the estimated magnitude has been normalized by it's energy, and where μ, which falls in the interval [0, 1], is a scale factor to account for a possible desired reduction in the chirp duration less than a pitch period (e.g., to avoid response overlap).

Applying the KFH phase dispersion solution in the synthesis requires that the synthetic system phase in Equation (9.54), $\psi_{kfh}(\omega; m)$ which is a continuous function of frequency, be sampled along the sine-wave frequency tracks $\omega_k(m)$

$$\psi_{k,kfh}(m) = \psi_{kfh}[\omega_k(m); m] \qquad (9.55)$$

where the subscript "k, kfh" denotes the KFH phase along the kth track. The solution in Equation (9.55) is used only where the approximate periodicity assumption holds, whereas in aharmonic regions the original system phase is maintained. Therefore, the KFH phase is assigned only to those tracks designated "harmonic"[19]. The original system phase is assigned to those tracks designated "aharmonic", i.e., noise-like. Thus the phase assignment for the kth sine wave is given by

$$\theta_k(m) = \Omega_k(m) + b_k(m)\psi_k(m) + [1 - b_k(m)]\psi_{k,kfh}(m) \qquad (9.56)$$

where $b_k(m)$, defined in Equation (9.43), takes on a value of zero for a harmonic track and unity for an aharmonic track, where $\Omega_k(m)$ is the excitation phase, $\psi_k(m)$ is the original phase, and $\psi_{k,kfh}(m)$ is the synthetic phase.

An example of dispersing a synthetic periodic waveform, with fixed pitch and fixed system spectral envelope is illustrated in Figure 9.16. Estimation of the spectral envelope of the processed and original waveforms in Figure 9.16d used the straight-line spectral smoothing technique (SEEVOC) in [Paul, 1981]. For the same peak level as the original waveform, the processed waveform has a larger rms value and so has a lower peak-to-rms ratio. The vocal tract phase is modified significantly, as illustrated in Figures 9.16b and 9.16c. In Figure 9.16d the magnitude of the dispersed waveform is compared with the original magnitude and the agreement is very close, a property that is important to maintaining intelligibility and minimizing perceived distortion.

Dynamic Range Compression of Real Signals: As illustrated in Figure 9.16, the KFH phase traverses a very large range (e.g., from 0 to 300 radians) over a bandwidth of 5000 Hz. This phase calculation can then be sensitive to small measurement errors in pitch or spectrum. It is straightforward to show that for unity spectral magnitude and a one-sample error in the pitch period, the resulting change in the phase at $\omega = \pi$ is $\pi/2$, a very large change in the phase over an analysis frame interval. Such changes in phase can introduce undesirable changes in the sine-wave frequency trajectory which is manifested as a "roughness" to the synthetic signal.

To reduce large frame-to-frame fluctuations in the KFH phase, both the pitch and the spectral envelope, used by the KFH solution, are smoothed in time over successive analysis frames. The strategy for adapting the degree of smoothing to signal characteristics is important for maintaining dispersion through rapidly changing speech

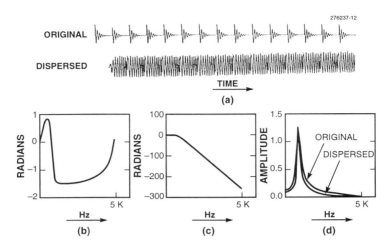

Figure 9.16 KFH phase dispersion using the sine-wave preprocessor: (a) Waveforms; (b) Original Phase; (c) Dispersed phase; (d) Spectral magnitudes. (Reprinted with permission from [Quatieri and McAulay, 1991], ©1991, IEEE)

events. In order that transitions from aharmonic to harmonic regions (and vice versa) do not severely bias the averaging process, the degree of smoothing is controlled by the harmonicity measure $V_h(m)$ and by spectral and pitch "derivatives" which reflect the rate at which these parameters are changing in time. Under the assumption that signal quality degrades when "unnatural" changes in phase occur during steady-state sounds, the degree of smoothing increases when the spectrum and pitch are varying slowly. Such a design results in little smoothing during signal state transitions or other rapidly-varying events. The importance of adaptive phase smoothing along sine-wave tracks for preserving speech quality warrants a more thorough description which is given in [Quatieri and McAulay, 1991].

A second important element in processing real signals is that of amplitude compression whose goal is to reduce envelope fluctuations in the waveform. Conventional amplitude compression methods require an estimate of the waveform envelope [Blesser, 1969]. It has been shown that in the context of sine-wave analysis/synthesis, the KFH phase relations can be used to compute the envelope of the dispersed waveform in the frequency domain and allow for a simple frequency-domain-based automatic gain control (AGC) and dynamic range compression (DRC), corresponding to "slow" and "fast" compression dynamics, respectively [Quatieri et al., 1991]. Figure 9.17 illustrates an example of processing a speech waveform in which two important changes have taken place. The peakiness with respect to a pitch period duration has been reduced via adaptive dispersion and the long-time envelope fluctuations have been

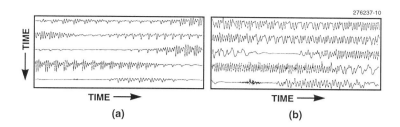

Figure 9.17 Comparison of original waveform and processed speech with combined dispersion and DRC: (a) Original; (b) Processed. (Reprinted with permission from [Quatieri and McAulay, 1991], ©1991, IEEE)

reduced by amplitude compression. The two waveforms have been peak normalized so that, since the processed waveform has a lower peak-to-rms ratio, it sounds louder than the original. In this case, after some mild clipping, about an 8 dB reduction in the peak-to-rms ratio was achieved. Although there is of course some loss in quality which is typical of this kind of processing, the resulting quality is generally acceptable in applications such as HF transmission [Quatieri et al., 1991].

9.4.3 Revisiting the Filter Bank-Based Approach

Section 2 of this chapter described the problem of phase dispersion encountered by the phase vocoder in time-scale modification. The phase vocoder therefore should be able to exploit the notion of phase coherence introduced in the previous section; in particular, the class of quasi-periodic signals may benefit from such an approach. In this section, however, a different class of short-duration aharmonic signals consisting of brief transient components that are closely spaced in time, such as from the closing of a stapler, will be investigated. The components of such signals are difficult to perceive aurally and any two such similar acoustic signals are difficult to discriminate. Thus signals of this kind may be enhanced by time-scale expansion, and in particular with the help of phase coherence imposed through the use of event onset times. It is shown that with appropriate subband phase coherence, the shape of their temporal envelope, which may play important role in auditory discrimination, can be preserved [Quatieri et al., 1994b, Quatieri et al., 1995, Quatieri et al., 1993].

Temporal Envelope. Temporal envelope is sometimes defined, typically in the context of bandpass signals, as the magnitude of the analytic signal representation[Oppenheim and Schafer, 1975]. Other definitions of temporal envelope have been proposed based on estimates of attack and release dynamics [Blesser, 1969]. One approach to time-scale modification, given the spectral envelope of a signal, is to select a Fourier-transform phase that results in a sequence with a time-scaled version of the original temporal envelope [Quatieri et al., 1993]. A close match to both the spectral envelope and modified temporal envelope, however, may not be consistent with the relationship between a sequence and its Fourier transform. Alternately, within the framework of the phase vocoder (Equations (9.9) and (9.17)), rather than attempting to maintain the temporal envelope over all time, a different approach is to maintain the subband phase relations at time instants that are associated with distinctive features of the envelope [Quatieri et al., 1994b, Quatieri et al., 1995, Quatieri et al., 1993]. As a stepping stone to the approach, the notion of *instantaneous invariance* is introduced.

Instantaneous Invariance. It is assumed that the temporal envelope of a waveform near a particular time instant $n = n_o$ is determined by the amplitude and phase of its subband components at that time [i.e., $a_k(n_o)$ and $\theta_k(n_o)$], and by the time rate of change of these amplitude and phase functions[20]. To preserve the temporal envelope in the new time scale near $n = \rho n_o$, these amplitude and phase relations are maintained at that time. Modification of amplitude and phase as in (9.15) does not maintain the phase relations; however, it does maintain the amplitudes (and relative amplitude and phase derivatives). The phase relations can be maintained by adding to each channel phase an offset, guaranteeing that the resulting phase trajectory takes on the desired phase at the specified time $n = \rho n_o$. Introduced in each channel is a phase correction that sets the phase of the modified filter output $\tilde{y}_k(n)$ at $n = \rho n_o$ to the phase at $n = n_o$ in the original time scale. Denoting the phase correction by ϕ_k, the modified channel signal becomes

$$\tilde{y}_k(n) = \tilde{a}_k(n)\cos[\rho\tilde{\theta}_k(n) + \phi_k], \qquad (9.57)$$

where $\phi_k = \theta_k(n_o) - \rho\tilde{\theta}_k(\rho n_o)$ and where $\tilde{a}_k(n)$ and $\tilde{\theta}_k(n)$ are the interpolated versions of the original amplitude and phase functions. An inconsistency arises, however, when preservation of the temporal envelope is desired at more than one time instant. One approach to resolving this inconsistency is to allow specific groups of subband components to contribute to different instants of time at which invariance is desired [Quatieri et al., 1994b, Quatieri et al., 1995, Quatieri et al., 1993].

The approach to invariance can be described by using the signal (in Figure 9.18a) that has a high- and low-frequency component, each with a different onset time. If all channels are "phase-aligned," as above, near the low-frequency event, the phase relations at the high-frequency event are changed and vice versa. For this signal, with two events of different frequency content, it is preferable to distribute the phase alignment over the two events; the high-frequency channels being phase-aligned at the first event and the low-frequency channels being phase-aligned at the second event. Equation (9.57) can then be applied to each channel group using the time instant for the respective event, thus aligning or *locking* the channel phases that most contribute to each event[21].

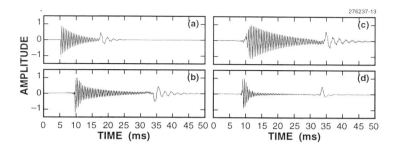

Figure 9.18 Time-scale expansion ($x2$) using subband phase correction; (a) Original; (b) Expansion with phase correction at 5ms; (c) with phase correction in clustered subbands; (d) without phase correction. (Reprinted with permission from [Quatieri et al., 1995], ©1995, IEEE)

Channels are assigned to time instants using the envelope of the filter bank outputs. Accordingly, the filter bank is designed such that each filter output reflects distinctive events that characterize the temporal envelope of the input signal. A perfect reconstruction filter bank with 21 uniformly spaced filters $h_k(n)$ was designed using a 2 ms prototype Gabor filter. Using the channel envelopes derived from the filter bank, channels are clustered according to their similarity in envelope across frequency. The *onset time* of an event is defined within each channel as the location of the maximum of the subband envelope $a_k(n)$ and is denoted by $n_o(k)$. It is assumed that the signal is of short duration with no more than two events and that only one onset time is assigned to each channel; more generally, multiple onset times would be required. A histogram of onset times is formed, and the average values of each of the two highest bins are selected as the event locations. These times are denoted by n_o^1 and n_o^2, and each of the k channels is assigned to n_o^1 or n_o^2 based on the minimum distance between $n_o(k)$ and the two possible event time instants. The distance is given by $D(p; k) = |n_o(k) - n_o^p|$ where $p = 1, 2$. The resulting two clusters of channels are denoted by $\{y_{k_p}^p(n)\}$ with $p = 1, 2$ and where for each p, k_p runs over a subset of the total number of bands. (For simplicity the subscript p will henceforth be dropped.)

Finally, based on the channel assignment a phase correction is introduced in each channel, making the phase of the modified filter output $\tilde{y}_k^p(n)$ at time $n = \rho n_o^p$ equal to the phase at the time instant $n = n_o^p$ in the original time scale. Denoting the phase correction for each cluster by ϕ_k^p, the modified channel signal becomes

$$\tilde{y}_k^p(n) = \tilde{a}_k^p(n)\cos[\rho\tilde{\theta}_k^p(n) + \phi_k^p], \tag{9.58}$$

where $\phi_k^p = \theta_k^p(n_o^p) - \rho\tilde{\theta}_k^p(\rho n_o^p)$ and where p refers to the first or second cluster.

Short-Time Processing. To process a waveform over successive frames, the filter bank modification is first applied to the windowed segment $z_l(n) = w(n - lL)x(n)$ where the frame length L is set to half the window length. The window $w(n)$ is chosen such that $\sum_l w(n - lL) = 1$, i.e., the overlapping windows form an identity. The two event time instants are saved and time-normalized with respect to the next frame. The procedure is repeated for frame $l + 1$. However, if the most recent event from frame l falls at least $L/4$ samples inside the current frame $l + 1$, then this event is designated the first event of frame $l + 1$[22]. With this condition, the second event time is found via the maximum of the histogram of the channel event occurrence times on frame $l + 1$ (excluding the previously chosen event time). Each channel is then assigned to a time instant based on the two event times and the measured occurrence times $n_o(k)$. In addition, a frame is also allowed to have no events by setting a histogram bin threshold below which a no-event condition is declared. In this case, channel phase offsets are selected to make the channel phases continuous across frame boundaries, i.e., the phase is allowed to "coast" from the previous frame.

Time-scale expansion can result in improved audibility of closely spaced components for a variety of such synthetic signals, as well as for sequences of actual complex acoustic signals of this kind (e.g., sums of rapidly damped sine waves) such as the sounds from mechanical impacts (e.g. a closing stapler), from percussion transients (e.g., tapping of a drum stick), and from biologics (e.g., dolphin clicks). An example of time-scale expansion of a sequence of transients from a closing stapler is shown in Figure 9.19, demonstrating the temporal and spectral fidelity in the time-scaled reconstruction.

In performing time-scale modification, a goal is to preserve the spectral envelope as well as the temporal envelope of the signal. Although for the signals demonstrated the original spectrum was approximately preserved in the time-scaled signal, an observed difference is the narrowing of resonant bandwidth, a change which is consistent with stretching the temporal envelope. This form of spectral preservation, however, does not always occur. In the case of a piano trill, for example, the closely-spaced events in time are also characterized by closely spaced frequency components. Our subband approach will approximately preserve the time-scaled temporal envelope of this complex signal, but harmonic smearing, due to the short (2 ms) response of the filter bank, becomes audible. A challenge remains in addressing this time-frequency resolution limitation. Further description of this approach, it's limitations, and ongoing work can be found in [Quatieri et al., 1994b] which also describes a means of using the subband approach in modifying a stochastic background component.

9.5 ADDITIVE DETERMINISTIC/STOCHASTIC MODEL

Sustained sounds such as speech vowels and steady musical tones from bowed strings and winds, though nearly periodic, have an aharmonic component that is a subtle

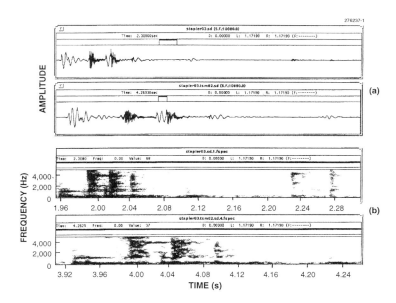

Figure 9.19 Time-scale expansion ($x2$) of a closing stapler using filter bank/overlap-add; (a) original and time-expanded waveform; (b) spectrograms of part (a). (Reprinted with permission from [Quatieri et al., 1995], ©1995, IEEE)

but essential part of the sound. This *additive* aharmonic component during sustained portions is distinct from the "pure" aharmonic sounds from certain speech fricatives and plosives, and musical attacks and percussive sounds. Sound analysis and synthesis is often deficient with regard to the accurate representation of these additive aharmonic components. Although the sine-wave model is applicable to speech and music signal representation, the harmonic and aharmonic components are sometimes difficult to distinguish and separate. One approach to separate these components, as in the previous section, assumes that they fall in separate time-varying bands. Although the adaptive harmonicity measure is effective in specifying this split-band cutoff frequency, it is however overly simple when the harmonic and aharmonic components are additively combined over the full band[23]. An alternative additive representation was developed by Serra and Smith [Serra, 1989, Serra and Smith, 1989]. This approach referred to as the "deterministic plus stochastic" sine-wave representation is the focus of this section.

9.5.1 Model

The deterministic component of the model consists of sinusoids with slowly-varying amplitude and frequency, or in musical terms the "partials" of the sound[24]. The stochastic component, sometimes referred to as the "residual," is then defined as the difference between the original and the deterministic part. In musical instruments this residual generally comprises the energy produced by the excitation mechanism (e.g., the bow in a string instrument) that is not transformed by the resonating body into stationary vibrations, plus any other energy component that is not sinusoidal in nature. In speech the residual corresponds to the turbulence generated at the glottis or some vocal tract constriction as well as plosive sounds. The sum of the two components results in the sound.

The deterministic/stochastic model therefore can be expressed as

$$s(t) = d(t) + e(t) \tag{9.59}$$

where $d(t)$ and $e(t)$ are the deterministic and stochastic components, respectively. The deterministic component $d(t)$ is of the form

$$d(t) = \sum_{l=1}^{L} A_l(t) \cos[\theta_l(t)]$$

where the phase is given by the integral of the instantaneous frequency $\omega_l(t)$

$$\theta(t) = \int_0^t \omega_l(\tau) d\tau \tag{9.60}$$

and where $\omega_l(t)$ are not necessarily harmonic and correspond to sustained sinusoidal components with slowly-varying amplitude and frequency trajectories. The deterministic component is therefore defined in the same way as in the baseline sinusoidal model except that now the sine waves are restricted to be "stable" thus modeling only the partials of the sound. In the baseline sinusoidal model, the spectral peaks need not correspond to such stable long-term trajectories. A mechanism for determining these stable sine components is described in the following section[25].

The stochastic component $e(t) = s(t) - d(t)$ can be thought of as anything not deterministic and is modeled as the output of a linear time-varying system $h(t, \tau)$ with a white-noise input $u(t)$

$$e(t) = \int_0^t h(t, \tau) u(t - \tau) d\tau \tag{9.61}$$

where when time invariant the filter $h(t, \tau)$ reduces to $h(t)$. This is a different approach to modeling a stochastic component than taken in the baseline sine-wave model where noise is represented as a sum of sine waves with random phase. In the frequency domain, however, the noise being the output of a white-noise driven linear filter, can be approximated by a Fourier transform with random phases and a smooth spectrum of the underlying linear filter [Serra, 1989]. A possible problem with this stochastic representation, as further discussed below, is that it is generated independently of the deterministic component and thus the two components may not "fuse" perceptually [Hermes, 1991]. Furthermore, not all aharmonic signals are accurately modeled by a stochastic signal; for example, sharp attacks in musical signals and plosives in speech may be better represented by a sum of coherent sine waves or the output of an impulse-driven linear system[26]. Nevertheless, this simplification leads to a useful representation for a variety of applications.

9.5.2 Analysis/Synthesis

The analysis/synthesis system which corresponds to the deterministic/stochastic model is similar to the baseline sine-wave analysis/synthesis system. The primary differences lie in the frequency matching stage for extraction of the deterministic component, in the subtraction operation to obtain the residual (stochastic component), and in the synthesis of the stochastic component.

Extraction of the deterministic component requires that frequency tracking take place based on the peaks of the STFT magnitude. Although the matching algorithm of section 3.3 can be used, this algorithm does not necessarily extract the "stable" sine components. In order to obtain the partials of the sound, Serra and Smith [Serra, 1989] developed a tracking algorithm based on prediction of tracks into the future, as well as based on past tracks, over multiple frames. In this algorithm, "frequency guides" (which is a generalization of the frequency matching window of section 3.3), advance in time through spectral peaks looking for slowly-varying frequencies according to constraint rules. When the signal is known to be harmonic, the tracker is assisted by constraining each frequency guide to search for a specific harmonic number. A unique feature of the algorithm is the generalization of the birth and death process by allowing each track to enter a "sleep" state and then reappear as part of a single track. This "peak continuation" algorithm is described in detail in [Serra, 1989]. The algorithm helps prevent the artificial breaking up of tracks, to eliminate spurious peaks, and to generate sustained sine-wave trajectories which is important in representing the time evolution of true "partials"[27].

With matched frequencies from the peak continuation algorithm, the deterministic component can be constructed using the linear amplitude and cubic phase interpolation of section 3.4. The interpolators use the peak amplitudes from the peak continuation algorithm and the measured phases at the matched frequencies. The residual component can then be obtained by subtraction of the synthesized deterministic signal from the measured signal. The method can be made flexible in defining the deterministic component; that is, the analysis parameters can be set so that the deterministic component comprises a desired number of partials. The resulting residual relies on how strict a condition is imposed on selecting partials. In addition, the attack portion of a signal is better preserved in the residual using the following subtraction algorithm

[Serra, 1989]

$$\begin{aligned}e(n) &= max[s(n), d(n)] - s(n), s(n) > 0 \\ &= min[s(n), d(n)] - d(n), s(n) \leq 0\end{aligned} \quad (9.62)$$

which prevents the residual from having a larger amplitude than the original. The attack in the residual is also improved by decreasing the frame interval since the sharpest possible attack in the deterministic component is determined by these components [Serra, 1989]; shortening of the window may also help but at the expense of frequency resolution.

The residual is obtained then by subtracting the resulting deterministic component from the measured signal. This subtraction yields an accurate residual only when the original phase is preserved. Alternately it is possible to obtain a residual when the phase of the original signal is not preserved. In this suboptimal approach, Serra derives the deterministic component by ignoring the measured phase and integrating linear frequency trajectories from matched frequencies derived from the peak continuation algorithm. The reconstruction is similar to the "magnitude-only" synthesis of section 3.4. Disregarding phase[28], however, implies that waveform subtraction of the two signals is meaningless and therefore must be performed in the frequency domain using the spectral magnitude.

The first step in obtaining the stochastic component is to compute the STFT magnitude of the original, as well as of the deterministic component. The STFT's are computed with the same analysis window, FFT size, and frame interval. The STFT magnitude of the residual is then given by

$$|E(n, \omega)| = |X(n, \omega)| - |D(n, \omega)| \quad (9.63)$$

where $X(n, \omega)$ and $D(n, \omega)$ are the STFT of the original signal and deterministic component, respectively. Attaching the measured phase[29] $\angle X(n, \omega)$ to $|E(n, \omega)|$ and applying an inverse STFT yields a short-time stochastic component for each frame. The complete "magnitude-only" analysis/synthesis system is illustrated in Figure 9.20.

In either approach to computing the residual, i.e., with or without the measured phase of the deterministic component, the residual is simplified by assuming it to be stochastic, represented by the output of a time-varying linear system as in Equation 9.61. In order to obtain a functional form for this stochastic process, a smooth function, $|\hat{E}(n, \omega)|$, is fit to the spectral magnitude of the residual, $|E(n, \omega)|$, as for example with linear predictive (all-pole) modeling or a line-segment approximation (SEEVOC [Paul, 1981]). A synthetic version of the process is then obtained by passing a white-noise sequence into a time-varying linear filter with the smooth residual spectral envelope[30].

One frame-based implementation of this time-varying linear filtering is to filter windowed blocks of white noise and overlap and add the outputs over consecutive frames. A time-varying impulse response of a linear system can be associated with

AUDIO SIGNAL PROCESSING BASED ON SINUSOIDAL ANALYSIS/SYNTHESIS 389

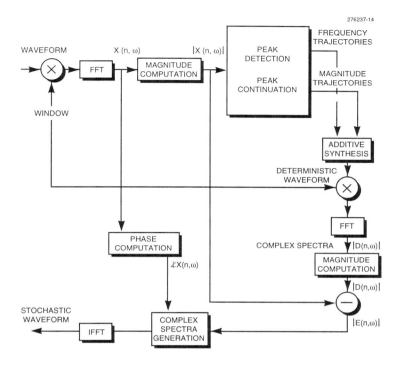

Figure 9.20 Block diagram of the deterministic plus stochastic system.

$\hat{E}(\omega, mP)$ and is given by the inverse Fourier transform of $\hat{E}(\omega; mP)^{1/2}$

$$h(n; mP) = 1/2\pi \int_0^{2\pi} B(\omega; mP)^{1/2} \exp[jn\omega]d\omega, \qquad (9.64)$$

which is a zero-phase response[31]. The synthetic stochastic signal over the mth frame is then given by

$$\hat{e}(n; mP) = h(n; mP) * [w(n - mP)u(n)], \qquad (9.65)$$

where $u(n)$, a white-noise input, is multiplied by the sliding analysis window with a frame interval of P samples. Because the window $w(n)$ and frame interval P are designed so that $\sum_m w(n - mP) = 1$. the overlapping sequences $\hat{e}(n; mP)$ can be summed to form the synthesized background

$$\hat{b}(n) = \sum_m b(n; mP). \qquad (9.66)$$

When the residual is stationary, the underlying impulse response is fixed so that as m becomes large, $h(n; mP)$ is approximately a time-invariant response $h(n)$. For large n, therefore,

$$\hat{b}(n) \approx h(n) * e(n), \qquad (9.67)$$

and thus the stochastic signal is approximately the output of a time-invariant linear filter.

9.5.3 Applications

Separation of Sound Components. The decomposition system can be applied to a wide range of sounds [Serra, 1989]. In the analysis of a guitar string, for example, the deterministic portion includes all the stable modes of vibration of the string. The residual includes the finger-noise, the attack portion of the sound, nonlinear components of the string vibration, plus other "unstable" components of the sound such as reverberation and tape hiss. In analyzing a flute sound, the residual is very prominent. Its main component is the air produced by the performer that is not transformed into periodic vibrations of the flute. Analysis of a piano tone illustrates how much noise is (surprisingly) present in a normal piano sound. The residual is a very important component of the sound and includes the noise that the fingers make when playing and the transient attack produced by the piano action. An example of the decomposition of the attack portion of a piano tone is illustrated in Figure 9.21. The deterministic/stochastic analysis/synthesis system was also applied to a very raspy voice. In this case, only a few stable harmonics were present. High-frequency harmonics are nearly completely masked by the breath noise; the deterministic analysis is unable to find them and thus are transferred to the residual. In this case, some form of harmonic continuation of the deterministic component from high frequencies to low frequencies might aid in extracting the high-frequency deterministic component [Cheng et al., 1994].

Signal Modification. The decomposition approach has been applied successfully to speech and music modification [Serra, 1989] where modification is performed differently on the two deterministic/stochastic components. Consider, for example, time-scale modification. With the deterministic component, the modification is performed as with the baseline system; using Equation (9.27), sustained (i.e., "steady") sine waves are compressed or stretched. For the aharmonic component, the white noise input lingers over longer or shorter time intervals and is matched to impulse responses (per frame) that vary slower or faster in time.

In one approach to implement synthesis of the modified stochastic component, the window length and frame interval are modified according to the rate change. A new window $w'(n)$ and frame interval P' are selected such that $\sum_m w'(n - mP') = 1$, and the factor P'/P equals the desired rate change factor ρ, which is assumed rational.

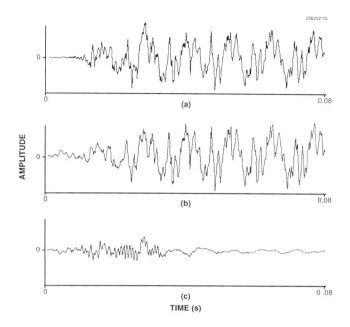

Figure 9.21 Decomposition example: (a) Attack of a piano tone; (b) Deterministic component; (c) Residual (Reprinted from [Serra, 1989], ©1989, with permission of the author)

The resulting time-scaled waveform is

$$\tilde{s}(n) = \sum_{m} h(n; mP') * [w'(n - mP')e'(n)], \qquad (9.68)$$

where $e'(n)$ is the white-noise input generated on the new time scale. As in the baseline system, when the response is stationary, for large n the synthesized residual approaches the output of a fixed linear filter $\tilde{b}(n) \approx h(n) * e'(n)$ where $h(n)$ is the time-invariant impulse response.

The advantage of separating out the additive stochastic component is that the character of noise-like sounds is not modified with the time scale; in particular, the noise may be stretched without the "tonality" that occurs in very large stretching of sine waves. On the other hand, the timbre of transient aharmonic sounds may be altered. In addition, component separation may suffer from a lack of "fusion," unlike sine-wave modification which models all components similarly. One approach to improve fusion of the two components is to exploit the property that for many sounds the stochastic component is in "synchrony" with the deterministic component. In speech, for example, the amplitude of the noise component is known to be modulated by the glottal air flow. Improved fusion can thus be obtained by temporal shaping of the stochastic component with the temporal envelope of the glottal air flow [Laroche et al., 1993a]. This is further discussed in chapter 7.

9.6 SIGNAL SEPARATION USING A TWO-VOICE MODEL

This section describes a sinusoidal-based approach to extracting two combined voices[32], as occurs with two simultaneous speakers or a musical duet. The goal is to separate and resynthesize the signal components while retaining as much of the original material as possible. In music analysis it may be desired, for example, to extract the violin part from a monaural recording of a violin and cello duet; while in speech enhancement, a low-level speaker may be sought in the presence of a loud interfering talker.

9.6.1 Formulation of the Separation Problem

The sinusoidal speech model for the single-voice case is easily generalized to the two-voice case. A waveform generated by two simultaneous voices can be represented by a sum of two sets of sine waves each with time-varying amplitudes, frequencies, and phases

$$x(n) = x_a(n) + x_b(n)$$

where

$$x_a(n) = \sum_{l=1}^{L_a} a_l(n) \cos[\theta_{a,l}(n)]$$

$$x_b(n) = \sum_{l=1}^{L_b} b_l(n) \cos[\theta_{b,l}(n)] \qquad (9.69)$$

where the sequences, $x_a(n)$ and $x_b(n)$ denote voice A and voice B, respectively. The amplitudes and phases associated with voice A are denoted by $a_l(n)$ and $\theta_{a,l}(n)$ and the frequencies are given by $\omega_{a,l}(n) = \dot{\theta}_{a,l}(n)$. A similar parameter set is associated with voice B. If the excitation is periodic, a two-voice harmonic model can be used where the frequencies associated with voice A and voice B are multiples of two underlying fundamental frequencies, $\omega_a(n)$ and $\omega_b(n)$, respectively. In the steady-state case where the excitation and system characteristics are assumed fixed over the analysis time interval, the model of Equation (9.69) is expressed as

$$s(n) = x_a(n) + x_b(n)$$

where

$$x_a(n) = \sum_{l=1}^{L_a} a_l \cos[\omega_{a,l} n + \phi_{a,l}]$$

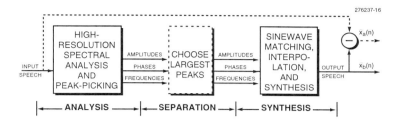

Figure 9.22   Two-voice, separation using sine-wave analysis/synthesis and peak-picking. (Reprinted with permission from [Quatieri and Danisewicz, 1990], ©1990, IEEE)

$$x_b(n) = \sum_{l=1}^{L_b} b_l \cos[\omega_{b,l} n + \phi_{b,l}] \qquad (9.70)$$

which is a useful model on which to base sine-wave analysis.

Using the model Equations (9.69) and (9.70), it is possible, as in the single voice case, to reconstruct the two-voice waveform with the baseline analysis-synthesis system illustrated in Figure 9.5. In order to obtain an accurate representation of the waveform, the number of sine waves in the underlying model is chosen to account for the presence of two voices. The presence of two voices also requires that the analysis window length be chosen to resolve frequencies more closely spaced than in the single-voice case. Due to the requirement of time resolution, however, the analysis window length is chosen to give adequate frequency resolution for the lower-pitch voice. The reconstruction, nevertheless, yields synthetic speech that is again nearly indistinguishable from the original two-voice waveform [McAulay and Quatieri, 1986b].

The capability to recover the summed waveform via the analysis-synthesis system of Figure 9.5 suggests the scheme in Figure 9.22 for recovering a desired waveform $x_b(n)$ which is of lower intensity than an interfering voice $x_a(n)$. The largest peaks of the summed spectra (the number of peaks is equal to or less than the number required to represent a single waveform) are chosen and are used to reconstruct the larger of the two waveforms. This waveform estimate is then subtracted from the combined waveform to form an estimate of the lower passage. The largest peaks of the summed spectra, however, do not necessarily represent the peaks of the spectra of the larger waveform; i.e., they will in general contain information about both passages. The parameters which form the basis for the reconstruction of the summed waveforms do not necessarily form the basis for reconstructing the individual speech waveforms. A problem with this technique, described below, is that closely spaced frequencies associated with different voices may be seen as one peak by the peak-picking process.

Alternatively, the frequency sets might be obtained by estimating a fundamental frequency for each voice and then sampling at these locations. This method is akin to *comb filtering* which extracts a waveform by processing the sum with a filter derived by placing its resonances about multiples of an assumed fundamental frequency [Shields, 1970]. Although these methods use more accurate frequency estimates than from peak-picking the summed STFTM, the accuracy of the corresponding amplitudes and phases is limited, as before, by the tendency of frequencies of the two waveforms to often be closely spaced.

Therefore, although the summed waveform $x(n) = x_a(n) + x_b(n)$ is well represented by peaks in the STFT of $x(n)$, the sine-wave amplitudes and phases of the individual waveforms are not easily extracted from these values. To look at this problem more closely, let $s_p(n)$ represent a windowed speech segment extracted from a time-shifted version of the sum of two sequences

$$s_p(n) = w(n)[x_a(n+pL) + x_b(n+pL)] \quad ; \quad \frac{-(N-1)}{2} < n < \frac{N-1}{2} \quad (9.71)$$

where the analysis window $w(n)$ is non zero over he interval $-(N-1)/2 < n < (N-1)/2$. With the model Equation (9.70), the Fourier transform of $s_p(n)$, denoted by $S_p(w)$, for $\omega > 0$, is given by the of scaled and shifted versions of the transform of the analysis window $W(\omega)$

$$\begin{aligned}
S_p(w) &= (1/2) \sum_{l=1}^{L_a} a_l \exp(j\phi_{a,l}) W(\omega - \omega_{a,l}) \\
&+ (1/2) \sum_{l=-1}^{L_a} a_l \exp(-j\phi_{a,l}) W(\omega + \omega_{a,l}) \\
&+ (1/2) \sum_{l=1}^{L_b} b_l \exp(j\phi_{b,l}) W(\omega - \omega_{b,l}) \\
&+ (1/2) \sum_{l=-1}^{L_b} b_l \exp(-j\phi_{b,l}) W(\omega + \omega_{b,l}) \quad (9.72)
\end{aligned}$$

where $W(\omega)$ denotes the Fourier transform of the time-domain window $w(n)$ and where for simplicity the time shift of the analysis frame in Equation (9.72) is assumed zero.

The success of extracting sine-wave parameters by peak-picking depends on the properties of the Fourier transform of the analysis window $W(\omega)$. The effective bandwidth of $W(\omega)$ is inversely proportional to N, the duration of the analysis window. Longer window lengths give rise to narrower spectral main lobes. If the spacing between the shifted versions of $W(\omega)$ in Equation (9.72) is such that the main lobes do not overlap, a reasonable strategy for extracting the model frequencies and performing the separation is the method of *peak-picking* . For the case of summed speech waveforms, however, this constraint is not often met since the analysis window cannot be made arbitrarily large. Even when the frequencies are known a priori when the frequencies are closely spaced, accurate estimates of the sine-wave amplitude and phases are generally not obtained.

Figure 9.23 illustrates an example where the frequencies are spaced closely enough to prevent accurate separation by the above methods. Figures 9.23a and 9.23b depict the STFTM of two steady state vowels over a 25 msec interval. The vowels have roughly equal intensity and belong to two voices with dissimilar fundamental frequencies. The STFT magnitude and phase of the summed waveforms appears in Figure 9.23c. A subset of the main lobes of the Fourier transform of the analysis windows overlap and add such that they merge to form a single composite lobe (Since the addition is complex, lobes may destructively interfere as well.). When the peak-picking strategy is applied to the STFTM of the summed speech waveform, the process may allot a single frequency to represent these composite structures. For this reason, the harmonic frequency sampling strategy will also have difficulty in recovering the individual sine-wave amplitude and phase parameters.

One approach to extracting the underlying amplitude and phase of the STFT of $x_a(n)$ and $x_b(n)$ is to detect the presence of overlap and then use the structure of the analysis window in the frequency domain to help in the separation [Parsons and Weiss, 1975]. Figure 9.23 shows that "features" in the STFTM of $x(n)$ are not, however, reliable in detecting the presence of a a single composite lobe formed by two overlapping lobes. Unique characteristics in the phase of the STFT (depicted by dotted lines in the Figure 9.23) of overlapping lobes are also difficult to determine. For example, lobe 2 in the summed spectra is characterized by both magnitude symmetry and a flat phase characteristic which characterizes either voice A or voice B. Thus any technique for separation relying on such features will be prone to error.

The discussion of the previous section suggests that the linear combination of the shifted and scaled Fourier transforms of the analysis window in Equation (9.72) must be explicitly accounted for in achieving separation. The (complex) scale factor applied to each such transform corresponds to the desired sine-wave amplitude and phase, and the location of each transform is the desired sine-wave frequency. Parameter estimation is difficult, however, due to the nonlinear dependence of the sine-wave representation on phase and frequency.

An alternate approach to separation first assumes a priori frequency knowledge, and performs a least squares fit to the summed waveform with respect to the unknown sine-wave parameters which can be written as

$$minimize \sum_n w(n)[x_a(n) + x_b(n) - s(n)]^2 \quad (9.73)$$

where the minimization takes place with respect to the unknown sine-wave amplitudes, frequencies, and phases of Equation (9.70) and where $w(n)$ is the analysis window. In the next section, the solution to Equation (9.73) is shown to be equivalent to solving for

396 APPLICATIONS OF DSP TO AUDIO AND ACOUSTICS

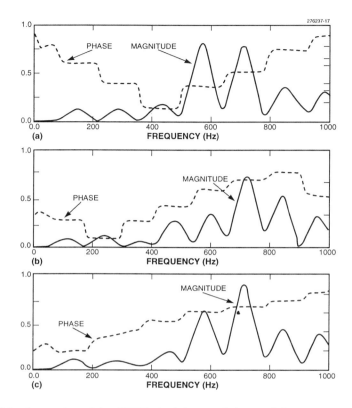

Figure 9.23 Properties of the STFT of $x(n) = x_a(n) + x_b(n)$: (a) STFT magnitude and phase of $x_a(n)$; (b) STFT magnitude and phase of $x_b(n)$; (c) STFT magnitude and phase of $x_a(n) + x_b(n)$.

the sine-wave amplitudes and phases via the linear relationships suggested by Equation (9.72). In section 6.4, this estimation problem will be simplified by constraining the frequencies to be harmonically related.

9.6.2 Analysis and Separation

In this section, the nonlinear problem of forming a least squares solution for the sine-wave amplitudes, phases, and frequencies is transformed into a linear problem. This is accomplished by assuming the sine-wave frequencies are known *apriori*, and by solving for the real and imaginary components of the quadrature representation of the sine waves, rather than solving for the sine-wave amplitudes and phases. The

AUDIO SIGNAL PROCESSING BASED ON SINUSOIDAL ANALYSIS/SYNTHESIS

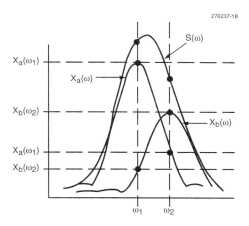

Figure 9.24 Least-squared error solution for two sine waves. (Reprinted with permission from [Quatieri and Danisewicz, 1990], ©1990, IEEE)

previous section suggests that these parameters can be obtained by exploiting the linear dependence of the STFT on scaled and shifted versions of the Fourier transform of the analysis window. This section begins with a solution based on this observation, and then show that the parameters derived by this approach represent the sine-wave parameters chosen by forming a least squares fit to the summed speech waveforms.

Figure 9.24 illustrates how the main lobes of two shifted versions of the Fourier transform of the analysis window, $W(\omega)$, typically overlap when they are centered at two closely spaced frequencies ω_1 and ω_2, corresponding to voice A and voice B, respectively, each consisting of a single frequency[33]. Fig. 9.24 suggests a strategy for separation by solving the following linear equations

$$\begin{bmatrix} 1 & W(\Delta\omega) \\ W(\Delta\omega) & 1 \end{bmatrix} \begin{bmatrix} S_a(\omega_1) \\ S_b(\omega_2) \end{bmatrix} = \begin{bmatrix} S(\omega_1) \\ S(\omega_2) \end{bmatrix} \quad (9.74)$$

where $S_a(\omega_1)$ and $S_b(\omega_2)$ denote the samples of the STFTs at known frequencies ω_1 and ω_2 and $\Delta\omega$ is the distance in frequency between them. The amplitudes and phases of $S_a(\omega_1)$ and $S_b(\omega_2)$ represent the unknown parameters of the two underlying sine waves. The STFT of the sum is denoted by $S(\omega)$. The Fourier transform of the analysis window is denoted by $W(\omega)$ with normalization $W(0) = 1$. Since the window transform is real, the matrix in the left side of Equation (9.74) is real; however, the STFT of the waveform is complex, so that the complex solution to Equation (9.74) can be obtained by solving separately the real and imaginary parts of the matrix equation. Equation (9.72) is not exact since the contribution from the Fourier transforms of the analysis window centered at $-\omega_1$ and $-\omega_2$ has not been included (In practice, the signal

398   APPLICATIONS OF DSP TO AUDIO AND ACOUSTICS

to be transformed is real and so both positive and negative frequency contributions will exist.). For simplicity, this contribution is assumed negligible.

Since from Equation (9.72), the STFT of a sum of sinusoids is a sum of shifted and scaled versions of $W(\omega)$, the two-lobe case of Figure 9.24 can be simply extended to the case where there are L overlapping lobes. Specifically, a relation can be written which reflects the linear dependence of the STFT on all L lobes [Quatieri and Danisewicz, 1990].

$$\begin{aligned} \mathbf{H}\underline{\alpha} &= 2Re[S(\underline{\omega})] \\ \mathbf{H}\underline{\beta} &= -2Im[S(\underline{\omega})] \end{aligned} \quad (9.75)$$

where,

$$\mathbf{H} = \begin{bmatrix} W(0) & W(\omega_1-\omega_2) & W(\omega_1-\omega_3) & . & W(\omega_1-\omega_L) \\ W(\omega_2-\omega_1) & W(0) & W(\omega_2-\omega_3) & . & W(\omega_2-\omega_L) \\ W(\omega_3-\omega_1) & W(\omega_3-\omega_2) & W(0) & . & W(\omega_3-\omega_L) \\ W(\omega_4-\omega_1) & W(\omega_4-\omega_2) & W(\omega_4-\omega_3) & . & W(\omega_4-\omega_L) \\ . & . & . & . & . \\ . & . & . & . & . \\ W(\omega_L-\omega_1) & W(\omega_L-\omega_2) & W(\omega_L-\omega_3) & . & W(0) \end{bmatrix} \quad (9.76)$$

and where the sinusoidal frequency vector $\underline{\omega}$ is given by

$$\underline{\omega} = [\omega_1, \omega_2, \omega_3, \ldots, \omega_L]^T \; ; \; \omega_1 \leq \omega_2 \leq \omega_3 \leq \ldots \omega_L$$

consisting of frequencies from both voice A and voice B.

The vectors $\underline{\alpha}$ and $\underline{\beta}$ consist of estimates of the unknown parameters of Equation (9.75) but in quadrature form [Danisewicz, 1987]

$$\hat{s}(n) = \sum_{l=0}^{L} \alpha_l \cos(\omega_l n) + \sum_{l=0}^{L} \beta_l \sin(\omega_l n) \quad (9.77)$$

with $\alpha_l = \hat{a}_l \cos(\hat{\phi}_l)$ and $\beta_l = -\hat{a}_l \sin(\hat{\phi}_l)$ with $L = L_a + L_b$. Equation (9.77) can also be expressed in terms of polar coordinates

$$\hat{s}(n) = \sum_{l=1}^{L} c_l \cos(\omega_l n + \hat{\phi}_l); \; c_l = \sqrt{\alpha_l^2 + \beta_l^2} \quad \hat{\phi}_l = \tan^{-1}(\beta_l/\alpha_l)$$

For voice separation, Equation (9.77) can be partitioned since the partitioning of the frequency vector $\underline{\omega}$ is assumed known a priori

$$\hat{s}(n) = \sum_{l=1}^{L_a} \hat{a}_l \cos(\omega_{a,l} n + \hat{\phi}_{a,l}) + \sum_{l=1}^{L_b} \hat{b}_l \cos(\omega_{b,l} n + \hat{\phi}_{b,l}) \quad (9.78)$$

and thus solution to the matrix Equation (9.75) yields the sine-wave amplitudes and phases of the two underlying speech components.

The preceding analysis views the problem of solving for the sine-wave amplitudes and phases in the frequency domain. Alternatively, the problem can be viewed in the time domain. It has been shown that [Quatieri and Danisewicz, 1990], for suitable window lengths, the vectors $\underline{\alpha}$ and $\underline{\beta}$ that satisfy Equation (9.75) also approximate the vectors that minimize the weighted mean square distance between the speech frame and the steady state sinusoidal model for summed vocalic speech with the sinusoidal frequency vector $\underline{\omega}$. Specifically, the following minimization is performed with respect to $\underline{\alpha}$ and $\underline{\beta}$

$$minimize \sum_{n=-(N-1)/2}^{(N-1)/2} w(n)[\hat{s}(n) - s(n)]^2 \qquad (9.79)$$

where the form of $\hat{s}(n)$ is given in Equation (9.78). The error weighting in the LSE problem Equation (9.78) is the analysis window that is used to obtain the STFT. Thus the solution in Equation (9.77) can be arrived at by two apparently different approaches; in the frequency domain, by investigating the linear dependence of the STFT on scaled and shifted versions of the Fourier transform of the analysis window, or, in the time domain, by the waveform minimization given in Equation (9.78). These two interpretations have analogies in the one-voice case where least-squares minimization in the time domain leads to the solution, developed in section 3, which chooses sine-wave amplitudes and phases at peaks in the STFT.

Figure 9.25 gives an example of the STFTM of two summed frames of vocalic speech and Figure 9.26 shows the corresponding **H** matrix. Although the **H** matrix has values that occur off of the main diagonal, these values fall off rapidly as the distance from the main diagonal increases. This property reflects the condition that overlap among the main lobes of scaled and shifted versions of the Fourier transform of the window occurs primarily between neighboring lobes of different voices (The analysis window is assumed long enough so that main lobes of a single voice do not overlap.). Occasionally, however, the **H** matrix will have a broader diagonal arising when the voices are low in pitch and the window lengths are short in duration.

9.6.3 The Ambiguity Problem

As frequencies of voice A come arbitrarily close to those of voice B, the conditioning of the **H** matrix deteriorates to where the matrix becomes singular [Quatieri and Danisewicz, 1990]. For these cases, solving the LSE problem does not permit separation. In detecting these cases, the spacing between neighboring frequencies is monitored. A single sinusoid is used to represent two sinusoids whose frequencies are closely spaced, e.g., less than 25 Hz apart. Close frequencies which satisfy this criterion are then combined as single entries in the LSE Equations (9.75) to (9.77).

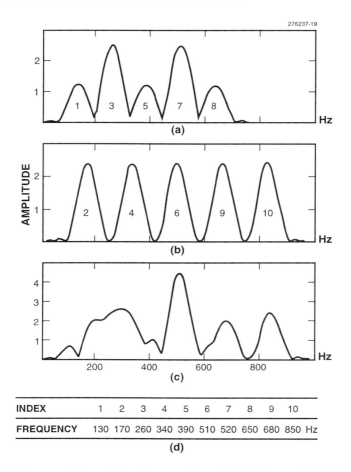

Figure 9.25 Demonstration of two-lobe overlap: (a) STFT magnitude of $x_a(n)$; (b) STFT magnitude $x_b(n)$; (c) STFT magnitude $x_a(n)+x_b(n)$; (d) Sine-wave frequencies. (Reprinted with permission from [Quatieri and Danisewicz, 1990], ©1990, IEEE)

Figure 9.27 illustrates such an example where a speaker B is 20 db below a second speaker A [34]. One lobe is missing in the reconstructed STFTM of each speaker. The monitoring procedure detected the presence of two frequencies which are close enough to cause ill-conditioning of the **H** matrix. These frequencies, merged as one in the LSE solution, were not used in the reconstruction. One strategy for resolving these ambiguities is to interpolate over the ill-conditioned regions from reliable parameter estimates in surrounding frames[35] [Quatieri and Danisewicz, 1990]. This interpolation

	1	2	3	4	5	6	7	8	9	10
1	.99	.52	.01	.00	.00	.00	.00	.00	.00	.00
2	.52	1.0	−.02	.00	.00	.00	.00	.00	.00	.00
3	.01	−.02	1.0	.01	.01	.00	.00	.00	.00	.00
4	.00	.00	.01	1.0	.34	.00	.00	.00	.00	.00
5	.00	.00	.01	.34	1.00	.00	.01	.00	.00	.00
6	.00	.00	.00	.00	.00	1.0	.96	.01	.00	.00
7	.00	.00	.00	.00	.01	.96	1.0	.01	.00	.00
8	.00	.00	.00	.00	.00	.01	.01	1.0	.70	.00
9	.00	.00	.00	.00	.00	.00	.00	.70	1.0	.00
10	.00	.00	.00	.00	.00	.00	.00	.00	.00	1.0

Figure 9.26 **H** matrix for the example in Figure 9.25 (with values quantized to two significant digits). (Reprinted with permission from [Quatieri and Danisewicz, 1990], ©1990, IEEE)

used with the solution of Equation (9.75), and together with an estimate of the pitch of the two voices (see below), has resulted in good separation of numerous *all-voiced* summed voices of about the same level [Quatieri and Danisewicz, 1990]; with different levels, either measured frequencies or *apriori* pitch is required.

Working in the music context, Maher proposed an alternate "multistrategy" approach to resolving the ambiguity problem [Maher, 1989, Maher, 1990]. The two fundamental frequencies of a duet are first estimated and used to generate the harmonic series of the two voices. The minimum spacing between adjacent partials is calculated. When a partial is at least 50 Hz away from every other partial, the component is considered "clean" and no "collision repair" occurs. A Kaiser window with a 6 dB bandwidth of 40 Hz is used. The criterion is changed appropriately if the window size is changed. If two partials are separated by less than 50 Hz but more than 25 Hz, the above least-squared error solution Equation (9.75) is applied. The 25 Hz condition assures a nonsingular solution. If two partials are separated by less than 25 Hz, Maher has proposed a number of possibilities for doing the separation. The first is to analyze the two closely-spaced frequencies in terms of a beating pattern, the amplitude modulation frequency being the frequency difference. However, if the collision is less than two or three beat periods, ($< 3/|\omega_1 - \omega_2|$), estimates of the beating parameters are not reliable. In this case, Maher interpolates in frequency, rather than in time, over the ill-conditioned region using reliable parameter estimates from neighboring harmonics. This approach, together with pitch estimation, has been

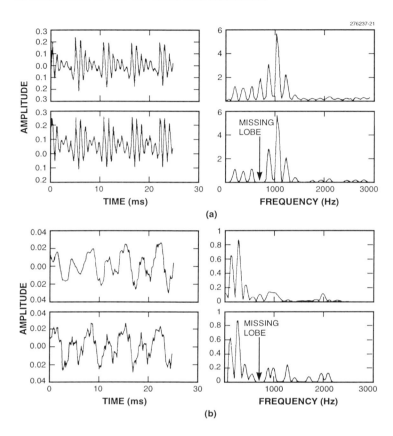

Figure 9.27 Demonstration of ill conditioning of the **H** matrix: (a) Speaker A (upper) compared to estimate of speaker A (lower); (b) Speaker B (upper) compared to estimate of speaker B (lower). (Reprinted with permission from [Quatieri and Danisewicz, 1990], ©1990, IEEE)

applied to separation of a number of musical duets including clarinet with bassoon, and tuba with trumpet [Maher, 1989, Maher, 1990].

9.6.4 Pitch and Voicing Estimation

It was noted above that reliable pitch estimation is necessary for two-voice separation using the solution of Equation (9.75). Under a harmonic assumption, since the function in Equation (9.73) is nonlinear in the fundamental frequencies, a simple closed-form solution for pitch based on a least squares approach does not exist. Under certain conditions, however, by exploiting pitch continuity the two fundamental frequencies can be tracked in time by using estimates on each analysis frame as initial estimates in a refinement procedure for the next frame [Quatieri and Danisewicz, 1990]. In particular, if the analysis frames are closely spaced, then pitch changes slowly across two consecutive frames k and $k + 1$. The pitch estimate obtained on frame k can then be used as the initial guess for estimating the pitch on frame $k + 1$. A grid search can be used as a means by which the tracking procedure be initialized. The iterative method of steepest descent [Widrow and Stearns, 1985] is then used for updating the pitch estimate on each frame. On each analysis frame, the method of steepest descent updates an initial pitch pair estimate by adding to the estimate a scaled error gradient with respect to the unknown pitch pair. The error signal for the update is the weighted least mean squared difference between the reconstructed waveform and the summed speech waveform. For a given pitch pair, the reconstructed waveform is obtained by using the amplitudes and phases that result from the solution to the LSE problem, and thus the error surface over which minimization occurs is itself a minimum for each pitch pair.

Although this pitch extraction algorithm has been applied successfully on a variety of two-voiced speech passages, the method suffers from a number of limitations including susceptibility to matrix conditioning problems and lapses from stationarity where the periodic model breaks down. Other sine-wave based approaches to the two-voice pitch estimation have been explored [Maher, 1989, Naylor and Boll, 1987]. Naylor and Porter [Naylor and Boll, 1987], for example, adapted an autoregressive spectral estimation algorithm by exploiting narrow spectral peaks in the estimate. A clustering algorithm was developed to group spectral peaks which are harmonically related to candidate pitch values. In spite of the many attempts, however, the two-pitch estimation problem is still largely unsolved with closely-spaced harmonics or large intensity differences in the two voices. Finally there is the problem of separation when the voices take on the many forms of aharmonicity described in the previous sections. In these cases, there is no known adequate solution to the two-voice separation problem.

9.7 FM SYNTHESIS

FM synthesis, first introduced by Chowning [Chowning, 1973] in a music synthesis context, is a simple and elegant approach to efficiently represent a complex sound. Although a diversion from the remainder of this chapter, FM synthesis is essential in any treatment of sine-wave representations of audio signals. With this technique it is possible to generate a sum of sine waves by FM modulating a single sine wave. Although not necessarily a physical model of sound production, it provides a perceptual model for a large class. Indeed, the perceptual accuracy of FM synthesis was sufficient to provide the basis of many electronic music synthesizers for almost two decades.

9.7.1 Principles

Basic Model. Chowning's FM model [Chowning, 1973] is given by

$$x(n) = A\sin[\omega_c n + I\sin(\omega_m n)] \quad (9.80)$$

where ω_c and ω_m are the carrier and modulation frequencies, respectively, A is the amplitude, and I is the index of modulation. The instantaneous frequency is given by the phase derivative

$$\omega(n) = \omega_c - I\omega_m \cos(\omega_m n) \quad (9.81)$$

The maximum instantaneous frequency deviation $\Delta\omega_{max}$ is therefore given by $\Delta\omega_{max} = I\omega_m$. When the modulation index I is nonzero, side frequencies occur above and below the carrier ω_c, and the number of side frequencies increases with increasing I.

These relations are expressed through the following trigonometric identity

$$\begin{aligned}\sin(\theta + a\sin(b)) &= J_0(a)\sin(\theta) \\ &+ \sum_{k=1}^{\infty} J_k(a)[\sin(\theta + kb) + (-1)^k \sin(\theta - kb)] \end{aligned} \quad (9.82)$$

where $J_k(a)$ is the kth Bessel function and where $\theta = \omega_c n$, $b = \omega_m n$, and $a = I$. Equation (9.82) shows that the side frequencies occur a distance $k\omega_m$ to the right and left of the carrier and that the amplitudes of the carrier and side frequencies are determined by the Bessel functions whose argument is the modulation index I. As the index of the side frequency increases, the modulation index must also increase for it to have significant amplitude, leading to a total bandwidth approximately equal to twice the sum of the maximum frequency deviation and the modulating frequency [Chowning, 1973]

$$\begin{aligned} BW &\approx 2(\Delta\omega_{max} + \omega_m) \\ &= 2(I+1)\omega_m \end{aligned} \quad (9.83)$$

One way to achieve a harmonic series is to set $\theta = b$. Then the trigonometric identity becomes

$$\begin{aligned}\sin(\theta + a\sin(\theta)) &= [J_0(a) - J_2(a)]\sin(\theta) \\ &+ [J_1(a) + J_3(a)]\sin(2\theta) \\ &+ [J_2(a) - J_4(a)]\sin(3\theta) \\ &+ [J_3(a) + J_5(a)]\sin(4\theta)\ldots \end{aligned} \quad (9.84)$$

so that with $\theta = \omega_c n = \omega_m n$, a harmonic series results where the amplitudes are sums of Bessel functions. More generally, when the ratio of carrier-to-modulation frequency is rational then FM synthesis results in harmonic spectra.

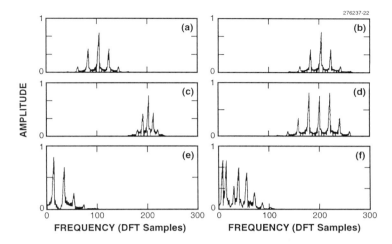

Figure 9.28 FM Synthesis with different carrier and modulation frequencies: (a) $\omega_c = 1000$, $\omega_m = 200$, and $I = 1.0$; (b) $\omega_c = 2000$, $\omega_m = 200$, and $I = 1.0$; (c) $\omega_c = 2000$, $\omega_m = 100$, and $I = 1.0$; (d) $\omega_c = 2000$, $\omega_m = 200$, and $I = 1.5$; (e) $\omega_c = 100$, $\omega_m = 200$, and $I = 1.5$; (f) $\omega_c = 200$, $\omega_m = 500/\sqrt{10}$, and $I = 1.5$.

Examples of rational carrier/modulation ratios are shown in Figures 9.28(a-d) for a variety of carrier and modulation frequencies, and modulation indices. The figure also illustrates how the position of the carrier, the bandwidth, and "fundamental frequency" (i.e., the difference between the carrier and modulation frequency) of the harmonic series can be manipulated. An interesting effect seen in Equation (9.82) is that when the carrier frequency is low, e.g., 100Hz, some side frequencies to the left of the carrier are negative in value, but become reflected back as positive frequencies with a phase shift of π (i.e., $\sin(-a) = \sin(a + \pi)$). For harmonic spectra, these reflected frequencies add to the positive frequencies with the result of potential greater complexity, but still preserving the harmonic nature of the spectrum. This effect is shown in Figure 9.28e.

The previous examples illustrate the special case of harmonic spectra. Aharmonic spectra result when the ratio of the carrier to modulation is irrational; e.g., $\omega_c/\omega_m = \sqrt{2}$. There is no "fundamental frequency" for the spectra, with aharmonic character arising from reflected side frequencies that do not fall at positive frequency locations (see Figure 9.28f).

Generalizations. An important generalization of the basic FM model is the introduction of dynamics. From Equation (9.83) an interesting property of FM synthesis is that the bandwidth of the spectrum increases with increasing modulation index I. Therefore, making the modulation index a function of time will allow spectra with

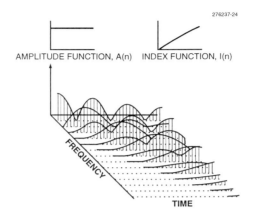

Figure 9.29 Spectral dynamics of FM synthesis with linearly changing modulation index $I(n)$. (Reprinted with permission from [Moorer, 1977], ©1977, IEEE)

dynamically changing bandwidth. As I increases, more energy goes to the sidelobes and the tone becomes less "tonal" and more "tinny" [Moorer, 1977]. By controlling I dynamically, a time-varying spectra and a resulting richness is introduced to the sound (see Figure 9.29). The resulting generalization is expressed as

$$x(n) = A(n)\sin[\omega_c n + I(n)\sin(\omega_m n)] \qquad (9.85)$$

where a time-varying amplitude envelope has also been introduced. Since the amplitude of a particular spectral component depends on Bessel functions, the amplitude change of components will depend on the specific rate of change of these functions. Further complexity is introduced by reflected side components that complicate the time-varying spectral shape.

Even more interesting sounds can be made by more complex usage of the FM formulas. With frequency modulation one might select more than one modulating waveform, or perhaps different waveforms than sinusoids. In addition, a complex amplitude modulation can be imposed. For example, one possibility is revealed in the trigonometric relation

$$e^{a\cos(b)}\sin[\theta + a\sin(b)] = \sum_{k=0}^{\infty} \frac{a^k}{k,}\sin[\theta + kb] \qquad (9.86)$$

where the sine-wave amplitudes are monotonic and a function the modulation index $a = I$. With appropriate selection of parameters this relation can yield spectra that are more "full" in comparison to Equation (9.82) which may yield a more "sparse"

AUDIO SIGNAL PROCESSING BASED ON SINUSOIDAL ANALYSIS/SYNTHESIS

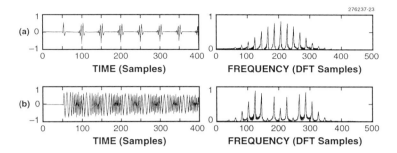

Figure 9.30 Comparison of Equation (9.82) and (9.86) for parameter settings $\omega_c = 2000$, $\omega_m = 200$, and $I = 5.0$: (a) Equation (9.86); (b) Equation (9.82). (Both waveforms and spectra are illustrated.)

spectrum (i.e., some partials may be low in amplitude). Figure 9.30 illustrates this difference for the same parameter settings in Equation (9.82) and (9.86).

A generalization of Equation (9.86) is given by

$$e^{a\cos(c)} \sin[\theta + a\sin(b) + \phi] \qquad (9.87)$$

which has a quite complicated trigonometric expression. In dynamic form this expression allows movement from sparse to full spectra, as well as harmonic and aharmonic spectra with possible *mixtures* of these two spectral classes [Moorer, 1977]. The phase angle allows another degree of freedom by influencing the manner in which reflected sidebands are combined.

9.7.2 Representation of Musical Sound

In Chowning's original work [Chowning, 1973], he explored three different classes of musical signals: brass tones, woodwind tones, and percussive-like sounds. An important characteristic of these musical sounds, in addition to their spectral composition, is the amplitude envelope of the sound and its relation to its instantaneous bandwidth.

In brass tones, all harmonics are generally present, the higher harmonics tend to increase with increasing intensity, and the rise and fall times for a typical attack and release is rapid with possible overshoot of the steady state. In creating this sound, therefore, the index of modulation $I(n)$ (hence, indirectly the bandwidth) changes in direct proportion to the amplitude envelope. To create a harmonic series ω_c is set equal to ω_m as in Equation (9.84). Figure 9.31 illustrates an example of a fast attack and decay envelope $A(n)$ (and hence rapidly-varying $I(n)$) for a trumpet-like sound which begins as nearly a pure sine wave, quickly evolves into a more complex spectrum, and ends with a narrow bandwidth [Moorer, 1977].

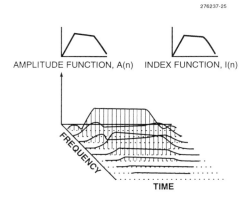

Figure 9.31 Spectral dynamics of trumpet-like sound using FM synthesis. (Reprinted with permission from [Moorer, 1977], ©1977, IEEE)

In woodwinds, on the other hand, odd harmonics can be present and the bandwidth may decrease as the attack increases, higher frequencies first becoming prominent. Odd harmonics can be created by making the modulation frequency exceed the carrier. For example, the frequency modulation $\omega_m = 2\omega_c$ gives only odd harmonics

$$\begin{aligned}\sin(\theta + a\sin(2\theta)) &= [J_0(a) + J_1(a)]\sin(\theta) \\ &+ [J_1(a) - J_2(a)]\sin(3\theta) \\ &+ [J_2(a) + J_3(a)]\sin(5\theta) \\ &+ [J_3(a) - J_4(a)]\sin(7\theta)...\end{aligned} \quad (9.88)$$

This signal representation with only odd harmonics is an approximate model for a clarinet; as with a uniform tube closed at one end and open at the other. In order to capture the time-varying envelope and bandwidth, one applies a $A(n)$ with a fast attack and slow release, and also makes the modulation index $I(n)$ inversely proportional to this envelope, thus emulating the decreasing bandwidth as a function of time.

In percussive sounds, such as a bell, gong, drum and other nonperiodic sounds, spectral components are typically aharmonic and can be simulated by forming an irrational relation between ω_c and ω_m (e.g., $\omega_m = \sqrt{2}\omega_c$). In addition, the envelope is characterized by a sharp (almost instantaneous) attack and rapid decay, and the bandwidth moves from wide to narrow. Bell-like sounds, for example, can be made by making the modulation index proportional to an amplitude envelope which has exponential decay. For a drum-like sound, the envelope decay is even more rapid than the bell, and also has a quick overshoot giving a reduced initial bandwidth, followed by a widening and then narrowing of the bandwidth.

9.7.3 Parameter Estimation

It has been assumed that AM and FM control functions can be selected, based on experience and musical knowledge, to create a variety of instrumental-like sounds with a specific *timbre*. Nevertheless, a more formal approach is desired in the AM-FM analysis and synthesis. Justice [Justice, 1979] addressed the problem of finding an analysis technique which can yield the parameters of a given FM signal; he also investigated the use of the FM synthesis model in representing a signal which consists of a sum of sine waves.

In exploring the approach of Justice, consider first Chowning's basic FM model

$$x(n) = A\cos[\omega_c n + I\cos(\omega_m n)] \tag{9.89}$$

where for convenience the cosine has replaced the sine in Equation (9.80), and suppose that the signal $s(n)$ is generated by this FM process. The first step is to write $s(n)$ in analytic form as

$$\tilde{s}(n) = s(n) + j\hat{s}(n) \tag{9.90}$$

where $\hat{s}(n)$ is the Hilbert transform of $s(n)$. In polar form Equation (9.90) is written as

$$\tilde{s}(n) = a(n)\exp[j\phi(n)] \tag{9.91a}$$

with $\phi(n)$ given as

$$\phi(n) = \omega_c n + I\cos(\omega_m n) \tag{9.91b}$$

The goal is to determine the three model parameters ω_c, I, and ω_m.

Justice proposed to first extract the linear phase term $\omega_c n$ by fitting a least-squares straight line through $\phi(n)$. The estimate $\hat{\omega}_c n$ can then be subtracted from $\phi(n)$ to obtain an estimate of the modulating function

$$\begin{aligned} s_m(n) &= \phi(n) - \hat{\omega}_c n \\ &= I\cos(\omega_m n) \end{aligned} \tag{9.92}$$

This technique can be thought of as the time-domain dual of sine-wave-based onset estimation of section 7 since the estimated slope is the carrier frequency which represents the frequency around which the spectrum of $s(n)$ is situated. In onset-time estimation, a straight line whose slope corresponds to the onset time of the system response is fit to the measured sine-wave phases in frequency [McAulay and Quatieri, 1986a].

The analytic signal representation of $s_m(n)$ is next constructed to obtain a signal of the form $\tilde{s}_m(n) = a_1(n)\exp[j\phi_1(n)]$ from which an estimate of I and also $\omega_m n$ by fitting a straight line to $\phi_1(n)$ is obtained. If the original signal does not follow the assumed FM model then a *phase residual* will result from the least-squared error

410 APPLICATIONS OF DSP TO AUDIO AND ACOUSTICS

process. In that case the modulator is itself modeled as modulated and the "nested" least-squared error approach is repeated.

This method when activated beyond the first nested modulator assumes a more general class of nested modulators. For example, a simple modulated modulator is given by

$$x(n) = A\cos[\omega_c n + I\cos(\omega_{m_1} n + I_1 \cos(\omega_{m_2} n))] \quad (9.93a)$$

and from the trigonometric identity in Equation (9.82)

$$x(n) = A\cos[\omega_c n + I\sum_k J_k(I_1)\cos(\omega_{m_1} + k\omega_{m_2})n))] \quad (9.93b)$$

Further analysis of this representation is quite complex, but it can be shown to encompass a large class of signals[36].

In addition to FM signals with nested modulators, Justice also considered a class of signals modeled by a harmonic sum of sine waves with a slowly varying amplitude, typical of many speech and music sounds

$$s(n) = a(n)\sum_{k=0}^{N} c_k \cos[k\Omega n] \quad (9.94)$$

This signal can be put in analytic form which can be expressed approximately as[37]

$$\tilde{s}(n) = e(n)\exp[\phi(n)] \quad (9.95a)$$

with

$$e(n) = a(n)|\sum_{k=0}^{N} c_k \exp[jk\Omega n]| \quad (9.95b)$$

and

$$\phi(n) = arctan[\sum_k c_k \sin(k\Omega n)/\sum_k c_k \cos(k\Omega n)] \quad (9.95c)$$

The envelope $e(n)$ of the resulting signal in general is not equal to the original envelope $a(n)$ but will "follow" $a(n)$ due to the periodicity of the second term of $e(n)$ in Equation (9.95). Now at this point we could assume a model for the phase $\phi(n)$ in the form of a nested modulator with a resulting phase residual. An alternative, as argued by Justice [Justice, 1979], is to note that $\phi(n)$ is a periodic function and express it as a Fourier series expansion; i.e.,

$$s(n) = e(n)\cos[\Omega n + \sum_k f_k \sin(k\Omega n)] \quad (9.96)$$

where the parameters of the modulator are given by f_k, thus providing a representation for FM synthesis.

9.7.4 Extensions

Although not corresponding necessarily to physical reality, the sound perception of FM synthesis can be quite realistic. There are, however, limitations of the approach. In brass and woodwind instruments, often a beating (AM) on a particular harmonic is observed. In attempting to capture this effect in the trumpet, for example, a small constant is added to the modulating frequency [Chowning, 1973]. If for example .5Hz is added, then reflected lower side frequencies do not fall on positive harmonics and so the resulting frequencies, which are closely spaced to the positive harmonics, produce beating. Although beating can be emulated through this method, there is little control over the precise beating patterns.

More generally, FM synthesis is said to not completely capture musical sound. Some of the unnatural character of FM synthesis may be attributed to the lack of vibrato[38]. Although this may capture tremolo to some extent, harmonic vibrato is not necessarily captured, a problem that remains unsolved in the context of FM synthesis. Maher and Beauchamp [Maher and Beauchamp, 1990] has suggested that since sine-wave analysis/synthesis [McAulay and Quatieri, 1986b] allows control over AM/FM on each harmonic (it can be taken in and out), the sine-wave representation may give a handle on where the FM synthesis is failing. For example, one can modify sine-wave synthesis to emulate the signal constructed from FM synthesis, particularly during sustained sounds where FM synthesis gives an unnaturally constant behavior of the synthesized tones. The importance of vibrato in musical sound construction by FM synthesis might then be determined. Another approach to improve FM synthesis may lie in the FM synthesis models proposed by Justice that offer more complexity than Chowning's original FM model.

9.8 CONCLUSIONS

A sine-wave model for an arbitrary signal class resulted in sine-wave analysis/synthesis applicable to a variety of problems in speech and music sound processing, including signal modification, separation, and interpolation. Tailoring the sine-wave representation, however, to specific signal classes can improve performance. A source/filter phase model for quasi-periodic signals led to a means to preserve sine-wave phase coherence through a pitch onset model. As a consequence, it was shown that phase coherence could also be introduced into the phase vocoder, an early form of sine-wave analysis/synthesis, to maintain the temporal envelope of certain transformed signals. The sine-wave analysis/synthesis was also tailored to signals with additive harmonic and aharmonic components by introducing a deterministic/stochastic model. Finally, a particular compact representation of a sum of sine waves, FM synthesis, was reviewed in the context of music analysis/synthesis. Although describing the many successes as well as limitations of sine-wave analysis/synthesis, this chapter, being of finite length, was not able to cover all extensions, refinements, and applications of the approach; nor was it able to adequately address a variety of unsolved problems. A natural extension of the models of section 4 and 5, for example, is to introduce phase coherence in the framework of the deterministic/stochastic model, thus preserving waveform shape during quasi-periodic segments, while keeping the noise-like quality of the stochastic component; an approach to this integration will be given in chapter 7.

One of the more challenging unsolved problems is the representation of transient events, such as attacks in musical percussive sounds and plosives in speech, which are neither quasi-periodic nor random. The residual which results from the deterministic/stochastic model generally contains everything which is not "deterministic," i.e., everything that is not sine-wave-like. Treating this residual as stochastic when it contains transient events, however, can alter the timbre of the sound, as for example in time-scale expansion. A possible approach to improve the quality of such transformed sounds is to introduce a second layer of decomposition where transient events are separated and transformed with appropriate phase coherence as developed in section 4.4. One recent method performs a wavelet analysis on the residual to estimate and remove transients in the signal [Hamdy et al., 1996]; the remainder is a broadband noise-like component.

Yet another unsolved problem is the separation of two voices that contain closely spaced harmonics or overlapping harmonic and aharmonic components. The time-varying nature of sine-wave parameters, as well as the synchrony of movement of these parameters within a voice [Bregman, 1990], may provide the key to solving this more complex separation problem. Section 6.3 revealed, for example, the limitation of assuming constant sine-wave amplitude and frequency in analysis and as a consequence proposed a generalization of Equation (9.75) based on linear sine-wave amplitude and frequency trajectories as a means to aid separation of sine waves with closely-spaced frequencies.

These and other unsolved problems often reflect the inherent tradeoff of time-frequency resolution in sine-wave analysis. Sine-wave analysis is therefore likely to benefit from multiresolution analysis/synthesis. For example, phase manipulations used in stretching low-frequency sine waves should perhaps use longer (in time) subband filters, as well as longer analysis windows and frame intervals, than used for high frequency events, since the rate of change of low-frequency phase is far smaller than for high-frequency phase. An added benefit of this frequency adaptivity is that narrow (in frequency) filters give more sine-like outputs for closely-spaced frequencies; while wide filters give better temporal resolution for closely-spaced temporal events. Ellis [Ellis, 1992], for example, has used a multi-resolution front-end in a sine-wave context in attempting to track synchronous events in two-voice signal separation. Ghitza [Ghitza, 1986] and more recently Anderson [Anderson, 1996] have exploited auditory spectral masking with constant-Q filters in attempting to reduce the number of sine waves required in sine-wave synthesis.

Finally, there remain a variety of application areas not addressed within this chapter. Use of a sum of sine-wave representation in low-rate speech coding, for example, is a vast area which warrants its own exposition [McAulay and Quatieri, 1992, McAulay and Quatieri, 1987]. Sine-wave analysis/synthesis has also been applied to signal enhancement including interference suppression and signal modification to improve signal audibility in underwater sound [Quatieri et al., 1994a, Quatieri et al., 1992]. Sine-wave-based enhancement is also being explored through signal manipulation for the hearing impaired, as for example in signal compensation for recruitment of loudness [Rutledge, 1989] and for enhancing speech in noise [Kates, 1994]. Other applications exploit the capability of sine-wave analysis/synthesis to blend signal operations, such as joint time-scale and pitch modification [Quatieri and McAulay, 1992], signal splicing [Serra and Smith, 1989, McMillen, 1994], and coding [McAulay and Quatieri, 1992, McAulay and Quatieri, 1987]; these applications include prosody manipulation in speech synthesis [Banga and Garcia-Mateo, 1995, Macon and Clements, 1996] and joint time-scale modification and speech coding for playback of stored speech in voice mail. Clearly, the generality of the sine-wave model and the flexibility of the resulting analysis/synthesis structures make sine-wave-based processing an important tool to be used in an ever-expanding set of signal processing problems.

Acknowledgements

This work was sponsored by the Department of the Air Force. Opinions, interpretations, conclusions, and recommendations are those of the authors and are not necessarily endorsed by the United States Air Force.

Notes

1. The term "harmonic" is used to refer to sinusoidal components that are nearly integral multiples of the fundamental frequency and are not necessarily exactly harmonic. The term "partials" refers to slowly-varying sinusoidal components that have frequencies with arbitrary value. This terminology is widely used in the speech and music literature [Moorer, 1977].

2. An approximate derivative can be obtained by first differencing the discrete-time unwrapped phase.

3. The filters $h_k(n)$ can be designed to satisfy a perfect reconstruction constraint $\sum_k h_k(n) = \delta(n)$ where $\delta(n)$ is the unit sample sequence. One sufficient condition for perfect reconstruction is that the length of $h(n)$ be less than the frequency sampling factor, i.e., $S < R$ [Nawab and Quatieri, 1988a]. With this perfect reconstruction constraint, the signal $x(n)$ can be recovered as $x(n) = \sum_k x(n) * h_k(n) = x(n) * \sum_k h_k(n) = x(n)$.

4. In applications, it is often advantageous to express each analysis output in terms of the channel phase derivative $\dot{\theta}(n, \omega_k)$, and initial phase offset which for a single sine wave are given by $\dot{\theta}(n, \omega_k) = (\omega_o - \omega_k)$ and $\theta(0, \omega_k) = \theta_o$, respectively [Flanagan and Golden, 1966]. The unwrapped phase can be obtained by integration of the phase derivative which is added to the carrier phase $\omega_k n$. Since for a single sine-like input $\dot{\theta}(n, \omega_k)$ is slowly-varying, this representation is particularly useful in speech compression.

5. Noninteger rate change can be performed by combined interpolation and decimation.

414   APPLICATIONS OF DSP TO AUDIO AND ACOUSTICS

6. Pitch estimation is performed using a sine-wave based approach [McAulay and Quatieri, 1990]. A fundamental frequency is obtained such that a harmonic set of sine waves is a "best fit" to the measured set of sine waves. The accuracy of the harmonic fit becomes a measure of the degree to which the analyzed waveform segment is periodic; i.e., a measure of "harmonicity".

7. An alternate approach, developed in a speech coding context [McAulay and Quatieri, 1986a, McAulay and Quatieri, 1992], uses a harmonically-dependent set of sine waves with a random phase modulation. Yet another related technique [Marques and Almeida, 1988], represents the signal by a sum of adjacent narrowband sines of uniformly-spaced center frequency, with random amplitude and frequency modulation.

8. The idea of applying a cubic polynomial to interpolate the phase between frame boundaries was independently proposed by [Almeida and Silva, 1984b] for use in their harmonic sine-wave synthesizer.

9. An alternative synthesis method is motivated by setting the matching window of the frequency tracker to zero. In this case, the synthesis of Equation (9.26) can be shown to be equivalent to first generating constant-amplitude and frequency sine waves, weighting the sum of these sine waves with a triangular window of length twice the frame width, and then overlapping and adding the windowed segments from successive frames [McAulay and Quatieri, 1986b]. Consequently, an FFT-based *overlap-and-add* synthesis can be formulated by filling FFT buffers with complex sine-waves, Fourier transform inverting the FFT buffers, and adding triangularly windowed short-time segments. This implementation can be particularly important in applications where computational efficiency is important [McAulay and Quatieri, 1988].

10. Since each harmonic is a multiple of the time-varying fundamental, higher harmonics vibrato have a larger bandwidth than lower harmonics. With rapid pitch vibrato, the temporal resolution required for frequency estimation increases with harmonic number. One approach to improve resolution time-warps the waveform inversely to pitch to remove vibrato [Ramalho, 1994]. This approach may also be useful in reducing channel cross-talk within the phase vocoder.

11. The time-bandwidth product, constraining the minimum analysis filter bandwidth to be inversely proportional to the observation time interval, must also be confronted.

12. Mcadams [McAdams, 1984] hypothesizes that the tracing of resonant amplitude by frequency modulation contributes to the distinctness of the sound in the presence of competing sound sources.

13. This linear evolution model may be even more important in the context of the signal separation problem described in section 6.

14. An iterative approach was also developed by George [George, 1991] for improving the estimation of low-level sine waves in spectra of wide dynamic range. A least-squared error minimization of sine-wave parameters was formulated as an analysis-by-synthesis procedure, successively subtracting each sine wave estimate from the original signal in order of its magnitude.

15. Multi-resolution analysis can be provided by the constant-Q property of the wavelet transform [Mallat and Hwang, 1992]. Although a short window for temporal tracking reduces frequency resolution, the human ear may not require as high a frequency resolution in perceiving high frequencies as for low frequencies. Ellis [Ellis, 1992, Ellis et al., 1991] exploited this property of auditory perception in developing a constant-Q analysis within the sine-wave framework for tracking signal fine structure for signal separation.

16. In adding synthetic harmonic and aharmonic components, it is important that the two components "fuse" perceptually [Hermes, 1991]; i.e., that the two components are perceived as emanating from the same sound source. Through informal listening, sine-wave phase randomization appears to yield a noise component that "fuses" with the harmonic component of the signal.

17. Estimation of the absolute onset times can be performed using a least-squared error approach to finding the unknown $n_o(m)$ [McAulay and Quatieri, 1986a]. Although this method can yield onset times to within a few samples, this slight inaccuracy is enough to generate a "rough" quality to the synthesis. It is interesting to note that this approach to onset estimation will be seen in section 7 as the frequency-domain dual to the time-domain estimation of the carrier frequency required in FM synthesis.

18. A computationally efficient FFT overlap-add implementation of the synthesis has been formulated by George [George, 1991].

19. Schroeder also derived an approach to "optimally" flatten a harmonic series [Schroeder, 1986, Schroeder, 1970a]. This method however requires exact harmonicity and can be shown to be a special case of the KFH phase dispersion formula.

20. A more formal approach requires a strict definition of temporal envelope

21. A recently proposed alternate approach to reducing dispersion in the phase vocoder introduces phase locking by replacing a channel phase with the phase of the weighted average of itself and its two adjacent neighbors [Puckette, 1995]. Thus the strongest of the three channels dominates the channel phase. This and other methods of phase locking are described in chapter 7.

22. The notion of synchronizing pulses over adjacent frames in overlap-add for time-scale modification was first introduced by Roucos and Wilgus in the speech context [Roucos and Wilgus, 1985]. This method relies on cross-correlation of adjacent frames to align pulses and not on phase synchronization of a subband decomposition.

23. Griffin and Lim [Griffin, 1987] generalized this split-band approach to multibands; multibands, however, do not adequately model *additively* combined harmonic and aharmonic components.

24. Recall that a partial refers to a sinusoidal component of a sound that usually corresponds to a mode of vibration of the producing sound system and is not necessarily harmonic.

25. As an alternative deterministic/stochastic separation scheme, Therrien [Therrien et al., 1994] has introduced an adaptive ARMA model for sample-by-sample tracking of sine-wave amplitude and frequencies of the deterministic signal. This component is subtracted from the original signal and parameters of the resulting residual are also adaptively estimated using a ARMA representation. This technique is being applied to signal modification to synthetically expand limited training data for signal classification [Therrien et al., 1994].

26. Alternatively, a more realistic music attack or speech plosive may be obtained by *splicing* into the synthesized deterministic component using the method described in section 3.6. Splicing of the actual attack has been shown to significantly improve the sound quality for a number of musical sounds including the piano and marimba [Serra, 1989]. Hybrid sounds can also be created by matching the sine-wave phases of the attack of one sound with the phases of the deterministic component of a second sound at a splice point.

27. A sine-wave frequency tracker has also been developed using hidden Markov modeling of the time evolution of sine-wave frequencies over multiple frames. This approach is particularly useful is tracking crossing frequency trajectories which can occur in complex sounds[Depalle et al., 1993].

28. Disregarding the measured phase also implies that the deterministic component, as well as its transformations, will suffer from waveform dispersion.

29. The residual, being assumed stochastic, is characterized by second-order statistics; i.e., the specific phase of the residual is of no importance. Nevertheless, this selection must be made carefully since spectral phase significantly influences the temporal properties of the signal.

30. An alternate approach to stochastic synthesis introduced by Serra [Serra, 1989] appends a random phase to the envelope $|\hat{E}(n,\omega)|$ and applies a time-domain window to the inverse STFT since the random phase may cause a splattering of the signal outside of the a desired short-time interval. An overlap and add procedure as in Equation (9.66) is then performed. Although this method is similar in spirit to the convolutional approach, the phase correlation of the resulting signals differ because the operations of windowing and convolution do not commute. The perceptual differences in the two approachs requires further study. Another alternative is to find a sequence whose STFT approximates in a least-squared error sense the STFT constructed with the spectral magnitude of the residual and with random phase [Griffin and Lim, 1984a]. This approach results in a STFT phase which may be close to the random phase, but which also meets the desired short-time constraint.

31. A minimun-phase version of the filter can also be constructed.

32. The term *voice* refers to a single speaker or single musical instrument.

33. Note that the stationarity assumption results in a spectrum consisting of identically shaped (window) pulses placed at the sine-wave frequencies. Most signals of interest however are generally nonstationary (e.g., the frequencies may change over the window extent due to amplitude and frequency modulation), and so the window transform may deviate from this fixed shape. Naylor and Porter [Naylor and Boll, 1986] have developed an extension of the approach of this section that accounts for the deviation from the ideal case.

34. The frequency estimates used in the solution Equation (9.75) were obtained by peak-picking the STFTM of each separate waveform. A 4096 FFT was found to give sufficient frequency resolution for adequate separation. The Gauss Siedel iterative method [Strang, 1980] was then used in solving Equation (9.75). Convergence of this algorithm is guaranteed for positive definite matrices, a property of the matrices in the least squares problem. The vector obtained by sampling the STFT at the sine-wave frequencies was used as an initial guess in the iterative algorithm.

35. An alternate approach is to impose continuity constraints *prior* to estimation, thus utilizing the sine waves that would be eliminated *aposteriori* due to ill-conditioning. For example, a linear model for each frequency trajectory can be shown to lead to a generalization of Equation (9.77). Such an approach may lead to more robust separation with the presence of closely spaced frequencies.

36. Consider signals of the form $x(n) = \cos(\omega n + \phi(n))$ with $\phi(n) = \sum_{k=1}^{N} I_k \cos(\omega_k n)$. Then the expansion of $x(n)$ takes the form $x(n) = \sum_{n_1} \sum_{n_2} \cdots \sum_{n_N} J_{n_1}(I_1) J_{n_2}(I_2) ... J_{n_N}(I_N) \cos[(\omega + n_1\omega_1 + n_2\omega_2 + ... + n_N\omega_N)n]$ which yields very complex spectra [Justice, 1979].

37. We assume for convenience that the analytic form of the signal equals its quadrature representation.

38. Capturing vibrato through adjusting the carrier/modulation frequency ratio requires very precise control which is difficult to achieve.

10 PRINCIPLES OF DIGITAL WAVEGUIDE MODELS OF MUSICAL INSTRUMENTS

Julius O. Smith III

Center for Computer Research in Music and Acoustics (CCRMA)
Music Department, Stanford University
Stanford, California 94305
http://www-ccrma.stanford.edu

jos@ccrma.stanford.edu

Abstract: Basic principles of digital waveguide modeling of musical instruments are presented in a tutorial introduction intended for graduate students in electrical engineering with a solid background in signal processing and acoustics. The vibrating string is taken as the principal illustrative example, but the formulation is unified with that for acoustic tubes. Modeling lossy stiff strings using delay lines and relatively low-order digital filters is described. Various choices of wave variables are discussed, including velocity waves, force waves, and root-power waves. Signal scattering at an impedance discontinuity is derived for an arbitrary number of waveguides intersecting at a junction. Various computational forms are discussed, including the Kelly-Lochbaum, one-multiply, and normalized scattering junctions. A relatively new three-multiply normalized scattering junction is derived using a two-multiply transformer to normalize a one-multiply scattering junction. Conditions for strict passivity of the model are discussed. Use of commutativity of linear, time-invariant elements to greatly reduce computational cost is described. Applications are summarized, and models of the clarinet and bowed-string are described in some detail. The reed-bore and bow-string interactions are modeled as nonlinear scattering junctions attached to the bore/string acoustic waveguide.

10.1 INTRODUCTION

Music synthesizers are beginning to utilize physical models of musical instruments in their sound-generating algorithms. The thrust of this trend is to obtain maximum expressivity and sound quality by providing all of the responsiveness of natural musical instruments. This is happening at a time when most synthesizers are based on "sampling" of the acoustic waveform. While sample-playback instruments sound great on the notes that were recorded, they tend to lack the expressive range of natural instruments.

Another potential application for physical models of sound production is in *audio compression*. Compression ratios can be enormous when coding parameters of a physical model of the sound source. High quality audio compression techniques such as used in MPEG are presently based on psychoacoustically motivated *spectral models* which yield up to an order of magnitude of "transparent" compression (see the chapter by Brandenburg or [Bosi et al., 1996a]). Physical models, on the other hand, can achieve much higher compression ratios for specific sounds. By combining model-based and spectral-based compression techniques, large average compression ratios can be achieved at very high quality levels.

The Musical Instrument Digital Interface (MIDI) format provides an example of the profound compression ratios possible for certain sounds by encoding only synthesizer control parameters. For example, two or three bytes of MIDI data can specify an entire musical note. In future audio compression standards, a compressed audio stream will be able to switch among a variety of compression formats. When arbitrary decompression algorithms can be included in the compressed-audio data stream, model-based compression will be fully enabled. In terms of existing standards, for example, one could extend MIDI to provide for MPEG-2 audio segments and "instrument definitions" (synthesis algorithms) written in Java (performance issues aside for the moment). In this context, instrument definitions serve a role analogous to "outline fonts" in a page description language such as PostScript, while sampled audio segments are more like "bit-map fonts." General, self-defining, instrument-based, audio synthesis scripts have been in use since the 1960s when the Music V program for computer music was developed [Mathews, 1969].

10.1.1 Antecedents in Speech Modeling

The original Kelly-Lochbaum (KL) speech model employed a ladder-filter with delay elements in physically meaningful locations, allowing it to be interpreted as a discrete-time, traveling-wave model of the vocal tract [Kelly and Lochbaum, 1962]. Assuming a reflecting termination at the lips, the KL model can be transformed via elementary manipulations to modern ladder/lattice filters [Smith, 1986b]. The early work of Kelly and Lochbaum appears to have been followed by two related lines of development: *articulatory speech synthesis* and *linear predictive coding* (LPC) of speech.

The most elaborate physical models for speech production are developed in the field of articulatory speech synthesis [Keller, 1994, Deller Jr. et al., 1993]. While they represent the forefront of our understanding of speech production, they are generally too complex computationally to yield practical speech coding algorithms at present. However, there have been ongoing efforts to develop low bit-rate speech coders based on simplified articulatory models [Schroeter and Sondhi, 1994, Flanagan et al., 1980]. The main barrier to obtaining practical speech coding methods has been the difficulty of estimating vocal-tract shape given only the speech waveform.

LPC is speech coding technique which has enjoyed widespread usage and intensive study [Atal and Hanauer, 1971, Markel and Gray, 1976, Campbell Jr. et al., 1990, Deller Jr. et al., 1993]. The allpole filter used in LPC synthesis is often implemented as a ladder or lattice digital filter [Markel and Gray, 1976] which is not far from having a physical interpretation. However, as normally implemented, the delay in samples from the filter input to its output equals the filter order, while the delay from the output back to the input is zero. This non-physical, asymmetric distribution of delay precludes building upon the ladder/lattice filter as a physical modeling element; for example, branching the filter to add a nasal tract is not immediately possible in the LPC synthesis model [Lim and Lee, 1996] while it is straightforward in the Kelly-Lochbaum model [Cook, 1990]. The benefits of moving all of the reverse delays to the forward path are that (1) the sampling rate can be reduced by a factor of two, and (2) the reflection coefficients can be uniquely computed from the autocorrelation function of the speech waveform by means of orthogonal polynomial expansions. While these advantages make sense for practical *coding* of speech, they come at the price of giving up the physical model.

Instead of a physical model, the LPC signal model is better regarded as a *source-filter* signal representation. The *source* is typically taken to be either a periodic impulse train, corresponding to voiced speech, or white noise, for unvoiced speech. The *filter* implements vocal tract resonances, or *formants* of speech, and part of this filter can be interpreted as arising from a physical model for the vocal tract consisting of a piecewise cylindrical acoustic tube [Markel and Gray, 1976, Rabiner and Schafer, 1978b]. However, the same filter must also represent the glottal pulse shape, since it is driven by impulses in place of physically accurate glottal pulses. Since an LPC filter encodes both vocal-tract and glottal-pulse characteristics, it is not an explicit physical model of either. However, it remains closely related structurally to the Kelly-Lochbaum model which does have a physical interpretation.

Another way to characterize the LPC filter is as an autoregressive (AR) *spectral envelope model* [Kay, 1988]. The error minimized by LPC (time-waveform prediction error) forces the filter to model parametrically the *upper* spectral envelope of the speech waveform [Makhoul, 1975]. Since the physical excitation of the vocal tract is not spectrally flat, the filter obtained by whitening the prediction error is not a physical model of the vocal tract. (It would be only if the glottal excitation were an impulse or white noise.) However, by factoring out a rational approximation to the spectral envelope of the glottal excitation, and accounting for the lip radiation transfer function, a vocal tract model can be pursued via post-processing of the LPC spectral model.

There have been some developments toward higher quality speech models retaining most of the simplicity of a source-filter model such as LPC while building on a true physical interpretation: A frequency-domain model of the vocal tract in terms of chain scattering matrices supports variable tube-section length and frequency-dependent losses in each section [Sondhi and Schroeter, 1987]. A two-mass model of the vocal cords has been widely used [Ishizaka and Flanagan, 1972]. An extended derivative of the Kelly-Lochbaum model, adding a nasal tract, neck radiation, and internal damping, has been used to synthesize high-quality female singing voice [Cook, 1990]. Sparse acoustic tubes, in which many reflection coefficients are constrained to be zero, have been proposed [Frank and Lacroix, 1986]. Also, conical (rather than the usual cylindrical) tube segments and sparsely distributed interpolating scattering junctions have been proposed as further refinements [Välimäki and Karjalainen, 1994b].

While coding algorithms for physical speech models are typically much more expensive computationally than LPC, the voice quality obtainable can increase significantly, as demonstrated by Cook. Also, when true physical modeling components are developed, they tend to be more modular and amenable to rule-based transformations. As an example, a true vocal-tract model can be easily scaled to change its length, pointing the way to simple, rule-based transformations as a function of gender, age, or singing technique. Similarly, vocal excitations can be parametrized in richer detail in ways which are understood physically; these should make it possible to directly parametrize a broader range of vocal textures such as "breathiness," "hardness" of glottal closure, and other important attributes of voice quality.

Since computational power has increased enormously in personal computers, and since there is a strong need for maximal compression of multimedia, especially over the internet, it seems reasonable to expect future growth in the development of model-based sound synthesis. Model-based image generation and algorithmic sound synthesis are already under consideration for the MPEG-4 compression standard.

10.1.2 Physical Models in Music Synthesis

At the time of this writing, use of physical models in music synthesizers is only just beginning.[1] Historically, physical models of musical instruments led to prohibitively expensive synthesis algorithms, analogous to articulatory speech models [Ruiz, 1969, Chaigne and Askenfelt, 1994]. More recently, "digital waveguide" models of musical instruments have been developed which are more analogous to acoustic tube models. For an overview of recent research in this area, see [Smith, 1996].

Physical models of musical instruments promise the highest quality in imitating natural instruments. Because the "virtual" instrument can have the same control parameters as the real instrument, expressivity of control is unbounded. Also, as in the case of speech, audio compression algorithms based on generative models promise huge compression ratios.

Digital waveguide models are essentially discrete-time models of *distributed* media such as vibrating strings, bores, horns, plates, and the like. They are often combined with models of *lumped* elements such as masses and springs. Lumped modeling is the main focus of *wave digital filters* as developed principally by Fettweis [1986], and they are also based on a a scattering theoretic formulation [Belevitch, 1968] which simplifies interfacing to waveguide models. For realizability of lumped models with feedback, wave digital filters also incorporate short, unit-sample, waveguide sections called "unit elements," but these are ancillary to the main development. Digital waveguide models are more closely related to "unit element filters" which were developed much earlier in microwave engineering [Rhodes et al., 1973]. As a result of the availability of unit elements in wave digital filters, some authors describe digital waveguide filters as a special case of wave digital filters. This has led to some confusion in the literature such as, for example, incorrectly assuming the existence of frequency warping in the Kelly-Lochbaum model. It can be said that a digital waveguide model may be obtained via simple *sampling* of the traveling waves in a unit-element filter, while a wave digital filter, on the other hand, is typically derived via the *bilinear transformation* of the scattering-theoretic formulation of an RLC (or mass-spring-dashpot) circuit, with some additional special techniques for avoiding delay-free loops.

It turns out that digital waveguide models of many musical instruments (such as winds, strings, and brasses) enjoy much greater efficiency than acoustic-tube speech models as a result of their relative simplicity as acoustic waveguides. This is because the vocal tract is a highly variable acoustic tube while strings, woodwind bores, and horns are highly uniform. As a result, there is very little *scattering* in digital waveguides used to model these simple one-dimensional acoustic waveguides. In fact, we normally approximate them as little more than delay lines; scattering junctions are rarely used except, e.g., in a high quality tone-hole model or string excitation.

Since a delay line can be implemented in software by a single fetch, store, and pointer update for each sample of output, a lossless waveguide simulation requires $O(1)$ computations per sample of output in contrast with $O(N)$ computations using conventional physical modeling methods, where N is the number of samples along the waveguide. If the delay line used in a CD-quality string or bore model is, say, $N = 500$ samples long, (corresponding to a pitch of $44100/500 = 88$ Hz), computational requirements relative to numerical integration of the wave equation on a grid are reduced by *three orders of magnitude.* As a result, for very simple physical models, several CD-quality voices can be sustained in real time on a single DSP chip costing only a few dollars.[2]

10.1.3 Summary

The remainder of this chapter is divided into two parts, addressing theory and applications, respectively. Part I reviews the underlying theory of digital waveguide modeling, starting with the wave equation for vibrating strings. Transverse waves on a string are taken as the primary example due to the relative clarity of the underlying physics, but the formulation for strings is unified with that of acoustic tubes. In particular, the use of transverse *force* and *velocity* waves for strings exactly parallels the use of pressure and longitudinal volume-velocity waves in acoustic tubes.

Longitudinal stress waves in strings are taken as an example leading to signal scattering formulas identical to those encountered in acoustic tubes. The well known Kelly-Lochbaum, one-multiply, and normalized scattering junctions are derived, as well as a lesser known transformer-normalized one-multiply junction. General ways of ensuring passivity of junction computations in the presence of round-off error are discussed. Scattering relations are also derived for the case of N lossless waveguides meeting at a lumped load impedance, illustrating in part how the digital waveguide formulation interfaces naturally to other kinds of physical simulation methodologies.

In addition to ideal strings, lossy and dispersive (stiff) strings are considered. Most generally, the approach holds for all wave equations which admit solutions in the form of non-scattering traveling waves having arbitrary frequency-dependent attenuation and dispersion. The important principle of *lumping* losses and dispersion, via commutativity, at discrete points in the waveguide, replacing more expensive distributed losses, is discussed. Finally, Part II describes digital waveguide models of single-reed woodwinds and bowed strings.

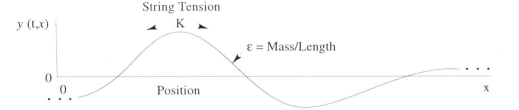

Figure 10.1 The ideal vibrating string.

Part I — Theory

10.2 THE IDEAL VIBRATING STRING

The *wave equation* for the ideal (lossless, linear, flexible) vibrating string, depicted in Fig. 10.1, is given by

$$Ky'' = \epsilon \ddot{y} \qquad (10.1)$$

where

$$
\begin{aligned}
K &\triangleq \text{string tension} & y &\triangleq y(t,x) \\
\epsilon &\triangleq \text{linear mass density} & \dot{y} &\triangleq \tfrac{\partial}{\partial t} y(t,x) \\
y &\triangleq \text{string displacement} & y' &\triangleq \tfrac{\partial}{\partial x} y(t,x)
\end{aligned} \qquad (10.2)
$$

where "$\triangleq$" means "is defined as." The wave equation is fully derived in [Morse, 1981] and in most elementary textbooks on acoustics. It can be interpreted as a statement of Newton's second law, *"force = mass × acceleration,"* on a microscopic scale. Since we are concerned with transverse vibrations on the string, the relevant restoring force (per unit length) is given by the string tension times the curvature of the string (Ky''); the restoring force is balanced at all times by the inertial force per unit length of the string which is equal to mass density times transverse acceleration ($\epsilon \ddot{y}$).

The same wave equation applies to any perfectly elastic medium which is displaced along one dimension. For example, the air column of a clarinet or organ pipe can be modeled using the one-dimensional wave equation by substituting air-pressure deviation for string displacement, and longitudinal volume velocity for transverse string velocity. We refer to the general class of such media as *one-dimensional waveguides*. Extensions to two and three dimensions (and more, for the mathematically curious[3]), are also possible [Van Duyne and Smith, 1995].

For a physical string model, at least three coupled waveguide models should be considered, corresponding to the horizontal and vertical transverse wave polarizations, as well as longitudinal waves. For bowed strings, torsional waves should also be considered, since they affect the bow-string friction force and provide an important loss mechanism for transverse waves [McIntyre et al., 1983]. In the piano, for key ranges in which the hammer strikes three strings simultaneously, *nine* coupled waveguides are required per key for a complete simulation (not including torsional waves); however, in a practical, high-quality, virtual piano, one waveguide per coupled string (modeling only the vertical, transverse plane) suffices quite well. It is difficult to get by with less than the correct number of strings, however, because their detuning determines the entire amplitude envelope as well as beating and aftersound effects [Weinreich, 1977].

10.2.1 The Finite Difference Approximation

In the musical acoustics literature, the normal method for creating a computational model from a differential equation is to apply the so-called *finite difference approximation* (FDA) in which differentiation is replaced by a finite difference [Strikwerda, 1989, Chaigne, 1992]. For example

$$\dot{y}(t,x) \approx \frac{y(t,x) - y(t-T,x)}{T} \tag{10.3}$$

and

$$y'(t,x) \approx \frac{y(t,x) - y(t,x-X)}{X} \tag{10.4}$$

where T is the time sampling interval to be used in the simulation, and X is a spatial sampling interval. These approximations can be seen as arising directly from the definitions of the partial derivatives with respect to t and x. The approximations become exact in the limit as T and X approach zero. To avoid a delay error, the second-order finite-differences are defined with a compensating time shift:

$$\ddot{y}(t,x) \approx \frac{y(t+T,x) - 2y(t,x) + y(t-T,x)}{T^2} \tag{10.5}$$

$$y''(t,x) \approx \frac{y(t,x+X) - 2y(t,x) + y(t,x-X)}{X^2} \tag{10.6}$$

The odd-order derivative approximations suffer a half-sample delay error while all even order cases can be compensated as above.

General Properties of the FDA. To understand the properties of the finite difference approximation in the frequency domain, we may look at the properties of its s-plane to z-plane mapping

$$s = \frac{1 - z^{-1}}{T} \tag{10.7}$$

The FDA does not alias, since the mapping $s = 1 - z^{-1}$ is one-to-one. Setting T to 1 for simplicity and solving the FDA mapping for z gives

$$z = \frac{1}{1-s} \tag{10.8}$$

We see that "analog dc" ($s = 0$) maps to "digital dc" ($z = 1$) as desired, but higher frequencies unfortunately map inside the unit circle rather than onto the unit circle in the z plane. Solving for the image in the z plane of the $j\omega$ axis in the s plane gives

$$z = \frac{1}{1-j\omega} = \frac{1+j\omega}{1+\omega^2} \tag{10.9}$$

From this it can be checked that the FDA maps the $j\omega$ axis in the s plane to the circle of radius $1/2$ centered at the point $z = 1/2$ in the z plane. Under the FDA, analog and digital frequency axes coincide well enough at very low frequencies (high sampling rates), but at high frequencies relative to the sampling rate, artificial *damping* is introduced as the image of the $j\omega$ axis diverges away from the unit circle. Consider, for example, an undamped mass-spring system. There will be a complex conjugate pair of poles on the $j\omega$ axis in the s plane. After the FDA, those poles will be inside the unit circle, and therefore damped in the digital counterpart. The higher the resonance frequency, the larger the damping. It is even possible for unstable s-plane poles to be mapped to stable z-plane poles.

FDA of the Ideal String. Substituting the FDA into the wave equation gives

$$K\frac{y(t, x+X) - 2y(t, x) + y(t, x-X)}{X^2} = \epsilon \frac{y(t+T, x) - 2y(t, x) + y(t-T, x)}{T^2}$$

which can be solved to yield the following recursion for the string displacement:

$$y(t+T, x) = \frac{KT^2}{\epsilon X^2}[y(t, x+X) - 2y(t, x) + y(t, x-X)]$$
$$+ 2y(t, x) - y(t-T, x)$$

In a practical software implementation, it is common to set $T = 1$, $X = (\sqrt{K/\epsilon})T$, and evaluate on the integers $t = nT = n$ and $x = mX = m$ to obtain the difference equation

$$y(n+1, m) = y(n, m+1) + y(n, m-1) - y(n-1, m) \tag{10.10}$$

Thus, to update the sampled string displacement, past values are needed for each point along the string at time instants n and $n - 1$. Then the above recursion can be carried out for time $n + 1$ by iterating over all m along the string.

Perhaps surprisingly, it will be shown in a later section that the above recursion is *exact* at the sample points in spite of the apparent crudeness of the finite difference approximation.

When more terms are added to the wave equation, corresponding to complex losses and dispersion characteristics, more terms of the form $y(n - l, m - k)$ appear in (10.10). This approach to numerical simulation was used in early computer simulation of musical vibrating strings [Ruiz, 1969], and it is still in use today [Chaigne, 1992, Chaigne and Askenfelt, 1994].

10.2.2 Traveling-Wave Solution

It can be readily checked that the lossless 1D wave equation $Ky'' = \epsilon\ddot{y}$ is solved by any fixed string shape which travels to the left or right with speed $c \triangleq \sqrt{K/\epsilon}$. If we denote right-going traveling waves in general by $y_r(t - x/c)$ and left-going traveling waves by $y_l(t + x/c)$, where y_r and y_l are arbitrary twice-differentiable functions, then the general class of solutions to the lossless, one-dimensional, second-order wave equation can be expressed as

$$y(x,t) = y_r(t - x/c) + y_l(t + x/c) \tag{10.11}$$

Note that we have, by definition, $\ddot{y}_r = c^2 y_r''$ and $\ddot{y}_l = c^2 y_l''$, showing that the wave equation is satisfied for all traveling wave shapes y_r and y_l. However, the derivation of the wave equation itself assumes the string slope is much less than 1 at all times and positions [Morse, 1981]. The traveling-wave solution of the wave equation was first published by d'Alembert in 1747 [Lindsay, 1973].

An example of the appearance of the traveling wave components shortly after plucking an infinitely long string at three points is shown in Fig. 10.2.

10.3 SAMPLING THE TRAVELING WAVES

To carry the traveling-wave solution into the "digital domain," it is necessary to *sample* the traveling-wave amplitudes at intervals of T seconds, corresponding to a sampling rate $f_s \triangleq 1/T$ samples per second. For CD-quality audio, we have $f_s = 44.1$ kHz. The natural choice of *spatial sampling interval* X is the distance sound propagates in one temporal sampling interval T, or $X \triangleq cT$ meters.

Formally, sampling is carried out by the change of variables

$$\begin{aligned} x &\to x_m = mX \\ t &\to t_n = nT \end{aligned}$$

Substituting into the traveling-wave solution of the wave equation gives

$$y(t_n, x_m) = y_r(t_n - x_m/c) + y_l(t_n + x_m/c) \tag{10.12}$$

PRINCIPLES OF DIGITAL WAVEGUIDE MODELS OF MUSICAL INSTRUMENTS   427

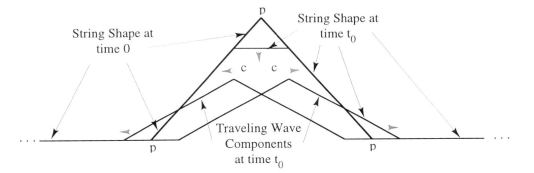

Figure 10.2   An infinitely long string, "plucked" simultaneously at *three* points, labeled "p" in the figure, so as to produce an initial triangular displacement. The initial displacement is modeled as the sum of two identical triangular pulses which are exactly on top of each other at time 0. At time t_0 shortly after time 0, the traveling waves centers are separated by $2ct_0$ meters, and their sum gives the trapezoidal physical string displacement at time t_0 which is also shown. Note that only three short string segments are in motion at that time: the flat top segment which is heading to zero where it will halt forever, and two short pieces on the left and right which are the leading edges of the left- and right-going traveling waves. The string is not moving where the traveling waves overlap at the same slope. When the traveling waves fully separate, the string will be at rest everywhere but for two half-amplitude triangular pulses heading off to plus and minus infinity at speed c.

$$= y_r(nT - mX/c) + y_l(nT + mX/c)$$
$$= y_r[(n-m)T] + y_l[(n+m)T]$$

Since T multiplies all arguments, let's suppress it by defining

$$y^+(n) \triangleq y_r(nT) \qquad y^-(n) \triangleq y_l(nT) \qquad (10.13)$$

This new notation also introduces a "+" superscript to denote a traveling-wave component propagating to the right, and a "−" superscript to denote propagation to the left. This notation is similar to that used for acoustic tubes [Markel and Gray, 1976].

The term $y_r[(n-m)T] = y^+(n-m)$ can be thought of as the output of an m-sample delay line whose input is $y^+(n)$. Similarly, the term $y_l[(n+m)T] \triangleq y^-(n+m)$ can be thought of as the *input* to an m-sample delay line whose *output* is $y^-(n)$. This can be seen as the lower "rail" in Fig. 10.3. Note that the position along the string, $x_m = mX = mcT$ meters, is laid out from left to right in the diagram, giving a physical interpretation to the horizontal direction in the diagram. Finally, the left- and right-going traveling waves must be summed to produce a physical output according to the formula

$$y(t_n, x_m) = y^+(n-m) + y^-(n+m) \qquad (10.14)$$

We may compute the physical string displacement at any spatial sampling point x_m by simply adding the upper and lower rails together at position m along the delay-line pair. In Fig. 10.3, "transverse displacement outputs" have been arbitrarily placed at $x = 0$ and $x = 3X$. The diagram is similar to that of well known ladder and lattice digital filter structures, except for the delays along the upper rail, the absence of scattering junctions, and the direct physical interpretation. We could proceed to ladder and lattice filters as in [Markel and Gray, 1976] by (1) introducing a perfectly reflecting (rigid or free) termination at the far right, and (2) commuting the delays rightward from the upper rail down to the lower rail [Smith, 1986b]. The absence of scattering junctions is due to the fact that the string has a uniform wave impedance. In acoustic tube simulations, such as for voice or wind instruments (discussed in a later section), lossless scattering junctions are used at changes in cross-sectional tube area and lossy scattering junctions are used to implement tone holes. In waveguide bowed-string synthesis (also discussed in a later section), the bow itself creates an active, time-varying, and nonlinear scattering junction on the string at the bowing point.

A more compact simulation diagram which stands for either sampled or continuous simulation is shown in Fig. 10.4. The figure emphasizes that the ideal, lossless waveguide is simulated by a *bidirectional delay line*, and that bandlimited spatial interpolation may be used to construct a displacement output for an arbitrary x not a multiple of cT, as suggested by the output drawn in Fig. 10.4. Similarly, bandlimited interpolation across time serves to evaluate the waveform at an arbitrary time not an integer multiple of T [Smith and Gossett, 1984, Laakso et al., 1996].

PRINCIPLES OF DIGITAL WAVEGUIDE MODELS OF MUSICAL INSTRUMENTS   429

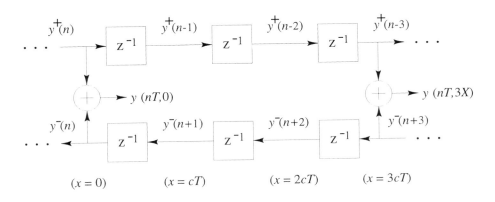

Figure 10.3 Digital simulation of the ideal, lossless waveguide with observation points at $x = 0$ and $x = 3X = 3cT$.

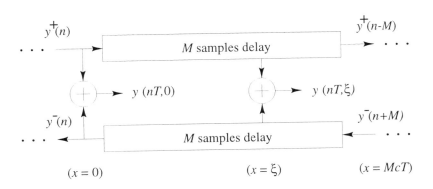

Figure 10.4 Conceptual diagram of interpolated digital waveguide simulation.

Any ideal, one-dimensional waveguide can be simulated in this way. It is important to note that the simulation is *exact* at the sampling instants, to within the numerical precision of the samples themselves. To avoid aliasing associated with sampling, we require all waveshapes traveling along the string to be initially *bandlimited* to less than half the sampling frequency. In other words, the highest frequencies present in the signals $y_r(t)$ and $y_l(t)$ may not exceed half the temporal sampling frequency $f_s \triangleq 1/T$; equivalently, the highest spatial frequencies in the shapes $y_r(x/c)$ and $y_l(x/c)$ may not exceed half the spatial sampling frequency $\nu_s \triangleq 1/X$.

10.3.1 Relation to Finite Difference Recursion

It is interesting to compare the digital waveguide simulation technique to the recursion produced by the finite difference approximation (FDA) applied to the wave equation. Recall from (10.10) that the time update recursion for the ideal string digitized via the FDA is given by

$$y(n+1, m) = y(n, m+1) + y(n, m-1) - y(n-1, m) \tag{10.15}$$

To compare this with the waveguide description, we substitute the traveling-wave decomposition $y(n,m) = y^+(n-m) + y^-(n+m)$ (which is exact in the ideal case at the sampling instants) into the right-hand side of the FDA recursion above and see how good is the approximation to the left-hand side $y(n+1,m) = y^+(n+1-m) + y^-(n+1+m)$. Doing this gives

$$\begin{aligned} y(n+1,m) &= y(n,m+1) + y(n,m-1) - y(n-1,m) \quad (10.16)\\ &= y^+(n-m-1) + y^-(n+m+1)\\ &\quad + y^+(n-m+1) + y^-(n+m-1)\\ &\quad - y^+(n-m-1) - y^-(n+m-1)\\ &= y^-(n+m+1) + y^+(n-m+1)\\ &= y^+[(n+1)-m] + y^-[(n+1)+m]\\ &\triangleq y(n+1,m) \end{aligned}$$

Thus, we obtain the result that the FDA recursion is also *exact* in the lossless case. This is surprising since the FDA introduces artificial damping when applied to lumped, mass-spring systems, as discussed earlier.

The last identity above can be rewritten as

$$\begin{aligned} y(n+1,m) &\triangleq y^+[(n+1)-m] + y^-[(n+1)+m] \quad (10.17)\\ &= y^+[n-(m-1)] + y^-[n+(m+1)] \end{aligned}$$

which says the displacement at time $n+1$, position m, is the superposition of the right-going and left-going traveling wave components at positions $m-1$ and $m+1$, respectively, from time n. In other words, the physical wave variable can be computed for the next time step as the sum of incoming traveling wave components from the left and right. This picture also underscores the lossless nature of the computation.

10.4 ALTERNATIVE WAVE VARIABLES

We have thus far considered discrete-time simulation of transverse *displacement* y in the ideal string. It is equally valid to choose *velocity* $v \triangleq \dot{y}$, *acceleration* $a \triangleq \ddot{y}$, *slope* y', or perhaps some other derivative or integral of displacement with respect to time or position. Conversion between various time derivatives can be carried out by means *integrators* and *differentiators*. Since integration and differentiation are linear operators, and since the traveling wave arguments are in units of time, the conversion formulas relating y, v, and a hold also for the traveling wave *components* $y^{\pm}, v^{\pm}, a^{\pm}$.

Digital filters can be designed to give arbitrarily accurate differentiation and integration by finding an optimal, complex, rational approximation to $H(e^{j\omega}) = (j\omega)^k$ over the interval $-\omega_{\max} \leq \omega \leq \omega_{\max}$, where k is an integer corresponding to the degree of differentiation or integration, and $\omega_{\max} < \pi$ is the upper limit of human hearing. For small guard bands $\delta \triangleq \pi - \omega_{\max}$, the filter order required for a given error tolerance is approximately inversely proportional to δ. Methods for digital filter design given an arbitrary desired frequency response can be found in [Rabiner and Gold, 1975, Parks and Burrus, 1987, Laakso et al., 1996, Gutknecht et al., 1983, Smith, 1983, Beliczynski et al., 1992].

10.4.1 Spatial Derivatives

In addition to time derivatives, we may apply any number of *spatial derivatives* to obtain yet more wave variables to choose from. The first spatial derivative of string displacement yields *slope waves:*

$$\begin{aligned} y'(t,x) &\triangleq \frac{\partial}{\partial x} y(t,x) \\ &= y'_r(t-x/c) + y'_l(t+x/c) \\ &= -\frac{1}{c}\dot{y}_r(t-x/c) + \frac{1}{c}\dot{y}_l(t+x/c) \end{aligned} \quad (10.18)$$

or, in discrete time,

$$\begin{aligned} y'(t_n, x_m) &\triangleq y'(nT, mX) \\ &= y'_r[(n-m)T] + y'_l[(n+m)T] \\ &\triangleq y'^{+}(n-m) + y'^{-}(n+m) \\ &= -\frac{1}{c}\dot{y}^{+}(n-m) + \frac{1}{c}\dot{y}^{-}(n+m) \end{aligned} \quad (10.19)$$

$$\triangleq -\frac{1}{c}v^+(n-m) + \frac{1}{c}v^-(n+m)$$
$$= \frac{1}{c}\left[v^-(n+m) - v^+(n-m)\right]$$

>From this we may conclude that $v^- = cy'^-$ and $v^+ = -cy'^+$. That is, traveling slope waves can be computed from traveling velocity waves by dividing by c and negating in the right-going case. Physical string slope can thus be computed from a velocity-wave simulation in a digital waveguide by *subtracting* the upper rail from the lower rail and dividing by c.

By the wave equation, *curvature waves*, $y'' = \ddot{y}/c^2$, are simply a scaling of acceleration waves.

In the field of acoustics, the state of a vibrating string at any instant of time t_0 is normally specified by the displacement $y(t_0, x)$ and velocity $\dot{y}(t_0, x)$ for all x [Morse, 1981]. Since displacement is the *sum* of the traveling displacement waves and velocity is proportional to the *difference* of the traveling displacement waves, one state description can be readily obtained from the other.

In summary, all traveling-wave variables can be computed from any one, as long as both the left- and right-going component waves are available. Alternatively, any *two* linearly independent *physical* variables, such as displacement and velocity, can be used to compute all other wave variables. Wave variable conversions requiring differentiation or integration are relatively expensive since a large-order digital filter is necessary to do it right. Slope and velocity waves can be computed from each other by simple scaling, and curvature waves are identical to acceleration waves to within a scale factor.

In the absence of factors dictating a specific choice, *velocity waves* are a good overall choice because (1) it is numerically easier to perform digital integration to get displacement than it is to differentiate displacement to get velocity, (2) slope waves are immediately computable from velocity waves. Slope waves are important because they are proportional to force waves.

10.4.2 Force Waves

Referring to Fig. 10.5, at an arbitrary point x along the string, the vertical force applied at time t *to* the portion of string to the left of position x *by* the portion of string to the right of position x is given by

$$f_l(t, x) = K\sin(\theta) \approx K\tan(\theta) = Ky'(t, x) \qquad (10.20)$$

assuming $|y'(t, x)| \ll 1$, as is assumed in the derivation of the wave equation. Similarly, the force applied *by* the portion to the left of position x *to* the portion to the right is given by

$$f_r(t, x) = -K\sin(\theta) \approx -Ky'(t, x) \qquad (10.21)$$

PRINCIPLES OF DIGITAL WAVEGUIDE MODELS OF MUSICAL INSTRUMENTS   433

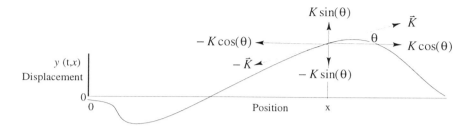

Figure 10.5   Transverse force propagation in the ideal string.

These forces must cancel since a nonzero net force on a massless point would produce infinite acceleration.

Vertical force waves propagate along the string like any other transverse wave variable (since they are just slope waves multiplied by tension K). We may choose either f_l or f_r as the string force wave variable, one being the negative of the other. It turns out that to make the description for vibrating strings look the same as that for air columns, we have to pick f_r, the one that *acts to the right*. This makes sense intuitively when one considers longitudinal pressure waves in an acoustic tube: a compression wave traveling to the right in the tube pushes the air in front of it and thus acts to the right. We therefore define the *force wave variable* to be

$$f(t,x) \triangleq f_r(t,x) = -Ky'(t,x) \qquad (10.22)$$

Note that a negative slope pulls up on the segment to the right. Substituting from (10.18), we have

$$f(t,x) = \frac{K}{c}\left[\dot{y}_r(t-x/c) - \dot{y}_l(t+x/c)\right] \qquad (10.23)$$

where $K/c \triangleq K/\sqrt{K/\epsilon} = \sqrt{K\epsilon}$. This is a fundamental quantity known as the *wave impedance* of the string (also called the *characteristic impedance*), denoted as

$$R \triangleq \sqrt{K\epsilon} = \frac{K}{c} = \epsilon c \qquad (10.24)$$

The wave impedance can be seen as the geometric mean of the two resistances to displacement: tension (spring force) and mass (inertial force).

The digitized traveling force-wave components become

$$\begin{aligned} f^+(n) &= Rv^+(n) \\ f^-(n) &= -Rv^-(n) \end{aligned} \qquad (10.25)$$

which gives us that the right-going force wave equals the wave impedance times the right-going velocity wave, and the left-going force wave equals *minus* the wave impedance times the left-going velocity wave. Thus, in a traveling wave, force is always *in phase* with velocity (considering the minus sign in the left-going case to be associated with the direction of travel rather than a 180 degrees phase shift between force and velocity). Note also that if the left-going force wave were defined as the string force acting to the left, the minus sign would disappear. The fundamental relation $f^+ = Rv^+$ is sometimes referred to as the mechanical counterpart of *Ohm's Law*, and R in c.g.s. units can be called *acoustical ohms* [Kolsky, 1963].

In the case of the *acoustic tube* [Morse, 1981, Markel and Gray, 1976], we have the analogous relations

$$\begin{aligned} p^+(n) &= R_t u^+(n) \\ p^-(n) &= -R_t u^-(n) \end{aligned} \quad (10.26)$$

where $p^+(n)$ is the right-going traveling *longitudinal pressure wave* component, $p^-(n)$ is the left-going pressure wave, and $u^{\pm}(n)$ are the left and right-going *volume velocity waves*. In the acoustic tube context, the wave impedance is given by

$$R_t = \frac{\rho c}{A} \quad \text{(Acoustic Tubes)} \quad (10.27)$$

where ρ is the mass per unit volume of air, c is sound speed in air, and A is the cross-sectional area of the tube. Note that if we had chosen *particle velocity* rather than volume velocity, the wave impedance would be $R_0 = \rho c$ instead, the wave impedance in open air. Particle velocity is appropriate in open air, while volume velocity is the conserved quantity in acoustic tubes or "ducts" of varying cross-sectional area [Morse and Ingard, 1968].

10.4.3 Power Waves

Basic courses in physics teach us that *power* is work per unit time, and *work* is a measure of *energy* which may be defined as force times distance. Therefore, power is in physical units of force times distance per unit time, or force times velocity. It therefore should come as no surprise that *traveling power waves* are defined for strings as

$$\begin{aligned} \mathcal{P}^+(n) &\triangleq f^+(n)v^+(n) \\ \mathcal{P}^-(n) &\triangleq -f^-(n)v^-(n) \end{aligned} \quad (10.28)$$

From the elementary relations $f^+ = Rv^+$ and $f^- = -Rv^-$, we also have

$$\begin{aligned} \mathcal{P}^+(n) &= R[v^+(n)]^2 = [f^+(n)]^2/R \\ \mathcal{P}^-(n) &= R[v^-(n)]^2 = [f^-(n)]^2/R \end{aligned} \quad (10.29)$$

Thus, both the left- and right-going components as defined are *nonnegative*. The sum of the traveling powers at a point gives the total power at that point in the waveguide:

$$\mathcal{P}(t_n, x_m) \triangleq \mathcal{P}^+(n-m) + \mathcal{P}^-(n+m) \tag{10.30}$$

If we had left out the minus sign in the definition of left-going power waves, the sign of the traveling power would indicate its direction of travel, and the sum of left- and right-going components would instead be the *net* power flow.

Power waves are important because they correspond to the actual ability of the wave to do work on the outside world, such as on a violin bridge at the end of a string. Because energy is conserved in closed systems, power waves sometimes give a simpler, more fundamental view of wave phenomena, such as in conical acoustic tubes. Also, implementing nonlinear operations such as rounding and saturation in such a way that signal power is not increased gives suppression of limit cycles and overflow oscillations, as discussed in the later section on signal scattering.

10.4.4 Energy Density Waves

The vibrational energy per unit length along the string, or *wave energy density* [Morse, 1981] is given by the sum of potential and kinetic energy densities:

$$W(t,x) \triangleq \frac{1}{2} K y'^2(t,x) + \frac{1}{2} \epsilon \dot{y}^2(t,x) \tag{10.31}$$

Sampling across time and space, and substituting traveling wave components, one can show in a few lines of algebra that the *sampled* wave energy density is given by

$$W(t_n, x_m) \triangleq W^+(n-m) + W^-(n+m) \tag{10.32}$$

where

$$
\begin{aligned}
W^+(n) &= \frac{\mathcal{P}^+(n)}{c} = \frac{f^+(n)v^+(n)}{c} = \epsilon\left[v^+(n)\right]^2 = \frac{\left[f^+(n)\right]^2}{K} \\
W^-(n) &= \frac{\mathcal{P}^-(n)}{c} = -\frac{f^-(n)v^-(n)}{c} = \epsilon\left[v^-(n)\right]^2 = \frac{\left[f^-(n)\right]^2}{K}
\end{aligned}
\tag{10.33}
$$

Thus, traveling power waves (energy per unit time) can be converted to energy density waves (energy per unit length) by simply dividing by c, the speed of propagation. Quite naturally, the *total wave energy* in the string is given by the integral along the string of the energy density:

$$\mathcal{E}(t) = \int_{x=-\infty}^{\infty} W(t,x)dx \approx \sum_{m=-\infty}^{\infty} W(t, x_m) X \tag{10.34}$$

In practice, of course, the string length is finite, and the limits of integration are from the x coordinate of the left endpoint to that of the right endpoint, e.g., 0 to L.

10.4.5 Root-Power Waves

It is sometimes helpful to *normalize* the wave variables so that signal power is uniformly distributed numerically. This can be especially helpful in fixed-point implementations.

From (10.29), it is clear that power normalization is given by

$$\begin{aligned} \tilde{f}^+ &\triangleq f^+/\sqrt{R} & \tilde{f}^- &\triangleq f^-/\sqrt{R} \\ \tilde{v}^+ &\triangleq v^+\sqrt{R} & \tilde{v}^- &\triangleq v^-\sqrt{R} \end{aligned} \qquad (10.35)$$

where we have dropped the common time argument '(n)' for simplicity. As a result, we obtain

$$\begin{aligned} \mathcal{P}^+ &= f^+ v^+ &&= \tilde{f}^+ \tilde{v}^+ \\ &= R(v^+)^2 &&= (\tilde{v}^+)^2 \\ &= (f^+)^2/R &&= (\tilde{f}^+)^2 \end{aligned} \qquad (10.36)$$

and

$$\begin{aligned} \mathcal{P}^- &= -f^- v^- &&= -\tilde{f}^+ \tilde{v}^+ \\ &= R(v^-)^2 &&= (\tilde{v}^-)^2 \\ &= (f^-)^2/R &&= (\tilde{f}^-)^2 \end{aligned} \qquad (10.37)$$

The normalized wave variables $\tilde{f}^\pm$ and $\tilde{v}^\pm$ behave physically like force and velocity waves, respectively, but they are scaled such that either can be squared to obtain instantaneous signal power. Waveguide networks built using normalized waves have many desirable properties such as the obvious numerical advantage of uniformly distributing signal power across available dynamic range in fixed-point implementations. Another is that only in the normalized case can the wave impedances be made *time varying* without modulating signal power [Gray and Markel, 1975, Smith, 1987]. In other words, use of normalized waves eliminates "parametric amplification" effects: signal power is decoupled from parameter changes.

10.5 SCATTERING AT AN IMPEDANCE DISCONTINUITY

When the wave impedance changes, signal *scattering* occurs, i.e., a traveling wave impinging on an impedance discontinuity will partially reflect and partially transmit at the junction in such a way that energy is conserved. This is a classical topic in transmission line theory [Main, 1978], and it is well covered for acoustic tubes in a variety of references [Markel and Gray, 1976, Rabiner and Schafer, 1978b]. However, for completeness, elementary scattering relations will be outlined here for the case of *longitudinal* force and velocity waves in an ideal string or rod. (In solids, force waves are referred to as *stress* waves [Kolsky, 1963].) Longitudinal compression waves in strings and rods behave like longitudinal pressure waves in acoustic tubes.

A single waveguide section between two partial sections is shown in Fig. 10.6. The sections are numbered 0 through 2 from left to right, and their wave impedances are

PRINCIPLES OF DIGITAL WAVEGUIDE MODELS OF MUSICAL INSTRUMENTS 437

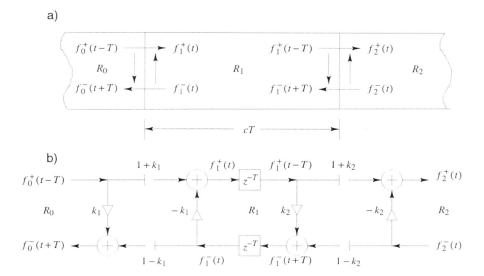

Figure 10.6 A waveguide section between two partial sections. a) Physical picture indicating traveling waves in a continuous medium whose wave impedance changes from R_0 to R_1 to R_2. b) Digital simulation diagram for the same situation. The section propagation delay is denoted as z^{-T}. The behavior at an impedance discontinuity is characterized by a lossless splitting of an incoming wave into transmitted and reflected components.

R_0, R_1, and R_2, respectively. Such a rod might be constructed, for example, using three different materials having three different densities. In the ith section, there are two force traveling waves: f_i^+ traveling to the right at speed c, and f_i^- traveling to the left at speed c. To minimize the numerical dynamic range, velocity waves may be chosen instead when $R_i > 1$.

As in the case of transverse waves (10.25), the traveling longitudinal plane waves in each section satisfy

$$\begin{aligned} f_i^+(t) &= R_i v_i^+(t) \\ f_i^-(t) &= -R_i v_i^-(t) \end{aligned} \qquad (10.38)$$

where the wave impedance is now $R_i = \sqrt{Y\rho}$, with ρ being the mass density, and Y being the *Young's modulus* of the medium (defined as the stress over the strain, where strain means displacement). If the wave impedance R_i is constant, the shape of a traveling wave is not altered as it propagates from one end of a section to the other. In this case we need only consider f_i^+ and f_i^- at one end of each section as a function of time. As shown in Fig. 10.6, we define $f_i^{\pm}(t)$ as the traveling force-wave component at the *extreme left* of section i. Therefore, at the extreme right of section i, we have the traveling waves $f_i^+(t-T)$ and $f_i^-(t+T)$, where T is the travel time from one end of a section to the other.

For generality, we may allow the wave impedances to vary with time. A number of possibilities exist which satisfy (10.38) in the time-varying case. For the moment, we will assume the traveling waves at the extreme right of section i are still given by $f_i^+(t-T)$ and $f_i^-(t+T)$. This definition, however, implies the velocity varies inversely with the wave impedance. As a result, signal energy, being the product of force times velocity, is "pumped" into or out of the waveguide by a changing wave impedance. Use of normalized waves $\tilde{f}_i^{\pm}$ avoids this. However, normalization increases the required number of multiplications, as we will see shortly.

As before, the physical force and velocity at the left end of section i are obtained by summing the left- and right-going traveling wave components:

$$\begin{aligned} f_i &= f_i^+ + f_i^- \\ v_i &= v_i^+ + v_i^- \end{aligned} \qquad (10.39)$$

Let $f_i(t,x)$ denote the force at position x and time t in section i, where x is measured from the extreme left of section i ($0 \leq x \leq cT$). Then we have

$$f_i(t,x) \triangleq f_i^+(t-x/c) + f_i^-(t+x/c)$$

within section i. In particular, at the left and right boundaries of section i, we have

$$\begin{aligned} f_i(t,0) &= f_i^+(t) + f_i^-(t) & (10.40) \\ f_i(t,cT) &= f_i^+(t-T) + f_i^-(t+T), & (10.41) \end{aligned}$$

respectively, as labeled in Fig. 10.6b.

10.5.1 The Kelly-Lochbaum and One-Multiply Scattering Junctions

At the impedance discontinuity, the force and velocity must be continuous, i.e.,

$$f_{i-1}(t, cT) = f_i(t, 0) \tag{10.42}$$
$$v_{i-1}(t, cT) = v_i(t, 0)$$

where force and velocity are defined as positive to the right on both sides of the junction. Equations (10.38), (10.39), and (10.42) imply the following *scattering equations* (a derivation is given in the next section for the more general case of N waveguides meeting at a junction):

$$f_i^+(t) = [1 + k_i(t)] f_{i-1}^+(t - T) - k_i(t) f_i^-(t)$$
$$f_{i-1}^-(t + T) = k_i(t) f_{i-1}^+(t - T) + [1 - k_i(t)] f_i^-(t) \tag{10.43}$$

where

$$k_i(t) \triangleq \frac{R_i(t) - R_{i-1}(t)}{R_i(t) + R_{i-1}(t)} \tag{10.44}$$

is called the ith *reflection coefficient*. (For generality, we allow the wave impedances, hence the reflection coefficients, to vary with time; time variation induced by wave-impedance variations remains "physical" and therefore does not invalidate the many desirable passivity properties.) Since $R_i(t) \geq 0$, we have $k_i(t) \in [-1, 1]$. It can be shown that if $|k_i| > 1$, then either R_i or R_{i-1} is negative, and this implies an active (as opposed to passive) medium. Correspondingly, lattice and ladder recursive digital filters are *stable* if and only if all reflection coefficients are bounded by 1 in magnitude.

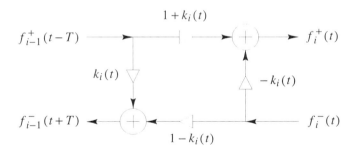

Figure 10.7 The Kelly-Lochbaum scattering junction.

The scattering equations are illustrated in Figs. 10.6b and 10.7. In linear predictive coding of speech, this structure is known as the *Kelly-Lochbaum scattering junction*, and it is one of several types of scattering junction used to implement lattice and ladder digital filter structures.

By factoring out $k_i(t)$ in each equation of (10.43), we can write

$$\begin{aligned} f_i^+(t) &= f_{i-1}^+(t-T) + f_\Delta(t) \\ f_{i-1}^-(t+T) &= f_i^-(t) + f_\Delta(t) \end{aligned} \quad (10.45)$$

where

$$f_\Delta(t) \triangleq k_i(t) \left[f_{i-1}^+(t-T) - f_i^-(t) \right] \quad (10.46)$$

Thus, only *one multiplication* is actually necessary to compute the transmitted and reflected waves from the incoming waves in the Kelly-Lochbaum junction. This computation is shown in Fig. 10.8, and it is known as the *one-multiply scattering junction* [Markel and Gray, 1976].

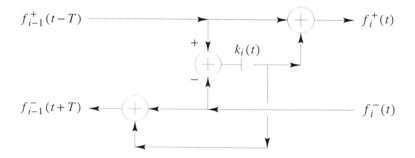

Figure 10.8 The one-multiply scattering junction.

Another one-multiply form is obtained by organizing (10.43) as

$$\begin{aligned} f_i^+(t) &= f_i^-(t) + \alpha_i(t)\tilde{f}_d(t) \\ f_{i-1}^-(t+T) &= f_i^+(t) - \tilde{f}_d(t) \end{aligned} \quad (10.47)$$

where

$$\begin{aligned} \alpha_i(t) &\triangleq 1 + k_i(t) \\ \tilde{f}_d(t) &\triangleq f_{i-1}^+(t-T) - f_i^-(t) \end{aligned} \quad (10.48)$$

As in the previous case, only one multiplication and three additions are required per junction. This one-multiply form generalizes more readily to junctions of more than two waveguides, as we'll see in a later section.

PRINCIPLES OF DIGITAL WAVEGUIDE MODELS OF MUSICAL INSTRUMENTS

A scattering junction well known in the LPC speech literature but not described here is the so-called *two-multiply* junction [Markel and Gray, 1976] (requiring also two additions). This omission is because the two-multiply junction is not valid as a general, *local*, physical modeling building block. Its derivation is tied to the reflectively terminated, cascade waveguide chain. In cases where it applies, however, it can be the implementation of choice; for example, in DSP chips having a fast multiply-add instruction, it may be possible to implement the two-multiply, two-add scattering junction using only two instructions.

10.5.2 Normalized Scattering Junctions

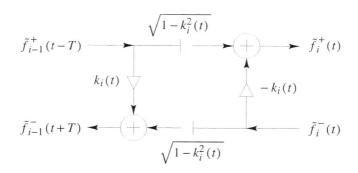

Figure 10.9 The normalized scattering junction.

Using (10.35) to convert to *normalized waves* $\tilde{f}^{\pm}$, the Kelly-Lochbaum junction (10.43) becomes

$$\tilde{f}_i^+(t) = \sqrt{1-k_i^2(t)}\,\tilde{f}_{i-1}^+(t-T) - k_i(t)\tilde{f}_i^-(t)$$
$$\tilde{f}_{i-1}^-(t+T) = k_i(t)\tilde{f}_{i-1}^+(t-T) + \sqrt{1-k_i^2(t)}\,\tilde{f}_i^-(t) \qquad (10.49)$$

as diagrammed in Fig. 10.9. This is called the *normalized scattering junction* [Markel and Gray, 1976], although a more precise term would be the "normalized-wave scattering junction."

It is interesting to define $\theta_i \triangleq \sin^{-1}(k_i)$, always possible for passive junctions since $-1 \leq k_i \leq 1$, and note that the normalized scattering junction is equivalent to a *2D rotation*:

$$\tilde{f}_i^+(t) = \cos(\theta_i)\tilde{f}_{i-1}^+(t-T) - \sin(\theta_i)\tilde{f}_i^-(t)$$
$$\tilde{f}_{i-1}^-(t+T) = \sin(\theta_i)\tilde{f}_{i-1}^+(t-T) + \cos(\theta_i)\tilde{f}_i^-(t) \qquad (10.50)$$

where, for conciseness of notation, the time-invariant case is written.

While it appears that scattering of normalized waves at a two-port junction requires four multiplies and two additions, it is possible to convert this to three multiplies and three additions using a two-multiply "transformer" to power-normalize an ordinary one-multiply junction.

Transformer Normalization. The *transformer* is a lossless two-port defined by [Fettweis, 1986]

$$f_i^+(t) = g_i f_{i-1}^+(t - T)$$
$$f_{i-1}^-(t + T) = \frac{1}{g_i} f_i^-(t) \qquad (10.51)$$

The transformer can be thought of as a device which steps the wave impedance to a new value without scattering; instead, the traveling signal power is redistributed among the force and velocity wave variables to satisfy the fundamental relations $f^\pm = \pm R v^\pm$ (10.25) at the new impedance. An impedance change from R_{i-1} on the left to R_i on the right is accomplished using

$$g_i \triangleq \sqrt{\frac{R_i}{R_{i-1}}} = \sqrt{\frac{1 + k_i(t)}{1 - k_i(t)}} \qquad (10.52)$$

as can be quickly derived by requiring $(f_{i-1}^+)^2/R_{i-1} = (f_i^+)^2/R_i$. The parameter g_i can be interpreted as the "turns ratio" since it is the factor by which force is stepped (and the inverse of the velocity step factor).

The Three-Multiply Normalized Scattering Junction. Figure 10.10 illustrates a *three-multiply normalized scattering junction* [Smith, 1986b]. The one-multiply junction of Fig. 10.8 is normalized by a transformer. Since the impedance discontinuity is created *locally* by the transformer, all wave variables in the delay elements to the left and right of the overall junction are at the same wave impedance. Thus, using transformers, all waveguides can be normalized to the same impedance, e.g., $R_i \equiv 1$.

It is important to notice that g_i and $1/g_i$ may have a large dynamic range in practice. For example, if $k_i \in [-1 + \epsilon, 1 - \epsilon]$, the transformer coefficients may become as large as $\sqrt{2/\epsilon - 1}$. If ϵ is the "machine epsilon," i.e., $\epsilon = 2^{-(n-1)}$ for typical n-bit two's complement arithmetic normalized to lie in $[-1, 1)$, then the dynamic range of the transformer coefficients is bounded by $\sqrt{2^n - 1} \approx 2^{n/2}$. Thus, while transformer-normalized junctions trade a multiply for an add, they require up to 50% more bits of dynamic range within the junction adders.

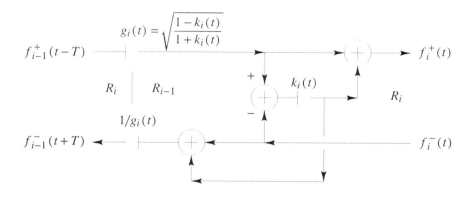

Figure 10.10 A three-multiply normalized scattering junction.

10.5.3 Junction Passivity

In fixed-point implementations, the round-off error and other nonlinear operations should be confined when possible to *physically meaningful* wave variables. When this is done, it is easy to ensure that signal power is not increased by the nonlinear operations. In other words, nonlinear operations such as rounding can be made *passive*. Since signal power is proportional to the square of the wave variables, all we need to do is make sure amplitude is never increased by the nonlinearity. In the case of rounding, *magnitude truncation*, sometimes called "rounding toward zero," is one way to achieve passive rounding. However, magnitude truncation can attenuate the signal excessively in low-precision implementations and in scattering-intensive applications such as the digital waveguide mesh [Van Duyne and Smith, 1993]. Another option is *error power feedback* in which case the cumulative round-off error power averages to zero over time.

A valuable byproduct of passive arithmetic is the suppression of *limit cycles* and *overflow oscillations*. Formally, the signal power of a conceptually infinite-precision implementation can be viewed as a Lyapunov function bounding the squared amplitude of the finite-precision implementation.

To formally show that magnitude truncation is sufficient to suppress overflow oscillations and limit cycles in waveguide networks built using structurally lossless scattering junctions, we can look at the signal power entering and leaving the junction. A junction is passive if the power flowing away from it does not exceed the power flowing into it. The total power flowing away from the ith junction is bounded by the incoming power if

$$\underbrace{\frac{[f_i^+(t)]^2}{R_i(t)} + \frac{[f_{i-1}^-(t+T)]^2}{R_{i-1}(t)}}_{\text{outgoing power}} \leq \underbrace{\frac{[f_{i-1}^+(t-T)]^2}{R_{i-1}(t)} + \frac{[f_i^-(t)]^2}{R_i(t)}}_{\text{incoming power}} \qquad (10.53)$$

Let $\hat{f}$ denote the finite-precision version of f. Then a *sufficient* condition for junction passivity is

$$\left|\hat{f}_i^+(t)\right| \leq \left|f_i^+(t)\right| \qquad (10.54)$$

$$\left|\hat{f}_{i-1}^-(t+T)\right| \leq \left|f_{i-1}^-(t+T)\right| \qquad (10.55)$$

Thus, if the junction computations do not increase either of the output force amplitudes, no signal power is created. An analogous conclusion is reached for velocity scattering junctions.

Passive Kelly-Lochbaum and One-Multiply Junctions. The Kelly-Lochbaum and one-multiply scattering junctions are *structurally lossless* [Vaidyanathan, 1993] because they can be computed exactly in terms of only one parameter k_i (or α_i), and all quantizations of the parameter within the allowed interval $[-1, 1]$ (or $[0, 2]$) correspond to lossless scattering.[4] The structural losslessness of the one-multiply junction has been used to construct a numerically stable, one-multiply, sinusoidal digital oscillator [Smith and Cook, 1992].

In the Kelly-Lochbaum and one-multiply scattering junctions, because they are structurally lossless, we need only double the number of bits at the output of each multiplier, and add one bit of extended dynamic range at the output of each two-input adder. The final outgoing waves are thereby *exactly* computed before they are finally rounded to the working precision and/or clipped to the maximum representable magnitude.

For the Kelly-Lochbaum scattering junction, given n-bit signal samples and m-bit reflection coefficients, the reflection and transmission multipliers produce $n + m$ and $n + m + 1$ bits, respectively, and each of the two additions adds one more bit. Thus, the intermediate word length required is $n + m + 2$ bits, and this must be rounded *without amplification* down to n bits for the final outgoing samples. A similar analysis gives also that the one-multiply scattering junction needs $n + m + 2$ bits for the extended precision intermediate results before final rounding and/or clipping.

Passive Four-Multiply Normalized Junctions. Unlike the structurally lossless cases, the (four-multiply) normalized scattering junction has *two* parameters, $s_i \triangleq k_i$ and $c_i \triangleq \sqrt{1 - k_i^2}$, and these can "get out of synch" in the presence of quantization.

Specifically, let $\hat{s}_i \triangleq s_i - \epsilon_s$ denote the quantized value of s_i, and let $\hat{c}_i \triangleq c_i - \epsilon_c$ denote the quantized value of c_i. Then it is no longer the case in general that $\hat{s}_i^2 + \hat{c}_i^2 = 1$. As a result, the normalized scattering junction is not structurally lossless in the presence of coefficient quantization. A few lines of algebra shows that a passive rounding rule for the normalized junction must depend on the sign of the wave variable being computed, the sign of the coefficient quantization error, and the sign of at least one of the two incoming traveling waves. We can assume one of the coefficients is exact for passivity purposes, so assume $\epsilon_s = 0$ and define $\hat{c}_i = \lfloor \sqrt{1 - s_i^2} \rfloor$, where $\lfloor x \rfloor$ denotes largest quantized value less than or equal to x. In this case we have $\epsilon_c \geq 0$. Therefore,

$$\hat{f}_i^+ = \hat{c}_i f_{i-1}^+ - s_i f_i^- = f_i^+ - \epsilon_c f_{i-1}^+$$

and a passive rounding rule which guarantees $|\hat{f}_i^+| \leq |f_i^+|$ need only look at the sign bits of $\hat{f}_i^+$ and f_{i-1}^+.

Passive Three-Multiply Normalized Junctions. The three-multiply normalized scattering junction is easier to "passify." While the transformer is not structurally lossless, its simplicity allows it to be made passive simply by using magnitude truncation on both of its coefficients as well as on its output wave variables. (The transformer is passive when the product of its coefficients has magnitude less than or equal to 1.) Since there are no additions following the transformer multiplies, double-precision adders are not needed. However, precision and a half is needed in the junction adders to accommodate the worst-case increased dynamic range. Since the one-multiply junction is structurally lossless, the overall junction is passive if magnitude truncation is applied to g_i, $1/g_i$, the outgoing transformer waves, and the waves leaving the one-multiply junction. In other words, the three-multiply normalized scattering junction is passive as long as the transformer coefficients and the four computed wave variables are not amplified by numerical round-off. Again these are sufficient but not necessary conditions, and magnitude truncation will generally result in extra damping.

In summary, a general means of obtaining passive waveguide networks is to compute exact results internally within each junction, and apply *saturation* (clipping on overflow) and *magnitude truncation* (truncation toward zero) to the final outgoing wave variables. Because the Kelly-Lochbaum and one-multiply junctions are structurally lossless, exact intermediate results are obtainable using extended internal precision. For the four-multiply normalized scattering junction, a passive rounding rule can be developed based on two sign bits. For the three-multiply normalized scattering junction, it is sufficient to apply magnitude truncation to the transformer coefficients and all four outgoing wave variables.

10.6 SCATTERING AT A LOADED JUNCTION OF N WAVEGUIDES

In this section, scattering relations will be derived for the general case of N waveguides meeting at a *load*. When a load is present, the scattering is no longer lossless, unless the load itself is lossless (i.e., its impedance has a zero real part). For $N > 2$, v_i^+ will denote a velocity wave traveling *into* the junction, and will be called an "incoming" velocity wave as opposed to "right-going."[5]

Consider first the *series* junction of N waveguides containing transverse force and velocity waves. At a series junction, there is a common velocity while the forces sum. For definiteness, we may think of N ideal strings intersecting at a single point, and the intersection point can be attached to a lumped load impedance $R_J(s)$, as depicted in Fig. 10.11 for $N = 4$. The presence of the lumped load means we need to look at the wave variables in the frequency domain, i.e., $V(s) = \mathcal{L}\{v\}$ for velocity waves and $F(s) = \mathcal{L}\{f\}$ for force waves, where $\mathcal{L}\{\cdot\}$ denotes the Laplace transform. In the discrete-time case, we use the z transform instead, but otherwise the story is identical.

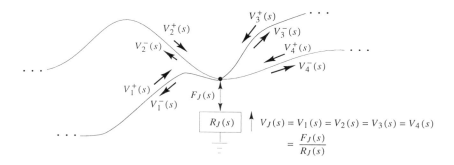

Figure 10.11 Four ideal strings intersecting at a point to which a lumped impedance is attached. This is a series junction for transverse waves.

The physical constraints at the *series* junction are

$$V_1(s) = V_2(s) = \cdots = V_N(s) \triangleq V_J(s) \quad (10.56)$$
$$F_1(s) + F_2(s) + \cdots + F_N(s) = V_J(s) R_J(s) \triangleq F_J(s) \quad (10.57)$$

where the reference direction for the load force F_J is taken to be opposite that for the F_i. (It can be considered the "equal and opposite reaction" force at the junction.) For a wave traveling into the junction, force is positive pulling up, acting toward the junction. When the load impedance $R_J(s)$ is zero, giving a free intersection point, the junction reduces to the unloaded case, and signal scattering is energy preserving. In general, the loaded junction is *lossless* when re $\{R_J(j\omega)\} \equiv 0$, and *memoryless* when im $\{R_J(j\omega)\} \equiv 0$.

The *parallel* junction is characterized by

$$F_1(s) = F_2(s) = \cdots = F_N(s) \triangleq F_J(s) \tag{10.58}$$
$$V_1(s) + V_2(s) + \cdots + V_N(s) = F_J(s)/R_J(s) \triangleq V_J(s) \tag{10.59}$$

For example, $F_i(s)$ could be pressure in an acoustic tube and $V_i(s)$ the corresponding volume velocity. In the parallel case, the junction reduces to the unloaded case when the load impedance $R_J(s)$ goes to infinity.

The scattering relations for the series junction are derived as follows, dropping the common argument '(s)' for simplicity:

$$R_J V_J = \sum_{i=1}^{N} F_i = \sum_{i=1}^{N}(F_i^+ + F_i^-) \tag{10.60}$$

$$= \sum_{i=1}^{N}(R_i V_i^+ - R_i \underbrace{V_i^-}_{V_J - V_i^+}) \tag{10.61}$$

$$= \sum_{i=1}^{N}(2R_i V_i^+ - R_i V_J) \tag{10.62}$$

where R_i is the wave impedance in the ith waveguide, a real, positive constant. Bringing all terms containing V_J to the left-hand side, and solving for the junction velocity gives

$$V_J = 2\left(R_J + \sum_{i=1}^{N} R_i\right)^{-1} \sum_{i=1}^{N} R_i V_i^+ \tag{10.63}$$

$$\triangleq \sum_{i=1}^{N} A_i(s) V_i^+(s) \tag{10.64}$$

where

$$A_i(s) \triangleq \frac{2R_i}{R_J(s) + R_1 + \cdots + R_N} \tag{10.65}$$

Finally, from the basic relation $V_J = V_i = V_i^+ + V_i^-$, the outgoing velocity waves can be computed from the junction velocity and incoming velocity waves as

$$V_i^-(s) = V_J(s) - V_i^+(s) \tag{10.66}$$

In the unloaded case, $R_J(s) = 0$, and we can return to the time domain and define

$$\alpha_i = \frac{2R_i}{R_1 + \cdots + R_N} \qquad (10.67)$$

These we call the *alpha parameters*, and they are analogous to those used to characterize "adaptors" in wave digital filters [Fettweis, 1986]. For unloaded junctions, the alpha parameters obey

$$0 \le \alpha_i \le 2 \qquad (10.68)$$

and

$$\sum_{i=1}^{N} \alpha_i = 2 \qquad (10.69)$$

In the unloaded case, the series junction scattering relations are given (in the time domain) by

$$v_J(t) = \sum_{i=1}^{N} \alpha_i v_i^+(t) \qquad (10.70a)$$

$$v_i^-(t) = v_J(t) - v_i^+(t) \qquad (10.70b)$$

The alpha parameters provide an interesting and useful parametrization of waveguide junctions. They are explicitly the coefficients of the incoming traveling waves needed to compute junction velocity for a series junction (or junction force or pressure at a parallel junction), and losslessness is assured provided only that the alpha parameters be nonnegative and sum to 2. Having them sum to something less than 2 simulates a "resistive load" at the junction.

Note that in the lossless, equal-impedance case, in which all waveguide impedances have the same value $R_i = R$, (10.67) reduces to

$$\alpha_i = \frac{2}{N} \qquad (10.71)$$

When, furthermore, N is a power of two, we have that there are *no multiplies* in the scattering computation (10.70a). This fact has been used to build multiply-free reverberators and other structures using digital waveguide meshes [Smith, 1987, Van Duyne and Smith, 1995, Savioja et al., 1995].

10.7 THE LOSSY ONE-DIMENSIONAL WAVE EQUATION

In any real vibrating string, there are energy losses due to yielding terminations, drag by the surrounding air, and internal friction within the string. While losses in solids generally vary in a complicated way with frequency, they can usually be well approximated by a small number of odd-order terms added to the wave equation. In the simplest case, force is directly proportional to transverse string velocity, independent of frequency. If this proportionality constant is μ, we obtain the modified wave equation

$$Ky'' = \epsilon\ddot{y} + \mu\dot{y} \qquad (10.72)$$

Thus, the wave equation has been extended by a "first-order" term, i.e., a term proportional to the first derivative of y with respect to time. More realistic loss approximations would append terms proportional to $\partial^3 y/\partial t^3$, $\partial^5 y/\partial t^5$, and so on, giving frequency-dependent losses.

It can be checked that, for small displacements, the following modified traveling wave solution satisfies the lossy wave equation:

$$y(t,x) = e^{-(\mu/2\epsilon)x/c} y_r(t - x/c) + e^{(\mu/2\epsilon)x/c} y_l(t + x/c) \qquad (10.73)$$

The left-going and right-going traveling-wave components decay *exponentially* in their respective directions of travel.

Sampling these exponentially decaying traveling waves at intervals of T seconds (or $X = cT$ meters) gives

$$y(t_n, x_m) = g^{-m} y^+(n - m) + g^m y^-(n + m) \qquad (10.74)$$

where $g \triangleq e^{-\mu T/2\epsilon}$.

The simulation diagram for the lossy digital waveguide is shown in Fig. 10.12.

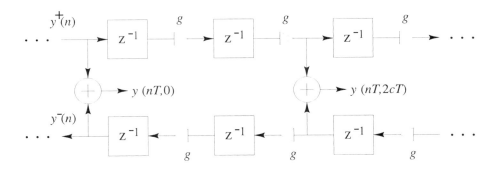

Figure 10.12 Discrete simulation of the ideal, lossy waveguide.

Again the discrete-time simulation of the decaying traveling-wave solution is an *exact* implementation of the continuous-time solution at the sampling positions and instants, even though losses are admitted in the wave equation. Note also that the losses which are *distributed* in the continuous solution have been consolidated, or *lumped*, at discrete intervals of cT meters in the simulation. The loss factor $g = e^{-\mu T/2\epsilon}$ *summarizes* the distributed loss incurred in one sampling interval. The lumping of distributed losses does not introduce an approximation error at the sampling points. Furthermore, bandlimited interpolation can yield arbitrarily accurate reconstruction between samples. The only restriction is again that all initial conditions and excitations be bandlimited to below half the sampling rate.

10.7.1 Loss Consolidation

In many applications, it is possible to realize vast computational savings in digital waveguide models by *commuting* losses out of unobserved and undriven sections of the medium and consolidating them at a minimum number of points. Because the digital simulation is linear and time invariant (given constant medium parameters K, ϵ, μ), and because linear, time-invariant elements commute, the diagram in Fig. 10.13 is exactly equivalent (to within numerical precision) to the previous diagram in Fig. 10.12.

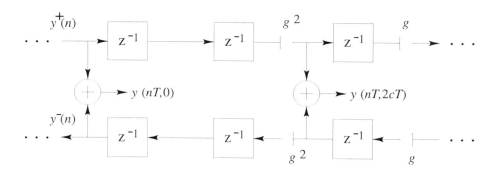

Figure 10.13 Discrete-time simulation of the ideal, lossy waveguide. Each per-sample loss factor g may be "pushed through" delay elements and combined with other loss factors until an input or output is encountered which inhibits further migration. If further consolidation is possible on the other side of a branching node, a loss factor can be pushed *through* the node by pushing a copy into each departing branch. If there are other *inputs* to the node, the *inverse* of the loss factor must appear on each of them. Similar remarks apply to pushing backwards through a node.

10.7.2 Frequency-Dependent Losses

In nearly all natural wave phenomena, losses increase with frequency. Distributed losses due to air drag and internal bulk losses in the string tend to increase with frequency. Similarly, air absorption increases with frequency, adding loss for sound waves in acoustic tubes or open air [Morse and Ingard, 1968].

The solution of a lossy wave equation containing higher odd-order derivatives with respect to time yields traveling waves which propagate with frequency-dependent attenuation. Instead of scalar factors g distributed throughout the diagram, we obtain lowpass filters having frequency-response per sample denoted by $G(\omega)$. If the wave equation (10.1) is modified by adding terms proportional to $\partial^3 y/\partial t^3$ and $\partial^5 y/\partial t^5$, for instance, then $G(\omega)$ is generally of the form

$$G(\omega) = g_0 + g_2 \omega^2 + g_4 \omega^4$$

where the g_i are constants depending on the constant coefficients in the wave equation. These per-sample loss filters may also be consolidated at a minimum number of points in the waveguide without introducing an approximation error in the linear, time-invariant case.

In an efficient digital simulation, lumped loss factors of the form $G^k(\omega)$ are approximated by a rational frequency response $\hat{G}_k(e^{j\omega T})$. In general, the coefficients of the optimal rational loss filter are obtained by minimizing $||G^k(\omega) - \hat{G}_k(e^{j\omega T})||$ with respect to the filter coefficients or the poles and zeros of the filter. To avoid introducing frequency-dependent delay, the loss filter should be a *zero-phase, finite-impulse-response* (FIR) filter [Rabiner and Gold, 1975]. Restriction to zero phase requires the impulse response $\hat{g}_k(n)$ to be finite in length (i.e., an FIR filter) and it must be symmetric about time zero, i.e., $\hat{g}_k(-n) = \hat{g}_k(n)$. In most implementations, the zero-phase FIR filter can be converted into a causal, *linear phase* filter by reducing an adjacent delay line by half of the impulse-response duration.

10.8 THE DISPERSIVE ONE-DIMENSIONAL WAVE EQUATION

Stiffness in a vibrating string introduces a restoring force proportional to the fourth derivative of the string displacement [Morse, 1981]:

$$\epsilon \ddot{y} = K y'' - \kappa y''''$$

where, for a cylindrical string of radius a and Young's modulus Y, the moment constant κ is equal to $\kappa = Y\pi a^4/4$.

At very low frequencies, or for very small κ, we return to the non-stiff case. At very high frequencies, or for very large κ, we approach the *ideal bar* in which stiffness is the only restoring force. At intermediate frequencies, between the ideal string and bar, the stiffness contribution can be treated as a correction term [Cremer, 1984]. This is the region of most practical interest because it is the principal operating region for strings, such as piano strings, whose stiffness has audible consequences (an inharmonic, stretched overtone series). The first-order effect of stiffness is to increase the wave propagation speed with frequency:

$$c(\omega) \triangleq c_0 \left(1 + \frac{\kappa \omega^2}{2K c_0^2}\right)$$

where c_0 is the wave travel speed in the absence of stiffness. Since sound speed depends on frequency, traveling waveshapes will "disperse" as they propagate along the string. That is, a traveling wave is no longer a static shape moving with speed c and expressible as a function of $t \pm x/c$. In a stiff string, the high frequencies propagate *faster* than the low-frequency components. As a result, a traveling velocity step, such as would be caused be a hammer strike, "unravels" into a smoother velocity step with high-frequency "ripples" running out ahead.

In a digital simulation, a frequency-dependent speed of propagation can be implemented in a lumped fashion using *allpass filters* which have a non-uniform delay versus frequency.

Since the temporal and spatial sampling intervals are related by $X = cT$, this must generalize to $X = c(\omega)T \Rightarrow T(\omega) = X/c(\omega) = c_0 T_0 / c(\omega)$, where $T_0 = T(0)$ is the size of a unit delay in the absence of stiffness. Thus, a unit delay z^{-1} may be replaced by

$$z^{-1} \to z^{-c_0/c(\omega)} \qquad \text{(for frequency-dependent wave velocity)}$$

That is, each delay element becomes an allpass filter which approximates the required delay versus frequency. A diagram appears in Fig. 10.14, where $H_a(z)$ denotes the allpass filter which provides a rational approximation to $z^{-c_0/c(\omega)}$.

For computability of the string simulation in the presence of scattering junctions, there must be at least one sample of pure delay along each uniform section of string. This means for at least one allpass filter in Fig. 10.14, we must have $H_a(\infty) = 0$ which implies $H_a(z)$ can be factored as $z^{-1} H'_a(z)$, where $H'_a(z)$ is a causal, stable allpass. In a systolic VLSI implementation, it is desirable to have at least one real delay from the input to the output of *every* allpass filter, in order to be able to pipeline the computation of all of the allpass filters in parallel. Computability can be arranged in practice by deciding on a minimum delay, (e.g., corresponding to the wave velocity at a maximum frequency), and using an allpass filter to provide excess delay beyond the minimum.

Because allpass filters are linear and time invariant, they commute like gain factors with other linear, time-invariant components. Fig. 10.15 shows a diagram equivalent to Fig. 10.14 in which the allpass filters have been commuted and consolidated at two points. For computability in all possible contexts (e.g., when looped on itself), a

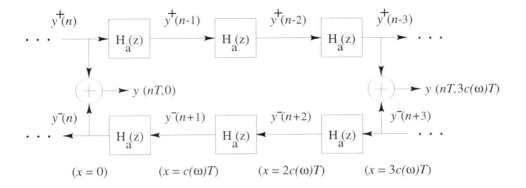

Figure 10.14 Section of a stiff string where allpass filters play the role of unit delay elements.

single sample of delay is pulled out along each rail. The remaining transfer function, $H_c(z) = zH_a^3(z)$ in the example of Fig. 10.15, can be approximated using any allpass filter design technique [Laakso et al., 1996, Lang and Laakso, 1994, Yegnanarayana, 1982]. Alternatively, both gain and dispersion for a stretch of waveguide can be provided by a single filter which can be designed using any general-purpose filter design method which is sensitive to frequency-response phase as well as magnitude; examples include equation error methods (such as used in the Matlab invfreqz() function [Smith, 1983, pp. 48–50]), and Hankel norm methods [Gutknecht et al., 1983, Beliczynski et al., 1992].

In the case of a lossless, stiff string, if $H_c(z)$ denotes the consolidated allpass transfer function, it can be argued that the filter design technique used should minimize the *phase-delay error*, where phase delay is defined by

$$P_c(\omega) \triangleq -\frac{\angle H_c\left(e^{j\omega T}\right)}{\omega} \qquad \text{(Phase Delay)}$$

Minimizing the Chebyshev norm of the phase-delay error, $||P_c(\omega) - c_0/c(\omega)||_\infty$, approximates minimization of the error in *mode tuning* for the freely vibrating string [Smith, 1983, pp. 182–184]. Since the stretching of the overtone series is typically what we hear most in a stiff, vibrating string, the worst-case phase-delay error seems a good choice in such a case. However, psychoacoustic experiments are necessary to determine the error tolerance and the relative audibility of different kinds of error behaviors.

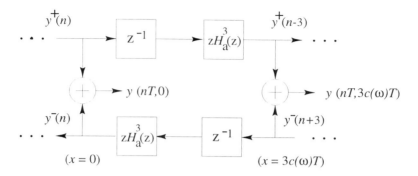

Figure 10.15 Section of a stiff string where the allpass delay elements are consolidated at two points, and a sample of pure delay is extracted from each allpass chain.

Alternatively, a lumped allpass filter can be designed by minimizing *group delay*,

$$D_c(\omega) \triangleq -\frac{d\angle H_c\left(e^{j\omega T}\right)}{d\omega} \qquad \text{(Group Delay)}$$

The group delay of a filter gives the delay experienced by the amplitude *envelope* of a narrow frequency band centered at ω, while the phase delay applies to the "carrier" at ω, or a sinusoidal component at frequency ω [Papoulis, 1977]. As a result, for proper *tuning* of overtones, phase delay is what matters, while for precisely estimating (or controlling) the *decay* time in a lossy waveguide, group delay gives the effective filter delay "seen" by the exponential decay envelope.

To model stiff strings, the allpass filter must supply a phase delay which *decreases* as frequency increases. A good approximation may require a fairly high-order filter, adding significantly to the cost of simulation. To a large extent, the allpass order required for a given error tolerance increases as the number of lumped frequency-dependent delays is increased. Therefore, increased dispersion consolidation is accompanied by larger required allpass filters, unlike the case of resistive losses.

Part II — Applications

We will now review selected applications in digital waveguide modeling, specifically single-reed woodwinds (such as the clarinet), and bowed strings (such as the violin). In these applications, a sustained sound is synthesized by the interaction of the digital waveguide with a *nonlinear* junction causing spontaneous, self-sustaining oscillation in response to an applied mouth pressure or bow velocity, respectively. This type of nonlinear oscillation forms the basis of the Yamaha "VL" series of synthesizers ("VL" standing for "virtual lead").

10.9 SINGLE-REED INSTRUMENTS

A simplified model for a single-reed woodwind instrument is shown in Fig. 10.16.

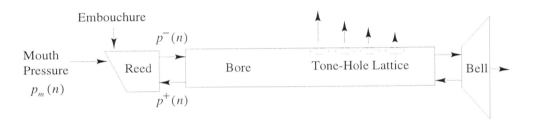

Figure 10.16 A schematic model for woodwind instruments.

If the bore is cylindrical, as in the clarinet, it can be modeled quite simply using a bidirectional delay line [Smith, 1986a, Hirschman, 1991]. If the bore is conical, such as in a saxophone, it can still be modeled as a bidirectional delay line, but interfacing to it is slightly more complex, especially at the mouthpiece [Benade, 1988, Gilbert et al., 1990, Smith, 1991, Välimäki and Karjalainen, 1994a, Scavone, 1997] Because the main control variable for the instrument is air pressure in the mouth at the reed, it is convenient to choose pressure wave variables.

To first order, the bell passes high frequencies and reflects low frequencies, where "high" and "low" frequencies are divided by the wavelength which equals the bell's diameter. Thus, the bell can be regarded as a simple "cross-over" network, as is used to split signal energy between a woofer and tweeter in a loudspeaker cabinet. For a clarinet bore, the nominal "cross-over frequency" is around 1500 Hz [Benade, 1976]. The flare of the bell lowers the cross-over frequency by decreasing the bore characteristic impedance toward the end in an approximately non-reflecting manner

[Berners and Smith, 1994]; it serves a function analogous to that of a transformer coupling of two electrical transmission lines.

Tone holes can also be treated as simple cross-over networks. However, it is more accurate to utilize measurements of tone-hole acoustics in the musical acoustics literature and convert the "transmission matrix" description, often used in the acoustics literature, to the traveling-wave formulation by a simple linear transformation.[6] For typical fingerings, the first few open tone holes jointly provide a bore termination [Benade, 1976]. Either the individual tone holes can be modeled as (interpolated) scattering junctions, or the whole ensemble of terminating tone holes can be modeled in aggregate using a single reflection and transmission filter, like the bell model. The tone-hole model can be simply a lossy two-port junction, to model only the internal bore loss characteristics, or a three-port junction, when it is desired also to model accurately the transmission characteristics to the outside air. Since the tone-hole diameters are small compared with most audio wavelengths, the reflection and transmission coefficients can be implemented to a reasonable approximation as constants, as opposed to crossover filters as in the bell. Taking into account the inertance of the air mass in the tone hole, the tone hole can be modeled as a two-port loaded junction having load impedance given by the air-mass inertance [Fletcher and Rossing, 1993, Välimäki et al., 1993]. At a higher level of accuracy, adapting transmission-matrix parameters from the best available musical acoustics literature [Keefe, 1982, Keefe, 1990] yields first-order reflection and transmission filters in the s-plane, and second-order digital approximations give very good approximations in the z-plane [Scavone, 1997, Scavone and Smith, 1997, Smith and Scavone, 1997]. Digital waveguide models of tone holes are elaborated further in [Scavone, 1997, Välimäki, 1995]. Outstanding issues for tone hole models include nonlinear fluid dynamics effects such as vortex shedding and flow separation at the tone hole [Hirschberg et al., 1995, Hirschberg et al., 1991]. For simple practical implementations, the bell model can be used unchanged for all tunings, as if the bore were being cut to a new length for each note and the same bell were attached.

Since the length of the clarinet bore is only a quarter wavelength at the fundamental frequency, (in the lowest, or "chalumeau" register), and since the bell diameter is much smaller than the bore length, most of the sound energy traveling into the bell reflects back into the bore. The low-frequency energy that makes it out of the bore radiates in a fairly omnidirectional pattern. Very high-frequency traveling waves do not "see" the enclosing bell and pass right through it, radiating in a more directional beam. The directionality of the beam is proportional to how many wavelengths fit along the bell diameter; in fact, many wavelengths away from the bell, the radiation pattern is proportional to the two-dimensional spatial Fourier transform of the exit aperture (a disk at the end of the bell) [Morse and Ingard, 1968].

The theory of the single reed is described in [McIntyre et al., 1983]. In the digital waveguide clarinet model described below [Smith, 1986a], the reed is modeled as

PRINCIPLES OF DIGITAL WAVEGUIDE MODELS OF MUSICAL INSTRUMENTS   457

a signal- and embouchure-dependent nonlinear reflection coefficient terminating the bore. Such a model is possible because the reed mass is neglected. The player's embouchure controls damping of the reed, reed aperture width, and other parameters, and these can be implemented as parameters on the contents of the lookup table or nonlinear function.

10.9.1 Clarinet Overview

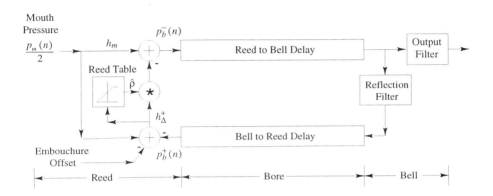

Figure 10.17 Waveguide model of a single-reed, cylindrical-bore woodwind, such as a clarinet.

A diagram of the basic clarinet model is shown in Fig. 10.17. The delay-lines carry left-going and right-going *pressure* samples p_b^+ and p_b^- (respectively) which sample the traveling pressure-wave components within the bore.

The reflection filter at the right implements the bell or tone-hole losses as well as the round-trip attenuation losses from traveling back and forth in the bore. The bell output filter is highpass, and power complementary with respect to the bell reflection filter [Vaidyanathan, 1993].

At the far left is the reed mouthpiece controlled by *mouth pressure* p_m. Another control is *embouchure*, changed in general by modifying the *reflection-coefficient* function $\rho(h_\Delta^+)$, where $h_\Delta^+ \triangleq p_b^-/2 - p_b^+$. A simple choice of embouchure control is an offset in the reed-table address. Since the main feature of the reed table is the pressure-drop where the reed begins to open, a simple embouchure offset can implement the effect of biting harder or softer on the reed, or changing the reed stiffness.

10.9.2 Single-Reed Theory

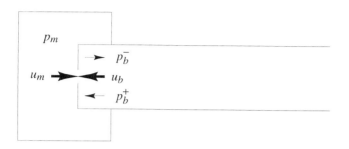

Figure 10.18 Schematic diagram of mouth cavity, reed aperture, and bore.

A simplified diagram of the clarinet mouthpiece is shown in Fig. 10.18. The pressure in the mouth is assumed to be a constant value p_m, and the bore pressure p_b is defined located at the mouthpiece. Any pressure drop $p_\Delta = p_m - p_b$ across the mouthpiece causes a flow u_m into the mouthpiece through the reed-aperture impedance $R_m(p_\Delta)$ which changes as a function of p_Δ since the reed position is affected by p_Δ.

The fundamental equation governing the action of the reed is *continuity of volume velocity*, i.e.,

$$u_b + u_m = 0 \qquad (10.75)$$

where

$$u_m(p_\Delta) \triangleq \frac{p_\Delta}{R_m(p_\Delta)} \qquad (10.76)$$

and

$$u_b \triangleq u_b^+ + u_b^- = \frac{p_b^+ - p_b^-}{R_b} \qquad (10.77)$$

is the volume velocity corresponding to the incoming pressure wave p_b^+ and outgoing pressure wave p_b^-. (The physical pressure in the bore at the mouthpiece is of course $p_b = p_b^+ + p_b^-$.) The wave impedance of the bore air-column is denoted R_b (computable as the air density times sound speed c divided by cross-sectional area).

In operation, the mouth pressure p_m and incoming traveling bore pressure p_b^+ are given, and the reed computation must produce an outgoing bore pressure p_b^- which satisfies (10.75), i.e., such that

$$0 = u_m + u_b = \frac{p_\Delta}{R_m(p_\Delta)} + \frac{p_b^+ - p_b^-}{R_b}, \qquad (10.78)$$

$$p_\Delta \triangleq p_m - p_b = p_m - (p_b^+ + p_b^-)$$

PRINCIPLES OF DIGITAL WAVEGUIDE MODELS OF MUSICAL INSTRUMENTS

Solving for p_b^- is not immediate because of the dependence of R_m on p_Δ which, in turn, depends on p_b^-. A graphical solution technique was proposed [Friedlander, 1953, Keller, 1953, McIntyre et al., 1983] which, in effect, consists of finding the intersection of the two terms of the equation as they are plotted individually on the same graph, varying p_b^-. This is analogous to finding the operating point of a transistor by intersecting its operating curve with the "load line" determined by the load resistance.

It is helpful to normalize (10.78) as follows: Define $G(p_\Delta) = R_b u_m(p_\Delta) = R_b p_\Delta / R_m(p_\Delta)$, and note that $p_b^+ - p_b^- = 2p_b^+ - p_m - (p_b^+ + p_b^- - p_m) \triangleq p_\Delta - p_\Delta^+$, where $p_\Delta^+ \triangleq p_m - 2p_b^+$. Then (10.78) can be multiplied through by R_b and written as $0 = G(p_\Delta) + p_\Delta - p_\Delta^+$, or

$$G(p_\Delta) = p_\Delta^+ - p_\Delta, \qquad p_\Delta^+ \triangleq p_m - 2p_b^+ \tag{10.79}$$

The solution is obtained by plotting $G(x)$ and $p_\Delta^+ - x$ on the same graph, finding the point of intersection at (x, y) coordinates $(p_\Delta, G(p_\Delta))$, and computing finally the outgoing pressure wave sample as

$$p_b^- = p_m - p_b^+ - p_\Delta(p_\Delta^+) \tag{10.80}$$

An example of the qualitative appearance of $G(x)$ overlaying $p_\Delta^+ - x$ is shown in Fig. 10.19.

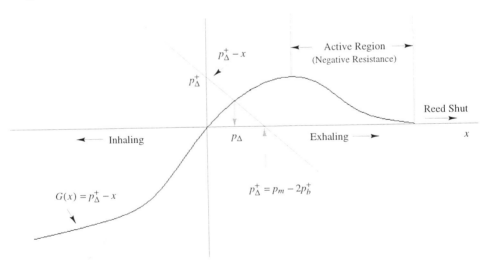

Figure 10.19 Normalized reed impedance $G(p_\Delta) \triangleq R_b u_m(p_\Delta)$ overlaid with the "bore load line" $p_\Delta^+ - p_\Delta = R_b u_b$.

Scattering-Theoretic Formulation of the Reed. Equation (10.78) can be solved for p_b^- to obtain

$$p_b^- = \frac{1-r}{1+r}p_b^+ + \frac{r}{1+r}p_m \quad (10.81)$$

$$= \rho p_b^+ + \frac{1-\rho}{2}p_m \quad (10.82)$$

$$= \frac{p_m}{2} - \rho\frac{p_\Delta^+}{2} \quad (10.83)$$

where

$$\rho(p_\Delta) \triangleq \frac{1 - r(p_\Delta)}{1 + r(p_\Delta)}, \qquad r(p_\Delta) \triangleq \frac{R_b}{R_m(p_\Delta)} \quad (10.84)$$

We interpret $\rho(p_\Delta)$ as a *signal-dependent reflection coefficient*.

Since the mouthpiece of a clarinet is nearly closed, $R_m \gg R_b$ which implies $r \approx 0$ and $\rho \approx 1$. In the limit as R_m goes to infinity relative to R_b, (10.82) reduces to the simple form of a rigidly capped acoustic tube, i.e., $p_b^- = p_b^+$.

Computational Methods. Since finding the intersection of $G(x)$ and $p_\Delta^+ - x$ requires an expensive iterative algorithm with variable convergence times, it is not well suited for real-time operation. In this section, fast algorithms based on precomputed nonlinearities are described.

Let h denote *half-pressure* $p/2$, i.e., $h_m \triangleq p_m/2$ and $h_\Delta^+ \triangleq p_\Delta^+/2$. Then (10.83) becomes

$$p_b^- = h_m - \rho(p_\Delta) \cdot h_\Delta^+ \quad (10.85)$$

Subtracting this equation from p_b^+ gives

$$p_b^+ + \rho h_\Delta^+ - h_m = p_b^+ - p_b^- = p_\Delta - p_\Delta^+$$

$$\Rightarrow \rho h_\Delta^+ = \underbrace{h_m - p_b^+}_{h_\Delta^+} + p_\Delta - 2h_\Delta^+ = p_\Delta - h_\Delta^+$$

$$\Rightarrow \rho = \frac{p_\Delta}{h_\Delta^+} - 1 \quad (10.86)$$

The last expression above can be used to precompute ρ as a function of $h_\Delta^+ \triangleq h_m - p_b^+ = p_m/2 - p_b^+$. Denoting this newly defined function as

$$\hat{\rho}(h_\Delta^+) = \rho(p_\Delta(h_\Delta^+)) \quad (10.87)$$

(10.85) becomes

$$p_b^- = h_m - \hat{\rho}(h_\Delta^+) \cdot h_\Delta^+ \quad (10.88)$$

This is the form chosen for implementation in Fig. 10.17. The control variable is mouth half-pressure h_m, and $h_\Delta^+ = h_m - p_b^+$ is computed from the incoming bore pressure using only a single subtraction. The table is indexed by h_Δ^+, and the result of the lookup is then multiplied by h_Δ^+. Finally, the result of the multiplication is subtracted from h_m to give the outgoing pressure wave into the bore. The cost of the reed simulation is only two subtractions, one multiply, and one table lookup per sample.

Because the table contains a coefficient rather than a signal value, it can be more heavily quantized both in address space and word length than a direct lookup of a signal value such as $p_\Delta(p_\Delta^+)$ or the like. A direct signal lookup, though requiring much higher resolution, would eliminate the multiply associated with the scattering coefficient.

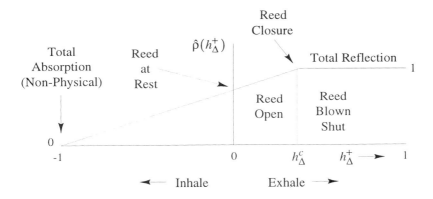

Figure 10.20 Simple, qualitatively chosen reed table for the digital waveguide clarinet.

In the field of computer music, it is customary to use simple piecewise linear functions for functions other than signals at the audio sampling rate, e.g., for amplitude envelopes, FM-index functions, and so on [Roads and Strawn, 1985, Roads, 1989]. Along these lines, good initial results were obtained using the simplified *qualitatively* chosen table

$$\hat{\rho}(h_\Delta^+) = \begin{cases} 1 - m(h_\Delta^c - h_\Delta^+), & -1 \leq h_\Delta^+ < h_\Delta^c \\ 1, & h_\Delta^c \leq h_\Delta^+ \leq 1 \end{cases} \quad (10.89)$$

depicted in Fig. 10.20 for $m = 1/(h_\Delta^c + 1)$. The corner point h_Δ^c is the smallest pressure difference giving reed closure.[7] Embouchure and reed stiffness correspond to the choice of offset h_Δ^c and slope m. For simplicity, an additive offset for shifting the curve laterally is generally used as an embouchure parameter. Brighter tones are obtained by increasing the curvature of the function as the reed begins to open; for example, one can use $\hat{\rho}^k(h_\Delta^+)$ for increasing $k \geq 1$.

Another approach is to replace the table-lookup contents by a piecewise polynomial approximation. While less general, good results have been obtained in practice [Cook, 1992, Cook, 1996]. For example, one of the SynthBuilder clarinet patches employs this technique using a cubic polynomial [Porcaro et al., 1995].

An intermediate approach between table lookups and polynomial approximations is to use interpolated table lookups. Typically, linear interpolation is used, but higher order polynomial interpolation can also be considered [Laakso et al., 1996].

Practical Details. To finish off the clarinet example, the remaining details of the SynthBuilder clarinet patch "Clarinet2.sb" are described.

The input mouth pressure is summed with a small amount of white noise, corresponding to turbulence. For example, 0.1% is generally used as a minimum, and larger amounts are appropriate during the attack of a note. Ideally, the turbulence level should be computed automatically as a function of pressure drop p_Δ and reed opening geometry [Flanagan and Ishizaka, 1976, Verge, 1995]. It should also be lowpass filtered as predicted by theory.

Referring to Fig. 10.17, the reflection filter is a simple one-pole with transfer function

$$H(z) = \frac{1 + a_1(t)}{1 + a_1(t)z^{-1}} \qquad (10.90)$$

where $a_1(t) = v(t) - 0.642$, $v(t) = A_v \sin(2\pi f_v t)$, A_v is vibrato amplitude (e.g., 0.03), and f_v is vibrato frequency (e.g., 5 Hz). Further loop filtering occurs as a result of using simple linear interpolation of the delay line. (There is only one delay line in the actual implementation since the lower delay line of Fig. 10.17 can be commuted with the reflection filter and combined with the upper delay line, ignoring the path to the output filter since a pure delay of less than a period in the final output sound is inconsequential.) There is no transmission filter or tone-hole modeling in this simple patch.

Legato note transitions are managed using two delay line taps and cross-fading from one to the other during a transition [Jaffe and Smith, 1995, Smith, 1996].

10.10 BOWED STRINGS

A schematic block diagram for bowed strings is shown in Fig. 10.21. The bow divides the string into two sections, so the bow model is a nonlinear two-port, in contrast with the reed which was a nonlinear one-port terminating the bore at the mouthpiece. In the case of bowed strings, the primary control variable is bow velocity, so velocity waves are the natural choice for the delay lines.

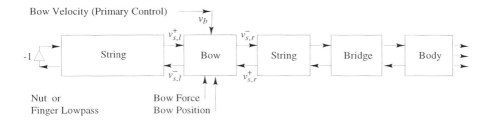

Figure 10.21   A schematic model for bowed-string instruments.

The theory of bow-string interaction is described in [Cremer, 1984, Friedlander, 1953, Keller, 1953, McIntyre and Woodhouse, 1979, McIntyre et al., 1983]. The basic operation of the bow is to reconcile the bow-string friction curve with the string state and string wave impedance. In a bowed string simulation as in Fig. 10.21, a velocity input (which is injected equally in the left- and right-going directions) must be found such that the transverse force of the bow against the string is balanced by the reaction force of the moving string. If bow-hair dynamics are neglected, the bow-string interaction can be simulated using a memoryless table lookup or segmented polynomial in a manner similar to single-reed woodwinds.

10.10.1   Violin Overview

A more detailed diagram of the digital waveguide implementation of the bowed-string instrument model is shown in Fig. 10.22 [Smith, 1986a]. The right delay-line pair carries left-going and right-going velocity waves samples $v_{s,r}^+$ and $v_{s,r}^-$, respectively, which sample the traveling-wave components within the string to the right of the bow, and similarly for the section of string to the left of the bow. The '+' superscript refers to waves traveling *into* the bow.

String velocity at any point is obtained by adding a left-going velocity sample to the right-going velocity sample immediately opposite in the other delay line, as indicated in Fig. 10.22 at the bowing point. The reflection filter at the right implements the losses at the bridge, bow, nut or finger-terminations (when stopped), and the round-trip attenuation/dispersion from traveling back and forth on the string. To a very good degree of approximation, the nut reflects incoming velocity waves (with a sign inversion) at all audio wavelengths. The bridge behaves similarly to a first order, but there are additional (complex) losses due to the finite bridge driving-point impedance (necessary for transducing sound from the string into the resonating body).

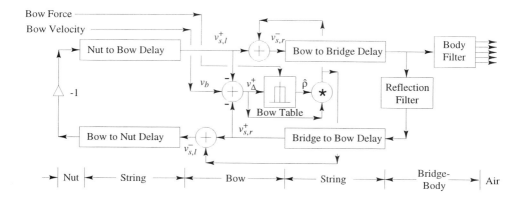

Figure 10.22 Waveguide model for a bowed string instrument, such as a violin.

Figure 10.22 is drawn for the case of the lowest note. For higher notes the delay lines between the bow and nut are shortened according to the distance between the bow and the finger termination. The bow-string interface is controlled by *differential velocity* v_Δ^+ which is defined as the bow velocity minus the current string velocity. Other controls include *bow force* and *angle* which are changed by modifying the reflection-coefficient $\rho(v_\Delta^+)$. Bow position is changed by taking samples from one delay-line pair and appending them to the other delay-line pair. Delay-line interpolation can be used to provide continuous change of bow position.

10.10.2 The Bow-String Scattering Junction

A derivation analogous to that for the single reed is possible for the simulation of the bow-string interaction. The final result is as follows.

$$v_{s,r}^- = v_{s,l}^+ + \hat{\rho}(v_\Delta^+) \cdot v_\Delta^+$$
$$v_{s,l}^- = v_{s,r}^+ + \hat{\rho}(v_\Delta^+) \cdot v_\Delta^+$$

where $v_{s,r}$ denotes transverse velocity on the segment of the bowed string to the *right* of the bow, and $v_{s,l}$ denotes velocity waves to the *left* of the bow. In addition we have $v_\Delta^+ \triangleq v_b - (v_{s,r}^+ + v_{s,l}^+)$, where v_b is bow velocity, and

$$\hat{\rho}(v_\Delta^+) = \frac{r(v_\Delta(v_\Delta^+))}{1 + r(v_\Delta(v_\Delta^+))}$$

The impedance ratio is defined as $r(v_\Delta) = 0.25 R_b(v_\Delta)/R_s$, where $v_\Delta = v_b - v_s$ is the velocity of the bow minus that of the string, $v_s = v_{s,l}^+ + v_{s,l}^- = v_{s,r}^+ + v_{s,r}^-$ is the string velocity in terms of traveling waves, R_s is the wave impedance of the string, and $R_b(v_\Delta)$ is the friction coefficient for the bow against the string, i.e., bow force $F_b(v_\Delta) = R_b(v_\Delta) \cdot v_\Delta$. (Force and velocity point in the same direction when they have the same sign.)

Nominally, $R_b(v_\Delta)$ is constant (the so-called static coefficient of friction) for $|v_\Delta| \leq v_\Delta^c$, where v_Δ^c is both the capture and break-away differential velocity. For $|v_\Delta| > v_\Delta^c$, $R_b(v_\Delta)$ falls quickly to a low dynamic coefficient of friction. It is customary in the bowed-string physics literature to assume that the dynamic coefficient of friction continues to approach zero with increasing $|v_\Delta| > v_\Delta^c$ [McIntyre et al., 1983, Cremer, 1984]. However, plausible bowed-string behavior can also be obtained using a simpler, idealized friction model [Guettler, 1992], in which the bow alternates between acting as a "velocity source" (while stuck to the string) and as a "force source" (while slipping against the string).

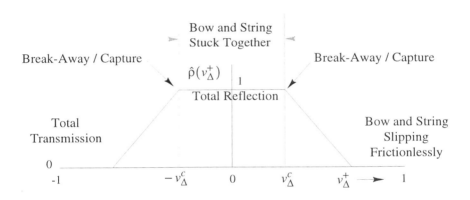

Figure 10.23 Simple, qualitatively chosen bow table for the digital waveguide violin.

Figure 10.23 illustrates a simplified, piecewise linear bow table $\hat{\rho}(v_\Delta^+)$. The flat center portion corresponds to a fixed reflection coefficient "seen" by a traveling wave encountering the bow stuck against the string, and the outer sections of the curve give a smaller reflection coefficient corresponding to the reduced bow-string interaction force while the string is slipping under the bow. The notation v_Δ^c at the corner point denotes the capture or break-away differential velocity. Note that hysteresis (which explains the "pitch flattening effect" when bow force is too heavy [McIntyre and Woodhouse, 1979]) is neglected.

10.11 CONCLUSIONS

Starting with the traveling-wave solution to the wave equation and sampling across time and space, we obtained a modeling framework known as the "digital waveguide" approach. Its main feature is computational economy in the context of a true physical model. Successful computational models have been obtained for the singing voice, several musical instruments of the string, wind, brass, and percussion families, and more are under development.

Physics-based synthesis can provide extremely high quality and expressivity in a very compact algorithm. Such computational models can provide extremely low bit rates at very high quality levels for certain sounds. In addition to data compression applications, such models can also provide a foundation for the future evolution of musical instruments, moving it from the real world of wood and metal into the "virtual world" where formerly impossible modifications are easily tried out.

Notes

1. The Yamaha VL1, introduced in 1994, appears to be the first synthesizer based on physical modeling principles. Korg introduced a related product in 1995. Since then, simplified software-only implementations have appeared.

2. As a specific example, the SynthBuilder virtual electric guitar patch implements 6 "steel" strings, a stereo flanger, amplifier distortion and feedback, and on-chip vibrato, in real time at a 22 kHz sampling rate, on a single Motorola DSP56001 DSP chip, clocked at 25 MHz, with 8 kWords of zero-wait-state SRAM [Porcaro et al., 1995].

3. Note that a "spring reverb" is essentially a one-dimensional structure, "plate reverbs" are primarily two-dimensional, and concert halls are three-dimensional. Since each of these increases in dimensionality is associated with a significant increase in quality, it is reasonable to conjecture that simulated reverberation in four dimensions or more may be even better.

4. Here it is assumed that $-k_i$ and $1 \pm k_i$ in the Kelly-Lochbaum junction can be computed exactly from k_i in the number system being used. This is the case in two's complement arithmetic as is typically used in practice.

5. In the acoustic tube literature which involves only a cascade chain of acoustic waveguides, x^+ is taken to be traveling to the *right* along the axis of the tube [Markel and Gray, 1976]. In classical network theory [Belevitch, 1968] and in circuit theory, velocity (current) at the terminals of an N-port device is by convention taken to be positive when it flows *into* the device.

6. For example, given $[p_1, u_1]^T = M[p_2, u_2]^T$ where p_1 and u_1 are the pressure and volume velocity on one side of a tone hole in an acoustic tube, p_2 and u_2 are the corresponding variables on the other side of the tone hole, and M is the transmission matrix, the transmission matrix formulation is easily converted to a scattering matrix formulation by replacing p_i by $p_i^+ + p_i^-$, replacing u_i by $(p_i^+ - p_i^-)/R_i$, where R_i is the wave impedance at position i in the tube, and solving for $[p_1^-, p_2^-]$ in terms of $[p_1^+, p_2^+]$, where p_i^+ are the known incoming traveling waves, and p_i^- are the unknown traveling waves which are scattered outward by the tone hole.

7. For operation in fixed-point DSP chips, the independent variable $h_\Delta^+ \triangleq p_m/2 - p_b^+$ is generally confined to the interval $[-1, 1)$. Note that having the table go all the way to zero at the maximum negative pressure $h_\Delta^+ = -1$ is not physically reasonable (0.8 would be more reasonable), but it has the practical benefit that when the lookup-table input signal is about to clip, the reflection coefficient goes to zero, thereby opening the feedback loop.

References

[Abraham and Box, 1979] Abraham, B. and Box, G. E. P. (1979). Bayesian analysis of some outlier problems in time series. *Biometrika*, 66(2):229–36.

[Adams, 1986] Adams, R. W. (1986). Design and Implementation of an Audio 18-bit Analog-to-Digital Converter Using Oversampling Techniques. *J. Audio Eng. Society*, 34(3):143–166.

[Akaike, 1974] Akaike, H. (1974). A new look at statistical model identification. *IEEE Trans. Autom. Control*, 19(6):716–723.

[Akansu and Haddad, 1992] Akansu, A. N. and Haddad, R. A. (1992). *Multiresolution Signal Decomposition*. Academic Press, inc.

[Allen, 1975] Allen, J. (1975). Computer Architecture for Signal Processing. *Proc. IEEE*, 63(4):624–633.

[Allen, 1982] Allen, J. (1982). Application of the Short-Time Fourier Transform to Speech Processing and Spectral Analysis. *Proc. IEEE ICASSP-82*, pages 1012–1015.

[Allen, 1991] Allen, J. (1991). Overview of text-to-speech systems. In Furui, S. and Sondhi, M., editors, *Advances in Speech Signal Processing*, chapter 23, pages 741–790. Marcel Dekker.

[Allen and Rabiner, 1977] Allen, J. B. and Rabiner, L. R. (1977). A unified approach to short-time Fourier analysis and synthesis. *Proc. IEEE*, 65(11):1558–1564.

[Alles, 1977] Alles, H. G. (1977). A Modular Approach to Building Large Digital Synthesis Systems. *Computer Music Journal*, 1(4):10–13.

[Alles, 1980] Alles, H. G. (1980). Music Synthesis Using Real Time Digital techniques. *Proc. IEEE*, 68(4):436–449.

[Alles, 1987] Alles, H. G. (1987). A Portable Digital Sound-Synthesis System. In Roads, C. and Strawn, J., editors, *Foundations of computer music*, pages 244–250. MIT Press. (Also appeared in Computer Music Journal, vol. 1, no. 4, 1977).

[Alles and di Giugno, 1977] Alles, H. G. and di Giugno, P. (1977). The 4B: A One-Card 64 Channel Digital Synthesizer. *Computer Music Journal*, 1(4):7–9.

[Almeida and Silva, 1984a] Almeida, L. and Silva, F. (1984a). Variable-frequency synthesis: an improved harmonic coding scheme. *Proc. IEEE ICASSP-84*, pages 27.5.1–27.5.4.

[Almeida and Silva, 1984b] Almeida, L. and Silva, F. (1984b). Variable-frequency synthesis: An improved harmonic coding scheme. In *Proc. IEEE Int. Conf. Acoustics, Speech, and Signal Processing.*, San Diego, CA.

[Alonso, 1979] Alonso, S. A. (1979). Musical synthesis envelope control techniques. U.S. Patent 4,178,822.

[Ammitzboll, 1987] Ammitzboll, K. (1987). Resonant peak control. U.S. Patent 4,689,818.

[Anderson and Moore, 1979] Anderson, B. D. O. and Moore, J. B. (1979). *Optimal Filtering*. Prentice-Hall, Englewood Cliffs, NJ.

[Anderson, 1996] Anderson, D. (1996). Speech analysis and coding using a multi-resolution sinusoidal transform. In *Proc. IEEE Int. Conf. Acoustics, Speech, and Signal Processing.*, pages 1037–1040, Atlanta, GA.

[Andreas et al., 1996] Andreas, D. C., Dattorro, J., and Mauchly, J. W. (1996). Digital Signal Processor for audio applications. U.S. Patent 5,517,436.

[ANSI, 1987] ANSI (1987). ANSI S3.22-1987: Specification of Hearing Aid Characteristics. Technical report, American National Standards Institute.

[Arakawa et al., 1986] Arakawa, K., Fender, D., Harashima, H., Miyakawa, H., and Saitoh, Y. (1986). Separation of a non-stationary component from the EEG by a nonlinear digital filter. *IEEE Trans. Biomedical Engineering*, 33(7):724–726.

[Arfib, 1991] Arfib, D. (1991). Analysis, transformation, and resynthesis of musical sounds with the help of a time-frequency representation. In de Poli, G., Piccialli, A., and Roads, C., editors, *Representation of musical signals*, pages 87–118. M.I.T Press, Cambridge.

[Arfib and Delprat, 1993] Arfib, D. and Delprat, N. (1993). Musical transformations using the modification of time-frequency images. *Computer Music J.*, 17(2):66–72.

[Asta et al., 1980] Asta, V., A. Chauveau, G. D. G., and Kott, J. (1980). Il sistema di sintesi digitale in tempo reale 4X (The real time digital synthesis system 4X). *Automazione e Strumentazione*, XXVII(2):119–133. (In Italian).

[Atal and Hanauer, 1971] Atal, B. S. and Hanauer, L. S. (1971). Speech analysis and synthesis by linear prediction of the speech wave. *J. Acoustical Soc. of America*, 50:637–655.

[ATSC, 1995] ATSC (1995). Digital audio compression (AC-3) standard. Doc. A/52/10, U.S. Advanced Television Systems Committee.

[Axon and Davies, 1949] Axon, P. E. and Davies, H. (1949). A study of frequency fluctuations in sound recording and reproducing systems. *Proc. IEE, Part III*, page 65.

[Baer et al., 1993] Baer, T., Moore, B., and Gatehouse, S. (1993). Spectral contrast enhancement of speech in noise for listeners with sensorineural hearing impairment: Effects on intelligibility, quality and response times. *J. Rehab. Res. and Devel*, 30:49–72.

[Banga and Garcia-Mateo, 1995] Banga, E. and Garcia-Mateo, C. (1995). Shape-invariant pitch-synchronous text-to-speech conversion. In *Proc. IEEE Int. Conf. Acoustics, Speech, and Signal Processing.*, volume 1, pages 656–659, Detroit, Michigan.

[Barnett and Lewis, 1984] Barnett, V. and Lewis, T. (1984). *Outliers in statistical data, 2nd Ed.* Chichester:Wiley.

[Barrière et al., 1989] Barrière, J.-B., Freed, A., Baisnée, P.-F., and Baudot, M.-D. (1989). A Digital Signal Multiprocessor and its Musical Application. In *Proc. of the 15th Intl. Computer Music Conference*. Computer Music Association.

[Barron and Marshall, 1981] Barron, M. and Marshall, A. H. (1981). "Spatial Impression Due to Early Lateral Reflection in Concert Halls: The Derivation of a Physical Measure". *J. Sound and Vibration*, 77(2):211–232.

[Baudot, 1987] Baudot, M.-D. (1987). Hardware design of a digital mixer for musical applications. In *Proc. AES 83rd convention*. Audio Eng. Society. Preprint 2506.

[Bauer and Seitzer, 1989a] Bauer, D. and Seitzer, D. (1989a). Frequency domain statistics of high quality stereo signals. In *Proc. of the 86th. AES-Convention*. Preprint 2748.

[Bauer and Seitzer, 1989b] Bauer, D. and Seitzer, D. (1989b). Statistical properties of high quality stereo signals in the time domain. In *Proc. IEEE Int. Conf. Acoust., Speech and Signal Proc*, pages 2045 – 2048.

[Baumgarte et al., 1995] Baumgarte, F., Ferekidis, C., and Fuchs, H. (1995). A nonlinear psychoacoustic model applied to the ISO MPEG Layer 3 coder. In *Proc. of the 99th. AES-Convention*. Preprint 4087.

[Bech, 1995] Bech, S. (1995). "Perception of Reproduced Sound: Audibility of Individual Reflections in a Complete Sound Field, II". In *Proc. Audio Eng. Soc. Conv.* Preprint 4093.

[Beckman and Cook, 1983] Beckman, R. J. and Cook, R. D. (1983). Outlier. . . . . s. *Technometrics*, 25(2):119–165.

[Beerends, 1989] Beerends, J. G. (1989). A stochastic subharmonic pitch model. In *Pitches of simultaneous complex tones*, chapter 5, pages 84–100. PhD. dissertation, Technical University Eindhoven.

[Beerends, 1995] Beerends, J. G. (1995). Measuring the quality of speech and music codecs, an integrated psychoacoustic approach. *Contribution to the 98th AES Convention, Paris, February 1995, preprint 3945*.

[Beerends and Stemerdink, 1992] Beerends, J. G. and Stemerdink, J. A. (1992). A perceptual audio quality measure based on a psychoacoustic sound representation. *J. Audio Eng. Soc.*, 40(12):963–978.

[Beerends and Stemerdink, 1994a] Beerends, J. G. and Stemerdink, J. A. (1994a). Modelling a cognitive aspect in the measurement of the quality of music codecs. *Contribution to the 96th AES Convention, Amsterdam, February 1994, preprint 3800*.

[Beerends and Stemerdink, 1994b] Beerends, J. G. and Stemerdink, J. A. (1994b). A perceptual speech quality measure based on a psychoacoustic sound representation. *J. Audio Eng. Soc.*, 42(3):115–123.

[Beerends et al., 1996] Beerends, J. G., van den Brink, W. A. C., and Rodger, B. (1996). The role of informational masking and perceptual streaming in the measurement of music codec quality. *Contribution to the 100th AES Convention, Copenhagen, May 1996, preprint 4176*.

[Beerends94dec, 1994] Beerends94dec (1994). *Correlation between the PSQM and the subjective results of the ITU-T 8 kbit/s 1993 speech codec test*. ITU-T Study Group 12. COM 12-31.

[Begault, 1992] Begault, D. R. (1992). "Perceptual effects of synthetic reverberation on three-dimensional audio systems". *J. Audio Eng. Soc.*, 40:895–904.

[Begault, 1994] Begault, D. R. (1994). *3-D Sound for Virtual Reality and Multimedia*. Academic Press, Cambridge, MA.

[Belevitch, 1968] Belevitch, V. (1968). *Classical Network Theory*. Holden Day, San Francisco.

[Beliczynski et al., 1992] Beliczynski, B., Kale, I., and Cain, G. D. (1992). Approximation of FIR by IIR digital filters: An algorithm based on balanced model reduction. *IEEE Trans. Acoustics, Speech, Signal Processing*, 40(3):532–542.

[Benade, 1976] Benade, A. H. (1976). *Fundamentals of Musical Acoustics*. Oxford University Press, Oxford and New York. (Reprinted by Dover Publications, New York, 1990).

[Benade, 1988] Benade, A. H. (1988). Equivalent circuits for conical waveguides. *J. Acoustical Soc. of America*, 83(5):1764–1769.

[Bennett, 1948] Bennett, W. R. (1948). Spectra of quantized signals. *Bell System Technical Journal*, pages 446–471.

[Beranek, 1986] Beranek, L. L. (1986). *Acoustics*. American Institute of Physics, New York, NY.

[Beranek, 1992] Beranek, L. L. (1992). "Concert hall acoustics". *J. Acoust. Soc. Am.*, 92(1):1–39.

[Berger et al., 1994] Berger, J., Coifman, R. R., and Goldberg, M. J. (1994). Removing noise from music using local trigonometric bases and wavelet packets. *J. Audio Eng. Soc.*, 42(10):808–818.

[Berger, 1971] Berger, T. (1971). *Rate Distortion Theory*. Englewood Cliffs.

[Bernardo and Smith, 1994] Bernardo, J. and Smith, A. (1994). *Bayesian Theory*. John Wiley & Sons.

[Berners and Smith, 1994] Berners, D. P. and Smith, J. O. (1994). On the use of Schrodinger's equation in the analytic determination of horn reflectance. In *Proc. 1994 Int. Computer Music Conf., Århus*, pages 419–422. Computer Music Association.

[Berouti et al., 1979] Berouti, M., Schwartz, R., and Makhoul, J. (1979). Enhancement of speech corrupted by additive noise. In *Proc. IEEE Int. Conf. Acoust., Speech, Signal Processing*, pages 208–211.

[Billings and Voon, 1986] Billings, S. A. and Voon, W. S. F. (1986). A prediction-error and stepwise-regression estimation algorithm for non-linear systems. *Int. J. Control*, 44(3):803–822.

[Bisgaard, 1993] Bisgaard, N. (1993). Digital feedback suppression: Clinical experiences with profoundly hearing impaired. In Beilin, J. and Jensen, G., editors, *Recent Developments in Hearing Instrument Technology: 15th Danavox Symposium*, pages 370–384.

[Blauert, 1983] Blauert, J. (1983). *Spatial Hearing*. MIT Press, Cambridge, MA.

[Blauert and Cobben, 1978] Blauert, J. and Cobben, W. (1978). "Some consideration of binaural cross-correlation analysis". *Acustica*, 39(2):96–104.

[Blauert and Tritthart, 1975] Blauert, J. and Tritthart, P. (1975). Ausnutzung von Verdeckungseffekten bei der Sprachcodierung (Exploitation of Masking Effects in Speech Coding). In *Proc. DAGA '75*, pages 377 – 380. (in German).

[Blesser, 1969] Blesser, B. (1969). Audio dynamic range compression for minimum perceived distortion. *IEEE Trans. Audio and Electroacoustics*, Vol. AU-17(1):22–32.

[Blesser and Kates, 1978] Blesser, B. and Kates, J. M. (1978). Digital Processing in Audio Signals. In Oppenheim, A. V., editor, *Applications of Digital Signal Processing*, pages 29–116. Prentice-Hall.

[Blesser, 1978] Blesser, B. A. (1978). Digitization of Audio: A Comprehensive Examination of Theory, Implementation and Current Practice. *J. Audio Eng. Society*, 26(10):739–771.

[Blood, 1980] Blood, W. (1980). *The MECL System Design Handbook*. Motorola Semiconductor.

[Boddie et al., 1981] Boddie, J., Daryanani, G., Eldumiati, I., Gadenz, R., Thompson, J., and Waters, S. (1981). Digital Signal Processor: Architecture and Performance. *Bell System Techn. J.*, 60(7):1449–1462.

[Boddie et al., 1986] Boddie, J. R., Hays, W. P., and Tow, J. (1986). The Architecture, Instruction Set and Development Support for the WE DSP-32 Digital Signal Processor. In *Proc. IEEE Int. Conf. Acoust., Speech and Signal Proc. 1986*, pages 421–424.

[Boers, 1980] Boers, P. (1980). Formant enhancement of speech for listeners with sensorineural hearing loss. Technical report, Inst. voor Perceptie Onderzoek.

[Boll, 1979] Boll, S. F. (1979). Suppression of acoustic noise in speech using spectral subtraction. *IEEE Trans. Acoust., Speech, Signal Processing*, 27(2):113–120.

[Boll, 1991] Boll, S. F. (1991). Speech enhancement in the 1980s: noise suppression with pattern matching. In Furui, S. and Sondhi, M., editors, *Advances in Speech Signal Processing*, chapter 10, pages 309–325. Marcel Dekker.

[Boothroyd et al., 1988] Boothroyd, A., Springer, N., Smith, L., and Schulman, J. (1988). Amplitude compression and profound hearing loss. *J. Speech and Hearing Res.*, 31:362–376.

[Borish, 1984] Borish, J. (1984). "Extension of the Image Model to Arbitrary Polyhedra". *J. Acoust. Soc. Am.*, 75(6):1827–1836.

[Bosi et al., 1996a] Bosi, M., Brandenburg, K., Quackenbush, S., Fielder, L., Akagin, K., Fuchs, H., Dietz, M., Herre, J., Davidson, G., and Oikawa, Y. (1996a). ISO / IEC MPEG-2 advanced audio coding. *Audio Eng. Soc. Convention*, Preprint 4382. 36 pages. See also ISO/IEC International Standard IS 13818-7 entitled "MPEG-2 Advanced Audio Coding," April, 1997.

[Bosi et al., 1996b] Bosi, M., Brandenburg, K., Quackenbush, S., Fielder, L., Akagiri, K., Fuchs, H., Dietz, M., Herre, J., Davidson, G., and Oikawa, Y. (1996b). ISO/IEC MPEG-2 Advanced Audio Coding. In *Proc. of the 101st AES-Convention*. Preprint 4382.

[Box and Tiao, 1973] Box, G. and Tiao, G. (1973). *Bayesian Inference in Statistical Analysis*. Addison-Wesley.

[Box and Jenkins, 1970] Box, G. E. P. and Jenkins, G. M. (1970). *Time Series Analysis, Forecasting and Control*. Holden-Day.

[Box and Tiao, 1968] Box, G. E. P. and Tiao, G. C. (1968). A Bayesian approach to some outlier problems. *Biometrika*, 55:119–129.

[Braida et al., 1979] Braida, L., Durlach, N., Lippmann, R., Hicks, B., Rabinowitz, W., and Reed, C. (1979). Hearing aids - a review of past research on linear amplification, amplitude compression and frequency lowering. In *ASHA Monogr. 19*. Am. Speech-Lang.-Hearing Assn.

[Brandenburg, 1987] Brandenburg, K. (1987). OCF – a new coding algorithm for high quality sound signals. In *Proc. IEEE Int. Conf. Acoust., Speech and Signal Proc*, pages 141–144.

[Brandenburg, 1988] Brandenburg, K. (1988). High quality sound coding at 2.5 bits/sample. In *Proc of the 84th. AES-Convention*. Preprint 2582.

[Brandenburg, 1991] Brandenburg, K. (1991). ASPEC coding. In *Proc of the 10th. International conference of the AES*, pages 81 – 90. Audio Eng. Society.

[Brandenburg and Bosi, 1997] Brandenburg, K. and Bosi, M. (1997). Overview of MPEG audio: Current and future standards for low bit-rate audio coding. *J. Audio Eng. Soc.*, 45(1/2):4 – 21.

[Brandenburg and Henke, 1993] Brandenburg, K. and Henke, R. (1993). Near lossless coding of high quality digital audio: First results. In *Proc. IEEE Int. Conf. Acoust., Speech and Signal Proc*, volume 1, pages 193–196.

[Brandenburg et al., 1991] Brandenburg, K., Herre, J., Johnston, J. D., Mahieux, Y., and Schroeder, E. F. (1991). ASPEC: Adaptive spectral perceptual entropy coding of high quality music signals. In *Proc of the 90th. AES-Convention*. Preprint 3011.

[Brandenburg and Johnston, 1990] Brandenburg, K. and Johnston, J. (1990). Second generation perceptual audio coding: The hybrid coder. In *Proc. of the 88th. AES-Convention*, Montreaux. preprint 2937.

[Brandenburg et al., 1982] Brandenburg, K., Langenbucher, G. G., Schramm, H., and Seitzer, D. (1982). A digital signal processor for real time adaptive transform coding of audio signals up to 20 khz bandwidth. In *Proc. of the ICCC*, pages 474–477.

[Brandenburg and Sporer, 1992] Brandenburg, K. H. and Sporer, T. (1992). NMR and masking flag: Evaluation of quality using perceptual criteria. In *Proceedings of the AES 11th international conference, Portland, Oregon USA*, pages 169–179.

[Branderbit, 1991] Branderbit, P. (1991). A standardized programming system and three-channel compression hearing aid. *Hear. Inst.*, 42:6,13.

[Bregman, 1990] Bregman, A. (1990). *Auditory Scene Analysis: The Perceptual Organization of Sound*. The MIT Press, Cambridge, MA.

[Brey et al., 1987] Brey, R., Robinette, M., Chabries, D., and Christiansen, R. (1987). Improvement in speech intelligibility in noise employing an adaptive filter with normal and hearing-impaired subjects. *J. Rehab. Res. and Devel.*, 24:75–86.

[Brigham, 1974] Brigham, E. O. (1974). *The Fast Fourier Transform*. Prentice-Hall, Inc., Englewood Cliffs, NJ.

[Brillinger, 1981] Brillinger, D. R. (1981). *Time Series Data Analysis and Theory*. Holden-Day, expanded edition.

[Bristow-Johnson, 1995] Bristow-Johnson, R. (1995). A detailed analysis of a time-domain formant-corrected pitch-shifting algorithm. *J. Audio Eng. Soc.*, 43(5):340–352.

[Bunnell, 1990] Bunnell, H. T. (1990). On enhancement of spectral contrast in speech for hearing-impaired listeners. *J. Acoust. Soc. Am.*, 88(6):2546–2556.

[Burkhard and Sachs, 1975] Burkhard, M. and Sachs, R. (1975). Anthropometric manikin for acoustic research. *J. Acoust. Soc. Am.*, 58:214–222.

[Bustamante and Braida, 1987] Bustamante, D. and Braida, L. (1987). Principal-component amplitude compression for the hearing impaired. *J. Acoust. Soc. Am.*, 82:1227–1242.

[Bustamante et al., 1989] Bustamante, D., Worrell, T., and Williamson, M. (1989). Measurement of adaptive suppression of acoustic feedback in hearing aids. *Proc. 1989 Int. Conf. Acoust. Speech and Sig. Proc.*, pages 2017–2020.

[Byrd and Yavelow, 1986] Byrd, D. and Yavelow, C. (1986). The Kurzweil 250 digital synthesizer. *Computer Music J.*, 10(1):64–86.

[Campbell Jr. et al., 1990] Campbell Jr., J. P., Tremain, T. E., and Welch, V. C. (1990). The Proposed Federal Standard 1016 4800 bps Voice Coder: CELP. *Speech Technology Magazine*, pages 58–64.

[Canagarajah, 1991] Canagarajah, C. N. (1991). A single-input hearing aid based on auditory perceptual features to improve speech intelligibility in noise. In *Proc. IEEE Workshop Appl. of Signal Processing to Audio and Acoustics*, Mohonk Mountain House, New Paltz, NY.

[Canagarajah, 1993] Canagarajah, C. N. (1993). *Digital Signal Processing Techniques for Speech Enhancement in Hearing Aids*. PhD thesis, University of Cambridge.

[Candy and Temes, 1992] Candy, J. C. and Temes, G. C. (1992). Oversampling Methods for A/D and D/A Conversion. In Candy, J. C. and Temes, G. C., editors, *Oversampling Delta-Sigma Data Converters*, pages 1–29. IEEE Press.

[Cappé, 1991] Cappé, O. (1991). *Noise reduction techniques for the restoration of musical recordings* (text in French). PhD thesis, Ecole Nationale Supérieure des Télécommunications, Paris.

[Cappé, 1994] Cappé, O. (1994). Elimination of the musical noise phenomenon with the Ephraim and Malah noise suppressor. *IEEE Trans. Speech and Audio Processing*, 2(2):345–349.

[Cappé and Laroche, 1995] Cappé, O. and Laroche, J. (1995). Evaluation of short-time spectral attenuation techniques for the restoration of musical recordings. *IEEE Trans. on Speech and Audio Processing*, 3(1):84–93.

[Cappé et al., 1995] Cappé, O., Laroche, J., and Moulines, E. (1995). Regularized estimation of cepstrum envelope from discrete frequency points. In *Proc. IEEE Workshop Appl. of Signal Processing to Audio and Acoustics*, Mohonk Mountain House, New Paltz, NY.

[Carlson, 1988] Carlson, E. (1988). An output amplifier whose time has come. *Hearing Instr.*, 39:30–32.

[Carrey and Buckner, 1976] Carrey, M. J. and Buckner, I. (1976). A system for reducing impulsive noise on gramophone reproduction equipment. *The Radio Electronic Engineer*, 50(7):331–336.

[Cavaliere, 1991] Cavaliere, S. (1991). New Generation Architectures for Music and Sound Processing. In Poli, G. D., Piccialli, A., and Roads, C., editors, *Representations of Musical Signals*, pages 391–411. MIT Press.

[Cavaliere et al., 1992] Cavaliere, S., DiGiugno, G., and Guarino, E. (1992). MARS – The X20 device and the SM1000 board. In *Proc. of the Intl. Conf. on Computer Music*, pages 348–351.

[CCIRrec562, 1990] CCIRrec562 (1990). *Subjective assessment of sound quality*. ITU-R. Vol. X, Part1, Recommendation 562-3, Düsseldorf.

[CCITT86sg12con46, 1986] CCITT86sg12con46 (1986). *Objective evaluation of non-linear distortion effects on voice transmission quality*. ITU-T Study Group 12. Contribution 46.

[CCITTrecG728, 1992] CCITTrecG728 (1992). *Coding of speech at 16 kbit/s using low-delay code excited linear prediction*. ITU-T. Recommendation G.728.

[CCITTrecG729, 1995] CCITTrecG729 (1995). *Coding of speech at 8 kbit/s using code excited linear prediction*. ITU-T. Recommendation G.729.

[CCITTrecP48, 1989] CCITTrecP48 (1989). *Specification for an intermediate reference system*. ITU-T. Recommendation P.48.

[CCITTrecP80, 1994] CCITTrecP80 (1994). *Telephone transmission quality subjective opinion tests*. ITU-T. Recommendation P.80.

[CCITTsup13, 1989] CCITTsup13 (1989). *Noise Spectra*. ITU-T. Series P Recommendations Suppl. No. 13.

[CCITTsup3, 1989] CCITTsup3 (1989). *Models for predicting transmission quality from objective measurements*. ITU-T. Series P Recommendations Suppl. No. 3.

[Cellier, 1994] Cellier, C. (1994). Lossless audio bit rate reduction. In *Proc. AES UK Conference Managing the Bit Budget*, pages 107 – 122. Audio Eng. Society.

[Chabries et al., 1987] Chabries, D., Christiansen, R., Brey, R., Robinette, M., and Harris, R. (1987). Applications of adaptive digital signal processing to speech enhancement for the hearing impaired. *J. Rehab. Res. and Devel.*, 24:65–72.

[Chaigne, 1992] Chaigne, A. (1992). On the use of Finite Differences for Musical Synthesis. Application to Plucked Stringed Instruments. *J. d'Acoustique*, 5(2):181–211.

[Chaigne and Askenfelt, 1994] Chaigne, A. and Askenfelt, A. (1994). Numerical Simulations of Piano Strings. I. A Physical Model for a Struck String Using Finite Difference Methods. *J. Acoustical Soc. of America*, 95(2):1112–1118.

[Chen and Billings, 1989] Chen, S. and Billings, S. A. (1989). Modelling and analysis of non-linear time series. *Int. J. Control*, 50(5):2151–2171.

[Chen and Fang, 1992] Chen, Y.-H. and Fang, H.-D. (1992). Frequency-domain implementation of Griffiths-Jim adaptive beamformer. *J. Acoust. Soc. Am.*, 91:3354–3366.

[Cheng et al., 1994] Cheng, Y., O'Shaughnessy, D., and Mermelstein, P. (1994). Statistical recovery of wideband speech from narrowband speech. *IEEE Trans. on Speech and Audio Processing*, ASSP-2(4):544–548.

[Cheng and O'Shaughnessy, 1991] Cheng, Y. M. and O'Shaughnessy, D. (1991). Speech enhancement based conceptually on auditory evidence. *IEEE Trans. Signal Processing*, 39(9):1943–1954.

[Chowning, 1973] Chowning, J. (1973). The synthesis of complex audio spectra by means of frequency modulation. *Journal of the Audio Engineering Society*, 21(7):526–534. (Reprinted in Roads and Strawn, *Foundations of Computer Music*, pp. 6–29).

[Chowning, 1977] Chowning, J. M. (1977). Method of synthesizing a musical sound. U.S. Patent 4,018,121.

[Chui, 1992a] Chui, C. K. (1992a). *Wavelet Analysis and its Applications, Volume 1: An Introduction to Wavelets*. Academic Press, inc.

[Chui, 1992b] Chui, C. K. (1992b). *Wavelet Analysis and its Applications, Volume 2: Wavelets: A Tutorial in Theory and Applications*. Academic Press, inc.

[Clark et al., 1981a] Clark, D. W., Lampson, B. W., and Pier, K. A. (1981a). The Memory System of a High-Performance Personal Computer. *IEEE Transactions on Computers*, C-30(10):715–733.

[Clark et al., 1981b] Clark, G., Mitra, S., and Parker, S. (1981b). Block implementations of adaptive digital filters. *IEEE Trans. Acoust. Speech and Sig. Proc.*, ASSP-29:744–752.

[Clark et al., 1983] Clark, G., Parker, S., and Mitra, S. (1983). A unified approach to time- and frequency-domain realization of FIR adaptive digital filters. *IEEE Trans. Acoust. Speech and Sig. Proc.*, ASSP-31:1073–1083.

[Cole, 1993] Cole, W. (1993). Current design options and criteria for hearing aids. *J. Speech-Lang. Path. and Audiol. Monogr. Suppl. 1*, pages 7–14.

[Collins, 1993] Collins, M. (1993). Infinity: DSP sampling tools for Macintosh. *Sound on Sound*, 9(1):44–47.

[Colomes et al., 1994] Colomes, C., Lever, M., and Dehery, Y. F. (1994). A perceptual objective measurement POM system for the quality assessment of perceptual codecs. *Contribution to the 96th AES Convention, Amsterdam, February 1994, Preprint 3801*.

[Cook and Bernfeld, 1967] Cook, C. and Bernfeld, M. (1967). *Radar Signals*. Academic Press, New York, NY.

[Cook, 1990] Cook, P. R. (1990). *Identification of Control Parameters in an Articulatory Vocal Tract Model, with Applications to the Synthesis of Singing*. PhD thesis, Elec. Eng. Dept., Stanford University.

[Cook, 1992] Cook, P. R. (1992). A meta-wind-instrument physical model, and a meta-controller for real time performance control. In *Proc. 1992 Int. Computer Music Conf., San Jose*, pages 273–276. Computer Music Association.

[Cook, 1996] Cook, P. R. (1996). Synthesis toolkit in C++, version 1.0. In *SIGGRAPH Proceedings*. Assoc. Comp. Mach. See http://www.cs.princeton.edu/~prc/NewWork.html for a copy of this paper and the software.

[Cooper and Yates, 1994] Cooper, N. and Yates, G. (1994). Nonlinear input-output functions derived from the response of guinea-pig cochlear nerve fibres. *Hearing Res.*, 78:221–234.

[Cornelisse et al., 1991] Cornelisse, L., Gagneé, J.-P., and Seewald, R. (1991). Ear level recordings of the long-term average spectrum of speech. *Ear and Hearing*, 12:47–54.

[Cox, 1973] Cox, H. (1973). Resolving power and sensitivity to mismatch of optimum array processors. *J. Acoust. Soc. Am.*, 54:771–785.

[Cox et al., 1986] Cox, H., Zeskind, R., and Kooij, T. (1986). Practical supergain. *IEEE Trans. Acoust. Speech and Sig. Proc.*, ASSP-34:393–398.

[Cox et al., 1987] Cox, H., Zeskind, R., and Owen, M. (1987). Robust adaptive beamforming. *IEEE Trans. Acoust. Speech and Sig. Proc.*, ASSP-35:1365–1376.

[Cox et al., 1983] Cox, R., Crochiere, R., and Johnston, J. (1983). Real-time implementation of time domain harmonic scaling of speech for rate modification and coding. *IEEE Trans. Acoust., Speech, Signal Processing*, 31(1):258–271.

[Cremer, 1984] Cremer, L. (1984). *The Physics of the Violin*. MIT Press, Cambridge, MA.

[Crochiere, 1980] Crochiere, R. (1980). A weighted overlap-add method of short-time Fourier analysis/synthesis. *IEEE Trans. Acoust., Speech, Signal Processing*, ASSP-28(2):99–102.

[Crochiere and Rabiner, 1983] Crochiere, R. E. and Rabiner, L. R. (1983). *Multirate Digital Signal Processing*. Prentice-Hall, Englewood Cliffs, NJ.

[Danisewicz, 1987] Danisewicz, R. (1987). Speaker separation of steady state vowels. Master's thesis, Department of Electrical Engineering and Computer Science, Massachusetts Institute of Technology.

[Dattorro, 1987] Dattorro, J. (1987). Using digital signal processor chips in a stereo audio time compressor/expander. *Proc. 83rd AES Convention, New York*. preprint 2500 (M-6).

[Dattorro, 1997] Dattorro, J. (1997). "Effect Design Part I: Reverberator and Other Filters". *J. Audio Eng. Soc.*, 45(9):660–684.

[Davidson et al., 1990] Davidson, G., Fielder, L., and Antill, M. (1990). High-quality audio transform coding at 128 kbits/s. In *Proc. IEEE Int. Conf. Acoust., Speech and Signal Proc*, pages 1117 – 1120.

[Dehery, 1991] Dehery, Y. F. (1991). MUSICAM source coding. In *Proc. of the 10th. International AES Conference*, pages 71–80.

[Deller Jr. et al., 1993] Deller Jr., J. R., Proakis, J. G., and Hansen, J. H. (1993). *Discrete-Time Processing of Speech Signals*. Macmillan, New York.

[Dembo and Malah, 1988] Dembo, A. and Malah, D. (1988). Signal synthesis from modified discrete short-time transform. *IEEE Trans. Acoust., Speech, Signal Processing*, ASSP-36(2):168–181.

[Dempster et al., 1977] Dempster, A. P., Laird, N. M., and Rubin, D. B. (1977). Maximum likelihood from incomplete data via the EM algorithm. *Journal of the Royal Statistical Society, Series B*, 39(1):1–38.

[Dendrinos et al., 1991] Dendrinos, M., Bakamidis, S., and Carayannis, G. (1991). Speech enhancement from noise: A regenerative approach. *Speech Communication*, 10:45–57.

[Depalle, 1991] Depalle, P. (1991). *Analyse, Modélisation et synthèse des sons basées sur le modèle source-filtre (Analysis, Modelling and synthesis of sounds based on the source-filter model)*. PhD thesis, Université du Maine, Le Mans, France. (in French).

[Depalle et al., 1993] Depalle, P., Garcia, G., and Rodet, X. (1993). Analysis of sound for additive synthesis: Tracking of partials using hidden markov models. In *Proc. 1993 Workshop on Applications of Signal Processing to Audio and Acoustics*, Mohonk Mountain House, New Paltz, NY.

[Deutsch, 1982] Deutsch, D., editor (1982). *The Psychology of Music*. AP series in cognition and perception. Academic Press.

[Dillier et al., 1993] Dillier, N., Frölich, T., Kompis, M., Bögli, H., and Lai, W. (1993). Digital signal processing (DSP) applications for multiband loudness correction digital hearing aids and cochlear implants. *J. Rehab. Res. and Devel.*, 30:95–109.

[Dillon, 1985] Dillon, H. (1985). Earmolds and high frequency response modification. *Hear. Instr.*, 36:8–12.

[Dillon and Lovegrove, 1993] Dillon, H. and Lovegrove, R. (1993). Single-microphone noise reduction systems for hearing aids: A review and an evaluation. In Studebaker, G. and Hochberg, I., editors, *Acoustical Factors Affecting Hearing Aid Performance*, pages 353–372. Allyn and Bacon.

[Dimino and Parladori, 1995] Dimino, G. and Parladori, G. (1995). Entropy reduction in high quality audio coding. In *Proc. of the 99th. AES-Convention*. Preprint 4064.

[Doblinger, 1982] Doblinger, G. (1982). "Optimum" filter for speech enhancement using integrated digital signal processors. *Proc. 1982 IEEE Int. Conf. on Acoust. Speech and Sig. Proc.*, pages 168–171.

[Dolby, 1967] Dolby, R. M. (1967). An audio noise reduction system. *J. Audio Eng. Soc.*, 15(4):383–388.

[Dolson, 1986] Dolson, M. (1986). The phase vocoder: A tutorial. *Computer Music Journal*, 10(4):14–27.

[Duifhuis, 1980] Duifhuis, H. (1980). Level effects in psychophysical two-tone suppression. *J. Acoust. Soc. Am.*, 67:914–927.

[Durbin, 1959] Durbin, J. (1959). Efficient estimation of parameters in moving-average models. *Biometrika*, 46:306–316.

[Durlach et al., 1986] Durlach, N. I., Braida, L. D., and Ito, Y. (1986). Towards a model for discrimination of broadband signals. *J. Acoust. Soc. Am.*, 80:63–72.

[Dyrlund and Bisgaard, 1991] Dyrlund, O. and Bisgaard, N. (1991). Acoustic feedback margin improvements in hearing instruments using a prototype DFS (digital feedback suppression) system. *Scand. Audiol.*, 20:49–53.

[Eastty et al., 1995] Eastty, P., Cooke, C., Densham, R., Konishu, T., and Matsushige, T. (1995). The Hardware Behind a Large Digital Mixer. In *AES 99th convention*. Preprint 4124.

[Edler, 1988] Edler, B. (1988). Prädiktive Teilbandcodierung mit Vektorquantisierung für Audiosignale hoher Tonqualität (in German). In *ITG Fachbericht*, volume 106, pages 223–228.

[Edler, 1989] Edler, B. (1989). Coding of audio signals with overlapping block transform and adaptive window functions. *Frequenz*, 43:252 – 256. (in German).

[Edler, 1992] Edler, B. (1992). Aliasing reduction in sub-bands of cascaded filter banks with decimation. *Electronics Letters*, 28:1104 – 1105.

[Edler, 1995] Edler, B. (1995). *Äquivalenz von Transformation und Teilbandzerlegung in der Quellencodierung*. Dissertation, Universität Hannover. (in German).

[Efron and Jeen, 1992] Efron, A. and Jeen, H. (1992). Pre-whitening for detection in correlated plus impulsive noise. *Proc. IEEE Int. Conf. Acoust., Speech and Signal Proc*, II:469–472.

[Egan and Hake, 1950] Egan, J. and Hake, H. (1950). On the masking pattern of a simple auditory stimulus. *J. Acoust. Soc. Am.*, 22:622–630.

[Egolf, 1982] Egolf, D. (1982). Review of the acoustic feedback literature from a control theory point of view. In *The Vanderbilt Hearing-Aid Report*, Monographs in Contemporary Audiology, pages 94–103.

[Egolf et al., 1986] Egolf, D., Haley, B., and Larson, V. (1986). The constant-velocity nature of hearing aids: Conclusions based on computer simulations. *J. Acoust. Soc. Am.*, 79:1592–1602.

[Egolf et al., 1985] Egolf, D., Howell, H., Weaver, K., and Barker, S. (1985). The hearing aid feedback path: Mathematical simulations, experimental verification. *J. Acoust. Soc. Am.*, 78:1578–1587.

[Egolf et al., 1978] Egolf, D., Tree, D., and Feth, L. (1978). Mathematical predictions of electroacoustic frequency response of in situ hearing aids. *J. Acoust. Soc. Am.*, 63:264–271.

[Ellis, 1992] Ellis, D. (1992). A perceptual representation of sound. Master's thesis, Department of Electrical Engineering and Computer Science, Massachusetts Institute of Technology.

[Ellis et al., 1991] Ellis, D., Vercoe, B., , and Quatieri, T. (1991). A perceptual representation of audio for co-channel source separation. In *Proc. IEEE Workshop Appl. of Signal Processing to Audio and Acoustics*, Mohonk Mountain House, New Paltz, NY.

[Engebretson and French-St.George, 1993] Engebretson, A. and French-St.George, M. (1993). Properties of an adaptive feedback equalization algorithm. *J. Rehab. Res. and Devel.*, 30:8–16.

[Engebretson et al., 1990] Engebretson, A., O'Connell, M., and Gong, F. (1990). An adaptive feedback equalization algorithm for the CID digital hearing aid. *Proc. 12th Annual Int. Conf. of the IEEE Eng. in Medicine, Biology Soc.*, Part 5:2286–2287.

[Ephraim, 1992] Ephraim, Y. (1992). Statistical-model-based speech enhancement systems. *Proc. IEEE*, 80(10):1526–1555.

[Ephraim and Malah, 1983] Ephraim, Y. and Malah, D. (1983). Speech enhancement using optimal non-linear spectral amplitude estimation. In *Proc. IEEE Int. Conf. Acoust., Speech, Signal Processing*, pages 1118–1121, Boston.

[Ephraim and Malah, 1984] Ephraim, Y. and Malah, D. (1984). Speech enhancement using a minimum mean-square error short-time spectral amplitude estimator. *IEEE Trans. Acoust., Speech, Signal Processing*, 32(6):1109–1121.

[Ephraim and Malah, 1985] Ephraim, Y. and Malah, D. (1985). Speech enhancement using a minimum mean-square error log-spectral amplitude estimator. *IEEE Trans. Acoust., Speech, Signal Processing*, 33(2):443–445.

[Ephraim and Van Trees, 1995] Ephraim, Y. and Van Trees, H. L. (1995). A signal subspace approach for speech enhancement. *IEEE Trans. Speech and Audio Processing*, 3(4):251–266.

[Ephraim and VanTrees, 1993] Ephraim, Y. and VanTrees, H. (1993). A signal subspace approach for speech enhancement. *Proc. IEEE Int. Conf. Acoust., Speech and Signal Proc*, II:359–362.

[Erwood and Xydeas, 1990] Erwood, A. and Xydeas, C. (1990). A multiframe spectral weighting system for the enhancement of speech signals corrupted by acoustic noise. In *SIGNAL PROCESSING V: Theories and Applications*, pages 1107–1110. Elsevier.

[Esteban and Galand, 1977] Esteban, D. and Galand, C. (1977). Application of Quadrature Mirror Filters to Split Band Voice Coding Schemes. In *Proc. IEEE Int. Conf. Acoust., Speech and Signal Proc*, pages 191–195.

[ETSI91tm74, 1991] ETSI91tm74 (October 1991). *Global analysis of selection tests: Basic data*. ETSI/TM/TM5/TCH-HS. Technical Document 91/74.

[ETSIstdR06, 1992] ETSIstdR06 (1992). *Speech codec specifications*. ETSI/GSM. ETS R.06.

[Etter and Moschytz, 1994] Etter, W. and Moschytz, G. S. (1994). Noise reduction by noise-adaptive spectral magnitude expansion. *J. Audio Eng. Soc.*, 42(5):341–349.

[Evans et al., 1981] Evans, J., Johnson, J., and Sun, D. (1981). High resolution angular spectrum estimation techniques for terrain scattering analysis and angle of arrival estimation. *Proc. 1st ASSP Workshop on Spectral Estimation*, pages 5.3.1–5.3.10.

[Fabry, 1991] Fabry, D. (1991). Programmable and automatic noise reduction in existing hearing aids. In Studebaker, B. and Beck, editors, *The Vanderbilt Hearing Aid Report II*, pages 65–78. York Press.

[Fabry and Tasell, 1990] Fabry, D. and Tasell, D. V. (1990). Evaluation of an articulation-index based model for predicting the effects of adaptive frequency response hearing aids. *J. Speech and Hearing Res.*, 33:676–689.

[Fairbanks et al., 1954] Fairbanks, G., Everitt, W., and Jaeger, R. (1954). Method for time or frequency compression-expansion of speech. *IEEE Trans. Audio and Electroacoustics*, AU-2:7–12.

[Ferrara, 1980] Ferrara, E. (1980). Fast implementation of LMS adaptive filters. *IEEE Trans. Acoust. Speech and Sig. Proc.*, Vol ASSP-28:474–475.

[Fettweis, 1986] Fettweis, A. (1986). Wave digital filters: Theory and practice. *Proc. IEEE*, 74(2):270–327.

[Fielder, 1985] Fielder, L. D. (1985). Modulation Noise in Floating-Point Conversion Systems. *J. Audio Eng. Society*, 33(10):770–781.

[Fielder, 1989] Fielder, L. D. (1989). Human Auditory Capabilities and Their Consequences for Digital Audio Converter Design. In Pohlmann, K., editor, *Audio in Digital Times*, pages 45–62. Audio Engineering Society.

[Fielder et al., 1996] Fielder, L. D., Bosi, M., Davidson, G., Davis, M., Todd, C., and Vernon, S. (1996). AC-2 and AC-2: Low-Complexity Transform-Based Audio Coding. In Gilchrist, N. and Grewin, C., editors, *Collected Papers on Digial Audio Bit-Rate Reduction*, pages 54 – 72. Audio Eng. Society.

[Fisher, 1983] Fisher, J. A. (1983). Very Long Instruction Word Architectures and the ELI-512. In *Proc. 10th Annual Intl. Symposium on Computer Architecture*, pages 140–150.

[Fitzgibbons and Wightman, 1982] Fitzgibbons, P. and Wightman, F. (1982). Gap detection in normal and hearing-impaired listeners. *J. Acoust. Soc. Am.*, 72:761–765.

[Flanagan and Golden, 1966] Flanagan, J. and Golden, R. (1966). Phase vocoder. *Bell System Technical Journal*, 45(9):1493–1509.

[Flanagan and Ishizaka, 1976] Flanagan, J. L. and Ishizaka, K. (1976). Automatic generation of voiceless excitation in a vocal cord-vocal tract speech synthesizer. *IEEE Trans. Acoustics, Speech, Signal Processing*, 24(2):163–170.

[Flanagan et al., 1980] Flanagan, J. L., Ishizaka, K., and Shipley, K. L. (1980). Signal models for low bit-rate coding of speech. *J. Acoustical Soc. of America*, 68(3):780–791.

[Fletcher, 1940] Fletcher, N. (1940). Auditory patterns. *Rev. Modern Phys.*, 12(0):47–65.

[Fletcher and Rossing, 1993] Fletcher, N. H. and Rossing, T. D. (1993). *The Physics of Musical Instruments*. Springer Verlag, New York.

[Florentine and Buus, 1981] Florentine, M. and Buus, S. (1981). An excitation-pattern model for intensity discrimination. *J. Acoust. Soc. Am.*, 70:1646–1654.

[Florentine and Buus, 1984] Florentine, M. and Buus, S. (1984). Temporal gap detection in sensorineural and simulated hearing impairment. *J. Speech and Hear. Res.*, 27:449–455.

[Fortune and Preves, 1992] Fortune, T. and Preves, D. (1992). Hearing aid saturation and aided loudness discomfort. *J. Speech and Hearing Res.*, 35:175–185.

[Foster, 1986] Foster, S. (1986). "Impulse response measurement using Golay codes". In *Proc. IEEE Int. Conf. Acoust., Speech and Signal Proc*, volume 2, pages 929–932.

[Fowle, 1967] Fowle, E. (1967). The design of FM pulse-compression signals. *IEEE Trans. Information Theory*, 10:61–67.

[Frank and Lacroix, 1986] Frank, W. and Lacroix, A. (1986). Improved vocal tract models for speech synthesis. *Proc. IEEE Int. Conf. Acoust., Speech, Signal Processing*, 3:2011–2014.

[French and Steinberg, 1947] French, N. and Steinberg, J. (1947). Factors governing the intelligibility of speech sounds. *J. Acoust. Soc. Am.*, 19:90–119.

[French-St.George et al., 1993] French-St.George, Wood, M., and Engebretson, A. (1993). Behavioral assessment of adaptive feedback cancellation in a digital hearing aid. *J. Rehab. Res. and Devel.*, 30:17–25.

[Friedlander, 1953] Friedlander, F. G. (1953). On the oscillations of the bowed string. *Proc. Cambridge Philosophy Soc.*, 49:516–530.

[Frindle, 1992] Frindle, P. A. (1992). Digital to Analog Conversion Circuit with Dither and Overflow Prevention. U. S. Patent 5,148,163.

[Frindle, 1995] Frindle, P. A. (1995). Design Considerations for Successful Analogue Conversion Systems for Large Scale Digital Audio Console Applications. In *Proc. AES 99th convention*. Audio Eng. Society. Preprint 4126.

[Fuccio et al., 1988] Fuccio, M. L., Gadenz, R. N., Garen, C. J., Huser, J. M., Ng, B., Pekarich, S. P., , and Ulery, K. D. (1988). The DSP-32C: AT&T's Second-Generation Floating-Point Digital Signal Processor. *IEEE Micro*, 8(6):30–47.

[Fuchs, 1995] Fuchs, H. (1995). Improving MPEG audio coding by backward adaptive linear stereo prediction. In *Proc. of the 99th. AES-Convention*. Preprint 4086.

[Fujita, 1996] Fujita, Y. (1996). Waveform generating apparatus for musical instrument . U. S. Patent 5,553,011.

[Furst, 1946] Furst, U. R. (1946). Periodic variations of pitch in sound reproduction by phonographs. *Proc. IRE*, page 887.

[Gagné, 1988] Gagné, J.-P. (1988). Excess masking among listeners with a sensorineural hearing loss. *J. Acoust. Soc. Am.*, 83:2311–2321.

[Gardner, 1994] Gardner, W., editor (1994). *Cyclostationarity in communications and signal processing*. IEEE Press, New York.

[Gardner, 1992] Gardner, W. G. (1992). "A realtime multichannel room simulator". *J. Acoust. Soc. Am.*, 92(4 (A)):2395. http://sound.media.mit.edu/papers.html.

[Gardner, 1995] Gardner, W. G. (1995). "Efficient convolution without input-output delay". *J. Audio Eng. Soc.*, 43(3):127–136.

[Gardner and Griesinger, 1994] Gardner, W. G. and Griesinger, D. (1994). "Reverberation level matching experiments". In *Proceedings of the Sabine Centennial Symposium*, pages 263–266, Cambridge, MA. Acoustical Society of America.

[Gehringer et al., 1987] Gehringer, E. F., Siewiorek, D. P., and Segall, Z. (1987). *Parallel Processing - The Cm* Experience*. Digital Press.

[Gelfand and Smith, 1990] Gelfand, A. E. and Smith, A. F. M. (1990). Sampling-based approaches to calculating marginal densities. *J. Am. Statist. Assoc.*, 85:398–409.

[Geman and Geman, 1984] Geman, S. and Geman, D. (1984). Stochastic relaxation, Gibbs distributions and the Bayesian restoration of images. *IEEE Trans. Pattern Analysis and Machine Intelligence*, 6:721–741.

[George, 1991] George, E. (1991). *An Analysis-by-Synthesis Approach to Sinusoidal Modeling Applied to Speech and Music Signal Processing*. PhD thesis, Department of Electrical Engineering and Computer Science, Georgia Institute of Technology.

[George and Smith, 1992] George, E. B. and Smith, M. J. T. (1992). Analysis-by-synthesis/Overlap-add sinusoidal modeling applied to the analysis and synthesis of musical tones. *J. Audio Eng. Soc.*, 40(6):497–516.

[Gerbrands, 1981] Gerbrands, J. J. (1981). On the relationships between the SVD, KLT and PCA. *Pattern Recognition*, 14(1):375–381.

[Gerr and Allen, 1994] Gerr, N. and Allen, J. (1994). The generalized spectrum and spectral coherence of a harmonizable time series. *Digital Signal Processing*, 4(4):222–238.

[Gerzon, 1971] Gerzon, M. A. (1971). "Synthetic Stereo Reverberation". *Studio Sound*, 13:632–635.

[Gerzon, 1972] Gerzon, M. A. (1972). "Synthetic Stereo Reverberation". *Studio Sound*, 14:209–214.

[Gerzon, 1976] Gerzon, M. A. (1976). "Unitary (energy preserving) multichannel networks with feedback". *Electronics Letters*, 12(11):278–279.

[Ghitza, 1986] Ghitza, O. (1986). Speech analysis/synthesis based on matching the synthesized and the original representations in the auditory nerve level. In *Proc. IEEE Int. Conf. Acoustics, Speech, and Signal Processing.*, pages 1995–1998, Tokyo, Japan.

[Ghitza, 1994] Ghitza, O. (1994). Auditory models and human performance in tasks related to speech coding and speech recognition. *IEEE Trans. on Speech and Audio Processing*, 2:115–132.

[Giguère and Woodland, 1994a] Giguère, C. and Woodland, P. C. (1994a). A computational model of the auditory periphery for speech and hearing research. I. Ascending paths. *J. Acoust. Soc. Am.*, 95:331–342.

[Giguère and Woodland, 1994b] Giguère, C. and Woodland, P. C. (1994b). A computational model of the auditory periphery for speech and hearing research. II. Descending paths. *J. Acoust. Soc. Am.*, 95:343–349.

[Gilbert et al., 1990] Gilbert, J., Kergomard, J., and Polack, J. D. (1990). On the reflection functions associated with discontinuities in conical bores. *J. Acoustical Soc. of America*, 87(4):1773–1780.

[Glasberg et al., 1987] Glasberg, B., Moore, B., and Bacon, S. (1987). Gap detection and masking in hearing-impaired and normal-hearing subjects. *J. Acoust. Soc. Am.*, 81:1546–1556.

[Glinski and Roe, 1994] Glinski, S. and Roe, D. (1994). Spoken Language Recognition on a DSP Array Processor. *IEEE Trans. on Parallel and Distributed Systems*, 5(7):697–703.

[Godara, 1990] Godara, L. (1990). Beamforming in the presence of correlated arrivals using structured correlation matrix. *IEEE Trans. Acoust. Speech and Sig. Proc.*, 38:1–15.

[Godara, 1991] Godara, L. (1991). Adaptive beamforming in the presence of correlated arrivals. *J. Acoust. Soc. Am.*, 89:1730–1736.

[Godara, 1992] Godara, L. (1992). Beamforming in the presence of broadband correlated arrivals. *J. Acoust. Soc. Am.*, 92:2702–2708.

[Godara and Gray, 1989] Godara, L. and Gray, D. (1989). A structured gradient algorithm for adaptive beamforming. *J. Acoust. Soc. Am.*, 86:1040–1046.

[Godsill, 1993] Godsill, S. J. (1993). *The Restoration of Degraded Audio Signals*. PhD thesis, Cambridge University Engineering Department, Cambridge, England. Available in A5 book format (compressed Postscript http://www-com-serv.eng.cam.ac.uk/~sjg/thesis/a5thesis.zip) and A4 format (compressed Postscript http://www-com-serv.eng.cam.ac.uk/~sjg/thesis/thesis.zip).

[Godsill, 1994] Godsill, S. J. (1994). Recursive restoration of pitch variation defects in musical recordings. In *Proc. IEEE Int. Conf. Acoust., Speech and Signal Proc*, volume 2, pages 233–236, Adelaide, Australia.

[Godsill, 1997a] Godsill, S. J. (1997a). Bayesian enhancement of speech and audio signals which can be modelled as ARMA processes. *International Statistical Review*, 65(1):1–21.

[Godsill, 1997b] Godsill, S. J. (1997b). Robust modelling of noisy ARMA signals. In *Proc. IEEE Int. Conf. Acoust., Speech and Signal Proc*, volume 5, pages 3797–3800.

[Godsill and Rayner, 1992] Godsill, S. J. and Rayner, P. J. W. (1992). A Bayesian approach to the detection and correction of bursts of errors in audio signals. In *Proc. IEEE Int. Conf. Acoust., Speech and Signal Proc*, volume 2, pages 261–264, San Francisco, CA.

[Godsill and Rayner, 1993a] Godsill, S. J. and Rayner, P. J. W. (1993a). Frequency-domain interpolation of sampled signals. In *Proc. IEEE Int. Conf. Acoust., Speech and Signal Proc*, volume I, pages 209–212, Mineapolis, MN.

[Godsill and Rayner, 1993b] Godsill, S. J. and Rayner, P. J. W. (1993b). The restoration of pitch variation defects in gramophone recordings. In *Proc. IEEE Workshop Appl. of Signal Processing to Audio and Acoustics*, Mohonk Mountain House, New Paltz, NY.

[Godsill and Rayner, 1995a] Godsill, S. J. and Rayner, P. J. W. (1995a). A Bayesian approach to the restoration of degraded audio signals. *IEEE Trans. on Speech and Audio Processing*, 3(4):267–278.

[Godsill and Rayner, 1995b] Godsill, S. J. and Rayner, P. J. W. (1995b). Robust noise modelling with application to audio restoration. In *Proc. IEEE Workshop Appl. of Signal Processing to Audio and Acoustics*, Mohonk Mountain House, New Paltz, NY.

[Godsill and Rayner, 1996a] Godsill, S. J. and Rayner, P. J. W. (1996a). Robust noise reduction for speech and audio signals. In *Proc. IEEE Int. Conf. Acoust., Speech and Signal Proc*, volume 2, pages 625–628.

[Godsill and Rayner, 1996b] Godsill, S. J. and Rayner, P. J. W. (1996b). Robust treatment of impulsive noise in speech and audio signals. In *Bayesian Robustness*, volume 29, pages 331–342. IMS Lecture Notes - Monograph Series (1996) Volume 29.

[Godsill and Rayner, 1998] Godsill, S. J. and Rayner, P. J. W. (1998). Robust reconstruction and analysis of autoregressive signals in impulsive noise using the Gibbs sampler. Accepted to *IEEE Trans. on Speech and Audio Processing*. (to appear) Available as http://www-com-serv.eng.cam.ac.uk/~sjg/papers/95/impulse.ps.

[Golub and Van Loan, 1989] Golub, G. H. and Van Loan, C. F. (1989). *Matrix Computations*. The John Hopkins University Press.

[Goodman, 1965] Goodman, A. (1965). Reference zero levels for pure-tone audiometer. *ASHA*, 7:262–263.

[Goodman and Nash, 1982] Goodman, D. J. and Nash, R. D. (1982). Subjective quality of the same speech transmission conditions in seven different countries. *IEEE Trans. on Commun.*, 4:642–654.

[Gordon, 1985] Gordon, J. W. (1985). System Architectures for Computer Music. *Computing Surveys*, 17(2):191–233.

[Gordon and Strawn, 1985] Gordon, J. W. and Strawn, J. (1985). An Introduction to the Phase Vocoder. In Strawn, J., editor, *Digital audio signal processing : an anthology*. William Kaufmann.

[Gordon-Salant, 1986] Gordon-Salant, S. (1986). Recognition of natural and time/intensity altered cv's by young and elderly subjects with normal hearing. *J. Acoust. Soc. Am.*, 80:1599–1607.

[Gordon-Salant, 1987] Gordon-Salant, S. (1987). Effects of acoustic modification on consonant recognition in elderly hearing-impaired subjects. *J. Acoust. Soc. Am.*, 81:1199–1202.

[Gray and Markel, 1975] Gray, A. H. and Markel, J. D. (1975). A normalized digital filter structure. *IEEE Trans. Acoustics, Speech, Signal Processing*, ASSP-23(3):268–277.

[Gray and Markel, 1976] Gray, A. H. and Markel, J. D. (1976). Distance measures for speech processing. *IEEE Trans. on Acoust., Speech and Signal Processing*, 24:380–391.

[Gray et al., 1980] Gray, R. M., Buzo, A., Gray, A. H., and Matsuyama, Y. (1980). Distortion measures for speech processing. *IEEE Trans. on Acoust., Speech and Signal Processing*, 28:367–375.

[Green and Huerta, 1994] Green, D. and Huerta, L. (1994). Tests of retrocochlear function. In Katz, J., editor, *Handbook of Clinical Audiology*, pages 176–180. Williams and Wilkins, 4th edition edition.

[Greenberg, 1994] Greenberg, J. (1994). *Improved design of microphone-array hearing aids*. PhD thesis, Mass. Inst. Tech.

[Greenberg and Zurek, 1992] Greenberg, J. and Zurek, P. (1992). Evaluation of an adaptive beamforming method for hearing aids. *J. Acoust. Soc. Am.*, 91:1662–1676.

[Griesinger, 1989] Griesinger, D. (1989). "Practical processors and programs for digital reverberation". In *Proc. Audio Eng. Soc. 7th Int. Conf.*, pages 187–195, Toronto, Ontario, Canada. Audio Eng. Society.

[Griesinger, 1991] Griesinger, D. (1991). "Improving Room Acoustics Through Time-Variant Synthetic Reverberation". In *Proc. Audio Eng. Soc. Conv.* Preprint 3014.

[Griesinger, 1992] Griesinger, D. (1992). "IALF - Binaural measures of spatial impression and running reverberance". In *Proc. Audio Eng. Soc. Conv.* Preprint 3292.

[Griesinger, 1995] Griesinger, D. (1995). "How loud is my reverberation?". In *Proc. Audio Eng. Soc. Conv.* Preprint 3943.

[Griesinger, 1997] Griesinger, D. (1997). "The Psychoacoustics of Apparent Source Width, Spaciousness and Envelopment in Performance Spaces". *Acustica*, 83(4):721–731.

[Griffin, 1987] Griffin, D. (1987). *A Multiband Excitation Vocoder*. PhD thesis, Department of Electrical Engineering and Computer Science, Massachusetts Institute of Technology.

[Griffin and Lim, 1984a] Griffin, D. and Lim, J. (1984a). Signal Estimation from Modified Short-Time Fourier Transform. *IEEE Trans. on Acoustics, Speech, and Signal Processing*, ASSP-32(2):236–242.

[Griffin and Lim, 1988] Griffin, D. and Lim, J. (1988). Multiband-excitation vocoder. *IEEE Trans. Acoust., Speech, Signal Processing*, ASSP-36(2):236–243.

[Griffin and Lim, 1984b] Griffin, D. W. and Lim, J. S. (1984b). Signal estimation from modified short-time Fourier transform. *IEEE Trans. Acoust., Speech, Signal Processing*, 32(2):236–242.

[Griffiths and Jim, 1982] Griffiths, L. and Jim, C. (1982). An alternative approach to linearly constrained adaptive beamforming. *IEEE Trans. Antennas and Propagation*, AP-30:27–34.

[Guettler, 1992] Guettler, K. (1992). The bowed string computer simulated — some characteristic features of the attack. *Catgut Acoustical Soc. J.*, 2(2):22–26. Series II.

[Gutknecht et al., 1983] Gutknecht, M., Smith, J. O., and Trefethen, L. N. (1983). The Caratheodory-Fejer (CF) method for recursive digital filter design. *IEEE Trans. Acoustics, Speech, Signal Processing*, 31(6):1417–1426.

[Hagerman and Gabrielsson, 1984] Hagerman, B. and Gabrielsson, A. (1984). Questionnaires on desirable properties of hearing aids. Technical report, Karolinska Inst.

[Haggard et al., 1987] Haggard, M., Trinder, J., Foster, J., and Lindblad, A. (1987). Two-state compression of spectral tilt: Individual differences and psychoacoustical limitations to the benefit from compression. *J. Rehab. Res. and Devel.*, 25:193–206.

[Halka and Heute, 1992] Halka, U. and Heute, U. (1992). A new approach to objective quality-measures based on attribute-matching. *Speech Communication*, 11:15–30.

[Hall, 1980] Hall, D. E. (1980). *Musical acoustics: An introduction.* Wadsworth.

[Hamdy et al., 1996] Hamdy, K., Ali, M., and Tewfik, A. (1996). Low bit rate high quality audio coding with combined harmonic and wavelet representations. In *Proc. IEEE Int. Conf. Acoustics, Speech, and Signal Processing.*, Atlanta, GA.

[Hardam, 1990] Hardam, E. (1990). High quality time scale modification of speech signals using fast synchronized overlap add algorithms. *Proc. IEEE ICASSP-90*, pages 409–412.

[Harris, 1990] Harris, S. (1990). The Effects of Sampling Clock Jitter on Nyquist Sampling Analog-to-Digital Converters, and on Oversampling Delta-Sigma ADCs. *J. Audio Eng. Soc.*, 38(7/8):537–542.

[Harrison et al., 1981] Harrison, R., Aran, J.-M., and Erre, J.-P. (1981). Ap tuning curves from normal and pathological human and guinea pig cochleas. *J. Acoust. Soc. Am.*, 69:1374–1385.

[Hartmann, 1987] Hartmann, W. M. (1987). Digital waveform generation by fractional addressing. *J. Acoustical Society of America*, 82(6):1883–1891.

[Hastings, 1987] Hastings, C. (1987). A Recipe for Homebrew ECL. In Roads, C. and Strawn, J., editors, *Foundations of computer music*, pages 335–362. MIT Press. Originally appeared in *Computer Music Journal*, vol. 2, no. 1, 1978, pages 48–59.

[Hastings, 1970] Hastings, W. K. (1970). Monte Carlo sampling methods using Markov chains and their applications. *Biometrika*, 57:97–109.

[Hauser, 1991] Hauser, M. (1991). Principles of Oversampling A/D Conversion. *J. Audio Eng. Society*, 39(1/2):3–26.

[Hawkins and Yacullo, 1984] Hawkins, D. and Yacullo, W. (1984). Signal-to-noise ratio advantage of binaural hearing aids, directional microphones under different levels of reverberation. *J. Speech and Hearing Disorders*, 49:278–286.

[Hayashi and Kitawaki, 1992] Hayashi, S. and Kitawaki, N. (1992). An objective quality assessment method for bit-reduction coding of wideband speech. *J. Acoust. Soc. Am.*, 92:106–113.

[Hayes et al., 1980] Hayes, M., Lim, J., and Oppenheim, A. (1980). Signal reconstruction from phase or magnitude. *IEEE Trans. Acoust., Speech, Signal Processing*, ASSP-28(6):672–680.

[Haykin, 1991] Haykin, S. (1991). *Adaptive Filter Theory, Second Edition*. Prentice-Hall, Englewood Cliffs, NJ.

[Hedelin, 1981] Hedelin, P. (1981). A tone-oriented voice excited vocoder. In *Proc. IEEE Int. Conf. Acoustics, Speech, and Signal Processing.*, Atlanta, GA.

[Hellman, 1972] Hellman, R. P. (1972). Asymmetry of masking between noise and tone. *Perception and Psychophysics*, 11:241–246.

[Hermes, 1991] Hermes, D. (1991). Synthesis of breathy vowels: Some research methods. *Speech Communications*, 10:497–502.

[Herre, 1995] Herre, J. (1995). *Fehlerverschleierung bei Spektral Codier Audiosignalen*. PhD thesis, University of Erlangen-Nuremburg. (in German).

[Herre et al., 1992] Herre, J., Eberlein, E., and Brandenburg, K. (1992). Combined stereo coding. In *Proc. of the 93rd AES-Convention*, San Francisco. preprint 3369.

[Herre and Johnston, 1996] Herre, J. and Johnston, J. D. (1996). Enhancing the performance of perceptual audio coders by using temporal noise shaping (TNS). In *Proc. of the 101st AES-Convention*. Preprint 4384.

[Heyser, 1967] Heyser, R. C. (1967). "Acoustical measurements by time delay spectrometry". *J. Audio Eng. Soc.*, 15:370–382.

[Hicks and Godsill, 1994] Hicks, C. M. and Godsill, S. J. (1994). A 2-channel approach to the removal of impulsive noise from archived recordings. In *Proc. IEEE Int. Conf. Acoust., Speech and Signal Proc*, volume 2, pages 213–216, Adelaide, australia.

[Hidaka et al., 1995] Hidaka, T., Beranek, L. L., and Okano, T. (1995). "Interaural cross-correlation, lateral fraction, and low- and high-frequency sound levels as measures of acoustical quality in concert halls". *J. Acoust. Soc. Am.*, 98(2):988–1007.

[Hingley, 1994] Hingley, A. (1994). Cost Effective DSP for Audio Mixing. In *UK Conference: Digital Signal Processing*, pages 226–234. Audio Eng. Soc.

[Hirschman, 1991] Hirschman, S. (1991). *Digital Waveguide Modelling and Simulation of Reed Woodwind Instruments*. PhD thesis, Elec. Eng. Dept., Stanford University. Available as CCRMA Technical Report Stan-M-72, Music Dept., Stanford University, July 1991.

[Hirschberg et al., 1991] Hirschberg, A., Gilbert, J., Wijnands, A. P. J., and Houtsma, A. J. M. (1991). Non-linear behavior of single-reed woodwind musical instruments. *Nederlands Akoestisch Genootschap*, 107:31–43.

[Hirschberg et al., 1995] Hirschberg, A., Kergomard, J., and Weinreich, G., editors (1995). *Mechanics of Musical Instruments*. Springer-Verlag, Berlin.

[Hochberg et al., 1992] Hochberg, I., Boothroyd, A., Weiss, M., and Hellman, S. (1992). Effects of noise suppression on speech perception by cochlear implant users. *Ear and Hearing*, 13:263–271.

[Hodgson and Lade, 1988] Hodgson, W. and Lade, K. (1988). Digital technology in hearing instruments: Increased flexibility for fitters and wearers. *Hear. J.*, 4:28–32.

[Hoffman et al., 1994] Hoffman, M., Trine, T., Buckley, K., and Tasell, D. V. (1994). Robust adaptive microphone array processing for hearing aids: Realistic speech enhancement. *J. Acoust. Soc. Am.*, 96:759–770.

[Houtgast, 1977] Houtgast, T. (1977). Auditory-filter characteristics derived from direct-masking data and pulsation-threshold data with rippled-noise masker. *J. Acoust. Soc. Am.*, 62:409–415.

[Huber, 1964] Huber, P. (1964). Robust estimation of a location parameter. *Annals of Mathematical Statistics*, pages 73–101.

[Huber, 1981] Huber, P. (1981). *Robust Statistics*. Wiley and Sons.

[Humes and Jesteadt, 1989] Humes, L. E. and Jesteadt, W. (1989). Models of the additivity of masking. *J. Acoust. Soc. Am.*, 85:1285–1294.

[IEEE Computer Society Standards Committee, 1985] IEEE Computer Society Standards Committee (1985). *IEEE standard for binary floating-point arithmetic*. Institute of Electrical and Electronics Engineers.

[Intel, 1991] Intel (1991). *i860 Microprocessor Family Programmer Reference Manual*. Intel Corporation.

[Ishizaka and Flanagan, 1972] Ishizaka, K. and Flanagan, J. L. (1972). Synthesis of voiced sounds from a two-mass model of the vocal cords. *Bell System Tech. J.*, 51:1233–1268.

[ISO90, 1990] ISO90 (1990). *MPEG/Audio test report*. ISO/IEC/JTC1/SC2/WG11 MPEG. Document MPEG90/N0030.

[ISO91, 1991] ISO91 (1991). *MPEG/Audio test report*. ISO/IEC/JTC1/SC2/WG11 MPEG. Document MPEG91/010.

[ISO92st, 1993] ISO92st (1993). *Coding of moving pictures and associated audio for digital storage media at up to about 1.5 Mbit/s*. ISO/IEC JTC 1/SC 29/WG11. International Standard 11172.

[ISO94st, 1994] ISO94st (1994). *Generic coding of moving pictures and assiocated audio*. ISO/IEC JTC 1/SC 29/WG11. International Standard 13818.

[ITURrecBS1116, 1994] ITURrecBS1116 (1994). *Methods for the subjective assessment of small impairments in audiosystems including multichannel sound systems*. ITU-R. Recommendation BS.1116.

[ITURsg10con9714, 1997] ITURsg10con9714 (1997). *Report on the fourth meeting of ITU-R Task Group 10/4, objective perceptual audio quality assessment methods*. ITU-R Task Group 10/4. Contribution 10-4/14.

[ITURsg10con9719, 1997] ITURsg10con9719 (1997). *Report on the fifth meeting of ITU-R Task Group 10/4, objective perceptual audio quality assessment methods*. ITU-R Task Group 10/4. Contribution 10-4/19.

[ITURsg10cond9343, 1993] ITURsg10cond9343 (1993). *CCIR listening tests, network verification tests without commentary codecs, final report*. ITU-R Task Group 10/2. Delayed contribution 10-2/43.

[ITURsg10cond9351, 1993] ITURsg10cond9351 (1993). *PAQM measurements*. ITU-R Task Group 10/2. Delayed contribution 10-2/51.

[ITUTrecP861, 1996] ITUTrecP861 (1996). *Objective quality measurement of telephone-band (300-3400 Hz) speech codecs*. ITU-T. Recommendation P.861.

[ITUTsg12con9674, 1996] ITUTsg12con9674 (1996). *Review of validation tests for objective speech quality measures*. ITU-T Study Group 12. Document COM 12-74.

[ITUTsg12rep31.96, 1996] ITUTsg12rep31.96 (June 1996). *Study group 12 - report R 31*. ITU-T Study Group 12. Document COM 12-R 31.

[ITUTsg12sq2.93, 1993] ITUTsg12sq2.93 (July 1993). *Subjective test methodology for a 8 kbit/s speech coder*. ITU-T Study Group 12, Speech Quality Expert Group. Document SQ-2.93.

[ITUTsg12sq3.94, 1994] ITUTsg12sq3.94 (February 1994). *8 kbit/s selection phase (1993-94): global results of experiment 2: effect of tandeming and input levels.* ITU-T Study Group 12, Speech Quality Expert Group. Document SQ-2.93.

[Iwadare et al., 1992] Iwadare, M., Sugiyama, A., Hazu, F., Hirano, A., and T.Nishitani (1992). A 128 kb/s Hi-Fi Audio CODEC Based on Adaptive Transform Coding with Adaptive Block Size MDCT. *IEEE Journal on Selected Areas in Communications*, 10(0):138 – 144.

[Iwakami et al., 1995] Iwakami, N., Moriya, T., and Miki, S. (1995). High quality audio coding at less than 64 kbit/s by using transform-domain interleave vector quantization (TWINVW). In *Proc. IEEE Int. Conf. Acoust., Speech and Signal Proc*, pages 3095 – 3098.

[Jackson et al., 1968] Jackson, L. B., Kaiser, J. F., and McDonald, H. (1968). An Approach to the Implementation of Digital Filters. *IEEE Trans. on Audio and Electroacoustics*, AU-16(3):413–421.

[Jaffe, 1995] Jaffe, D. A. (1995). Ten criteria for evaluating synthesis techniques. *Computer Music J.*, 19(1):76–87.

[Jaffe and Smith, 1995] Jaffe, D. A. and Smith, J. O. (1995). Performance expression in commuted waveguide synthesis of bowed strings. In *Proc. 1995 Int. Computer Music Conf., Banff*, pages 343–346. Computer Music Association.

[Janssen et al., 1986] Janssen, A. J. E. M., Veldhuis, R., and Vries, L. B. (1986). Adaptive interpolation of discrete-time signals that can be modeled as AR processes. *IEEE Trans. Acoustics, Speech and Signal Processing*, ASSP-34(2):317–330.

[Jayant and Noll, 1982] Jayant, N. and Noll, P. (1982). *Digital Coding of Waveforms: Principles and Applications to Speech and Video*. Prentice-Hall, Englewood Cliffs, NJ.

[Jensen et al., 1995] Jensen, S. H., Hansen, P. C., Hansen, S. D., and Sørensen, J. A. (1995). Reduction of broad-band noise in speech by truncated QSVD. *IEEE Trans. Speech and Audio Processing*, 3(6):439–448.

[Jerger et al., 1989] Jerger, J., Jerger, S., Oliver, T., and Pirozzolo, F. (1989). Speech understanding in the elderly. *Ear and Hearing*, 10:79–89.

[Jesteadt et al., 1982] Jesteadt, W. J., Bacon, S. P., and Lehman, J. R. (1982). Forward masking as a function of frequency, masker level, and signal delay. *J. Acoust. Soc. Am.*, 71:950–962.

[Johnson and Schnier, 1988] Johnson, J. and Schnier, W. (1988). An expert system for programming a digitally controlled hearing instrument. *Hear. Instr.*, 39:24–26.

[Johnston, 1980] Johnston, J. D. (1980). A Filter Family Designed for Use in Quadrature Mirror Filter Banks. In *Proc. IEEE Int. Conf. Acoust., Speech and Signal Proc*, pages 291–294.

[Johnston, 1988] Johnston, J. D. (1988). Estimation of perceptual entropy using noise masking criteria. In *Proc. IEEE Int. Conf. Acoust., Speech and Signal Proc*, pages 2524–2527.

[Johnston, 1989a] Johnston, J. D. (1989a). Perceptual transform coding of wideband stereo signals. In *Proc. IEEE Int. Conf. Acoust., Speech and Signal Proc*, pages 1993–1996.

[Johnston, 1989b] Johnston, J. D. (1989b). Transform coding of audio signals using perceptual noise criteria. *IEEE Journal on Selected Areas in Communications*, 6:314–323.

[Johnston, 1996] Johnston, J. D. (1996). Audio coding with filter banks. In Akansu, A. N. and Smith, M. J. T., editors, *Subband and Wavelet Transforms*, pages 287–307. Kluwer Academic Publishers.

[Johnston and Brandenburg, 1992] Johnston, J. D. and Brandenburg, K. (1992). Wideband coding–Perceptual considerations for speech and music. In Furui, S. and Sondhi, M. M., editors, *Advances in speech signal processing*. Marcel Dekker, New York.

[Johnston et al., 1996] Johnston, J. D., Herre, J., Davis, M., and Gbur, U. (1996). MPEG-2 NBC audio – stereo and multichannel coding methods. In *Proc. of the 101st AES-Convention*. Preprint 4383.

[Johnstone et al., 1986] Johnstone, B., Patuzzi, R., and Yates, G. (1986). Basilar membrane measurements and the travelling wave. *Hear. Res.*, 22:147–153.

[Jones and Parks, 1988] Jones, D. and Parks, T. (1988). On the generation and combination of grains for music synthesis. *Computer Music J.*, 12(2):27–34.

[Jot, 1992a] Jot, J. M. (1992a). "An analysis/synthesis approach to real-time artificial reverberation". In *Proc. IEEE Int. Conf. Acoust., Speech and Signal Proc*, volume 2, pages 221–224.

[Jot, 1992b] Jot, J. M. (1992b). *Etude et réalisation d'un spatialisateur de sons par modèles physiques et perceptifs (Design and implementation of a sound spatializer based on physical and perceptual models, in French)*. PhD thesis, Telecom Paris.

[Jot and Chaigne, 1991] Jot, J. M. and Chaigne, A. (1991). "Digital delay networks for designing artificial reverberators". In *Proc. Audio Eng. Soc. Conv.* Preprint 3030.

[Jot et al., 1995] Jot, J. M., Larcher, V., and Warusfel, O. (1995). "Digital signal processing issues in the context of binaural and transaural stereophony". In *Proc. Audio Eng. Soc. Conv.* Preprint 3980.

[Jullien et al., 1992] Jullien, J. P., Kahle, E., Winsberg, S., and Warusfel, O. (1992). "Some results on the objective characterisation of room acoustical quality in both laboratory and real environments". In *Proc. Inst. of Acoust.*, Birmingham.

[Justice, 1979] Justice, J. (1979). Analytic signal processing in music computation. *IEEE Trans. on Acoustics, Speech, and Signal Processing*, ASSP-27(6):897–909.

[Kaegi et al., 1978] Kaegi, W., and Tempelaars, S. (1978). VOSIM—A New Sound Synthesis System. *J. Audio Eng. Soc.*, 26(6):418–24.

[Kahrs, 1988] Kahrs, M. (1988). The architecture of DSP.*: A DSP multiprocessor. In *Proc. IEEE Int. Conf. Acoust., Speech, Signal Processing*, pages 2552–2555.

[Kahrs and Killian, 1992] Kahrs, M. and Killian, T. (1992). Gnot Music: A Flexible Workstation for Orchestral Synthesis. *Computer Music Journal*, 16(3):48–56.

[Kalapathy, 1997] Kalapathy, P. (1997). Hardware-Software Interactions in MPACT. *IEEE Micro*, 17(2):20–26.

[Kalliojärvi et al., 1994] Kalliojärvi, K., Kontro, J., and Neuvo, Y. (1994). Novel Floating-Point A/D and D/A Conversion Methods. In *IEEE International Symposium on Circuits and Systems*, volume 2, pages 1–4.

[Karp, 1996] Karp, A. H. (1996). Bit reversal on uniprocessors. *SIAM Review*, 38(1):1–26.

[Karplus and Strong, 1983] Karplus, K. and Strong, A. (1983). Digital synthesis of plucked-string and drum timbres. *Computer Music J.*, 7(2):43–55.

[Kasparis and Lane, 1993] Kasparis, T. and Lane, J. (1993). Adaptive scratch noise filtering. *IEEE Trans. Consumer Electronics*, 39(4).

[Kates, 1986] Kates, J. (1986). Signal processing for hearing aids. *Hearing Instr.*, 37(2):19–21.

[Kates, 1988] Kates, J. (1988). A computer simulation of hearing aid response and the effects of ear canal size. *J. Acoust. Soc. Am.*, 83:1952–1963.

[Kates, 1989] Kates, J. (1989). Hearing-aid signal-processing system. U.S. Patent 4,852,175.

[Kates, 1990] Kates, J. (1990). A time-domain digital simulation of hearing aid response. *J. Rehab. Res. and Devel.*, 27:279–294.

[Kates, 1991a] Kates, J. (1991a). Feedback cancellation in hearing aids: Results from a computer simulation. *IEEE Trans. Sig. Proc.*, Vol.39:553–562.

[Kates, 1991b] Kates, J. (1991b). A time-domain digital cochlear model. *IEEE Trans. Signal Processing*, 39(12):2573–2592.

[Kates, 1992] Kates, J. (1992). On the use of coherence to measure distortion in hearing aids. *J. Acoust. Soc. Am.*, 91:2236–2244.

[Kates, 1993a] Kates, J. (1993a). Accurate tuning curves in a cochlear model. *IEEE Trans. Speech and Audio Proc.*, 1:453–462.

[Kates, 1993b] Kates, J. (1993b). Hearing aid design criteria. *J. Speech-Lang. Path. and Audiol. Monogr. Suppl. 1*, pages 15–23.

[Kates, 1993c] Kates, J. (1993c). Superdirective arrays for hearing aids. *J. Acoust. Soc. Am.*, 94:1930–1933.

[Kates, 1993d] Kates, J. (1993d). Toward a theory of optimal hearing aid processing. *J. Rehab. Res. and Devel.*, 30:39–48.

[Kates, 1994] Kates, J. (1994). Speech enhancement based on a sinusoidal model. *Journal of Speech and Hearing Research*, 37:449–464.

[Kates, 1995] Kates, J. (1995). Two-tone suppression in a cochlear model. *IEEE Trans. Speech and Audio Proc*, 3:396–406.

[Kates and Kozma-Spytek, 1994] Kates, J. and Kozma-Spytek, L. (1994). Quality ratings for frequency-shaped peak-clipped speech. *J. Acoust. Soc. Am.*, 95:3586–3594.

[Kates and Weiss, 1996] Kates, J. and Weiss, M. (1996). A comparison of hearing-aid processing techniques. *J. Acoust. Soc. Am.*, 99:3138–3148.

[Kay, 1988] Kay, S. M. (1988). *Modern Spectral Estimation*. Prentice-Hall, Inc., Englewood Cliffs, NJ.

[Kay, 1993] Kay, S. M. (1993). *Fundamentals of statistical signal processing: Estimation theory*. PH signal processing series. Prentice-Hall, Englewood Cliffs, NJ.

[Keefe, 1982] Keefe, D. H. (1982). Theory of the single woodwind tone hole. experiments on the single woodwind tone hole. *J. Acoustical Soc. of America*, 72(3):676–699.

[Keefe, 1990] Keefe, D. H. (1990). Woodwind air column models. *J. Acoustical Soc. of America*, 88(1):35–51.

[Keller, 1994] Keller, E. (1994). *Fundamentals of Speech Synthesis*. John Wiley and Sons, Inc., New York.

[Keller, 1953] Keller, J. B. (1953). Bowing of violin strings. *Comm. Pure Applied Math.*, 6:483–495.

[Kelly and Lochbaum, 1962] Kelly, J. L. and Lochbaum, C. C. (1962). Speech synthesis. *Proc. Fourth Int. Congress on Acoustics, Copenhagen*, pages 1–4. Paper G42.

[Kennedy et al., 1996] Kennedy, E., Levitt, H., Neuman, A., and Weiss, M. (1996). Consonant-vowel ratios for maximizing consonant recognition by hearing-impaired listeners.

[Key et al., 1959] Key, E., Fowle, E., and Haggarty, R. (1959). A method of pulse compression employing nonlinear frequency modulation. Technical report 207, Lincoln Laboratory, M.I.T.

[Kiang, 1980] Kiang, N.-S. (1980). Processing of speech by the auditory nervous system. *J. Acoust. Soc. Am.*, 68:830–835.

[Kiessling, 1993] Kiessling, J. (1993). Current approaches to hearing aid evaluation. *J. Speech-Lang. Path. and Audiol. Monogr. Suppl. 1*, pages 39–49.

[Killion, 1976] Killion, M. (1976). Noise of ears and microphones. *J. Acoust. Soc. Am.*, 59:424–433.

[Killion, 1981] Killion, M. (1981). Earmold options for wideband hearing aids. *J. Speech Hear. Disorders*, 46:10–20.

[Killion, 1988] Killion, M. (1988). Principles of high-fidelity hearing-aid amplification. In Sandlin, R., editor, *Handbook of Hearing Aid Amplification*, volume Volume I: Theoretical and Technical Considerations, pages 45–79. College-Hill Press.

[Killion, 1993] Killion, M. (1993). The K-Amp hearing aid: An attempt to present high fidelity for persons with impaired hearing. *Am. J. Audiol.*, 2:52–74.

[Kinzie, Jr. and Gravereaux, 1973] Kinzie, Jr., G. R. and Gravereaux, D. W. (1973). Automatic detection of impulse noise. *J. Audio Eng. Soc.*, 21(3):331–336.

[Klatt, 1980] Klatt, D. (1980). Software for a cascade/parallel formant synthesizer. *J. Acoust. Soc. Am.*, 67:971–995.

[Kleiner et al., 1993] Kleiner, M., Dalenback, B.-I., and Svensson, P. (1993). "Auralization - an overview". *J. Audio Eng. Soc.*, 41(11):861–875.

[Kloker, 1986] Kloker, K. (1986). Motorola DSP56000 Digital Signal Processor. *IEEE Micro*, 6(6):29–48.

[Kloker and Posen, 1988] Kloker, K. L. and Posen, M. P. (1988). Modulo arithmetic unit having arbitrary offset values. U.S. Patent 4,742,479.

[Knudsen, 1975] Knudsen, M. J. (1975). Real-Time Linear-Predictive Coding of Speech on the SPS-41 Triple-Microprocessor Machine. *IEEE Trans. Acoustics Speech and Signal Processing*, ASSP-23(1):140–145.

[Kochkin, 1992] Kochkin, S. (1992). Marketrak III identifies key factors in determining consumer satisfaction. *Hearing J.*, 45:39–44.

[Koilpillai and Vaidyanathan, 1991] Koilpillai, D. and Vaidyanathan, P. P. (1991). New Results on Cosine-Modulated FIR Filter Banks Satisfying Perfect Reconstruction. In *Proc. IEEE Int. Conf. Acoust., Speech and Signal Proc*, pages 1793 – 1796.

[Kollmeier et al., 1993] Kollmeier, B., Peisseig, J., and Hohnmann, V. (1993). Real-time multiband dynamic compression and noise reduction for binaural hearing aids. *J. Rehab. Res. and Devel.*, 30:82–94.

[Kolsky, 1963] Kolsky, H. (1963). *Stress Waves in Solids*. Dover, New York.

[Kompis and Dillier, 1994] Kompis, M. and Dillier, N. (1994). Noise reduction for hearing aids: Combining directional microphones with an adaptive beamformer. *J. Acoust. Soc. Am.*, 96:1910–1913.

[Kontro et al., 1992] Kontro, J., Kalliojärvi, K., and Neuvo, Y. (1992). Use of Short Floating Point Formats in Audio Applications. *IEEE Trans. on Consumer Electronics*, 38(3):200–207.

[Koo et al., 1989] Koo, B., Gibson, J. D., and Gray, S. D. (1989). Filtering of coloured noise for speech enhancement and coding. *Proc. IEEE Int. Conf. Acoust., Speech and Signal Proc*, pages 349–352.

[Koren, 1993] Koren, I. (1993). *Computer arithmetic algorithms*. Prentice Hall.

[Krahé, 1988] Krahé, D. (1988). *Grundlagen eines Verfahrens zur Datenreduktion bei qualitativ hochwertigen, digitalen Audiosignalen auf Basis einer adaptiven Transformationscodierung unter Berücksichtigung psychoakustischer Phänomene*. Dissertation, Universität Duisburg. (in German).

[Krasner, 1979] Krasner, M. A. (1979). Digital encoding of speech and audio signals based on the perceptual requirements of the auditory system. Technical Report 535, Massachusetts Institute of Technology, Lincoln Laboratory, Lexington.

[Kriz, 1975] Kriz, J. S. (1975). A 16-Bit A-D-A Conversion System for High-Fidelity Audio Research. *IEEE Trans. on ASSP*, ASSP-23(1):146–149.

[Kriz, 1976] Kriz, J. S. (1976). An Audio Analog-Digital-Analog Conversion System. In *55th Convention*. Preprint 1142.

[Krokstad et al., 1968] Krokstad, A., Strom, S., and Sorsdal, S. (1968). "Calculating the acoustical room response by the use of a ray tracing technique". *J. Sound and Vibration*, 8:118–125.

[Kronland-Martinet et al., 1987] Kronland-Martinet, R., , Morlet, J., and Grossmann, A. (1987). Analysis of sound patterns through wavelets transforms. *Int. J. Patt. Recog. Art. Int.*, 1(2):97–126.

[Kryter, 1962] Kryter, K. (1962). Methods for the calculation and use of the articulation index. *J. Acoust. Soc. Am.*, 34:1689–1697.

[Kubichek et al., 1989] Kubichek, R., Quincy, E., and Kiser, K. (1989). Speech quality assessment using expert pattern recognition techniques. IEEE Pacific Rim Conference on Communications Computers and Signal Processing.

[Kuk et al., 1989] Kuk, F., Tyler, R., Stubbing, P., and Bertschy, M. (1989). Noise reduction circuitry in ITE instruments. *Hearing Instr.*, 40:20–26ff.

[Kuttruff, 1991] Kuttruff, H. (1991). *Room Acoustics*. Elsevier Science Publishing Company, New York, NY.

[Laakso et al., 1996] Laakso, T. I., Välimäki, V., Karjalainen, M., and Laine, U. K. (1996). Splitting the Unit Delay—Tools for Fractional Delay Filter Design. *IEEE Signal Processing Magazine*, 13(1):30–60.

[Lagadec and Pelloni, 1983] Lagadec, R. and Pelloni, D. (1983). Signal enhancement via digital signal processing. In *Preprints of the AES 74th Convention*, New York.

[Landau, 1960] Landau, H. (1960). On the recovery of band-limited signals after instantaneous companding and subsequent bandlimiting. *Bell Sys. Tech. J.*, 39:351–364.

[Landau and Miranker, 1961] Landau, H. and Miranker, W. L. (1961). The recovery of distorted band-limited signals. *J. Math. Anal.Appl.*, 2:97–104.

[Lang and Laakso, 1994] Lang, M. and Laakso, T. I. (1994). Simple and robust method for the design of allpass filters using least-squares phase error criterion. *IEEE Trans. Circuits and Systems*, 41(1):40–48.

[Laroche, 1989] Laroche, J. (1989). A New Analysis/Synthesis System of Musical Signals Using Prony's Method: Application to Heavily Damped Percussive Sounds. In *Proc. IEEE Int. Conf. Acoustics, Speech, and Signal Processing.*, volume 3, pages 2053–2056, Glasgow, Scotland.

[Laroche, 1993] Laroche, J. (1993). Autocorrelation method for high quality time/pitch scaling. In *Proc. IEEE Workshop Appl. of Signal Processing to Audio and Acoustics*, Mohonk Mountain House, New Paltz, NY.

[Laroche, 1994] Laroche, J. (1994). Multichannel excitation/filter modeling of percussive sounds with application to piano. *IEEE Trans. on Acoustics, Speech, and Signal Processing*, ASSP-02(2):329–345.

[Laroche et al., 1993a] Laroche, J., Moulines, E., and Stylianou, Y. (1993a). HNS: Speech modification based on a harmonic + noise model. *Proc. IEEE ICASSP-93, Minneapolis*, pages 550–553.

[Laroche et al., 1993b] Laroche, J., Stylianou, Y., and Moulines, E. (1993b). HNM: A simple, efficient harmonic plus noise model for speech. *Proc. IEEE Workshop Appl. of Signal Processing to Audio and Acoustics*.

[Lawrence et al., 1983] Lawrence, R., Moore, B., and Glasberg, B. (1983). A comparison of behind-the-ear high-fidelity linear hearing aids and two-channel compression aids in the laboratory and in everyday life. *Brit. J. Audiol.*, 17:31–48.

[LeBrun, 1979] LeBrun, M. (1979). Digital waveshaping synthesis. *J. Audio Eng. Soc.*, 27(4):250–266.

[Lee, 1988] Lee, E. A. (1988). Programmable DSP Architectures: Part I. *IEEE ASSP Magazine*, 5(4):4–19.

[Lee, 1989] Lee, E. A. (1989). Programmable DSP Architectures: Part II. *IEEE ASSP Magazine*, 6(1):4–14.

[Lee, 1972] Lee, F. (1972). Time compression and expansion of speech by the sampling method. *J. Audio Eng. Soc.*, 20(9):738–742.

[Leek and Watson, 1984] Leek, M. R. and Watson, C. S. (1984). Learning to detect auditory pattern components. *J. Acoust. Soc. Am.*, 76:1037–1044.

[Lehnert and Blauert, 1992] Lehnert, H. and Blauert, J. (1992). "Principles of binaural room simulation". *Applied Acoustics*, 36:259–291.

[Leontaritis and Billings, 1985] Leontaritis, I. J. and Billings, S. A. (1985). Input-output parametric models for non-linear systems. part I:deterministic non-linear systems; part II: Stochastic non-linear systems. *Int. J. Control*, 41(2):303–344.

[Leontaritis and Billings, 1987] Leontaritis, I. J. and Billings, S. A. (1987). Model selection and validation methods for non-linear systems. *Int. J. Control*, 45(1):311–341.

[Levitt and Neuman, 1991] Levitt, H. and Neuman, A. (1991). Evaluation of orthogonal polynomial compression. *J. Acoust. Soc. Am.*, 90:241–252.

[Liberman and Dodds, 1984] Liberman, M. and Dodds, L. (1984). Single neuron labeling and chronic cochlear pathology III: Stereocilia damage and alterations of threshold tuning curves. *Hearing Res.*, 16:55–74.

[Lidbetter et al., 1988] Lidbetter, P., Bustance, D., and Boswell, G. (1988). Basic Concepts and Problems of Synchronization of Digital Audio Systems. *Proc. 84th AES Convention, Paris*. preprint 2605.

[Lidbetter, 1983] Lidbetter, P. S. (1983). Signal processing for the digital audio console. In *Proc. 6th European Conf. on Circuit Theory and Design*, pages 536–539.

[Lim and Lee, 1996] Lim, I. and Lee, B. G. (1996). Lossy pole-zero modeling of speech. *IEEE Trans. Speech and Audio Processing*, 4(2):80–88.

[Lim et al., 1978] Lim, J., Oppenheim, A., and Braida, L. (1978). Evaluation of an adaptive comb filtering method for enhancing speech degraded by white noise addition. *IEEE Trans. Acoust. Speech and Sig. Proc.*, ASSP-26:354–358.

[Lim, 1983] Lim, J. S., editor (1983). *Speech enhancement*. Prentice-Hall signal processing series. Prentice-Hall, Englewood Cliffs, NJ.

[Lim, 1986] Lim, J. S. (1986). Speech enhancement (preconference lecture). In *Proc. IEEE Int. Conf. Acoust., Speech, Signal Processing*, pages 3135–3142.

[Lim and Oppenheim, 1978] Lim, J. S. and Oppenheim, A. V. (1978). All-pole modelling of degraded speech. *IEEE Trans. Acoustics, Speech and Signal Processing*, ASSP-26(3).

[Lim and Oppenheim, 1979] Lim, J. S. and Oppenheim, A. V. (1979). Enhancement and bandwidth compression of noisy speech. *Proc. IEEE*, 67(12):1586–1604.

[Lin et al., 1987] Lin, K.-S., Frantz, G. A., and Simar, R. (1987). The TMS320 Family of Digital Signal Processors. *Proc. IEEE*, 75(9):1143–1159.

[Lindemann, 1987] Lindemann, E. (1987). DSP architectures for the digital audio workstation. In *AES 83rd convention*. Preprint 2498.

[Lindemann et al., 1991] Lindemann, E., Dechelle, F., Smith, B., and Starkier, M. (1991). The Architecture of the IRCAM Musical Workstation. *Computer Music Journal*, 15(3):41–49.

[Lindsay, 1973] Lindsay, R. B. (1973). *Acoustics: Historical and Philosophical Development*. Dowden, Hutchinson & Ross, Stroudsburg. Contains "Investigation of the Curve Formed by a Vibrating String, 1747," by Jean le Rond d'Alembert.

[Lippmann et al., 1981] Lippmann, R., Braida, L., and Durlach, N. (1981). Study of multichannel amplitude compression and linear amplification for persons with sensorineural hearing loss. *J. Acoust. Soc. Am.*, 69:524–534.

[Liu and Stanley, 1965] Liu, B. and Stanley, T. P. (1965). Error Bounds for Jittered Sampling. *IEEE Trans. on Automatic Control*, AC-10(4):449–454.

[Lockhart and Goodman, 1986] Lockhart, G. B. and Goodman, D. J. (1986). Reconstruction of missing speech packets by waveform substitution. *Signal Processing 3: Theories and Applications*, pages 357–360.

[Loy, 1981] Loy, D. G. (1981). Notes on the implementation of MUSBOX: A Compiler for the Systems Concepts Digital Synthesizer. *Computer Music Journal*, 5(1):34–50.

[Lufti, 1983] Lufti, R. A. (1983). Additivity of simultaneous masking. *J. Acoust. Soc. Am.*, 73:262–267.

[Lufti, 1985] Lufti, R. A. (1985). A power-law transformation predicting masking by sounds with complex spectra. *J. Acoust. Soc. Am.*, 77:2128–2136.

[Lynch et al., 1982] Lynch, T., Nedzelnitsky, V., and Peake, W. (1982). Input impedance of the cochlea in the cat. *J. Acoust. Soc. Am.*, 72:108–130.

[Lyon, 1981] Lyon, R. F. (1981). A Bit Serial VLSI Architecture Methodology for Signal Processing. In Gray, J. P., editor, *VLSI - 81*, pages 131–140. Academic Press.

[MacKay, 1992] MacKay, D. (1992). *Bayesian Methods for Adaptive Models*. PhD thesis, Calif. Inst. Tech.

[Macon and Clements,] Macon, M. and Clements, M. Phase dithering in sine-wave analysis/synthesis. *IEEE Signal Processing Letters*. (submitted for publication).

[Macon and Clements, 1996] Macon, M. and Clements, M. (1996). Speech concatenation and synthesis using an overlap-add sinusoidal model. In *Proc. IEEE Int. Conf. Acoustics, Speech, and Signal Processing.*, volume 1, pages 361–364, Atlanta, GA.

[Maher, 1989] Maher, R. (1989). *An Approach for the Separation of Voices in Composite Musical Signals*. PhD thesis, Department of Electrical Engineering and Computer Science, U. Illinois.

[Maher, 1990] Maher, R. (1990). Evaluation of a method for separating digitized duet signals. *J. Audio Eng. Soc.*, 38(12):956–979.

[Maher, 1994] Maher, R. (1994). A method for extrapolation of missing digital audio data. *J. Audio Eng. Soc.*, 42(5):350–357.

[Maher and Beauchamp, 1990] Maher, R. and Beauchamp, J. (1990). An investigation of vocal vibrato for synthesis. *Applied Acoustics*, 30:219–245.

[Mahieux et al., 1990] Mahieux, Y., Petit, J. P., and Charbonnier, A. (1990). Transform coding of audio signals at 64 kbit/s. In *Proc of GLOBECOM'90*.

[Main, 1978] Main, I. G. (1978). *Vibrations and Waves in Physics*. Cambridge University Press.

[Makhoul, 1975] Makhoul, J. (1975). Linear prediction: A tutorial review. *Proc. IEEE*, 63(4):561–580.

[Makhoul and El-Jaroudi, 1986] Makhoul, J. and El-Jaroudi, A. (1986). Time scale modification in medium to low rate speech coding. *Proc. IEEE ICASSP-86*, pages 1705–1708.

[Malah, 1979] Malah, D. (1979). Time-domain algorithms for harmonic bandwidth reduction and time scaling of speech signals. *IEEE Trans. on Acoustics, Speech, and Signal Processing*, ASSP-27(2):113–120.

[Malah and Flanagan, 1981] Malah, D. and Flanagan, J. (1981). Frequency scaling of speech signals by transform techniques. *Bell System Technical Journal*, 60(9):2107–2156.

[Mallat and Hwang, 1992] Mallat, S. and Hwang, W. (1992). Singularity detection and processing with wavelets. *IEEE Trans. Information Theory*, 38(2):617–643.

[Mallatt and Hwang, 1992] Mallatt, S. and Hwang, W. L. (1992). Singularity detection and processing with wavelets. *IEEE Trans. Info. Theory*, pages 617–643.

[Malvar, 1990] Malvar, H. S. (1990). Lapped transforms for efficient transform/subband coding. In *IEEE Trans. Acoust., Speech and Signal Processing*, volume 38, pages 969–978.

[Malvar, 1991] Malvar, H. S. (1991). Extended lapped transforms: Fast algorithms and applications. In *Proc. IEEE Int. Conf. Acoust., Speech and Signal Proc*, pages 1797–1800.

[Malvar, 1992] Malvar, H. S. (1992). *Signal Processing with Lapped Transforms*. Artech House, Norwood, MA.

[Mansour and Gray, 1982] Mansour, D. and Gray, A. (1982). Unconstrained frequency-domain adaptive filter. *IEEE Trans. Acoust. Speech and Sig. Proc.*, ASSP-30:726–734.

[Markel and Gray, 1976] Markel, J. D. and Gray, A. H. (1976). *Linear Prediction of Speech*. Springer Verlag, Berlin and New York.

[Marques and Almeida, 1987] Marques, J. and Almeida, L. (1987). Quasi-optimal analysis for sinusoidal representation of speech. *Proc. Gretsi*, pages 447–450.

[Marques and Almeida, 1988] Marques, J. and Almeida, L. (1988). Sinusoidal modeling of speech: Representation of unvoiced sounds with narrowband basis functions. In *Proc. EUSIPCO*.

[Marques and Almeida, 1989] Marques, J. and Almeida, L. (1989). Frequency-varying sinusoidal modeling of speech. *IEEE Trans. on Acoustics, Speech, and Signal Processing*, ASSP-37(5):763–765.

[Martens, 1982] Martens, J. (1982). A new theory for multitone masking. *J. Acoust. Soc. Am.*, 72:397–405.

[Martin et al., 1993] Martin, J., Maercke, D. V., and Vian, J.-P. (1993). "Binaural simulation of concert halls: a new approach for the binaural reverberation process". *J. Acoust. Soc. Am.*, 94(6):3255–3264.

[Martin, 1981] Martin, R. (1981). Robust methods for time series. *In Applied Time Series Analysis II, D.F. Findley, ed.*, pages 683–759.

[Massie, 1985] Massie, D. (1985). The Emulator II computer music environment. *Proc. of International Computer Music Conference*, pages 111–113.

[Massie, 1986] Massie, D. (1986). A survey of looping algorithms for sampled data musical instruments. In *Proc. IEEE Workshop Appl. of Signal Processing to Audio and Acoustics*, Mohonk Mountain House, New Paltz, NY.

[Mathews, 1969] Mathews, M. V. (1969). *The Technology of Computer Music*. MIT Press, Cambridge, MA.

[Mathews and Moore, 1970] Mathews, M. V. and Moore, F. R. (1970). GROOVE – A Program to Compose, Store and Edit Functions of Time. *Comm. of the ACM*, 13(7):715–721.

[Maxwell and Zurek, 1995] Maxwell, J. and Zurek, P. (1995). Reducing acoustic feedback in hearing aids. *IEEE. Trans. Speech and Audio Proc.*, 3:304–313.

[McAdams, 1984] McAdams, S. (1984). Spectral fusion, spectral parsing, and the formation of auditory images. STAN-M-22, CCRMA, Dept. of Music, Stanford, CA.

[McAulay and Quatieri, 1985] McAulay, R. and Quatieri, T. (1985). Speech analysis/synthesis based on a sinusoidal representation. Technical report 693, Lincoln Laboratory, M.I.T.

[McAulay and Quatieri, 1986a] McAulay, R. and Quatieri, T. (1986a). Phase modeling and its application to sinusoidal transform coding. In *Proc. IEEE Int. Conf. Acoustics, Speech, and Signal Processing.*, pages 1713–1715, Tokyo, Japan.

[McAulay and Quatieri, 1986b] McAulay, R. and Quatieri, T. (1986b). Speech analysis-synthesis based on a sinusoidal representation. *IEEE Trans. on Acoustics, Speech, and Signal Processing*, ASSP-34(4):744–754.

[McAulay and Quatieri, 1987] McAulay, R. and Quatieri, T. (1987). Multirate sinusoidal transform coding at rates from 2.4 kbps to 8 kbps. In *Proc. IEEE Int. Conf. Acoustics, Speech, and Signal Processing.*, pages 1645–1648, Dallas, TX.

[McAulay and Quatieri, 1988] McAulay, R. and Quatieri, T. (1988). Computationally efficient sine-wave synthesis and its application to sinusoidal transform coding. In *Proc. IEEE Int. Conf. Acoustics, Speech, and Signal Processing.*, pages 370–373, New York, NY.

[McAulay and Quatieri, 1990] McAulay, R. and Quatieri, T. (1990). Pitch estimation and voicing detection based on a sinusoidal speech model. In *Proc. IEEE Int. Conf. Acoustics, Speech, and Signal Processing.*, volume 1, pages 249–252, Albuquerque, NM.

[McAulay and Quatieri, 1992] McAulay, R. and Quatieri, T. (1992). Low rate speech coding based on the sinusoidal speech model. In *Advances in Speech Signal Processing*. Marcel Dekker, New York, NY.

[McAulay and Malpass, 1980] McAulay, R. J. and Malpass, M. L. (1980). Speech enhancement using a soft-decision noise suppression filter. *IEEE Trans. Acoust., Speech, Signal Processing*, 28(2):137–145.

[McClellan, 1988] McClellan, J. (1988). Parametric signal estimation. In Lim, J. S. and Oppenheim, A. V., editors, *Advanced Topics in Digital Signal Processing*. Prentice Hall, Englewood Cliffs, NJ.

[McCulloch and Tsay, 1994] McCulloch, R. E. and Tsay, R. S. (1994). Bayesian analysis of autoregressive time series via the Gibbs sampler. *Journal of Time Series Analysis*, 15(2):235–250.

[McDermott et al., 1992] McDermott, H., McKay, C., and Vandali, A. (1992). A new portable sound processor for the University of Melbourne/ Nucleus Limited multielectrode cochlear implant. *J. Acoust. Soc. Am.*, 91:3367–3371.

[McDonough, 1972] McDonough, R. (1972). Degraded performance of nonlinear array processors in the presence of data modeling errors. *J. Acoust. Soc. Am.*, 51:1186–1193.

[McIntyre et al., 1983] McIntyre, M. E., Schumacher, R. T., and Woodhouse, J. (1983). On the oscillations of musical instruments. *J. Acoustical Soc. of America*, 74(5):1325–1345.

[McIntyre and Woodhouse, 1979] McIntyre, M. E. and Woodhouse, J. (1979). On the fundamentals of bowed string dynamics. *Acustica*, 43(2):93–108.

[McMillen, 1994] McMillen, K. (1994). *A Technology Dossier on Fourier Analysis Resynthesis*. Oberheim Digital and G-WIZ Labs, Berkeley, CA.

[McNally, 1979] McNally, G. W. (1979). Microprocessor Mixing and Processing of Digital Audio Signals. *J. of the Audio Eng. Soc.*, 27(10):793–803.

[Meddis and Hewitt, 1991] Meddis, R. and Hewitt, M. J. (1991). Virtual pitch and phase sensitivity of a computer model of the auditory periphery. I: Pitch identification. *J. Acoust. Soc. Am.*, 89:2866–2882.

[Medwetsky and Boothroyd, 1991] Medwetsky, L. and Boothroyd, A. (1991). Effect of microphone placement on the spectral distribution of speech. *Am. Speech Lang. Hearing Assn.*

[Mercer, 1993] Mercer, K. J. (1993). *Identification of signal distortion models*. PhD thesis, University of Cambridge.

[Mick and Brick, 1980] Mick, J. and Brick, J. (1980). *Bit-slice microprocessor design*. McGraw-Hill.

[Miller, 1973] Miller, N. J. (1973). Recovery of singing voice from noise by synthesis. *Thesis Tech. Rep., ID UTEC-CSC-74-013, Univ. Utah.*

[Montgomery and Edge, 1988] Montgomery, A. and Edge, R. (1988). Evaluation of two speech enhancement techniques to improve intelligibility for hearing-impaired adults. *J. Speech and Hearing Res.*, 31:386–393.

[Montresor et al., 1990] Montresor, S., Valière, J. C., and Baudry, M. (1990). Détection et Suppression de Bruits Impulsionnels Appliqué à la Restauration d'Enregistrements Anciens (Detection and suppression of Impulsive noise applied to the restoration of old recordings). *Colloq. de Physique, C2, supplément au no. 2, Tome 51, Février*, pages 757–760.

[Monzingo and Miller, 1980] Monzingo, R. and Miller, T. (1980). *Introduction to Adaptive Arrays*. Wiley.

[Moore, 1987] Moore, B. (1987). Design and evaluation of a two-channel compression hearing aid. *J. Rehab. Res. and Devel.*, 24:181–192.

[Moore, 1989] Moore, B. (1989). *An Introduction to the Psychology of Hearing*. Academic Press, second edition.

[Moore et al., 1985] Moore, B., Glasberg, B., Hess, R., and Birchall, J. (1985). Effects of flanking noise bands on the rate of growth of loudness of tones in normal and recruiting ears. *J. Acoust. Soc. Am.*, 77:1505–1513.

[Moore, 1997] Moore, B. C. J. (1997). *An introduction to the psychology of hearing*. Academic Press, fourth edition.

[Moore, 1990a] Moore, F. (1990a). *Elements of Computer Music*. Prentice Hall, Englewood Cliffs, New Jersey.

[Moore, 1977a] Moore, F. R. (1977a). *Real Time Interactive Computer Music Synthesis*. PhD thesis, Stanford University, EE Dept.

[Moore, 1977b] Moore, F. R. (1977b). Table Lookup Noise for Sinusoidal Oscillators. *Computer Music Journal*, 1(2):26–29. pages 326–334.

[Moore, 1985] Moore, F. R. (1985). The FRMBox - A Modular Digital Music Synthesizer. In Strawn, J., editor, *Digital Audio Enginnering: An Anthology*, pages 95–108. William Kaufmann.

[Moore, 1990b] Moore, F. R. (1990b). *Elements of Computer Music*. Prentice Hall.

[Moorer, 1977] Moorer, J. (1977). Signal processing aspects of computer music: A survey. *Proceedings of IEEE*, 65:1108–1137.

[Moorer, 1976] Moorer, J. A. (1976). The synthesis of complex audio spectra by means of discrete summation formulae. *J. Audio Eng. Soc.*, 24(9):717–727.

[Moorer, 1978] Moorer, J. A. (1978). The use of the phase vocoder in computer music applications. *J. Audio Eng. Soc.*, 26(1).

[Moorer, 1979] Moorer, J. A. (1979). "About This Reverberation Business". *Computer Music Journal*, 3(2):3255–3264.

[Moorer, 1980a] Moorer, J. A. (1980a). Linked list of timed and untimed commands. U.S. Patent 4,497,023.

[Moorer, 1980b] Moorer, J. A. (1980b). Musical instrument and method for generating musical sound. U.S. Patent 4,215,617.

[Moorer, 1981] Moorer, J. A. (1981). Synthesizers I Have Known and Loved. *Computer Music Journal*, 5(1):4–12.

[Moorer, 1983] Moorer, J. A. (1983). The Audio Signal Processor: The next step in Digital Audio. In Blesser, B., Locanthi, B., and T. G. Stockham, J., editors, *Digital Audio*, pages 205–215. Audio Eng. Society.

[Moorer, 1985a] Moorer, J. A. (1985a). A Flexible Method for Synchronizing Parameter Updates for Real-Time Audio Signal Processors. In *AES 79th convention*. Preprint 2279.

[Moorer, 1985b] Moorer, J. A. (1985b). The Lucasfilm Digital Audio Facility. In Strawn, J., editor, *Digital Audio Enginnering: An Anthology*, pages 109–136. William Kaufmann.

[Moorer et al., 1986] Moorer, J. A., Abbott, C., Nye, P., Borish, J., and Snell, J. (1986). The Digital Audio Processing Station: A New Concept in Audio Postproduction. *J. Audio Engineering Society*, 34(6):454–462.

[Moorer and Berger, 1986] Moorer, J. A. and Berger, M. (1986). Linear-phase bandsplitting: Theory and applications. *J. Audio Eng. Soc.*, 34(3):143–152.

[Moorer et al., 1979] Moorer, J. A., Chauveau, A., Abbott, C., Eastty, P., and Lawson, J. (1979). The 4C Machine. *Computer Music Journal*, 3(3):16–24. pages 244–250.

[Morse, 1981] Morse, P. M. (1981). *Vibration and Sound*. American Institute of Physics, for the Acoustical Society of America, (516)349-7800 x 481. 1st ed. 1936, 4th ed. 1981.

[Morse and Ingard, 1968] Morse, P. M. and Ingard, K. U. (1968). *Theoretical Acoustics*. McGraw-Hill, New York. Reprinted by Princeton Univ. Press.

[Mosteller and Tukey, 1977] Mosteller, F. and Tukey, J. (1977). *Data Analysis and Regression*. Addison-Wesley, Reading, Mass.

[Moulines and Charpentier, 1990] Moulines, E. and Charpentier, F. (1990). Pitch-synchronous waveform processing techniques for text-to-speech synthesis using diphones. *Speech Communication*, 9(5/6):453–467.

[Moulines and Laroche, 1995] Moulines, E. and Laroche, J. (1995). Non parametric techniques for pitch-scale and time-scale modification of speech. *Speech Communication*, 16:175–205.

[Mourjopoulos et al., 1992] Mourjopoulos, J., Kokkinakis, G., and Paraskevas, M. (1992). Noisy audio signal enhancement using subjective spectra. In *Preprints of the AES 92nd Convention*, Vienna.

[MPEG, 1991] MPEG (1991). Report on the MPEG/AUDIO subjective listening tests. Doc. 91/331, ISO/IEC JTC1/SC29 WG11.

[MPEG, 1992] MPEG (1992). Coding of moving pictures and associated audio for digital storage media at up to 1.5 Mbit/s, part 3: Audio. International Standard IS 11172-3, ISO/IEC JTC1/SC29 WG11.

[MPEG, 1994a] MPEG (1994a). Information technology — generic coding of moving pictures and associated audio, part 3: Audio. International Standard IS 13818–3, ISO/IEC JTC1/SC29 WG11.

[MPEG, 1994b] MPEG (1994b). Report on the subjective testing of coders at low sampling frequencies. Doc. N0848, ISO/IEC JTC1/SC29 WG11.

[MPEG, 1996] MPEG (1996). Description of GCLs experiment for NBC core experiment 2. Doc. MPEG96/M0611, ISO/IEC JTC1/SC29 WG11.

[MPEG, 1997a] MPEG (1997a). MPEG–2 advanced audio coding, AAC. International Standard IS 13818–7, ISO/IEC JTC1/SC29 WG11.

[MPEG, 1997b] MPEG (1997b). Working draft of ISO/IEC 14496-3 MPEG-4 Audio V3.0. Doc. N1631, ISO/IEC JTC1/SC29 WG11.

[Musiek and Lamb, 1994] Musiek, F. and Lamb, L. (1994). Central auditory assessment: An overview. In Katz, J., editor, *Handbook of Clinical Audiology*, pages 197–211. Williams and Wilkins, 4th edition edition.

[Nakahashi and Ono, 1990] Nakahashi, T. and Ono, T. (1990). Analog to Digital Conversion System using dither. U.S. Patent 4,914,439.

[Narayan et al., 1983] Narayan, S., Peterson, A., and Narasimha, M. (1983). Transform domain LMS algorithm. *IEEE Trans. Acoust. Speech and Sig. Proc.*, ASSP-31:609–615.

[National Institutes of Health, 1989] National Institutes of Health (1989). National strategic research plan for hearing and hearing impairment and voice and voice disorders. Technical report, National Institutes of Health. NIH Publication 93-3443.

[Natvig, 1988] Natvig, J. E. (1988). Evaluation of six medium bit-rate coders for the pan-european digital mobile radio system. *IEEE J. on Select. Areas in Commun.*, 6(2):324–331.

[Nawab and Quatieri, 1988a] Nawab, H. and Quatieri, T. (1988a). Short-time fourier transform. In Lim, J. S. and Oppenheim, A. V., editors, *Advanced Topics in Signal Processing*. Prentice Hall, Englewood Cliffs, New Jersey.

[Nawab et al., 1983] Nawab, S., Quatieri, T., and Lim, J. (1983). Signal Reconstruction from Short-Time Fourier Transform Magnitude. *IEEE Trans. Acoust., Speech, Signal Processing*, ASSP-31(4):986–998.

[Nawab and Quatieri, 1988b] Nawab, S. H. and Quatieri, T. F. (1988b). Short-time Fourier transform. In Lim, J. S. and Oppenheim, A. V., editors, *Advanced topics in signal processing*, chapter 6. Prentice-Hall, Englewood Cliffs, NJ.

[Naylor and Boll, 1986] Naylor, J. and Boll, S. (1986). Simultaneous talker separation. In *Proc. 1986 Digital Signal Processing Workshop*, Chatham, MA.

[Naylor and Boll, 1987] Naylor, J. and Boll, S. (1987). Techniques for suppression of an interfering talker in co-channel speech. In *Proc. IEEE Int. Conf. Acoustics, Speech, and Signal Processing.*, pages 205–208, Dallas, Texas.

[Neuman et al., 1994] Neuman, A., Bakke, M., Hellman, S., and Levitt, H. (1994). The effect of compression ratio in a slow acting compression hearing aid: Paired-comparison judgments of quality. *J. Acoust. Soc. Am.*, 96:1471–1478.

[Neuman et al., 1995] Neuman, A., Bakke, M., Mackersie, C., Hellman, S., and Levitt, H. (1995). The effect of release time in compression hearing aids: Paired comparison judgments of quality. *J. Acoust. Soc. Am.*, 98:3182–3187.

[Neuman et al., 1985] Neuman, A., Mills, R., and Schwander, T. (1985). Noise reduction: Effects on consonant perception by normal hearing listeners. *Annual Convention of the Am. Speech-Lang.-Hearing Assn.*

[Neuman and Schwander, 1987] Neuman, A. and Schwander, T. (1987). The effect of filtering on the intelligibility and quality of speech in noise. *J. Rehab. Res. and Devel.*, 24:127–134.

[Niedźwiecki, 1994] Niedźwiecki, M. (1994). Recursive algorithm for elimination of measurement noise and impulsive disturbances from ARMA signals. *Signal Processing VII: Theories and Applications*, pages 1289–1292.

[Niedźwiecki and Cisowski, 1996] Niedźwiecki, M. and Cisowski, K. (1996). Adaptive scheme for elimination of broadband noise and impulsive disturbances from AR and ARMA signals. *IEEE Trans. on Signal Processing*, pages 528–537.

[Nieminen et al., 1987] Nieminen, A., Miettinen, M., Heinonen, P., and Nuevo, Y. (1987). Music restoration using median type filters with adaptive filter substructures. *In Digital Signal Processing-87, eds. V. Cappellini and A. Constantinides, North-Holland*.

[Nocerino et al., 1985] Nocerino, N., Soong, F. K., Rabiner, L. R., and Klatt, D. H. (1985). Comparative study of several distortion measures for speech recognition. *Speech Communication*, 4:317–331.

[Norton, 1988] Norton, J. P. (1988). *An Introduction to Identification*. Academic Press, London.

[Olson, 1957] Olson, H. (1957). *Acoustical Engineering*. Van Nostrand.

[Ono et al., 1983] Ono, H., Kanazaki, J., and Mizoi, K. (1983). Clinical results of hearing aid with noise-level-controlled selective amplification. *Audiology*, 22:494–515.

[Oppenheim and Schafer, 1975] Oppenheim, A. and Schafer, R. (1975). *Digital Signal Processing*. Prentice Hall, Englewood Cliffs, NJ.

[Oppenheim and Lim, 1981] Oppenheim, A. V. and Lim, J. S. (1981). The importance of phase in signals. *Proc. IEEE*, 69(5):529–541.

[Oppenheim and Schafer, 1989] Oppenheim, A. V. and Schafer, R. W. (1989). *Discrete-Time Signal Processing*. Prentice Hall, Englewood Cliffs, New Jersey.

[Oppenheim and Willsky, 1983] Oppenheim, A. V. and Willsky, A. S. (1983). *Signals and Systems*. Prentice Hall, Englewood Cliffs, New Jersey.

[ÓRuanaidh, 1994] ÓRuanaidh, J. J. K. (1994). *Numerical Bayesian methods in signal processing*. PhD thesis, University of Cambridge.

[ÓRuanaidh and Fitzgerald, 1993] ÓRuanaidh, J. J. K. and Fitzgerald, W. J. (1993). The restoration of digital audio recordings using the Gibbs' sampler. Technical Report TR I34 CUED INFENG, Cambridge University.

[ÓRuanaidh and Fitzgerald, 1994] ÓRuanaidh, J. J. K. and Fitzgerald, W. J. (1994). Interpolation of missing samples for audio restoration. *Electronic Letters*, 30(8).

[Paillard et al., 1992] Paillard, B., Mabilleau, P., Morisette, S., and Soumagne, J. (1992). PERCEVAL: Perceptual evaluation of the quality of audio signals. *J. Audio Eng. Soc.*, 40(1/2):21–31.

[Paliwal and Basu, 1987] Paliwal, K. K. and Basu, A. (1987). A speech enhancement method based on Kalman filtering. *Proc. IEEE Int. Conf. Acoust., Speech and Signal Proc*, pages 177–180.

[Papamichalis and Simar, 1988] Papamichalis, P. and Simar, R. (1988). The TMS320C30 Floating-Point Digital Signal Processor. *IEEE Micro*, 8(6):13–29.

[Papoulis, 1977] Papoulis, A. (1977). *Signal Analysis*. McGraw-Hill, New York.

[Papoulis, 1991] Papoulis, A. (1991). *Probability, random variables, and stochastic processes*. McGraw-Hill, New York, 3rd edition.

[Parks and Burrus, 1987] Parks, T. W. and Burrus, C. S. (1987). *Digital Filter Design*. John Wiley and Sons, Inc., New York.

[Parsons and Weiss, 1975] Parsons, T. and Weiss, M. (1975). Enhancing/intelligibility of speech in noisy or multi-talker environments. Final technical report, Rome Air Development Center.

[Patrick and Clark, 1991] Patrick, J. and Clark, G. (1991). The nucleus 22-channel cochlear implant system. *Ear and Hearing*, 12:3S–9S.

[Paul, 1981] Paul, D. (1981). The spectral envelope estimation vocoder. *IEEE Trans. on Acoustics, Speech, and Signal Processing*, ASSP-29(4):786–794.

[Pavlovic, 1987] Pavlovic, C. (1987). Derivation of primary parameters and procedures for use in speech intelligibility predictions. *J. Acoust. Soc. Am.*, 82:413–422.

[Pearsons et al., 1976] Pearsons, K., Bennett, R., and Fidell, S. (1976). Speech levels in various noise environments. Technical report, Bolt, Beranek, and Newman, Inc.

[Penner, 1980] Penner, M. J. (1980). The coding of intensity and the interaction of forward and backward masking. *J. Acoust. Soc. Am.*, 67:608–616.

[Penner and Shiffrin, 1980] Penner, M. J. and Shiffrin, R. M. (1980). Nonlinearities in the coding of intensity within the context of a temporal summation model. *J. Acoust. Soc. Am.*, 67:617–627.

[Perlmutter et al., 1977] Perlmutter, Y., Braida, L., Frazier, R., and Oppenheim, A. (1977). Evaluation of a speech enhancement system. *Proc. 1977 IEEE Int. Conf. on Acoust. Speech and Sig. Proc.*, Hartford, CT:212–215.

[Petersen and Boll, 1981] Petersen, T. L. and Boll, S. F. (1981). Acoustic noise suppression in the context of a perceptual model. In *Proc. IEEE Int. Conf. Acoust., Speech, Signal Processing*, volume 1086-1088.

[Pickles, 1988] Pickles, J. (1988). *An Introduction to the Physiology of Hearing*. Academic Press.

[Pitas and Venetsanopoulos, 1990] Pitas, I. and Venetsanopoulos, A. N. (1990). *Nonlinear Digital Filters*. Kluwer Academic Publishers.

[Platte and Rowedda, 1985] Platte, H. J. and Rowedda, V. (1985). A burst error concealment method for digital audio tape application. *AES preprint*, 2201:1–16.

[Plomp, 1978] Plomp, R. (1978). Auditory handicap of hearing impairment and the limited benefit of hearing aids. *J. Acoust. Soc. Am.*, 63:533–549.

[Plomp, 1988] Plomp, R. (1988). The negative effect of amplitude compression in multichannel hearing aids in the light of the modulation-transfer function. *J. Acoust. Soc. Am.*, 83:2322–2327.

[Pluvinage and Benson, 1988] Pluvinage, V. and Benson, D. (1988). New dimensions in diagnostics and fitting. *Hear. Instr.*, 39:28,30,39.

[Poirot et al., 1988] Poirot, G., Rodet, X., and Depalle, P. (1988). Diphone sound synthesis based on spectral envelopes and harmonic/noise excitation functions. *Proc. of International Computer Music Conference, Köln*.

[Pope and Rayner, 1994] Pope, K. J. and Rayner, P. J. W. (1994). Non-linear System Identification using Bayesian Inference. *Proc. IEEE Int. Conf. Acoust., Speech, Signal Processing*, pages 457–460.

[Porcaro et al., 1995] Porcaro, N., Scandalis, P., Smith, J. O., Jaffe, D. A., and Stilson, T. (1995). SynthBuilder—a graphical real-time synthesis, processing and performance system. In *Proc. 1995 Int. Computer Music Conf., Banff*, pages 61–62. Computer Music Association. See http://www-leland.stanford.edu/group/OTL/SynthBuilder.html for information on how to obtain and run SynthBuilder. See also http://www-ccrma.stanford.edu for related information.

[Portnoff, 1980] Portnoff, M. R. (1980). Time–Frequency Representation of Digital Signals and Systems Based on Short–Time Fourier Analysis. *IEEE Trans. Acoustics, Speech, Signal Processing*, ASSP-28:55–69.

[Portnoff, 1981] Portnoff, R. (1981). Time-scale modifications of speech based on short-time Fourier analysis. *IEEE Trans. Acoust., Speech, Signal Processing*, 29(3):374–390.

[Preis and Bloom, 1984] Preis, D. and Bloom, P. J. (1984). Perception of Phase Distortion in Anti-Alias Filters. *J. of the AES*, 32(11):842–847.

[Preis and Polchlopek, 1984] Preis, D. and Polchlopek, H. (1984). Restoration of nonlinearly distorted magnetic recordings. *J. Audio Eng. Soc.*, 32(12):79–86.

[Preves and Newton, 1989] Preves, D. and Newton, J. (1989). The headroom problem and hearing aid performance. *Hearing Journal*, 42:19–26.

[Priestley, 1981] Priestley, M. B. (1981). *Spectral Analysis and Time Series*. Academic Press.

[Priestley, 1988] Priestley, M. B. (1988). *Non-linear and Non-stationary Time Series Analysis*. Academic Press, London and San Diego.

[Princen et al., 1987] Princen, J., Johnson, A., and Bradley, A. (1987). Subband/transform coding using filter bank designs based on time domain aliasing cancellation. In *Proc. IEEE Int. Conf. Acoust., Speech and Signal Proc*, pages 2161–2164.

[Princen and Johnston, 1995] Princen, J. and Johnston, J. D. (1995). Audio coding with signal adaptive filterbanks. In *Proc. IEEE Int. Conf. Acoust., Speech and Signal Proc*, pages 3071 – 3074.

[Puckette, 1995] Puckette, M. (1995). Phase-locked vocoder. In *Proc. IEEE Workshop Appl. of Signal Processing to Audio and Acoustics*, Mohonk Mountain House, New Paltz, NY.

[Quackenbush et al., 1988] Quackenbush, S. R., III, T. P. B., and Clements, M. A. (1988). *Objective measures of speech quality*. Prentice Hall, New Jersey, Englewood Cliffs, NJ.

[Quatieri and Danisewicz, 1990] Quatieri, T. and Danisewicz, R. (1990). An approach to co-channel talker interference suppression using a sinusoidal model for speech. *IEEE Trans. on Acoustics, Speech, and Signal Processing*, ASSP-38(1):56–69.

[Quatieri et al., 1995] Quatieri, T., Dunn, R., , and Hanna, T. (1995). A subband approach to time-scale modification of complex acoustic signals. *Proc. IEEE Trans. Speech and Audio*, 3(6):515–519.

[Quatieri et al., 1993] Quatieri, T., Dunn, R., and Hanna, T. (1993). Time-scale modification with temporal envelope invariance. In *Proc. 1991 Workshop on Applications of Signal Processing to Audio and Acoustics*, Mohonk Mountain House, New Paltz, NY.

[Quatieri et al., 1994a] Quatieri, T., Dunn, R., and McAulay, R. (1994a). Signal enhancement in AM-FM interference. Technical Report TR-993, Lincoln Laboratory, M.I.T.

[Quatieri et al., 1992] Quatieri, T., Dunn, R., McAulay, R., and Hanna, T. (1992). Underwater signal enhancement using a sine-wave representation. In *Proc. IEEE Oceans92*, Newport, Rhode Island.

[Quatieri et al., 1994b] Quatieri, T., Dunn, R., McAulay, R., and Hanna, T. (1994b). Time-scale modification of complex acoustic signals in noise. Technical Report TR-990, Lincoln Laboratory, M.I.T.

[Quatieri et al., 1991] Quatieri, T., Lynch, J., M.L.Malpass, McAulay, R., and Weinstein, C. (1991). Speech processing for AM radio broadcasting. Technical Report TR-681, Lincoln Laboratory, M.I.T.

[Quatieri and McAulay, 1986] Quatieri, T. and McAulay, R. (1986). Speech transformations based on a sinusoidal representation. *IEEE Trans. on Acoustics, Speech, and Signal Processing*, ASSP-34(6):1449–1464.

[Quatieri and McAulay, 1989] Quatieri, T. and McAulay, R. (1989). Phase coherence in speech reconstruction for enhancement and coding applications. In *Proc. IEEE Int. Conf. Acoustics, Speech, and Signal Processing.*, pages 207–210, Glasgow, Scotland.

[Quatieri and McAulay, 1991] Quatieri, T. and McAulay, R. (1991). Peak-to-RMS Reduction Based on a Sinusoidal Model. *IEEE Trans. on Acoustics, Speech, and Signal Processing*, ASSP-39(2):273–288.

[Quatieri and McAulay, 1992] Quatieri, T. and McAulay, R. (1992). Shape-invariant time-scale and pitch modification of speech. *IEEE Trans. on Acoustics, Speech, and Signal Processing*, ASSP-40(3):497–510.

[Rabiner and Schafer, 1978a] Rabiner, L. and Schafer, R. (1978a). *Digital Processing of Speech Signals*. Prentice Hall, Englewood Cliffs, NJ.

[Rabiner, 1982] Rabiner, L. R. (1982). Digital techniques for changing the sampling rate of a signal. *Digital Audio; collected papers from the AES premier conference*, pages 79–89.

[Rabiner and Gold, 1975] Rabiner, L. R. and Gold, B. (1975). *Theory and Application of Digital Signal Processing*. Prentice-Hall, Inc., Englewood Cliffs, NJ.

[Rabiner and Schafer, 1978b] Rabiner, L. R. and Schafer, R. W. (1978b). *Digital Processing of Speech Signals*. Prentice-Hall, Englewood Cliffs, NJ.

[Rajan, 1994] Rajan, J. (1994). *Time Series Classification*. PhD thesis, University of Cambridge.

[Ramalho, 1994] Ramalho, M. (1994). *The Pitch Mode Modulation Model with Applications in Speech Processing*. PhD thesis, Rutgers University.

[Ramstadt and Tanem, 1991] Ramstadt, T. A. and Tanem, J. P. (1991). Cosine-modulated analysis-synthesis filterbank with critical sampling and perfect reconstruction. In *Proc. IEEE Int. Conf. Acoust., Speech and Signal Proc*, pages 1789–1792.

[Rao and Yip, 1990] Rao, K. R. and Yip, P. (1990). *Discrete cosine transform : algorithms, advantages, applications*. Academic Press.

[Rasmussen, 1943] Rasmussen, A. (1943). *Outlines of Neuro-Anatomy*. William C. Brown.

[Rayner and Godsill, 1991] Rayner, P. J. W. and Godsill, S. J. (1991). The detection and correction of artefacts in archived gramophone recordings. In *Proc. IEEE Workshop Appl. of Signal Processing to Audio and Acoustics*, Mohonk Mountain House, New Paltz, NY.

[Rhode, 1971] Rhode, W. (1971). Observations of the vibration of the basilar membrane in squirrel monkeys using the Mössbauer technique. *J. Acoust. Soc. Am.*, 49:1218–1231.

[Rhodes et al., 1973] Rhodes, J. D., Marston, P. C., and Youla, D. C. (1973). Explicit solution for the synthesis of two-variable transmission-line networks. *IEEE Trans. Circuit Theory*, CT-20(5):504–567.

[Rife and Vanderkooy, 1987] Rife, D. D. and Vanderkooy, J. (1987). "Transfer-Function Measurement using Maximum-Length Sequences". In *Proc. Audio Eng. Soc. Conv.* Preprint 2502.

[Roads, 1989] Roads, C., editor (1989). *The Music Machine*. MIT Press, Cambridge, MA.

[Roads, 1996] Roads, C. (1996). *The Computer Music Tutorial*. MIT Press, Cambridge, Massachusetts.

[Roads and Strawn, 1985] Roads, C. and Strawn, J., editors (1985). *Foundations of Computer Mu.* MIT Press, Cambridge, MA.

[Roberts, 1976] Roberts, L. G. (1976). Picture Coding Using Pseudo-Random Noise. In Jayant, N. S., editor, *Waveform Quantization and Coding*, pages 145–154. IEEE Press. Reprinted from IEEE Trans. on Information Theory, vol. IT-8, Feb. 1962.

[Rocchesso and Smith, 1994] Rocchesso, D. and Smith, J. O. (1994). "Circulant feedback delay networks for sound synthesis and processing". In *Proc. Int. Computer Music Conf.*, pages 378–381.

[Rocchesso and Smith, 1997] Rocchesso, D. and Smith, J. O. (1997). "Circulant and Elliptic Feedback Delay Networks for Artificial Reverberation". *IEEE Trans. Speech and Audio Processing*, 5(1):51–63.

[Rodet and Depalle, 1992] Rodet, X. and Depalle, P. (1992). Spectral envelopes and inverse FFT synthesis. *Proc. 93rd A.E.S Convention, San Francisco.* preprint 3393 (H-3).

[Rodet et al., 1989] Rodet, X., Potard, Y., and Barrière, J. (1989). The CHANT Project: From the Synthesis of the Singing Voice to Synthesis in General. In Roads, C., editor, *The Music Machine*, pages 449–465. MIT Press, Cambridge, MA.

[Roehrig, 1990] Roehrig, C. (1990). Time and pitch scaling of audio signals. *Proc. 89th AES Convention, Los Angeles.* preprint 2954 (E-1).

[Roesgen, 1986] Roesgen, J. P. (1986). The ADSP-2100 DSP Microprocessor. *IEEE Micro*, 6(6):49–59.

[Ross et al., 1974] Ross, M., Shaffer, H., Cohen, A., Freudberg, R., and Manley, H. (1974). Average magnitude difference function pitch extractor. *IEEE Trans. Acoust., Speech, Signal Processing*, 22:353–362.

[Rossum, 1992] Rossum, D. P. (1992). Digital sampling instrument for digital audio data. U.S. Patent 5,111,727.

[Rossum, 1994a] Rossum, D. P. (1994a). Digital sampling instrument. U.S. Patent 5,303,309.

[Rossum, 1994b] Rossum, D. P. (1994b). Digital sampling instrument employing cache memory. U.S. Patent 5,342,990.

[Rothweiler, 1983] Rothweiler, J. H. (1983). Polyphase Quadrature Filters – A new Subband Coding Technique. In *Proc. IEEE Int. Conf. Acoust., Speech and Signal Proc*, pages 1280–1283.

[Roucos and Wilgus, 1985] Roucos, S. and Wilgus, A. M. (1985). High quality time-scale modification of speech. *Proc. IEEE ICASSP-85, Tampa*, pages 493–496.

[Roys, 1978] Roys, H. E., editor (1978). *Disc Recording and Reproduction*. Dowden, Hutchinson and Ross, Inc., Stroudsburg, Pennsylvania.

[Ruiz, 1969] Ruiz, P. M. (1969). *A Technique for Simulating the Vibrations of Strings with a Digital Computer*. PhD thesis, Music Master Diss., Univ. Ill., Urbana.

[Rutledge, 1989] Rutledge, J. (1989). *Time-Varying, Frequency-Dependent Compensation for Recruitment of Loudness*. PhD thesis, Department of Electrical Engineering and Computer Science, Georgia Institute of Technology.

[Sabine, 1972] Sabine, W. C. (1972). "Reverberation". In Lindsay, R. B., editor, *Acoustics: Historical and Philosophical Development*. Dowden, Hutchinson, and Ross, Stroudsburg, PA. Originally published in 1900.

[Sachs and Kiang, 1968] Sachs, M. and Kiang, N. (1968). Two-tone inhibition in auditory nerve fibers. *J. Acoust. Soc. Am.*, 43:1120–1128.

[Samson, 1985] Samson, P. (1985). Architectural Issues in the Design of the System Concepts Digital Synthesizer. In Strawn, J., editor, *Digital Audio Enginnering: An Anthology*, pages 61–94. William Kaufmann.

[Samson, 1980] Samson, P. R. (1980). A General-Purpose Digital Synthesizer. *J. of the Audio Engineering Society*, 28(3):106–113.

[Sandy and Parker, 1982] Sandy, F. and Parker, J. (1982). Digital voice processor consortium interim report. Technical Report Report MTR-81w00159-01, The MITRE Corp.

[Sauvagerd, 1988] Sauvagerd, U. (1988). Bit-rate reduction for high quality audio signals using floating-point wave digital filters. In *Proc. of the ISCAS*, pages 2031 – 2034.

[Savioja et al., 1995] Savioja, L., Backman, J., Järvinen, A., and Takala, T. (1995). Waveguide mesh method for low-frequency simulation of room acoustics. *Proc. 15th Int. Conf. Acoustics (ICA-95), Trondheim, Norway*, pages 637–640.

[Scavone and Smith, 1997] Scavone, G. and Smith, J. O. (1997). Digital waveguide modeling of woodwind toneholes. In *Proc. 1997 Int. Computer Music Conf., Greece*. Computer Music Association.

[Scavone, 1997] Scavone, G. P. (1997). *An Acoustic Analysis of Single-Reed Woodwind Instruments with an Emphasis on Design and Performance Issues and Digital Waveguide Modeling Techniques*. PhD thesis, Music Dept., Stanford University. Available as CCRMA Technical Report No. STAN–M–100 or from ftp://ccrma-ftp.stanford.edu/pub/Publications/Theses/GaryScavoneThesis/.

[Schafer et al., 1981] Schafer, R. W., Mersereau, R. M., and Richards, M. A. (1981). Constrained iterative restoration algorithms. *Proc. IEEE*, 69:432–450.

[Scharf, 1964] Scharf, B. (1964). Partial masking. *Acustica*, 14:16–23.

[Scharf, 1970] Scharf, B. (1970). Critical Bands. In Tobias, J., editor, *Foundations of Modern Auditory Theory*, pages 159–202. Academic.

[Scharf and Buus, 1986] Scharf, B. and Buus, S. (1986). Stimulus, physiology, thresholds. In K.R. Boff, L. K. and Thomas, J., editors, *Handbook of Perception and Human Performance*, chapter 14, Basic sensory processes II. Wiley, New York.

[Scharf and Houtsma, 1986] Scharf, B. and Houtsma, A. J. M. (1986). Loudness, pitch, localization, aural distortion, pathology. In K.R. Boff, L. K. and Thomas, J., editors, *Handbook of Perception and Human Performance*, chapter 15, Basic sensory processes II. Wiley, New York.

[Schroeder and Voessing, 1986] Schroeder, E. F. and Voessing, W. (1986). High quality digital audio encoding with 3.0 bits/sample using adaptive transform coding. In *Proc. of the 80th. AES-Convention*. Preprint 2321.

[Schroeder, 1970a] Schroeder, M. (1970a). Synthesis of low-peak-factor signals and binary sequences with low autocorrelation. *IEEE Trans. Information Theory*, IT-16:85–89.

[Schroeder, 1986] Schroeder, M. (1986). *Number Theory in Science and Communication*. Springer Verlag, New York, NY, 2nd enlarged edition.

[Schroeder et al., 1967] Schroeder, M., Flanagan, J., and Lundry, E. (1967). Bandwidth compression of speech by analytic-signal rooting. *Proc. IEEE*, 55:396–401.

[Schroeder, 1954] Schroeder, M. R. (1954). "Die Statistischen Parameter der Frequenzkurven von Grossen Raumen". *Acustica*, 4:594–600. See Schroeder (1987) for English translation.

[Schroeder, 1962] Schroeder, M. R. (1962). "Natural Sounding Artificial Reverberation". *J. Audio Eng. Soc.*, 10(3).

[Schroeder, 1965] Schroeder, M. R. (1965). "New method of measuring reverberation time". *J. Acoust. Soc. Am.*, 37:409–412.

[Schroeder, 1970b] Schroeder, M. R. (1970b). "Digital simulation of sound transmission in reverberant spaces". *J. Acoust. Soc. Am.*, 47(2):424–431.

[Schroeder, 1987] Schroeder, M. R. (1987). "Statistical parameters of the frequency response curves of large rooms". *J. Audio Eng. Soc.*, 35(5):299–306. English translation of Schroeder (1954).

[Schroeder et al., 1979] Schroeder, M. R., Atal, B. S., and Hall, J. L. (1979). Optimizing digital speech coders by exploiting masking properties of the human ear. *J. Acoust. Soc. Am.*, 66:1647–1652.

[Schroeder and Kuttruff, 1962] Schroeder, M. R. and Kuttruff, K. H. (1962). "On frequency response curves in rooms. Comparison of experimental, theoretical, and

Monte Carlo results for the average frequency spacing between maxima". *J. Acoust. Soc. Am.*, 34(1):76–80.

[Schroeter and Sondhi, 1994] Schroeter, J. and Sondhi, M. M. (1994). Techniques for estimating vocal-tract shapes from the speech signal. *IEEE Trans. Speech and Audio Processing*, 2(1):133–150.

[Schuller, 1995] Schuller, G. (1995). A low delay filter bank for audio coding with reduced pre-echoes. In *Proc. of the 99th. AES-Convention*. Preprint 4088.

[Schwander and Levitt, 1987] Schwander, T. and Levitt, H. (1987). Effect of two-microphone noise reduction on speech recognition by normal-hearing listeners. *J. Rehab. Res. and Devel.*, 24:87–92.

[Scott and Gerber, 1972] Scott, R. and Gerber, S. (1972). Pitch-synchronous time-compression of speech. *Proceedings of the Conference for Speech Communication Processing*, pages 63–65.

[Segelken et al., 1992] Segelken, J. M., Wu, L. J., Lau, M. Y., Tai, K. L., Shively, R. R., and Grau, T. G. (1992). Ultra-Dense: An MCM-Based 3-D Digital Signal Processor. *IEEE Trans. on Components, Hybrids and Manufacturing Tech.*, 15(4):438–443.

[Segiguchi et al., 1983] Segiguchi, K., Ishizaka, K., Matsudaira, T. K., and Nakajima, N. (1983). A New Approach to High-Speed Digital Signal Processing Based on Microprogramming. *J. AES*, 31(7):517–522.

[Seitzer et al., 1988] Seitzer, D., Brandenburg, K., Kapust, R., Eberlein, E., Gerhäuser, H., Popp, H., and Schott, H. (1988). Real-time implementation of low complexity adaptive transform coding. In *Proc of the 84th. AES-Convention*. Preprint 2581.

[Seligson, 1970] Seligson, C. (1970). Comment on 'high-resolution frequency-wavenumber spectrum analysis'. *Proc. IEEE*, 58:947–949.

[Sellick et al., 1982] Sellick, P., Patuzzi, R., and Johnstone, B. (1982). Measurement of basilar membrane motion in the guinea pig using the Mossbauer technique. *J. Acoust. Soc. Am.*, 72:131–141.

[Seneff, 1982] Seneff, S. (1982). System to independently modify excitation and/or spectrum of speech waveform without explicit pitch extraction. *IEEE Trans. Acoust., Speech, Signal Processing*, ASSP-24:358–365.

[Serra, 1989] Serra, X. (1989). *A System for Sound Analysis/Transformation/Synthesis Based on a Deterministic Plus Stochastic Decomposition*. PhD thesis, CCRMA, Department of Music, Stanford University.

[Serra and Smith, 1989] Serra, X. and Smith, J. (1989). Spectral modeling synthesis: A sound analysis/synthesis system based on a deterministic plus stochastic decomposition. In *Proc. of Int. Computer Music Conf.*, pages 281–284, San Francisco, CA.

[Serra and Smith, 1990] Serra, X. and Smith, J. (1990). Spectral modeling synthesis: A sound analysis/synthesis system based on a deterministic plus stochastic decomposition. *Computer Music J.*, 14(4):12–24.

[Shan et al., 1985] Shan, T.-J., Wax, M., and Kailath, T. (1985). On spatial smoothing for direction-of-arrival estimation of coherent signals. *IEEE Trans. Acoust. Speech and Sig. Proc.*, ASSP-33:806–811.

[Shaw, 1974] Shaw, E. (1974). The external ear. In Keidel, W. and Neff, W., editors, *Handbook of Sensory Physiology*, pages 455–490. Springer.

[Sheingold, 1986] Sheingold, D. H., editor (1986). *Analog-digital conversion handbook, 3rd Ed.* Prentice-Hall.

[Shelton, 1989] Shelton, T. (1989). Synchronization in Digital Audio. In Pohlmann, K., editor, *Audio in Digital Times*, pages 241–250. Audio Engineering Society.

[Shields, 1970] Shields, V. (1970). Separation of additive speech signals by digital comb filtering. Master's thesis, Department of Electrical Engineering and Computer Science, Massachusetts Institute of Technology.

[Sigelman and Preves, 1987] Sigelman, J. and Preves, D. (1987). Field trials of a new adaptive signal processor hearing aid circuit. *Hearing J.*, 40:24–27.

[Simar et al., 1992] Simar, R., Koeppen, P., Leach, J., Marshall, S., Francis, D., Mekras, G., and Rosenstrauch, J. (1992). Floating-Point Processors Join Forces in Parallel Processing Architectures. *IEEE Micro*, 12(4):60–69.

[Simpson et al., 1990] Simpson, A., Moore, B., and Glasberg, B. (1990). Spectral enhancement to improve the intelligibility of speech in noise for hearing-impaired listeners. *Acta Otolaryngol. Suppl. 469*, pages 101–107.

[Sinclair, 1989] Sinclair, I. R. (1989). *Audio Electronics Reference Book*. BSP, Osney Mead, Oxford.

[Singhal, 1990] Singhal, S. (1990). High Quality Audio Coding using Multipulse LPC. In *Proc. IEEE Int. Conf. Acoust., Speech and Signal Proc*, pages 1101–1104.

[Sinha and Tewfik, 1993] Sinha, D. and Tewfik, A. H. (1993). Low bit-rate transparent audio compression using adapted wavelets. *IEEE Trans. Acoust,. Speech, and Signal Processing*, 41(12):3463–3479.

[Skinner, 1980] Skinner, M. (1980). Speech intelligibility in noise-induced hearing loss: Effects of high-frequency compensation. *J. Acoust. Soc. Am.*, 67:306–317.

[Smith et al., 1989] Smith, J., Jaffe, D., and Boyton, L. (1989). Music System Architecture on the NeXT Computer. In Pohlmann, K., editor, *Audio in Digital Times*, pages 301–312. Audio Engineering Society.

[Smith, 1983] Smith, J. O. (1983). *Techniques for Digital Filter Design and System Identification with Application to the Violin*. PhD thesis, Elec. Eng. Dept., Stanford University.

[Smith, 1985] Smith, J. O. (1985). A new approach to digital reverberation using closed waveguide networks. In *Proc. 1985 Int. Computer Music Conf., Vancouver*, pages 47–53. Computer Music Association. Also available in [Smith, 1987].

[Smith, 1986a] Smith, J. O. (1986a). Efficient simulation of the reed-bore and bow-string mechanisms. In *Proc. 1986 Int. Computer Music Conf., The Hague*, pages 275–280. Computer Music Association. Also available in [Smith, 1987].

[Smith, 1986b] Smith, J. O. (1986b). Elimination of limit cycles and overflow oscillations in time-varying lattice and ladder digital filters. In *Proc. IEEE Conf. Circuits and Systems, San Jose*, pages 197–299. Short conference version. Full version available in [Smith, 1987].

[Smith, 1987] Smith, J. O. (1987). Music applications of digital waveguides. Technical Report STAN–M–39, CCRMA, Music Dept., Stanford University. A compendium containing four related papers and presentation overheads on digital waveguide reverberation, synthesis, and filtering. CCRMA technical reports can be ordered by calling (415)723-4971 or by sending an email request to info@ccrma.stanford.edu.

[Smith, 1991] Smith, J. O. (1991). Waveguide simulation of non-cylindrical acoustic tubes. In *Proc. 1991 Int. Computer Music Conf., Montreal*, pages 304–307. Computer Music Association.

[Smith, 1996] Smith, J. O. (1996). Physical modeling synthesis update. *Computer Music J.*, 20(2):44–56. Available online at http://www-ccrma.stanford.edu/~jos/.

[Smith and Cook, 1992] Smith, J. O. and Cook, P. R. (1992). The second-order digital waveguide oscillator. In *Proc. 1992 Int. Computer Music Conf., San Jose*, pages 150–153. Computer Music Association. Available online at http://www-ccrma.stanford.edu/~jos/.

[Smith and Gossett, 1984] Smith, J. O. and Gossett, P. (1984). A flexible sampling-rate conversion method. In *Proc. Int. Conf. Acoustics, Speech, and Signal Processing, San Diego*, volume 2, pages 19.4.1–19.4.2, New York. IEEE Press. An expanded tutorial based on this paper and associated free software are available online at http://www-ccrma.stanford.edu/~jos/.

[Smith and Rocchesso, 1994] Smith, J. O. and Rocchesso, D. (1994). Connections between feedback delay networks and waveguide networks for digital reverberation. In *Proc. 1994 Int. Computer Music Conf., Århus*, pages 376–377. Computer Music Association.

[Smith and Scavone, 1997] Smith, J. O. and Scavone, G. (1997). The One-Filter Keefe Clarinet Tonehole. In *Proc. IEEE Workshop Appl. of Signal Processing to Audio and Acoustics*, Mohonk Mountain House, New Paltz, NY.

[Snell, 1977] Snell, J. (1977). Design of a Digital Oscillator That Will Generate up to 256 Low-distortion Sine Waves in Real Time. *Computer Music J.*, 1(2):4–25. reprinted in Foundations of Computer Music ed. J. Strawn & C. Roads 289-325, MIT Press 1985.

[Snell, 1989] Snell, J. M. (1989). Multiprocessor DSP Architectures and Implications for Software. In Pohlmann, K., editor, *Audio in Digital Times*, pages 327–336. Audio Engineering Society.

[Soede et al., 1993a] Soede, W., Berkhout, A., and Bilsen, F. (1993a). Development of a directional hearing instrument based on array technology. *J. Acoust. Soc. Am.*, 94:785–798.

[Soede et al., 1993b] Soede, W., Bilsen, F., and Berkhout, A. (1993b). Assessment of a directional microphone array for hearing-impaired listeners. *J. Acoust. Soc. Am.*, 94:799–808.

[Sohie and Kloker, 1988] Sohie, G. R. and Kloker, K. L. (1988). A Digital Signal Processor with IEEE Floating-Point Arithmetic. *IEEE Micro*, 8(6):49–67.

[Sondhi et al., 1981] Sondhi, M. M., Schmidt, C. E., and Rabiner, L. R. (1981). Improving the quality of a noisy speech signal. *The Bell System Technical Journal*, 60(8):1847–1859.

[Sondhi and Schroeter, 1987] Sondhi, M. M. and Schroeter, J. (1987). A hybrid time-frequency domain articulatory speech synthesizer. *IEEE Trans. Acoustics, Speech, Signal Processing*, ASSP-35(7):955–967.

[Sony, 1986] Sony (1986). DAE-1100A Digital Audio Editor: Service Manual. 4–876–860–05.

[Sony, 1989] Sony (1989). SDP-1000 Digital Audio Effector: Service Manual. 9–953–764–01.

[Sony, 1992] Sony (1992). Computer Audio/Video Semiconductor Data Book.

[Spath, 1991] Spath, H. (1991). *Mathematical Algorithms for Linear Regression*. Academic Press.

[Spenser, 1990] Spenser, P. S. (1990). *System Identification with Application to the Restoration of Archived Gramophone Recordings*. PhD thesis, University of Cambridge.

[Spenser and Rayner, 1989] Spenser, P. S. and Rayner, P. J. W. (1989). Separation of stationary and time-varying systems and its application to the restoration of gramophone recordings. *Proc. ISCAS89, Oregon*, 1:299–295.

[Spille, 1992] Spille, J. (1992). Messung der Vor- und Nachverdeckung bei Impulsen unter kritischen Bedingungen. Technical report, Thomson Consumer Electronics, Research and Development Laboratories, Hannover. unpublished.

[Sporer, 1998] Sporer, T. (1998). *Gehörangepasste Audiomesstechnik*. PhD thesis, University of Erlangen-Nuremburg. (To appear, in German).

[Srulovicz and Goldstein, 1983] Srulovicz, P. and Goldstein, J. L. (1983). A central spectrum model: a synthesis of auditory-nerve timing and place cues in monaural communication of frequency spectrum. *J. Acoust. Soc. Am.*, 73:1266–1276.

[Stadler and Rabinowitz, 1993] Stadler, R. and Rabinowitz, W. (1993). On the potential of fixed arrays for hearing aids. *J. Acoust. Soc. Am.*, 94:1332–1342.

[Stautner and Puckette, 1982] Stautner, J. and Puckette, M. (1982). "Designing multichannel reverberators". *Computer Music Journal*, 6(1):52–65.

[Stockham, 1972] Stockham, T. G. (1972). A-D and D-A Converters: Their effect on Digital Audio Fidelity. In Rabiner, L. and Rader, C., editors, *Digital Signal Processing*, pages 484–496. IEEE Press. Reprinted from 41st AES Convention, 1971.

[Stockham et al., 1975] Stockham, T. G., Cannon, T. M., and Ingebretsen, R. B. (1975). Blind deconvolution through digital signal processing. *Proc. IEEE*, 63(4):678–692.

[Stone and Moore, 1992] Stone, M. and Moore, B. (1992). Spectral feature enhancement for people with sensorineural hearing impairment: Effects on speech intelligibility and quality. *J. Rehab. Res. and Devel.*, 29:39–56.

[Strang, 1980] Strang, G. (1980). *Linear Algebra and Its Applications*. Academic Press, New York, NY.

[Strikwerda, 1989] Strikwerda, J. (1989). *Finite Difference Schemes and Partial Differential Equations.* Wadsworth and Brooks, Pacific Grove, CA.

[Suzuki and Misaki, 1992] Suzuki, R. and Misaki, M. (1992). Time-scale modification of speech signals using cross-correlation functions. *IEEE Trans. Consumer Elec.*, 38(3):357–363.

[Sylvestre and Kabal, 1992] Sylvestre, B. and Kabal, P. (1992). Time-scale modification of speech using an incremental time-frequency approach with waveform structure compensation. *Proc. IEEE ICASSP-92*, pages 81–84.

[Takamizawa et al., 1997] Takamizawa, Y., Iwadare, M., and Sugiyama, A. (1997). An efficient tonal component coding algorithm for MPEG-2 audio NBC. In *Proc. IEEE Int. Conf. Acoust., Speech and Signal Proc*, pages 331 – 334.

[Takao et al., 1986] Takao, K., Kikuma, N., and Yano, T. (1986). Toeplitzization of correlation matrix in multipath environment. *Proc. 1986 Int. Conf. on Acoust. Speech and Sig. Proc.*, Tokyo, Japan:1873–1876.

[Talambiras, 1976] Talambiras, R. P. (1976). Digital-to-Analog Converters: Some Problems in Producing High-Fidelity Signals. *Computer Design*, pages 63–69.

[Talambiras, 1985] Talambiras, R. P. (1985). Limitations on the Dynamic Range of Digitized Audio. In Strawn, J., editor, *Digital Audio Enginneering: An Anthology*, pages 29–60. William Kaufmann.

[Tellman et al., 1995] Tellman, E., Haken, L., and Holloway, B. (1995). Timbre morphing of sounds with unequal number of features. *J. Audio Eng. Soc.*, 43(9):678–689.

[Temerinac and Edler, 1993] Temerinac, M. and Edler, B. (1993). LINC: a common theory of transform and subband coding. *IEEE Transactions on Communications*, 41:266–274.

[Terhardt, 1979] Terhardt, E. (1979). Calculating virtual pitch. *Hearing Research*, 1:155–182.

[Tewksbury et al., 1978] Tewksbury, S. K., Meyer, F. C., Rollenhagen, D. C., Schoenwetter, H. K., and Souders, T. M. (1978). Terminology related to the performance of S/H, A/D, and D/A circuits. *IEEE Trans. Circuits and Systems*, CAS-25(7):419–426.

[Theile et al., 1987] Theile, G., Link, M., and Stoll, G. (1987). Low bit-rate coding of high quality audio signals. In *Proc of the 82nd. AES-Convention*.

[Therrien et al., 1994] Therrien, C., Cristi, R., and Allison, D. (1994). Methods for acoustic data synthesis. In *Proc. 1994 Digital Signal Processing Workshop*, Yosemite National Park, CA.

[Therrien, 1989] Therrien, C. W. (1989). *Decision, Estimation and Classification.* Wiley.

[Therrien, 1992] Therrien, C. W. (1992). *Discrete Random Signals and Statistical Signal Processing.* Prentice-Hall, Englewood Cliffs, NJ.

[Thiede and Kabot, 1996] Thiede, T. and Kabot, E. (1996). A new perceptual quality measure for bit rate reduced audio. *Contribution to the 100th AES Convention, Copenhagen, May 1996, preprint 4280.*

[Thornton, 1970] Thornton, J. E. (1970). *Design of a Computer: The Control Data 6600.* Scott, Foresman & Company, Glenview, IL.

[Tierney et al., 1971] Tierney, J., Rader, C. M., and Gold, B. (1971). A Digital Frequency Synthesizer. *IEEE Trans. Audio & Electroacoustics*, AU-19:48–56.

[Tong, 1990] Tong, H. (1990). *Non-linear Time Series.* Oxford Science Publications.

[Tong et al., 1979] Tong, Y., Black, R., Clark, G., Forster, I., Millar, J., and O'Loughlin, B. (1979). A preliminary report on a multiple-channel cochlear implant operation. *J. Laryngol. Otol.*, 93:679–695.

[Treurniet, 1996] Treurniet, W. (1996). Simulation of individual listeners with an auditory model. *Contribution to the 100th AES Convention, Copenhagen, May 1996, preprint 4154.*

[Troughton and Godsill, 1997] Troughton, P. T. and Godsill, S. J. (1997). Bayesian model selection for time series using Markov Chain Monte Carlo. In *Proc. IEEE Int. Conf. Acoust., Speech and Signal Proc*, volume 5, pages 3733–3736.

[Truax, 1994] Truax, B. (1994). Discovering inner complexity: Time shifting and transposition with a real-time granulation technique. *Computer Music J.*, 18(2):38–48.

[Tsoukalas et al., 1993] Tsoukalas, D., Paraskevas, M., and Mourjopoulos, J. (1993). Speech enhancement using psychoacoustic criteria. *Proc. IEEE Int. Conf. Acoust., Speech and Signal Proc*, II:359–362.

[Tukey, 1971] Tukey, J. W. (1971). *Exploratory Data Analysis.* Addison-Wesley.

[Uchiyama and Suzuki, 1986] Uchiyama, Y. and Suzuki, H. (1986). Electronic musical instrument forming tones by wave computation. U.S. Patent 4,616,546.

[Uchiyama and Suzuki, 1988] Uchiyama, Y. and Suzuki, H. (1988). Electronic musical instrument forming tones by wave computation. U.S. Patent 4,747,332.

[Vaidyanathan, 1993] Vaidyanathan, P. (1993). *Multirate Systems and Filter Banks.* PrenticeHall, Englewood Cliffs, NJ.

[Valière, 1991] Valière, J. C. (1991). *La Restauration d'Enregistrements Anciens par Traitement Numérique- Contribution à l'étude de Quelques techniques récentes (Restoration of old recording using digital techniques – Contribution to the study of some recent techniques).* PhD thesis, Université du Maine, Le Mans.

[Välimäki, 1995] Välimäki, V. (1995). *Discrete-Time Modeling of Acoustic Tubes Using Fractional Delay Filters.* PhD thesis, Report no. 37, Helsinki University of Technology, Faculty of Elec. Eng., Lab. of Acoustic and Audio Signal Processing, Espoo, Finland.

[Välimäki and Karjalainen, 1994a] Välimäki, V. and Karjalainen, M. (1994a). Digital waveguide modeling of wind instrument bores constructed of truncated cones. In *Proc. 1994 Int. Computer Music Conf., Århus*, pages 423–430. Computer Music Association.

[Välimäki and Karjalainen, 1994b] Välimäki, V. and Karjalainen, M. (1994b). Improving the Kelly-Lochbaum vocal tract model using conical tube sections and fractional delay filtering techniques. In *Proc. 1994 Int. Conf. Spoken Language Processing (ICSLP-94)*, volume 2, pages 615–618, Yokohama, Japan. IEEE Press.

[Välimäki et al., 1993] Välimäki, V., Karjalainen, M., and Laakso, T. I. (1993). Modeling of woodwind bores with finger holes. In *Proc. 1993 Int. Computer Music Conf., Tokyo*, pages 32–39. Computer Music Association.

[van de Plassche, 1994] van de Plassche, R. (1994). *Integrated Analog-to-Digital and Digital-to-Analog Converters.* Kluwer Academic Publishers.

[van der Waal and Veldhuis, 1991] van der Waal, R. G. and Veldhuis, R. N. J. (1991). Subband coding of stereophonic digital audio signals. In *Proc. IEEE Int. Conf. Acoust., Speech and Signal Proc*, pages 3601 – 3604.

[van Dijkhuizen et al., 1987] van Dijkhuizen, J., Anema, P., and Plomp, R. (1987). The effect of slope of the amplitude-frequency response on the masked speech-reception threshold of sentences. *J. Acoust. Soc. Am.*, 81:465–469.

[van Dijkhuizen et al., 1989] van Dijkhuizen, J., Festen, J., and Plomp, R. (1989). The effect of varying the amplitude-frequency response on the masked speech-reception threshold of sentences for hearing-impaired listeners. *J. Acoust. Soc. Am.*, 86:621–628.

[van Dijkhuizen et al., 1991] van Dijkhuizen, J., Festen, J., and Plomp, R. (1991). The effect of frequency-selective attenuation on the speech-reception threshold of sentences in conditions of low-frequency noise. *J. Acoust. Soc. Am.*, 90:885–894.

[Van Duyne and Smith, 1993] Van Duyne, S. A. and Smith, J. O. (1993). Physical modeling with the 2-D digital waveguide mesh. In *Proc. 1993 Int. Computer Music Conf., Tokyo*, pages 40–47. Computer Music Association.

[Van Duyne and Smith, 1995] Van Duyne, S. A. and Smith, J. O. (1995). The tetrahedral waveguide mesh: Multiply-free computation of wave propagation in free space. In *Proc. IEEE Workshop Appl. of Signal Processing to Audio and Acoustics*, pages 9a.6.1–4, New York. IEEE Press.

[Van Trees, 1968] Van Trees, H. L. (1968). *Detection, Estimation, and Modulation Theory, Part I*. J. Wiley & Sons.

[Vanderkooy and Lipshitz, 1984] Vanderkooy, J. and Lipshitz, S. (1984). Resolution below the Least Significant Bit in Digital Systems with Dither. *J. Audio Eng. Society*, 32:106–113. Correction, ibid., vol. 32, pp. 889, June 1987.

[Vanderkooy and Lipshitz, 1989] Vanderkooy, J. and Lipshitz, S. (1989). Digital dither: Signal processing with resolution far below the least significant bit. In Pohlmann, K., editor, *Audio in Digital Times*, pages 87–96. Audio Engineering Society.

[VanTrees, 1968] VanTrees, H. (1968). *Decision, Estimation and Modulation Theory, Part 1*. Wiley and Sons.

[Vary, 1983] Vary, P. (1983). On the enhancement of noisy speech. *Signal Processing II: Theories and Applications*, pages 327–330.

[Vary, 1985] Vary, P. (1985). Noise suppression by spectral magnitude estimation - Mechanism and theoretical limits. *Signal Processing*, 8(4):387–400.

[Vaseghi, 1988] Vaseghi, S. V. (1988). *Algorithms for Restoration of Archived Gramophone Recordings*. PhD thesis, University of Cambridge.

[Vaseghi and Frayling-Cork, 1992] Vaseghi, S. V. and Frayling-Cork, R. (1992). Restoration of old gramophone recordings. *J. Audio Eng. Soc.*, 40(10).

[Vaseghi and Rayner, 1988] Vaseghi, S. V. and Rayner, P. J. W. (1988). A new application of adaptive filters for restoration of archived gramophone recordings. *Proc. IEEE Int. Conf. Acoust., Speech and Signal Proc*, V:2548–2551.

[Vaseghi and Rayner, 1989] Vaseghi, S. V. and Rayner, P. J. W. (1989). The effects of non-stationary signal characteristics on the performance of adaptive audio restoration systems. *Proc. IEEE Int. Conf. Acoust., Speech and Signal Proc*, 1:377–380.

[Vaseghi and Rayner, 1990] Vaseghi, S. V. and Rayner, P. J. W. (1990). Detection and suppression of impulsive noise in speech communication systems. *IEE Proceedings, Part 1*, 137(1):38–46.

[Vaupelt, 1991] Vaupelt, T. (1991). *Ein Beitrag zur Transformationscodierung von Audiosignalen unter Verwendung der Methode der "Time Domain Aliasing Cancellation (TDAC)" und einer Signalkompandierung im Zeitbereich.* Dissertation, Universität Duisburg. (in German).

[Veldhuis, 1990] Veldhuis, R. (1990). *Restoration of Lost Samples in Digital Signals.* Prentice-Hall, Englewood Cliffs, NJ.

[Verge, 1995] Verge, M. P. (1995). *Aeroacoustics of Confined Jets with Applications to the Physical Modeling of Recorder-Like Instruments.* PhD thesis, Eindhoven University.

[Verhelst and Roelands, 1993] Verhelst, W. and Roelands, M. (1993). An Overlap-add Technique Based on Waveform Similarity (WSOLA) for High Quality Time-Scale Modification of Speech. *Proc. IEEE ICASSP-93, Minneapolis*, pages 554–557.

[Vernon, 1995] Vernon, S. (1995). Design and implementation of AC-3 coders. *IEEE Transactions on Consumer Electronics*, 41(3):754–759.

[Verschuure and Dreschler, 1993] Verschuure, J. and Dreschler, W. (1993). Present and future technology in hearing aids. *J. Speech-Lang. Path. and Audiol. Monogr. Suppl. 1*, pages 65–73.

[Vetterli and Kovačević, 1995] Vetterli, M. and Kovačević, J. (1995). *Wavelets and Subband Coding.* Prentice Hall, Englewood Cliffs, NJ.

[Victory, 1993] Victory, C. (1993). Comparison of signal processing methods for passive sonar data. Master's thesis, Department of Electrical Engineering and Computer Science, Naval Post Graduate School.

[Viergever, 1986] Viergever, M. (1986). Cochlear macromechanics - a review. In J. B. Allen and J. L. Hall and A. Hubbard and S. T. Neely and A. Tubis, editor, *Peripheral Auditory Mechanisms*, pages 63–72. Springer, Berlin.

[Villchur, 1973] Villchur, E. (1973). Signal processing to improve speech intelligibility in perceptive deafness. *J. Acoust. Soc. Am.*, 53:1646–1657.

[Waldhauer and Villchur, 1988] Waldhauer, F. and Villchur, E. (1988). Full dynamic range multiband compression in a hearing aid. *Hear. Journal*, 41:29–32.

[Walker et al., 1984] Walker, G., Byrne, D., and Dillon, H. (1984). The effects of multichannel compression/expansion amplification on the intelligibility of nonsense syllables in noise. *J. Acoust. Soc. Am.*, 76:746–757. .

[Wallraff, 1987] Wallraff, D. (1987). The DMX-1000 Signal-Processing Computer. In Roads, C. and Strawn, J., editors, *Foundations of computer music*, pages 225–243. MIT Press. Originally appeared in *Computer Music Journal*, vol. 3, no. 4, 1979, pages 44–49.

[Wang and Lim, 1982] Wang, D. L. and Lim, J. S. (1982). The unimportance of phase in speech enhancement. *IEEE Trans. Acoust., Speech, Signal Processing*, 30(4):679–681.

[Wang et al., 1992] Wang, S., Sekey, A., and Gersho, A. (1992). An objective measure for predicting subjective quality of speech coders. *IEEE J. on Select. Areas in Commun.*, 10(5):819–829.

[Waser and Flynn, 1982] Waser, S. and Flynn, M. J. (1982). *Introduction to Arithmetic for Digital Systems Designers*. Holt, Rinehart, Winston.

[Wawrzynek, 1986] Wawrzynek, J. (1986). A Reconfigurable Concurrent VLSI Architecture for Sound Synthesis. In S-Y Kung, R. O. and Nash, J., editors, *VLSI Signal Processing - II*, pages 385–396. IEEE Press.

[Wawrzynek, 1989] Wawrzynek, J. (1989). VLSI Models for Real-time Music Synthesis. In Mathews, M. and Pierce, J., editors, *Current Directions in Computer Music Research*, pages 113–148. MIT Press.

[Wawrzynek and von Eicken, 1989] Wawrzynek, J. and von Eicken, T. (1989). MIMIC, A Custom VLSI Parallel Processor for Musical Sound Synthesis. In Musgrave, G. and Lauther, U., editors, *Proceedings of the IFIP IC10/WG10.5 Working Conference on Very Large Scale Integration (VLSI-89)*, pages 389–398. Elsevier/North-Holland.

[Wawrzynek and Mead, 1985] Wawrzynek, J. C. and Mead, C. A. (1985). A bit serial architecture for sound synthesis. In Denyer, P. and Renshaw, D., editors, *VLSI Signal Processing: A Bit-serial approach*, pages 277–296. Addison-Wesley.

[Wayman and Wilson, 1988] Wayman, J. and Wilson, D. (1988). Some improvements on the synchronized-overlap-add method of time scale modification for use in real-time speech compression and noise filtering. *IEEE Trans. Acoust., Speech, Signal Processing*, 36(1):139–140.

[Weinreich, 1977] Weinreich, G. (1977). Coupled piano strings. *J. Acoustical Soc. of America*, 62(6):1474–1484. Also see *Scientific American*, vol. 240, p. 94, 1979.

[Weiss, 1987] Weiss, M. (1987). Use of an adaptive noise canceler as an input preprocessor for a hearing aid. *J. Rehab. Res. and Devel.*, 24:93–102.

[Weiss and Aschkenasy, 1975] Weiss, M. and Aschkenasy, E. (1975). Automatic detection and enhancement of speech signals. Technical report, Rome Air Devel. Ctr.

[Weiss et al., 1975] Weiss, M., Aschkenasy, E., and Parsons, T. (1975). Study and development of the INTEL technique for improving speech intelligibility. Technical report, Rome Air Devel. Ctr.

[Weiss and Neuman, 1993] Weiss, M. and Neuman, A. (1993). Noise reduction in hearing aids. In Studebaker, G. and Hochberg, I., editors, *Acoustical Factors Affecting Hearing Aid Performance*, pages 337–352. Allyn and Bacon.

[West, 1984] West, M. (1984). Outlier models and prior distributions in Bayesian linear regression. *Journal of the Royal Statistical Society, Series B*, 46(3):431–439.

[Widrow et al., 1975] Widrow, B., Glover, J. J., McCool, J., Williams, C., Hearn, R., Ziedler, J., Dong, E. J., and Goodlin, R. (1975). Adaptive noise canceling: Principles and applications. *Proc. IEEE*, 63:1692–1716.

[Widrow and Stearns, 1985] Widrow, B. and Stearns, S. (1985). *Adaptive Signal Processing*. Prentice Hall, Englewood Cliffs, NJ.

[Wiener, 1949] Wiener, N. (1949). *Extrapolation, Interpolation and Smoothing of Stationary Time Series with Engineering Applications*. MIT Press.

[Wightman and Kistler, 1989] Wightman, F. L. and Kistler, D. J. (1989). "Headphone simulation of free-field listening. I: Stimulus synthesis". *J. Acoust. Soc. Am.*, 85(2):858–867.

[Wilson et al., 1993] Wilson, B., Finley, C., Lawson, D., Wolford, R., and Zerbi, M. (1993). Design and evaluation of a continuous interleaved sampling (cis) processing strategy for multichannel cochlear implants. *J. Rehab. Res. and Devel.*, 30:110–116.

[Winckel, 1967] Winckel, F. (1967). *Music, Sound and Sensation*. Dover Publications, Inc., New York.

[Working Group on Communication Aids for the Hearing-Impaired, 1991] Working Group on Communication Aids for the Hearing-Impaired (1991). Speech-perception aids for hearing-impaired people: Current status and needed research. *J. Acoust. Soc. Am.*, 91:637–685.

[Wulf et al., 1981] Wulf, W. A., Levin, R., and Harbison, S. P. (1981). *HYDRA/C.mmp: An Experimental Computer System*. McGraw Hill.

[Yamasaki, 1983] Yamasaki, Y. (1983). Application of large amplitude dither to the quantization of wide range audio signals. *Journal of the Acoustical Society of Japan*, J-39(7):452–462. (In Japanese).

[Yang et al., 1992] Yang, X., Wang, K., and Shamma, S. A. (1992). Auditory representations of acoustic signals. *IEEE Trans. on Information Theory*, 38:824–839.

[Yegnanarayana, 1982] Yegnanarayana, B. (1982). Design of recursive group-delay filters by autoregressive modeling. *IEEE Trans. Acoustics, Speech, Signal Processing*, 30(4):632–637.

[Yund et al., 1987] Yund, E., Simon, H., and Efron, R. (1987). Speech discrimination with an 8-channel compression hearing aid and conventional aids in a background of speech-band noise. *J. Rehab. Res. and Devel.*, 24:161–180.

[Zelinski and Noll, 1977] Zelinski, R. and Noll, P. (1977). Adaptive transform coding of speech signals. *IEEE Trans. Acoust,. Speech, and Signal Processing*, 25:299–309.

[Zoelzer et al., 1990] Zoelzer, U., Fleige, N., Schonle, M., and Schusdziara, M. (1990). "Multirate Digital Reverberation System". In *Proc. Audio Eng. Soc. Conv.* Preprint 2968.

[Zwicker, 1977] Zwicker, E. (1977). Procedure for calculating loudness of temporally variable sounds. *J. Acoust. Soc. Am.*, 62:675–682.

[Zwicker, 1982] Zwicker, E. (1982). *Psychoakustik*. Springer-Verlag, Berlin Heidelberg New York. (in German).

[Zwicker and Fastl, 1990] Zwicker, E. and Fastl, H. (1990). *Psychoacoustics, Facts and Models*. Springer, Berlin.

[Zwicker and Feldtkeller, 1967] Zwicker, E. and Feldtkeller, R. (1967). *Das Ohr als Nachrichtenempfänger*. Hirzel-Verlag, Stuttgart. (in German).

[Zwicker and Zwicker, 1991] Zwicker, E. and Zwicker, U. T. (1991). Audio engineering and psychoacoustics: Matching signals to the final receiver, the human auditory system. *J. Audio Eng. Soc.*, 39(3):115–126.

Index

4A, 213
4B, 213
4C, 214
4X, 215

A/D
  Fixed point, 197
  Flash, 198
  counter (servo), 198
  floating point, 201
  integration, 198
AAC (*See* Advanced Audio Coding)
ACR (*See* Absolute Category Rating)
ADC (Analog to Digital Converter), 196
AGC (Automatic Gain Control), 255
AGC, 380
  hearing aid, 256
  input, 257
  output, 257
AMD 2901, 203, 216
AR (*See* Autoregressive)
ARMA (*See* Autoregressive moving average)
ASP, 203, 217
ATC (*See* Adaptive Transform Coding)
AT&T
  DSP-16, 221, 229
  DSP-32, 222, 228
  DSP-32C, 222
  DSP3, 232
Absolute Category Rating (ACR), 6
Absorptive filter, 102–103, 113, 116, 119, 122, 127–128, 130
Adaptive block switching, 58
Adaptive filter bank, 58
Adaptive high-pass filters, 263

Adaptive noise cancellation, 268
Adaptive parameter estimation, 136
Adaptive phase smoothing
  sine-wave analysis/synthesis, 380
Adaptive processing, 136
Addition rule in masking, 13
Additive synthesis, 346
Additivity of masking, 44
Advanced Audio Coding, 79
Akaike's Information Criterion (AIC), 189
Alias reduction, 57
All-pole model, 135
All-pole spectral modeling
  sine-wave residual, 388
Allen Organ Company, 319
Allpass feedback loop, 114, 116
Allpass filter, 105, 107–108, 111, 113–116, 121, 452
  lattice, 114
Allpass filters, 105
Allpass interpolation, 452
Alpha parameters, 448
Amplifier saturation, 251–252
Amplifier
  class-B, 252
  class-D, 252
Analog Devices
  2100, 221
  21000, 223
  SHARC, 223
Analog synthesizers, 312
Analogue restoration (*See* Restoration)
Analysis-by-synthesis, 65
Analysis/synthesis filterbank, 347

Analysis/synthesis, 51
Analytic signal
  definition, 409
Anti-aliasing filter, 196, 200
Array misalignment, 270
Articulation Index, 264
Articulatory speech synthesis, 419
Asymmetry of masking, 17
Attack time
  ANSI, 260
  compression, 255
Attack, 319, 322
Audio codec quality, 37
Audio quality, 1–2
Audio restoration (*See* Restoration)
Auditory scene analysis, 24, 26
Auditory system, 8, 10, 16
Auralization, 87
Automatic gain control, 255
Autoregressive (AR) model, 135, 142, 159, 164
Autoregressive (AR) model, Interpolation (*See* Restoration,Interpolation)
  Excitation energy, 148
  Model order, 136
Autoregressive Memoryless non-linearity (AR-MNL), 187
Autoregressive moving average (ARMA) model, 135, 164
Autoregressive non-linear AR model (AR-NAR), 188

BC coding (Backward Compatible), 77
Background noise, 7
Bandwidth
  hearing aid, 253
Bar vibrations, 451
Bark, 11, 14, 16, 19, 26, 37, 42
Basilar membrane, 240
Bayes' rule, 153
Bayesian methods (*See* Restoration)
Bernoulli model, 140
Big values, 65
Binaural impulse response, 86
Binaural processing, 36
Bit allocation, 63
  MPEG-1, 76
Bit reservoir, 67
Bit-reversal, 207
Blind identification, 184
Block companding, 64
Block convolution, 101
Block floating point, 64

Block-based processing, 136
Bowed strings
  bow force, 464
  bow-string interaction, 463
  friction force, 465
  scattering formulation, 464
  waveguide synthesis, 462
Breakages (*See* Restoration)
Breathiness
  example, 344
Buchla, 312

CDC 6000, 210
COPAS, 217
Caruso, 134
Cepstral distance, 30
Chaos measure, 61
Characteristic impedance, 433
Chipmunk effect, 322
Cholesky decomposition, 148
Chorusing, 298, 303
Circulant matrix, 126
Clarinet synthesis, 455
Clarity index, 98
Clicks (*See* Restoration)
Clipping, 340
Cochlea, 239–240, 259
Cochlear implants, 275
Cochlear partition, 240
Coding, 63
Cognitive modelling, 1, 7–8, 17, 22, 26, 29, 31, 37
Coherence function, 30
Coherence, 252
Comb filter, 105, 107–111, 113, 118, 121, 126–127, 394
  adaptive, 266
Commutativity simplifications, 450
Compatibility matrix, 73
Compression ratio, 255–256, 258
  auditory, 243, 259
Compression rule, 14
Compression threshold, 255–256
Compression, 10
  auditory, 243
  cochlear model based, 263
  feedforward, 257
  hearing aid, 255
  loudness-based, 262
  multi-channel, 261
  polynomial, 262
  principle component, 262
  single-channel, 256

INDEX 537

slow-acting, 262
syllabic, 257, 261
two-channel, 260, 263
two-stage, 261
lossy
    model-based, 418
Consonant-vowel ratio, 258
Continuously interleaved sampling, 276
Converter
    Analog-Digital, 196
    Digital-Analog, 196
    floating-point, 201
    oversampling, 199–200
    successive approximation, 198
Correction filter, 123, 128
Correlation matrix
    array, 272
Cosine-modulated filter banks, 56
Coupling channel, 72
Critical band, 42
Cross-correlation, 156
Cross-over distortion, 183
Crossbar, 222, 228, 231–232
Crossfade loop, 334

D'Alembert, 426
DAC (Digital to Analog Converter), 196
DCR (*See* Degradation Category Rating)
DCT (*Discrete Consine Transform*), 53
DCT (*Discrete Cosine Transform*), 79, 205
DFT (*Discrete Fourier Transform*), 53, 179, 206, 347, 353
DI (Disturbance Index), 35
DRC (Dynamic Range Compression), 380
DSP.*, 228
DX-7, 224
Data compression, 280
Deconvolution, 89
Degradation Category Rating (DCR), 6
Delay-and-sum beamforming, 269, 274
Detection of clicks (*See* Detection)
Detection, 141
    Probability of error, 141
    Bayesian, 154
    Clicks, 141, 144
    False Alarms, 140
    False detection, 143
    High-pass filter, 141
    Matched filter, 143
    Maximum *a posteriori*, 153
    Missed detection, 140
    Model-based, 142

Sequential, 153
Threshold selection, 141, 143
Deterministic plus stochastic signal model, 386
    time-scale modification, 391
Differentiation filters, 431
Digital Audio Broadcasting, 40
Digital waveguide network (DWN), 123, 125
Discrete Fourier Transform (DFT), 149
    Model for interpolation, 152
Dispersion, 451
Distortion
    harmonic, 252
    hearing aid, 259
    intermodulation, 252
    peak-related, 137
Dither, 199
    subtractive, 199
    triangular, 199
    nonsubtractive, 199
Downmix, 73
Drop sample tuning, 322
Dynamic Range Compression, 380
Dynamic range compression, 255
    sine-wave analysis/synthesis, 379
Dynamic range
    hearing aid, 251–252

E-mu, 320, 338, 340
ECL (*See* Emitter Coupled Logic)
ESC (escape) coding, 66
ETSI (European Telecommunications Standards Institute), 21, 25, 29
    speech codec quality, 21
Ear canal, 238, 250
Ear drum, 238
Ear, 238
    inner, 238
    outer, 238
Early decay time (EDT), 99
Echo density, 100–101, 107, 109–111, 113–114, 125–127
Embouchure modeling, 457, 461
Emitter Coupled Logic, 203
Emulator, 340
Energy decay curve (EDC), 94
Energy decay relief (EDR), 95, 99–100, 130
Enhancement (*See* Restoration)
Ensoniq
    ESP2, 226
Error power feedback, 443
Error resilience, 41
Excitation waveform

sine-wave model, 367
sine-wave phase, 367
sine-wave/pitch onset time, 368
Excitation, 8
Expectation-maximize (EM), 154, 194
Expert pattern recognition, 31
Expressivity, 314

FFT (*See* Fast Fourier Transform)
FM signal model
   Bessel function representation, 404
FM synthesis, 403–404
   model parameter estimation, 409
   musical sound, 407
   nested modulation, 410
   time-varying spectra, 405
FRMBox, 210, 223, 228
Fairlight Computer Music Instrument, 320, 340
Fast Fourier Transform (FFT), 14, 16, 101, 127, 205–206
Feedback cancellation, 254
Feedback delay network (FDN), 119–121, 123, 125–127
Feedback
   hearing aid, 249–250, 253
Fettweis, 421
Film sound tracks (*See* Sound recordings)
Filter bank, 50
   time-scale modification with phase coherence, 384
Filter design
   Hankel norm, 453
   differentiators, 431
   equation error, 453
   group delay error, 454
   integrators, 431
   phase error, 453
Finite differences, 424, 430
Finite impulse response (FIR) filter, 87, 101–102, 128
Flutter (*See* Restoration)
Flutter, 177
Force waves, 432
Formant, 302, 322
   speech, 245
Forward masking
   auditory, 246
Fourier transform, 285
Frame size (Layer 1, 2), 76
Frequency domain smearing, 9, 13
Frequency modulation, 178, 315
Frequency response envelope, 95, 100, 119, 123

G.729 speech codec, 34
GSM (Global System for Mobile communications), 21, 25, 29
GSM speech codecs, 21
Gap detection
   auditory, 243
Gaussian, 159
Gibbs Sampler, 149
Global degradation, 135
Glottal pulses, 245
Golden ear, 7
Gramophone disc recordings (*See* Sound recordings)
Griffiths-Jim array, 269–270, 272
Groove deformation, 183

Hair cells, 240, 243, 245, 259
Hammond organs, 336
Hankel norm, 453
Hanning window, 14, 16
Harmonicity, 315
Head-related transfer function (HRTF), 92, 102–103
Hearing aid acoustics, 251
Hearing aid cosmetics, 237
Hearing aid
   behind-the-ear, 249
   in-the-ear, 249
   linear, 248, 251
Hearing impairment, 236
Hearing loss, 236–237, 243, 247
   central, 247
   conductive, 239
   retrocochlear, 247
   simulated, 245
High-pass filter (*See* Detection)
High-pass filter, 156, 160
Hilbert transform
   definition, 409
Householder matrix, 125
Huffman coding, 65
Hybrid filter bank, 56
Hyperparameters, 182

IIR Filter (*See* Infinite Impulse Response Filter)
IRCAM
   4B, 213
   4C, 214
   4X, 215
   ISPW, 232
IRIS
   X-20, 225

INDEX 539

IS 13818-3, 77
ISO (International Standards Organization), 16–17, 25
ISO/MPEG audio quality, 15, 19
ISPW (IRCAM Signal Processing Workstation), 232
ITU (International Telecommunication Union), 1–2, 5, 25, 29, 35
ITU-T recommendation P.861, 1–2, 5, 33, 37
Impulse removal (*See* Restoration)
Impulse response, 3
Infinite Impulse Response (IIR) filter, 88, 128
Infinite Impulse Response Filter, 206
Information index, 30
Informational masking, 23
Inner loop, 65
Insertion loss
  hearing aid, 250
Instantaneous frequency, 288, 352
Integration filters, 431
Intensity coding, 71
Interaural cross-correlation coefficient (IACC), 93, 99, 111
Interaural intensity difference (IID), 92
Interaural time difference (ITD), 92
Intermodulation distortion, 183
Internal prepresentation calculation, 15
Internal representation measurement, 8
Internal representation, 1, 7, 9, 11, 14, 17–18, 20, 31
Internal sound representation, 2
International Telecommunication Union (*See* ITU)
Interpolation of clicks (*See* Restoration)
Interpolation, 141, 290, 295

Jitter, 202
  independent, 202
  locked, 202
  read-in, 202
  read-out, 202
  sampling, 202
Joint stereo coding, 68
Jupiter Systems, 336

Kaiser-Bessel-Derived window, 56
Kalman filter, 160, 164
Karhunen-Loève expansion
  noise-like signals, 354
Kelly-Lochbaum model
  acoustic tube, 418
  scattering junction, 439
Kurzweil, 320

LMS adaptation, 254, 268
LPC, 50, 204, 418–419
Ladder/lattice filters, 418, 428, 439
Layer 1, 75
Layer 2, 76
Layer 3, 76
Least squares AR-based (LSAR) interpolation, 145
Leslie Rotating Speaker, 336
Levinson-Durbin recursion, 145
Limit cycles, 443
Linear Gaussian model, 153
Linear Prediction (LP), 135
Linear predictive coding (*See* LPC)
Linear time-invariant (LTI) system, 156
Linear time-invariant system, 2, 5
Listening level, 7
Loaded waveguide junction, 446
Localized degradation, 135
Looping, 225, 331
Lossless prototype, 119, 122–123, 125, 127–128
Loudness discomfort level, 252
Loudness recruitment, 243
Loudness, 8, 14–17, 19, 32, 243
Lowpass filter, 92, 103

M-estimator, 136, 142
MDCT (Modified DCT), 54
MLT (Modulated Lapped Transform), 56
MOS (Mean Opinion Score), 6, 17, 20, 33–34
MPEG (Motion Picture Expert Group), 16–17, 25
MPEG-1 Audio, 75
MPEG-1 Layer 3, 76
MPEG-2 AAC, 79
MPEG-2 NBC, 79
MPEG-2 audio, 77
MPEG-4 Audio, 81
Magnetic tape (*See* Sound recordings)
Magnitude truncation, 443
Main-data-begin, 68
Marginal density, 149
Markov chain Monte Carlo (MCMC), 154, 164, 192, 194
Markov chain prior, 153
Masking (frequency domain), 42
Masking (time domain), 44
Masking of small clicks, 152
Masking properties, 164, 176
Masking
  auditory, 243, 264
Matched filter, 143, 158
Matrix, 73
Maximum likelihood (ML), 144, 163, 188

Maximum *a posteriori* (MAP) detection, 153
Maximum *a posteriori* (MAP), 144–145, 152, 159, 163
Mean Opinion Score (*See* MOS)
Median filter, 144
Mellotron, 337
Memory
  cache, 225
  sum, 212
Memoryless non-linearity, 186
Microphone array
  block frequency domain, 273
  multi-microphone, 271
  two-microphone, 269
Microphone arrays, 269
Microphone
  cardioid, 267, 269
  directional, 267
  hearing aid, 251
Microprocessors
  fixed-point, 220
Middle ear, 238
Minimum mean-squared error (MMSE), 144
Minimum variance estimation, 144
Minimum variance unbiased estimator, 145
Missing data (*See* Restoration, Interpolation)
Modal density, 97, 107, 109–110, 127
Model order selection, 189
Modelling of audio signals, 135, 137
Modulated Lapped Transform, 56
Moog, 312
Morpheus, 338
Motion Picture Expert Group (*See* MPEG)
Motorola
  56000, 204, 221, 223, 232–233
  96000, 223
Multi-channel coding, 73
Multi-electrode excitation, 275
Multiplication, 203
Multiplier
  Wallace tree, 203, 211
Multiresolution methods, 142
Munchkinize, 302
Music codec quality, 5–6
Musical Noise, 173

NBC (non backwards compatible), 79
NMR (*See* Noise to mask ratio)
NeXT, 221, 223
Nerve fibers
  afferent, 240
  efferent, 240

Nerve
  auditory, 240
New England Digital Synclavier, 320
Noise allocation, 64
Noise disturbance, 17–18, 26–27
Noise power, 199–200
Noise reduction (*See* Restoration)
Noise reduction, 163
Noise shaping, 200
Noise suppression, 275–276
  optimal filters for, 266
  single-channel, 263
Noise
  Additive, 137, 140
  Ambient noise, 163
  Bursts, 140
  Clicks, 137
  Clustering, 140
  Electrical circuit noise, 163
  Hiss, 163
  Localized, 137, 163
  Model, 153
  Musical, 173
  Non-stationary, 163
  Replacement, 137, 140
  Stationarity, 174
  Stationary, 163
  Suppression Rule, 165
  Thump noise, 156
  White, 163
Non-Gaussianity, 164
Non-linear AR (NAR) model, 187
Non-linear ARMA (NARMA) model, 185
Non-linear distortion (*See* Restoration)
Non-linearity, 158, 164
Non-stationarity, 164
Non-stationary (*See* Noise)
Non-uniform quantization, 64
Normal modes, 95
Normalized scattering junction, 441
Normalized waves, 436
Notch filters, 254

Objective quality assessment, 2
One-multiply scattering junction, 440
Organ of Corti, 240
Oscillator
  coupled-form, 208
  implementation, 208
  table lookup, 208
Ossicles, 239
Otitus media, 239

Otosclerosis, 239
Ototoxic drugs, 243
Outliers, 141, 152
Oval window, 239
Overflow oscillations, 443

PAQM (Perceptual Audio Quality Measure), 2, 5, 17, 36
PAQM
  validation on music, 20
  validation on speech, 20
PE (see Perceptual Entropy), 50
PERCEVAL (PERCeptual EVALuation method), 35
POM (Perceptual Objective Model), 35
PSOLA (Pitch Synchronous OverLap-Add), 299
PSQM (Perceptual Speech Quality Measure), 5, 7, 31, 37
Parameter update problem, 206, 212
Passive rounding, 445
Peak clipping, 252
Peak continuation algorithm
  sine waves, 387
Peak-to-rms ratio
  minimum, 377
Perceptual Audio Quality Measure (*See* PAQM)
Perceptual Coding, 205
Perceptual Entropy, 50
Perceptual Speech Quality Measure (*See* PSQM)
Perceptual characterization, 3
Perceptual coder: block diagram, 48
Perceptual coding, 39
Perceptual model 2 (MPEG), 61
Perceptual model, 59
Perceptual modelling, 1
Perfect reconstruction, 51
Permutation matrix, 126
Phase Vocoder, 205
Phase coherence, 309
  filter-bank-based onset times, 382
  filter-bank-based, 382
  shape invariance with sine waves, 369
  sine-wave-based onset times, 369
  sine-wave-based time-scale modification, 369
Phase dispersion
  optimal, 377
Phase unwrapping, 288
Phase vocoder, 286
  analysis/synthesis, 348
  phase dispersion, 351
  time-scale modification with phase coherence, 384

  time-scale modification, 350
Phasing, 298, 303
Physical modeling, 317
Piano
  component separation, 390
  synthesis, 452
Pinna, 238, 250
Pitch adjustment, 193
Pitch estimation
  sine-wave based, 414
  two voices, 403
Pitch period, 144, 146
Pitch shifting, 320
Pitch variation defects (*See* Restoration)
Pitch, 8, 11, 16, 19, 279
Polyphase filter bank, 53
Pops (*See* Restoration)
Posterior probability, 153
Power waves, 434
Pre-echo control, 62
Pre-echo, 45
Pre-whitening filter, 143
Precedence effect, 69
Prediction error filter, 142
Prediction, 72
Principal component analysis, 137
Prior information, 142
  Smoothness model, 182
Probe signal, 254
Prony's method, 365
Prosody, 315
Prototype window, 52
Pseudo-cepstrum, 265
Psychoacoustic model, 59
Psychoacoustics, 42

Quadrature mirror filter (QMF), 52
Quantization, 63
Quasi-periodic signals, 152

REDMASK, 265
RMS (Root Mean Square), 196
  value, 339
Random periodic process, 152
Rate distortion function, 41
Ray tracing, 92
Real time processing, 143
Recruitment compensation, 255–256
Recruitment, 243
Reflection coefficient, 439
Regularisation, 182
Relative Signal Level, 167, 172
Release time

ANSI, 260
compression, 258
Replacement of corrupted samples (*See*
    Restoration,Interpolation)
Resampling, 182, 290
Restoration, 133, 194
  Analogue de-clicking, 142
  Analogue methods, 134
  Background noise, 135, 163, 177
  Bayesian methods, 152, 164, 182
  Breakages, 135, 155
  Click removal, 137, 155
  Clicks, 135, 137
    Detection, 136
  Crackles, 135
  Flutter, 135
  Future trends, 133, 193–194
  Global defects, 135
  Interpolation, 144, 152
    Atypical excitation, 149
    Autoregressive, 144–145, 149,
    Model-based, 144
  Localized defects, 135
  Low frequency transients, 155, 160
    Model-based separation methods, 158, 160
    Template methods, 156, 158
  Noise reduction, 163
    Autoregressive model, 163
    Model-based, 163
  Non-linear distortions, 135, 182, 192
  Pitch variation defects, 177, 182
    Frequency domain, 179
  Scratches, 135, 155
  Statistical methods, 152, 155
  Stereo processing, 192
  Transform domain methods, 149
  Wow, 135
Reverberation, 303
  algorithm, 87
  description of, 88
  measurement, 89, 94
  time, 89, 94–95, 97–98, 100, 107–110, 113, 116,
    118–119, 122–123, 127–128, 130
Robust estimation, 136

SNR (*See* Signal to Noise Ratio)
SOLA (Synchronized OverLap Add), 298
SPL (Sound Pressure Level), 10
SPL, 13–14
SPS-41, 210
STFT (*See* Short-Time Fourier Transform)
Sample rate conversion, 182, 324

'Sinc' interpolation, 182
Sample-and-hold
  droop, 198
  jitter, 198
  slew rate limitations, 198
Sampling machines, 280
Sampling, 313
Samson Box, 211
Scaleable audio coding, 81
Scaled projection algorithm, 272, 275
Scalefactor select, 76
Scattering junction
  N waveguides, 446
  Kelly-Lochbaum, 439
  bow-string, 464
  frequency dependent, 456–457
  multiply free, 448
  nonlinear two-port, 462
  normalized, 441
  one-multiply, 440
  passive rounding, 443
  series connection, 448
  three-multiply normalized, 442
  two-multiply, 441
Scattering matrix, 125
Scattering, 436
Schroeder integral, 95–96, 99
Scratches (*See* Restoration)
Sensation level, 245
Shape-invariance, 304
Short window, 58
Short-Time Fourier Transform, 165, 286, 347
  Phase, 166
Short-Time Fourier processing, 163–164
Short-Time Spectral Attenuation (STSA), 136, 164,
    177
Signal separation, 159
Signal suppression
  array, 272
Signal to Noise Ratio (SNR), 89
Signal to Noise Ratio, 2
Signal to Noise Ratio), 196
Signal to Noise Ratio
  fixed point, 197
  floating point, 201
Silent intervals, 24, 27, 32–34, 37
Sine waves
  aharmonic, 344
  harmonic, 344
Sine-wave analysis/synthesis, 357
  magnitude-only reconstruction, 361
  multi-resolution, 412

overlap-add, 414
phase dispersion, 378
time-scale modification with phase coherence, 372
time-scale modification, 362
time-varying time-scale modification, 374
zero-phase reconstruction, 378
Sine-wave dithering
   frequency, 371
   phase, 371
Sine-wave representation
   amplitude estimation, 352
   birth-death process, 355
   definition, 352
   deterministic plus stochastic, 386
   frequency matching, 355
   harmonicity, 371
   phase estimation, 354
   phase unwrapping/interpolation, 355
   stochastic residual, 388
   time-scale modification with phase coherence, 371
Singing vibrato
   sine-wave analysis/synthesis, 364
Single-reed synthesis, 456, 458
Sinusoidal model, 136, 149, 178–179, 282
Sinusoidal modeling, 266
Smoothness model, 182
Sone density, 11
Sone, 14, 16, 37
Sony
   CXD1160, 221
   DAE-1000A, 216
   OXF, 226, 233
   SDP-1000, 221, 231
   reverb, 216
Sound Pressure Level (*See* SPL)
Sound extrapolation
   sine-wave analysis/synthesis, 363
Sound interpolation
   sine-wave analysis/synthesis, 363
Sound recordings
   Film sound tracks, 133, 137, 177
   Film sound tracks
      Optical, 155
   Gramophone discs, 133–134, 155, 177
      78 rpm, 163
      Laser reader, 156
   Magnetic tapes, 134, 163, 177
   Wax cylinders, 134, 163
Sound separation
   deterministic and stochastic sine waves, 390
   voiced and unvoiced speech, 390
Sound splicing
   sine-wave analysis/synthesis, 362
SoundDroid, 217
Source coding, 41
Source image method, 90, 92, 98, 101
Spaciousness, 93, 99, 103
Spectral Subtraction, 168, 194
Spectral contrast enhancement, 266
Spectral enhancement, 266
Spectral envelope, 302
Spectral subtraction, 264
   INTEL technique, 265, 276
Spectro-temporal weighting, 24, 27
Speech codec quality, 5, 21
Speech intelligibility, 236
Speech levels, 252
Speech reception threshold, 269
Speech signals
   sine-wave analysis/synthesis, 359
   sine-wave time-scale modification with phase coherence, 375
   sine-wave-based dynamic range compression, 380
Speech synthesis
   Kelly-Lochbaum acoustic tube model, 418
   articulatory, 419
   linear prediction, 419
Speech, 315, 322
Splicing method, 294
Spline interpolation, 160
Start window, 59
Stationarity, 136
   Short-term, 142
Stereo unmasking, 68
Stereocilia, 241
Stiff vibrating strings, 451
Stop window, 59
String
   numerical simulation, 426
   stiff, 451
   traveling-wave description, 426
   wave equation, 423
Structured correlation matrix, 273, 275
Sub-space methods, 164, 177
Subband coder, 50
Subjective quality assessment, 2
Subjective testing, 6
Superdirective array, 269, 274
Suppression Rule, 165, 167
   Ephraim and Malah, 174
Synchronization, 280

SynthBuilder, 462, 466
Synthesis
  Physical modelling, 209
  linear, 208
  nonlinear, 208
  subtractive, 208
  physical modeling, 466
  piano, 424
  source-filter modeling, 419
  strings, 423
  turbulence, 462
  waveguide, 420
  winds, 455
System characterization, 2

TDAC (*See* Time Domain Aliasing Cancellation)
TI (Texas Instruments)
  TMS320, 204, 220
  TMS320C30, 223
  TMS320C40, 223
TNS (Temporal Noise Shaping), 79
Tape saturation, 183
Tectorial membrane, 240
Telharmonium, 311
Temporal envelope
  definition, 381
Temporal noise shaping, 79
Temporal resolution
  auditory, 243
Text-to-speech synthesizer, 280
The masked threshold, 13
Threshold selection (*See* Detection)
Thumps (*See* Restoration)
Ticks (*See* Restoration)
Time Domain Aliasing Cancellation, 54
Time domain smearing, 9, 14
Time-frequency domain smearing, 10
Time-frequency tracking, 180
Tip-to-tail ratio, 241
Toeplitz equations, 145
Tonality, 60
Total harmonic distortion, 2
Tracing distortion, 183
Transform coder, 51
Transient smearing, 307
Transmission matrix, 456
Transparent, 2
Trauma
  acoustic, 243
Traveling waves
  Ohm's Law, 434
  acoustic tubes, 434

  choice of, 431
  dispersion, 451
  frequency-dependent losses, 451
  losses, 448
  nonlinear reflection, 457, 460
  normalized waves, 436
  passive rounding, 443
  power waves, 434
  reflection coefficient, 439
  scattering, 436
  transformer, 442
  vibrating strings, 426
  wave impedance, 433
Tremolo
  example, 344
Trumpet
  FM synthesis, 407
  example, 344
  sine-wave analysis/synthesis, 359
  sine-wave time-scale modification with phase coherence, 375
Tuning curve
  auditory, 240–241, 243
Turbulence synthesis, 462
Two-tone suppression
  auditory, 263
Two-voice
  model
    Fourier transform, 394
    least-squared error sine-wave solution, 395, 397
    sine waves, 392
  separation
    pitch estimation, 403
    sine waves, 399
    sine-wave ambiguity problem, 400
Tympanic membrane, 238

UPE (Universal Processing Element), 225
Unitary
  feedback loop, 119, 121–122, 127
  matrix, 118, 125–127

Variable rate, 67
Vector quantization, 65
Vent
  hearing aid, 249
Vibrato
  example, 344
Violin family synthesis, 462
Volterra series, 184

Wave digital filters, 421, 448

Wave equation
  ideal strings, 423
  lossy strings, 449
  stiff strings, 451
Wave impedance, 433
Wave variables, 431
  force waves, 432
  normalized waves, 436
  power waves, 434
  pressure waves, 434, 455, 457
  slope waves, 431
  velocity waves, 431, 462
Waveguide junction, 446
Waveguide mesh, 443, 448
Waveguide synthesis, 420
  bow force, 464
  bow-string friction, 465
  bow-string interaction model, 463
  bow-string table, 465
  bowed strings, 462
  development tools for, 462, 466
  legato transitions, 462
  noise injection, 462
  single-reed model, 455–456, 458
  woodwind bore termination model, 455
Waveguides, 451
  commutativity simplifications, 450
  dispersive, 451
  lossy, 448
  nonlinear termination, 460
  nonlinear two-port junction, 462
  polynomial nonlinearities, 462
  table-lookup nonlinearities, 461
Wavelets, 53, 137, 142, 149, 164, 177
Wavetable synthesis, 318
Wax cylinder recordings (*See* Sound recordings)
White Gaussian noise (WGN), 188
Wiener filter, 163–164
Window switching, 58
Windowing
  Hamming, 353
  pitch-adaptive, 353
Woodwinds
  bell filtering, 455
  finger-hole models, 456
  radiation characteristics, 456
  single-reed synthesis, 455
  synthesis, 455
  tone-hole modeling, 456
Wow (*See* Restoration)
Wow, 177

Yamaha
  DX-7, 224
  VL1 synthesizer, 466